Cell–Cell Interactions

METHODS IN MOLECULAR BIOLOGY™

John M. Walker, SERIES EDITOR

357. **Cardiovascular Proteomics:** *Methods and Protocols,* edited by *Fernando Vivanco, 2006*
356. **Cellomics:** *Methods and Protocols,* edited by *Ken Guiliano, D. Lansing Taylor, and Jeffrey Haskins, 2006*
355. **Plant Proteomics:** *Methods and Protocols,* edited by *Hervé Thiellement, Michel Zivy, Catherine Damerval, and Valerie Mechin, 2006*
354. **Plant–Pathogen Interactions:** *Methods and Protocols,* edited by *Pamela C. Ronald, 2006*
353. **DNA Analysis by Nonradioactive Probes:** *Methods and Protocols,* edited by *Elena Hilario and John. F. MacKay, 2006*
352. **Protein Engineering Protocols**, edited by *Kristian Müller and Katja Arndt, 2006*
351. ***C. elegans:*** *Methods and Applications,* edited by *Kevin Strange, 2006*
350. **Protein Folding Protocols,** edited by *Yawen Bai and Ruth Nussinov 2006*
349. **YAC Protocols,** *Second Edition,* edited by *Alasdair MacKenzie, 2006*
348. **Nuclear Transfer Protocols:** *Cell Reprogramming and Transgenesis,* edited by *Paul J. Verma and Alan Trounson, 2006*
347. **Glycobiology Protocols**, edited by *Inka Brockhausen-Schutzbach, 2006*
346. ***Dictyostelium discoideum*** **Protocols**, edited by *Ludwig Eichinger and Francisco Rivero-Crespo, 2006*
345. **Diagnostic Bacteriology Protocols**, *Second Edition,* edited by *Louise O'Connor, 2006*
344. **Agrobacterium Protocols, Second Edition:** *Volume 2,* edited by *Kan Wang, 2006*
343. **Agrobacterium Protocols, Second Edition:** *Volume 1,* edited by *Kan Wang, 2006*
342. **MicroRNA Protocols**, edited by *Shao-Yao Ying, 2006*
341. **Cell–Cell Interactions:** *Methods and Protocols,* edited by *Sean P. Colgan, 2006*
340. **Protein Design:** *Methods and Applications,* edited by *Raphael Guerois and Manuela López de la Paz, 2006*
339. **Microchip Capillary Electrophoresis:** *Methods and Protocols,* edited by *Charles Henry, 2006*
338. **Gene Mapping, Discovery, and Expression:** *Methods and Protocols,* edited by *M. Bina, 2006*
337. **Ion Channels:** *Methods and Protocols,* edited by *James D. Stockand and Mark S. Shapiro, 2006*
336. **Clinical Applications of PCR:** *Second Edition,* edited by *Y. M. Dennis Lo, Rossa W. K. Chiu, and K. C. Allen Chan, 2006*
335. **Fluorescent Energy Transfer Nucleic Acid Probes:** *Designs and Protocols,* edited by *Vladimir V. Didenko, 2006*
334. **PRINS and** ***In Situ*** **PCR Protocols:** *Second Edition,* edited by *Franck Pellestor, 2006*
333. **Transplantation Immunology:** *Methods and Protocols,* edited by *Philip Hornick and Marlene Rose, 2006*
332. **Transmembrane Signaling Protocols:** *Second Edition,* edited by *Hydar Ali and Bodduluri Haribabu, 2006*
331. **Human Embryonic Stem Cell Protocols**, edited by *Kursad Turksen, 2006*
330. **Embryonic Stem Cell Protocols, Second Edition,** *Vol. II: Differentiation Models,* edited by *Kursad Turksen, 2006*
329. **Embryonic Stem Cell Protocols, Second Edition,** *Vol. I***:** *Isolation and Characterization,* edited by *Kursad Turksen, 2006*
328. **New and Emerging Proteomic Techniques**, edited by *Dobrin Nedelkov and Randall W. Nelson, 2006*
327. **Epidermal Growth Factor:** *Methods and Protocols,* edited by *Tarun B. Patel and Paul J. Bertics, 2006*
326. ***In Situ*** **Hybridization Protocols,** *Third Edition,* edited by *Ian A. Darby and Tim D. Hewitson, 2006*
325. **Nuclear Reprogramming:** *Methods and Protocols,* edited by *Steve Pells, 2006*
324. **Hormone Assays in Biological Fluids,** edited by *Michael J. Wheeler and J. S. Morley Hutchinson, 2006*
323. **Arabidopsis Protocols,** *Second Edition,* edited by *Julio Salinas and Jose J. Sanchez-Serrano, 2006*
322. ***Xenopus*** **Protocols:** *Cell Biology and Signal Transduction,* edited by *X. Johné Liu, 2006*
321. **Microfluidic Techniques:** *Reviews and Protocols,* edited by *Shelley D. Minteer, 2006*
320. **Cytochrome P450 Protocols**, *Second Edition,* edited by *Ian R. Phillips and Elizabeth A. Shephard, 2006*
319. **Cell Imaging Techniques**, *Methods and Protocols,* edited by *Douglas J. Taatjes and Brooke T. Mossman, 2006*
318. **Plant Cell Culture Protocols**, *Second Edition,* edited by *Victor M. Loyola-Vargas and Felipe Vázquez-Flota, 2005*
317. **Differential Display Methods and Protocols**, *Second Edition,* edited by *Peng Liang, Jonathan Meade, and Arthur B. Pardee, 2005*
316. **Bioinformatics and Drug Discovery**, edited by *Richard S. Larson, 2005*
315. **Mast Cells:** *Methods and Protocols,* edited by *Guha Krishnaswamy and David S. Chi, 2005*
314. **DNA Repair Protocols:** *Mammalian Systems, Second Edition,* edited by *Daryl S. Henderson, 2006*
313. **Yeast Protocols:** *Second Edition,* edited by *Wei Xiao, 2005*
312. **Calcium Signaling Protocols:** *Second Edition,* edited by *David G. Lambert, 2005*
311. **Pharmacogenomics:** *Methods and Protocols,* edited by *Federico Innocenti, 2005*
310. **Chemical Genomics:** *Reviews and Protocols,* edited by *Edward D. Zanders, 2005*
309. **RNA Silencing:** *Methods and Protocols,* edited by *Gordon Carmichael, 2005*
308. **Therapeutic Proteins:** *Methods and Protocols,* edited by *C. Mark Smales and David C. James, 2005*
307. **Phosphodiesterase Methods and Protocols,** edited by *Claire Lugnier, 2005*
306. **Receptor Binding Techniques:** *Second Edition,* edited by *Anthony P. Davenport, 2005*
305. **Protein–Ligand Interactions:** *Methods and Applications,* edited by *G. Ulrich Nienhaus, 2005*
304. **Human Retrovirus Protocols:** *Virology and Molecular Biology,* edited by *Tuofu Zhu, 2005*
303. **NanoBiotechnology Protocols,** edited by *Sandra J. Rosenthal and David W. Wright, 2005*
302. **Handbook of ELISPOT:** *Methods and Protocols,* edited by *Alexander E. Kalyuzhny, 2005*

METHODS IN MOLECULAR BIOLOGY™

Cell–Cell Interactions

Methods and Protocols

Edited by

Sean P. Colgan

Center for Experimental Therapeutics and Reperfusion Injury
Brigham and Women's Hospital, Havard Medical School
Boston, MA

HUMANA PRESS ✳ TOTOWA, NEW JERSEY

999 Riverview Drive, Suite 208
Totowa, New Jersey 07512

www.humanapress.com

This publication is printed on acid-free paper. ∞
ANSI Z39.48-1984 (American Standards Institute) Permanence of Paper for Prined Library Materials.

Production Editor: Melissa Caravella

Cover design by Patricia F. Cleary

Cover illustration: From Fig. 2 in Chapter 3, "The Role of Junctional Adhesion Molecules in Interactions Between Vascular Cells," by Triantafyllos Chavakis and Valeria Orlova.

For additional copies, pricing for bulk purchases, and/or information about other Humana titles, contact Humana at the above address or at any of the following numbers: Tel.: 973-256-1699; Fax: 973-256-8341; E-mail: orders@humanapr.com; or visit our Website: www.humanapress.com

Printed in the United States of America. 10 9 8 7 6 5 4 3 2 1

eISBN 1-59745-113-4

ISSN 1064-3745

Library of Congress Cataloging-in-Publication Data

Cell–cell interactions : methods and protocols / edited by Sean P. Colgan.

p. ; cm. -- (Methods in molecular biology, ISSN 1064-3745 ; . 341)

Includes bibliographical references and index.

ISBN 978-1-61737-618-4 e-ISBN 978-1-59745-113-0

1. Cell interaction. 2. Cell junctions. 3. Epithelial cells.

[DNLM: 1. Cell Communication--physiology. 2. Endothelial Cells--physiology. 3. Epithelial Cells--physiology. 4. Tight Junctions--physiology. QU 375 C3925 2006] I. Colgan, Sean P. II. Series: Methods in molecular biology (Clifton, N.J.) ; v. 341.

QH604.2.C4418 2006

571.6--dc22

2005029552

Preface

Studies incorporating cell–cell interactions remain an exciting and ever-evolving discipline within the broader areas of biology. Although recent advances in genetics have provided much insight into gene- and disease-specific mechanisms, such models are often limited by tissue/organ complexity. The use of directed cell models to study the manner by which two or more cells interact has revolutionized our understanding of protein–protein interactions in a complex milieu and have provided an invaluable platform for the identification of new molecules.

The aim of *Cell–Cell Interactions: Methods and Protocols* is to provide a collection of diverse protocols, incorporating methods-based approaches both in vitro and in vivo. The authors of the various chapters are highly skilled experts in the development and utilization of cell–cell interaction assays. Their extensive expertise provides a detailed, step-by-step map of a reproducible protocol. Importantly, these protocols are readily adaptable for nearly any cell type and any organ system, and for this reason, I have tried to compile a diverse set of chapters incorporating numerous individual types of cells. It is my hope that this book will be both insightful to the beginner and inspiring to the experienced.

Sean P. Colgan

Contents

Contributors

MICHAEL ASCHNER • *Department of Pediatrics Pharmacology and The Kennedy Center, Vanderbilt University School of Medicine, Nashville, TN*

TANUSREE BHATTACHARYYA • *Faculty of Life Sciences, University of Manchester, Manchester, United Kingdom*

JOSEPH P. BRESSLER • *Department of Environmental Health Sciences, The Kennedy Krieger Institute, John Hopkins University Bloomberg School of Public Health, Baltimore, MD*

MATTHIAS BRÜWER • *Department of General Surgery, University of Münster, Münster, Germany*

GERALDINE CANNY • *Center for Experimental Therapeutics and Reperfusion Injury, Brigham and Women's Hospital, Harvard Medical School, Boston, MA*

TRIANTAFYLLOS CHAVAKIS • *Experimental Immunology Branch, National Cancer Institute, National Institutes of Health, Bethesda, MD*

NUNO CERCA • *The Channing Laboratory, Brigham and Women's Hospital, Harvard Medical School, Boston, MA*

NAN CHIANG • *Department of Anesthesiology, Perioperative, and Pain Medicine, Center for Experimental Therapeutics and Reperfusion Injury, Brigham and Women's Hospital, Harvard Medical School, Boston, MA*

ALEX C. CHIN • *Division of Gastrointestinal Pathology, Department of Pathology and Laboratory Medicine, Emory University School of Medicine, Atlanta, GA*

SEAN P. COLGAN • *Center for Experimental Therapeutics and Reperfusion Injury, Brigham and Women's Hospital, Harvard Medical School, Boston, MA*

DIONNE DANIELS • *Center for Experimental Therapeutics and Reperfusion Injury, Brigham and Women's Hospital, Harvard Medical School, Boston, MA*

HEIDI E. DE LUCA • *Gastrointestinal Cell Biology, Children's Hospital Boston, Harvard Medical School, Boston, MA*

BRADLEY M. DENKER • *Renal Division, Brigham and Women's Hospital, Harvard Medical School, Boston, MA*

ANA PAULA MARREILHA DOS SANTOS • *Faculty of Pharmacy, University of Lisbon, Lisbon, Portugal*

TOBIAS ECKLE • *Department of Anesthesiology and Intensive Care Medicine, Tübingen University Clinic, Tübingen, Germany*

HOLGER K. ELTZSCHIG • *Department of Anesthesiology and Intensive Care Medicine, Tübingen University Clinic, Tübingen, Germany*

JANICE P. EVANS • *Division of Reproductive Biology, Department of Biochemistry and Molecular Biology, John Hopkins University Bloomberg School of Public Health, Baltimore, MD*

VANESSA A. FITSANAKIS • *Department of Pediatrics, Vanderbilt University Medical Center, Nashville, TN*

ROBERT FUHLBRIGGE • *Harvard Skin Disease Research Center and Department of Dermatology, Harvard Medical School, Boston, MA*
GLENN T. FURUTA • *Division of Pediatric Gastroenterology and Center for Experimental Therapeutics and Reperfusion, Brigham and Women's Hospital, Children's Hospital Boston, Harvard Medical School, Boston, MA*
ALLISON J. GARDNER • *Division of Reproductive Biology, Department of Biochemistry and Molecular Biology, John Hopkins University Bloomberg School of Public Health, Baltimore, MD*
JUAN C. IBLA • *Department of Anesthesiology, Perioperative, and Pain Medicine, Center for Experimental Therapeutics and Reperfusion Injury, Brigham and Women's Hospital, Children's Hospital Boston, Harvard Medical School, Boston, MA*
KIMBERLY K. JEFFERSON • *Department of Microbiology and Immunology, Virginia Commonwealth University, Richmond, VA*
JOSEPH KHOURY • *Department of Anesthesiology, Perioperative, and Pain Medicine, Center for Experimental Therapeutics and Reperfusion Injury, Brigham and Women's Hospital, Children's Hospital Boston, Harvard Medical School, Boston, MA*
AIMEE LANDRY • *Department of Molecular Biology, Massachusetts General Hospital, Boston, MA*
WINSTON Y. LEE • *Division of Gastrointestinal Pathology, Department of Pathology and Laboratory Medicine, Emory University School of Medicine, Atlanta, GA*
WAYNE I. LENCER • *Gastrointestinal Cell Biology, Children's Hospital Boston, Harvard Medical School, Boston, MA*
YAW-CHYN LIM • *Department of Pathology and Physiology, Immunology Program, National University of Singapore, Singapore*
NANCY A. LOUIS • *Center for Experimental Therapeutics and Reperfusion Injury, Brigham and Women's Hospital, Harvard Medical School, Boston, MA*
FRANCIS W. LUSCINSKAS • *Department of Pathology, Center for Excellence in Vascular Biology, Brigham and Women's Hospital, Harvard Medical School, Boston, MA*
ALICE MAGER • *Department of Anesthesiology and Intensive Care Medicine, Tübingen University Clinic, Tübingen, Germany*
MICHELLE L. MATTER • *John A. Burns Medical School, University of Hawaii at Manoa, Honolulu, HI*
BETH A. MCCORMICK • *Department of Pediatric Gastroenterology and Nutrition, Massachusetts General Hospital and Department of Microbiology and Molecular Genetics, Harvard Medical School, Boston, MA*
ANITA J. MERRITT • *Faculty of Life Sciences, University of Manchester, Manchester, United Kingdom*
PETER NGO • *Fellow in Pediatric Gastroenterology, Center for Experimental Therapeutics and Reperfusion Injury, Brigham and Women's Hospital, Harvard Medical School, Boston, MA*

Asma Nusrat • *Department of Pathology and Laboratory Medicine, Epithelial Pathobiology Unit, Emory University School of Medicine, Atlanta, GA*
Nobuko Obara • *Department of Oral Anatomy, School of Dentistry, Health Sciences University of Hokkaido, Hokkaido, Japan*
Luisa Olivi • *The Kennedy Krieger Institute, John Hopkins University Bloomberg School of Public Health, Baltimore, MD*
Valeria Orlova • *Experimental Immunology Branch, National Cancer Institute, National Institutes of Health, Bethesda, MD; Department of Internal Medicine, University of Heidelberg, Heidelberg, Germany*
Charles A. Parkos • *Division of Gastrointestinal Pathology, Department of Pathology and Laboratory Medicine, Emory University School of Medicine, Atlanta, GA*
Jonathan A. Phillips • *Department of Gravitational Biology, Ames Research Center, Moffett Field, CA*
Punitha Ramalingam • *Center for Experimental Therapeutics and Reperfusion Injury, Brigham and Women's Hospital, Harvard Medical School, Boston, MA*
Joe W. Ramos • *Cancer Research Center of Hawaii, University of Hawaii at Manoa, Honolulu, HI*
Ernesto Sabath • *Renal Division, Brigham and Women's Hospital, Harvard Medical School, Boston, MA*
Robert Sackstein • *Harvard Skin Disease Research Center and Departments of Dermatology and Medicine, Brigham and Women's Hospital, Harvard Medical School, Boston, MA*
Anthea Scothern • *Faculty of Life Sciences, University of Manchester, Manchester, United Kingdom*
Charles N. Serhan • *Department of Anesthesiology, Perioperative, and Pain Medicine, Center for Experimental Therapeutics and Reperfusion Injury, Brigham and Women's Hospital, Harvard Medical School, Boston, MA*
Alexander Swidsinski • *Innere Klinik, Gastroenterologie, Charit, Berlin, Germany*
Markus Utech • *Department of Pathology and Laboratory Medicine, Epithelial Pathobiology Unit, Emory University School of Medicine, Atlanta, GA; USA Department of General Surgery, University of Münster, Münster, Germany*
Susan Voss • *Division of Gastrointestinal Pathology, Department of Pathology and Laboratory Medicine, Emory University School of Medicine, Atlanta, GA*
Thomas Weissmüller • *Department of Anesthesiology and Intensive Care Medicine, Tübingen University Clinic, Tübingen, Germany*
Genevieve B. Wortzman • *Division of Reproductive Biology, Department of Biochemistry and Molecular Biology, Johns Hopkins University Bloomberg School of Public Health, Baltimore, MD*
Ramnik Xavier • *Department of Medicine and Molecular Biology, Massachusetts General Hospital, Harvard Medical School, Boston, MA*

List of Color Plates

Color Plates follow p. 50

1

Blood–Brain Barrier and Cell–Cell Interactions

Methods for Establishing In Vitro Models of the Blood–Brain Barrier and Transport Measurements

Michael Aschner, Vanessa A. Fitsanakis, Ana Paula Marreilha dos Santos, Luisa Olivi, and Joseph P. Bressler

Summary

This chapter describes in vitro methods for studying the blood–brain barrier. These methods include a cell line and isolated brain microvessels. The rat brain endothelial cell line 4 (RBE4) express many properties that are expressed by brain endothelial cells in vivo. Tissue culture methods allow the investigator to design experiments for studying transporters and permeability that would be much more difficult in vivo. A method for making preparations of isolated brain microvessels also is described. These preparations are highly enriched and also can be used for studying transport in vitro, but their short life span is a limitation. Two methods are discussed for measuring transport in cell culture. In one method, permeability is measured across a cell monolayer. This method is useful for measuring luminal and abluminal transport. The second method is especially designed for measuring the families of efflux transporters. These in vitro methods will complement many of the in vivo techniques, and they may be used as screening for more timely and expensive experiments, and also reducing the need for experimental animals.

Key Words: Blood–brain barrier; RBE4 cells; brain microvessels; transporters; permeability; multidrug resistance.

1. Introduction

The blood–brain barrier (BBB) is a specialized capillary bed separating the blood from the central nervous system parenchyma. It regulates amino acid, peptide, and protein transport, allowing for tight control of the composition of the brain's interstitial fluid composition. It also assures that changes in blood metabolite and ion composition (e.g., nutrients) are not reflected within the brain's milieu by preventing charged and large molecules from entering the brain.

From: *Methods in Molecular Biology, vol. 341: Cell–Cell Interactions: Methods and Protocols*
Edited by: S. P. Colgan © Humana Press Inc., Totowa, NJ

Two main factors contribute to the BBB's restrictive nature, and they include a physical barrier comprised of tight junctions between the endothelial cells ***(1,2)*** and a plethora of proteins on the endothelium, which play a key role in regulating influx and efflux of nutrients, ions, and other materials between the two (blood and brain) compartments ***(3,4)***. Brain capillary endothelial cells are polarized into luminal (blood-facing) and abluminal (brain-facing) plasma membrane domains. Nutrients are transported from the luminal side to the abluminal side by transporters. Additionally, the BBB expresses efflux pumps belonging to the superfamily of adensoine-5′-triphosphate (ATP)-binding cassette (ABC) proteins that prevent a broad range of amphiphatic chemicals from entering the brain. Efflux pumps that have been discovered thus far in the BBB include multidrug resistance glycoprotein, also referred to as P-glycoprotein (and most recently referred to as ABCB1), and breast cancer resistance protein (and most recently referred to as ABCG2 ***[5]***). ABCB1 and ABCG2 are expressed on the luminal surface of brain endothelial cells in the BBB. The expression of members of the family of multidrug resistance-associated proteins (ABCC family) has been reported in cultures of bovine brain endothelial cells ***(6)*** but not in isolated human ***(7)*** or rat microvessels ***(8)*** ABCB1 and ABCG2 transport lipophilic and cationic compounds with overlap ***(9,10)***, whereas multidrug resistance proteins transport organic anions, glutathione, or glucuronide-conjugated compounds, as well as various nucleosides ***(11)***.

In vitro and in vivo methods have been described for studying permeability of the BBB. Obviously, the advantage of in vivo methods is that the integrity of the BBB is maintained, but these methods have distinct disadvantages when compared with in vitro methods, including cost and sensitivity. Also, the experimental design is more difficult when the experiments involve inhibitors and inducers of transporters. In an in vitro method, such as cell culture, it is relatively easy to modify the activity of transporters. Nonetheless, a distinct disadvantage of cultures of brain endothelial cells is that they do not fully mimic the properties of the BBB *in situ*. The most notable property that is not fully mimicked when using in vitro models of the BBB is the expression of a tight barrier (commonly measured as the electrical resistance ***[12]***). Accordingly, a combination of both in vivo and in vitro methods is needed.

This chapter discusses two in vitro models for studying permeability and transport; these include cultures of the rat brain endothelial cell line (RBE4) and isolated brain microvessels. We also describe methods for examining transport permeability. In addition to RBE4 cell line, other cell culture models are available. Primary cultures of brain endothelial cells, for example, commonly are used. Monolayer cultures of porcine and bovine brain endothelial cells have been reported to display higher electrical resistance than most cell lines ***(13–15)***. Nonetheless, a wide variation in electrical resistance has been observed in different preparations of primary cultures of endothelial cells from bovine and porcine brain from the same laboratory ***(13,14)***. Also, a practical issue in preparing primary cultures is the laboratory's proximity to the slaughterhouse. Poor cell yield will be obtained when brains are kept on ice for longer than 1 h. We refer the reader to manuscripts describing methods for preparing primary cultures of brain endothelial cells ***(14,15)***.

2. Materials

1. Minimal Essential Media (MEM), with Earle's salts (Gibco-BRL, Grand Island, NY; cat. no. 11095-080). MEM was modified to contain no Ca^{2+}, which can produce cell clumping as the result of interactions of extracellular matrix proteins.
2. Heat-inactivated horse serum (Gibco-BRL; cat. no. 26050-088). Repeated freeze–thaw cycles of serum should be avoided. To thaw frozen serum, place the bottle in a refrigerator for approx 24–48 h. If rapid thawing is needed, place in a water bath no warmer than 40°C. Once thawed, serum can be stored at 4°C for approx 1 wk; however, we suggest aliquoting into 50-mL tubes and storing at –20°C. These 50-mL tubes can then be thawed as needed at room temperature over a few hours.
3. Penicillin G (10,000 U/mL)–streptomycin sulfate (10,000 µg/mL; Gibco-BRL; cat. no. 15140-122). Storing penicillin/streptomycin in 10-mL aliquots at –20°C is suggested to avoid repeated freeze–thaw cycles and because it provides the correct amount for 1 L of media.
4. Fungizone® (Amphotericin B, 250 µg/mL; Gibco-BRL; cat. no. 15290-018): its use if optional to prevent fungal or yeast contaminations.
5. Trypan blue 0.4%. Trypan blue is used determine cell viability. Intact cells are able to exclude trypan blue, whereas dead or damaged cells retain the dye. This is a useful method when determining plating density needed for the experiments.

2.1. Cell Culture

1. RBE4 cells: Neurotech SA (Evry, France) graciously provided the RBE4 cells.
2. MEM, alpha medium, F-10 nutrient mixture, glutamine, penicillin, streptomycin, heat-inactivated fetal bovine serum, and geneticin (G418).
3. T-75 flasks, which must be collagen coated and are obtained from BD Bioscience (Bedford, MA). Experiments are routinely conducted after microscopic and visual confirmation of confluence and are performed in HEPES buffer of the following composition: 25 m*M* HEPES/NaOH, 122 m*M* NaCl, 3.3 m*M* KCl, 0.4 m*M* $MgSO_4$, 1.3 $CaCl_2$, 1.2 m*M* KH_2PO_4, 10 m*M* glucose, adjusted to pH 7.4, and at 25°C unless otherwise desired (for example, as in studies examining the effect of temperature-dependence on transport kinetics).

2.2. Efflux Transporters

1. Hank's Balanced Salt Solutions (HBSS; with Ca, Mg, and glucose) without phenol red. Verapamil is kept at 1 m*M* in HBSS in small aliquots at –20°C. Working dilution is 20 μ*M* in HBSS, and the final concentration is 10 μ*M*.
2. Calcein AM is made at 1 m*M* in dimethyl sulfoxide and kept in small aliquots in the freezer at –20°C. It is light sensitive. The final concentration is 2 μ*M* in HBSS.
3. MicroFluor 96-well black plates.
4. Phosphate-buffered saline with 1% Triton X-100 or an equivalent nonionic detergent. Store the buffer at room temperature.

2.3. Isolated Microvessels

1. Media 199.
2. Nytex nylon mesh (www.sefar.com).
3. Dextran.
4. Dounce homogenizer with a Teflon pestle (0.25-mm clearance). The homogenizer is purchased from a supply company and shaved by a machine shop.
5. Motorized homogenizer.
6. Dissecting tools, such as scissors and forceps.
7. Kim wipes.
8. 250-mL flat-bottom centrifuge tubes with lid.
9. Buchner funnel and fitting flask.
10. 3-mL syringe.
11. 0.25-mm glass beads.

3. Methods

3.1. Cell Culture

Our focus is on RBE4 cells in describing a cell culture method for studying the BBB. These cells were derived from rat brain microvascular endothelial cells immortalized with the plasmid pE1A-neo containing the E1A region of adenovirus 2 and a neomycin resistance gene ***(16–18)***. The RBE4 cell line has been used in our laboratory for a number of years with great success, and these cells have proven to be a useful tool for probing the characteristics of the BBB in vitro. RBE4 cells have been used extensively by others as well, representing a wide range of transporter studies, including multidrug transporters ***(19)*** serotonin (5-HT ***[20]***; L-3,4-dihydroxyphenylalanine or L-DOPA ***[21]***), organic cation transporter/carnitine transporter OCTN2 ***(22)***, and glucose transporter 1 ***(23)*** RBE4 cells also express the specific brain endothelial enzymes γ-glutamyltranspeptidase and alkaline phosphatase ***(16,24)***.

The RBE4 cell cultures are maintained in T-75 flasks coated with rat tail collagen I and subcultured into transparent six-well plates with porous nitrocellulose and collagen I-coated inserts (both from BD Biosciences, Bedford, MA) for transport studies. The flasks and/or plates are routinely monitored by inverse microscopy before their doubling rate, approximately every 20 to 24 h, and the media are replaced after each doubling. In reality, the media is replaced approximately every 48 h to prevent the unnecessary disruption of the RBE4 cell monolayers (*see* **Note 1**). RBE4 monolayers in the flasks are subcultured every 3 to 4 d just as each flask's monolayer becomes confluent at approx 1.5×10^5 cell/cm^2. During this fractional splitting, these newly confluent monolayers are washed with 3 mL of 0.125% trypsin, detached with 1 mL 0.25% trypsin, centrifuged at very low speed (~1000*g* for 1 min), and fractionated 1:3 or 1:4, depending on the quality of the monolayer and the time frame in which they are needed. This procedure is routinely conducted for each flask and for each type of media (either astrocyte-conditioned media [ACM] or normal alpha medium) at approx 10^5 cell/cm^2.

This method is a gentler way of manipulating the eukaryotic RBE4 cells and their collagen adherent proteins. Trituration, after the addition of trypsin, is accomplished more easily and with less shearing. The result is a more uniform and disbursed initial cell layer over the hydrated nitrocellulose collagen filters and a faster growing, uniformly confluent monolayer in the model chambers (*see* **Notes 2–4**).

The media specifications are largely dependent on the experimental design and the parameters that are to be tested. We have traditionally used two different types of medium for RBE4 growth. These include the traditional alpha-type media ***(16,18)*** and ACM, which is known to decrease transendothelial permeability by physically tightening the barrier and reducing transluminal diffusion ***(25)*** and thus representing a more amenable model system for probing issues related to transport properties of the BBB (*see* **Note 5**).

ACM is made just before use. It consists of equal parts (1:1, v/v) of fresh astrocyte growth media (MEM with Earl's salts containing 10% heat-inactivated horse serum, 100 U/mL penicillin, 100 µg/mL streptomycin, and 0.25 µg/mL Fungizone) and used astrocyte growth media siphoned from stock primary astrocyte colonies maintained in the laboratory.

3.2. Transcellular Transport

The RBE4 cells are grown to confluent monolayers on porous, collagen I-coated, nitrocellulose filters and transferred to six-well plates. Each well has the porous filters attached to the base acrylic-walled inserts. Each insert sits securely suspended in its respective well, creating a top well above the cell monolayer and a bottom layer below each filter. Permeability and flux derivations are based on a one-dimensional permeability mathematical model derived from Fick's law:

$$J_F(x,t) = -D\,(\delta C(x,t)/\delta x)$$

where D is the gross coefficient of diffusion, and C is the chemical expression of concentration of distributed mass. The minus sign insures a stable, thermodynamically sound system with a downhill concentration gradient.

Unidirectional flux (J) is calculated using the following equation:

$$J = \Delta Q/\Delta t^{*}A$$

where Q is the fraction of the original radioisotope on the basolateral side of the culture model to the original activity applied to the apical side of the system. This value, Q (nanomoles), is divided by time of the experiment (t; minutes), and area (A; cm^2) of the filter.

Concentration apparent permeability ($concP_{app}$) is calculated by dividing flux (J) by the initial concentration (C_a) used on the apical side of the model BBB system ***(26,27)***:

$$concP_{app} = J/Ca.$$

Results are presented as both $concP_{app}$ relative to percent control and as equivalent nanomoles over the time course of the assay.

The permeability or flux of $^{54}Mn^{2+}$ across confluent monolayers of RBE4 cells is quantified by liquid scintillation counting. The asymmetric timepoints chosen for measurement were 1, 5, 10, 30, 60, and 120 min. They proved to be spaced well enough apart for practical considerations and long enough in breadth between and in duration for Michaelis–Menton saturation kinetic calculations based on flux rate and concentration to be practicable as well.

The results are reported as mean ± SEM. All experiments are analyzed routinely per well, with $n \geq 9$ (one should use power analysis to calculate the number of wells and replicates needed to perform sound statistical analyses). In our laboratory, we routinely use InStat (GraphPad Software, El Camino Real, San Diego, CA) to analyze the data by one-way analysis of variance with a post-hoc test such as Tukey, when the number of groups is greater than two (to avoid multiple Student's *t*-tests). Results are considered to be statistically significantly from controls if the $p < 0.05$.

3.3. Efflux Pumps

The most direct method of assaying efflux pumps is a functional assay measuring intracellular accumulation of a radioactive or fluorescent substrate (probe) in the presence of a competing substrate. The competition will result in greater uptake of the probe. Alternatively, ATPase assays ***(28,29)*** have been described but are difficult to interpret because P-glycoprotein-associated ATPase activity is affected differently by different substrates. We prefer fluorescent probes in the functional assay because they are less expensive yet display the same level of sensitivity as radioactive probes. Instruments that have been used for measuring fluorescence include fluorescence-activated cell sorters, image analysis systems, spectrofluorometers, and multiwell plate fluorescent readers. The choice depends on the experimental design. In studying heterogeneous populations of cells, fluorescence-activated cell sorters and image analysis systems have the best potential of determining the cell type expressing the efflux pump. A spectrofluorometer (or image analysis system) would enable the investigator to examine the same population of cells over time. A multiwell plate fluorescent reader allows high-input screening and might be more readily available to the investigator than the other instruments. Because of availability, we use the multiwell plate fluorescent reader. Also, multiplate fluorescent readers can be used because many of the commonly used fluorescent substrates are strong emitters that can be easily detected.

Several probes have been reported for measuring *ABCB1*, *ABCG2*, and *ABCC* family. In assaying *ABCG2*, mitoxantrone, BODIPY-prazosin ***(30)***, and pheophorbide have been used ***(31)***. An issue with the first two is that both are also substrates for *ABCB1* ***(32,33)***, whereas pheophorbide is specific for *ABCG2* ***(31)***. Rhodamine 123 often is used as a specific substrate for P-glycoprotein, but at least one study found an interaction between rhodamine 123 and *ABCC1* ***(34)***. Also, rhodamine 123 is not transported by wild-type BCRP but transported by a mutated form of BCRP (Arg 482 to Thr) The mutation is commonly found in different cell lines ***(35)***. Carboxyfluorescein diacetate A often is used for studying many of the members of the *ABCC* family of efflux pumps.

Inhibitors of *ABCB1* and *ABCG2* are verapamil ***(36)*** and fumitremorgin C ***(37)***, respectively. The latter compound is not available currently for commercial use but can be obtained from the Natural Products Branch, Developmental Therapeutics Program, Division of Cancer Treatment and Diagnosis, and the National Cancer Institute. These drugs are often used as positive controls.

As we discussed previously, the underlying principal of the functional assay is that the putative substrate competitively inhibits efflux of the fluorescent probe, resulting in higher levels of probe inside the cells that is indicated by increased fluorescence. Most probes are lipophilic and may nonspecifically leak out of the cell, resulting in a lower intracellular fluorescence. To circumvent this problem, we have used calcein-acetoxymethylester (calcein-AM) to study *ABCB2* and members of the *ABCC* family. Upon entering the cell, the ester is cleaved by nonspecific esterases, and the calcein is impermeable and will not leak from the cell. We also would like to recommend caution when determining whether chemicals are substrates for efflux pumps. Many potential substrates are lipophilic and have a limited solubility in water.

The assay described here has been used for several cell lines. RBE4 cells, for example, are plated in 24-well dishes (collagen-coated wells) so that a confluent monolayer is achieved within 2 to 3 d. The investigator might be interested in coculturing endothelial cells with astrocytes to achieve higher levels of efflux

Table 1
Excitation and Emission Wavelengths of Commonly Used Probes

Probe	Excitation	Emission	Final concentration
Rhodamine 123	507	529	2.5 μ*M*
Calcein	493	515	1 μ*M*
BODIPY-prazosin	488	530	250 μ*M*
Pheophorbide	488	650	10 μ*M*

pumps (*see* **Subheading 3.1.** for culturing endothelial cells with astrocytes or in the presence of AC *[38]*). Co-cultures can be achieved either by plating astrocytes and endothelial cells on opposite sides of polycarbonate filters or by placing inserts containing cultures of astrocyte monolayers into endothelial cell monolayers in 24-well dishes for a few days. The efflux assay is performed immediately on the endothelial cells immediately after removing the insert.

Cultures are washed three times with HBSS and incubated at 37°C with 500 μL of verapamil and 500 μL of calcein-AM to achieve a final concentration of 10 μ*M* and 1 μ*M* of verapamil and calcein, respectively. In examining potential substrates for the efflux pumps, we routinely preincubate the cells with the substrate for 10 min at 37°C before adding calcein-AM. After 60 min with calcein-AM, the cells are again given three washes, and the intracellular calcein is extracted by incubating cells with 200 μL of phosphate-buffered saline/Triton X-100 for 30 min at 4°C. Because calcein-AM is light-sensitive, we shield the plates with aluminum foil. The extract is transferred to the microtiter plate, and the fluorescence is measured at an excitation of and emission as shown in **Table 1**. The data are expressed as relative fluorescence or computed as percent control (*see* **Note 6**).

3.4. Isolation of Microvessels From Bovine Brain

Isolated microvessels have been used for characterizing transporters and other brain capillary specific proteins and for studying signaling pathways. We use the term microvessel instead of capillary because the procedure isolates vessels that are 5 to 25 μm in diameter, which will include venules and arterioles, as well as capillaries. The isolated microvessels consist of endothelial cells, abluminally located pericytes, and their surrounding extracellular basement but are void of perivascular astrocytes that in vivo ensheath the microvessel ***(39)***. Whereas primary cultures of endothelial cells lose the expression of some of the genes are expressed in vivo and gain the expression of genes needed for sustenance and growth in culture, isolated microvessels are the best preparation of enriched endothelial cell protein. Indeed, typical fold enrichments of endothelial cell proteins (e.g., γ-glutatmyl peptidase and alkaline phosphatase) are approx 20 but might vary depending on the brain region ***(40)***. The reader should be aware of two problems in using isolated microvessels. First, isolated microvessels have a brief viability because they cannot maintain adequate levels of ATP ***(41)*** (*see* **Note 7**). The second problem is associated with transport measurements. In vivo, transport from blood to brain across the BBB is unidirectional. Most substrates start their journey from the luminal surface, cross to the abluminal surface, and enter the parenchyma. In isolated microvessels, transport is assayed by measuring uptake into the microvasculature and transport is bidirectional because both the luminal and abluminal surfaces are exposed to the substrate undergoing transport. Furthermore, the luminal surface might not be accessible to the substrate because the microvessels could have been damaged during the preparation. Nonetheless, amino acid and glucose transporters, and efflux pumps have been characterized in isolated microvessels ***(42)***. More recently, a method for studying transport has been described using fluorescent substrates with confocal microscopy ***(43)***. This approach has two advantages. First, undamaged capillaries can be selected and assayed. Second, a microscopic assay requires less material.

The procedure is used for isolating brain microvessels from rats ***(44)*** and cows ***(40)*** in our laboratory and likely has been applied to other mammals. When working with larger mammals, it is advised to work with only gray matter unless there is a specific reason for studying capillaries from white matter. It is important to keep brains at 4°C throughout the procedure. The basis of the procedure for isolating brain microvessels is that the integrity of brain capillaries is not compromised when brain is homogenized in a Dounce homogenizer. The pestle of the homogenizer is shaved by a machine shop to achieve a clearance of 0.25 mm. Because the microvessels are intact, but the remaining cells in the brain are disrupted, microvessels can be isolated as described in the next paragraph (*see* **Note 8**).

The following protocol is tailored for isolating brain microvessels from adult rats. Dissect the brains and place them on cellophane wrap on ice. Wipe the surface with a Kim wipe to remove meninges containing superficial arteries and membranes. Cut away the cerebellum and brain-stem and open up the two halves of the cortex to remove the choroid plexus from the ventricles. With forceps, remove arteries along the surface. Place the brain in a 250-mL beaker containing 50 mL of ice-cold Media 199 that was pregassed with 95% O_2/5% CO_2 containing 1% (bovine serum albumin [BSA]) solutions on ice. With four pairs of scissors, and two hands, mince the brain until it is the consistency of oatmeal. Divide the mince into four homogenization tubes and add 9 volumes of Media 199. Homogenize 20 strokes for 2 min slowly at 40 rpm with an electrical homogenizer so that there is no spillage.

Take the homogenate and dilute 1:10 with Media 199/BSA in flat-bottom centrifuge tubes that have closed lids. Shake the tubes for 1 min and centrifuge for 10 min at 1000*g*. Aspirate the supernatant and add 15% (wt/vol) dextran in Media 199 (*see* **Note 9**). Several investigators have substituted 25% BSA. Shake vigorously for 1 min and centrifuge for 10 min at 4000*g*. The myelin floats and the cell pellet containing microvessels and nuclei. If not shaken properly, myelin will not separate from the cell pellet. Aspirate (do not pour) the supernatant and wipe fat clinging to the walls of the centrifuge tube with Kim wipes. Resuspend the pellet thoroughly in ice-cold Media 199/1% BSA using a broken capillary pipet.

The microvessels are separated from larger vessels by filtering a nylon screen. Secure a piece of 118-µm mesh onto a 4-in. Buchner funnel with a rubber band. Wet the mesh and make a depression so liquid will flow to the center and not the edges. Add the suspension of microvessels to the filter and collect the microvessels that flow through. Fat and larger vessels will be retained. Keep washing the mesh with Media 199 while pushing aside fat and larger vessels with a spatula. Rinse the centrifuge tube making sure all of the microvessels have been removed. Approximately 10 mL of suspension will have resulted in at least 100 mL of flow-through. Centrifuge the flow-through at 4000*g* for 10 min and aspirate the supernatant. Resuspend the pellet with 4 mL of Media 199.

Microvessels are separated from nuclei through a column of glass beads. Plastic columns are prepared by cutting both ends of 3-mL syringe and securing one end with a small piece of 53-µm mesh that is secured with a rubber band. Fill the columns 1.5 cm with glass beads (0.25 mm) and wash column with 10 mL of Media 199. Add 1 mL of suspension to each column and allow the capillaries to be trapped by the glass beads. Squirt 2 mL of Media 199 into each column, let it flow until the media is collected, and repeat five times. Remove the column, add the glass beads and nylon mesh to 10 mL of Media 199 in a 50-mL tube, gently swirl, and allow the beads to settle and mesh, and transfer the fluid to a new centrifuge tube. Place a drop on a microscope slide and view under phase contrast. Small thin capillaries should be observed without contaminating nuclei. Wash the beads again and combine the washings in one tube. Centrifuge for 5 min at 500*g*. In addition to visually monitoring purity, we recommend comparing the specific activity of alkaline phosphatase or gamma-glutamyl transpeptidase between brain homogenate and microvessel homogenate. Alkaline phosphatase activity should be approx seven-fold greater and γ-glutamyl transpeptidase activity should be 20-fold greater in microvessels.

4. Notes

This section identifies major sources of problems associated with RBE4 cell culturing and notes how these problems can be identified and rectified. Specific attention is directed to the subculturing and culturing techniques along with a detailed section on troubleshooting.

1. A new subculturing technique has been used in our laboratory to negate the occurrence of aberrant confluence in some BBB model culture dish wells. The new method requires a wash step to clear any serum from the T-75 flask monolayer before trypsinization.
2. Important variables to consider when performing studies with cultures are the effects of confluency and time in culture. Unlike tumor cells, these cell cultures stop growing and dividing once confluent because of cell–cell interactions. It is strongly recommended that the effects of confluency be determined by treating cells at various states of confluency. In a related issue, the time the cells are maintained in culture can have profound effects on the genotype and phenotype of the cells. Thus, preliminary studies should examine temporal changes in expression of the product of interest, and subsequent studies must contain appropriate controls to account for temporal changes in expression during treatments.
3. Microbial contamination of cell culture is probably the most common problem encountered once the initial cell isolation is successful. Contamination may be the result of invasion by bacteria, yeast, fungi, molds, mycoplasma, viruses, or any combination of these. Although isolated contamination of an individual culture plate or two is not uncommon, widespread or continual contaminations (in more than two separate cultures) require investigation into the source of the microbes. Once contamination is suspected, the rapid and accurate identification of the offending class of microbe can aid in determining the source of contamination. We highly recommend if contamination of cells is suspected that a sample be sent to a certified microbiological laboratory for identification of the class of microbe. Any infected culture plates must be disposed of immediately. All surfaces, including microscopes, pipettors, incubator and refrigerator door handles, also should be cleaned with 70% ethanol.
4. Because the growth media contains antibiotics (penicillin and streptomycin), bacterial contamination is uncommon. The addition of gentamycin sulfate may be beneficial in controlling bacterial and mycoplasma contamination. When adding new antibiotics, we suggest beginning with the lowest recommended concentrations (available in Gibco-BRL catalog) and performing a dose-toxicity curve in a few noninfected plates to assess effects on growth rate and cytotoxicity. In general, the highest concentration of antibiotic that does not adversely inhibit cell growth should be used. In our experience, the most common sources of contamination are yeast and fungi. We have found the addition of 0.25 μg/mL Fungizone (Amphotericin B) effectively halts both yeast and fungal infections. We do not routinely add Fungizone to the growth media but use it only as necessary to control sporadic infections that seem to occur mainly during spring and summer. Cultures with yeast contamination can usually be readily identified by the examination of the media. The growth media appears cloudy and usually turns yellow as the result of acidification. The rapidly growing yeast consume the available glucose and convert

to anaerobic metabolism, which produces large quantities of lactic acid; it is this lactic acids which alters the pH (and thus the color) of the growth media. Microscopic examination of yeast-contaminated cultures reveals small round spores floating throughout the media. The source of yeast often is hard to identify and, thus, the list of possible suspects is long. Any item to be placed in the laminar flow hood should be cleaned with 70% ethanol, but special care must be taken when laboratory water baths are used to thaw or warm reagents for cell culture. The use of disposable plastic pipets can greatly reduce the possibility of contamination. Pipets should only be placed in the growth media once and used to feed one "stack" of culture plates. This prevents transfer of contaminates from a single plate to the growth media, which can then infect the entire culture. The Pasteur pipets used for removal of media should similarly be changed frequently. Fungus often is found growing directly on the cell layers, or even the walls of culture plates. It grows rapidly to form green to yellow sponge-like circles, and rapidly absorbs much of the growth media. When a fungal infection is discovered careful examination of the incubator is required, as this is often a reservoir of fungal spores. A thorough disinfection of the incubator with 70% ethanol is suggested. Mycoplasma contamination cannot be easily detected by visual inspection of the cultures. If cultures die before reaching confluency or are growing slowly, mycoplasma contamination may be suspected. Once again, we recommend sending samples to a certified microbiological laboratory for testing of mycoplasma. Alternatively, kits designed for mycoplasma detection are commercially available (MycoTest, Gibco-BRL, 15672-017 or Mycoplasma Detection Kit, Roche, 1296744).

5. The total time for the preparation of cultures is dependent upon the desired number of plates. Routinely splitting cells into 12 to 18 filters for the study of transport can take as long as 2 to 4 h. The changing of media ("feeding") and subculturing usually requires approx 1 to 2 h but will of course vary with the number of plates/T-75 flasks.
6. Because potential chemical substrates of efflux pumps are lipophilic and have limited solubility in water, care should be exercised in solubilizing the chemical. The functional assay could be conducted in 96-well plates for high-input screening. **Table 1** lists dyes that have been described for studying different transporters. The concentrations are those that have worked for us using the protocol described.
7. Isolation of Microvessels. Because microvessels become energy depleted, brains should not be kept on ice for longer than 60 min.
8. The most common cause for poor recovery is associated with homogenizing tubes that do not provide a 0.25-mm clearance. Homogenizing tubes should be discarded when poor recovery is noticed.
9. Dextran is made at 33% in water (w/v) with constant heating and stirring. A 2X (30%) working solution is made by diluting it 1 : 10 with 10X Media 199.

Acknowledgments

This review was partially supported by Public Health grants from the National Institutes of Health 10563 and 07331 (to M.A.), Department of

Defense DAMD17-01-1-0685 (to M.A.), Gerber Foundation (to J.P.B.), and National Institutes of Health Center Grant 03819 (to J.P.B.).

References

1. Reese, T. S. and Karnovsky, M. J. (1967) Fine structural localization of a blood–brain barrier to exogenous peroxidase. *J. Cell Biol.* **34,** 207–217.
2. Brightman, M. and Reese, T. (1969) Junctions between intimately apposed cell membranes in the vertebrate brain. *J. Cell Biol.* **40,** 648–677.
3. Vannucci, S. J., Maher, F., and Simpson, I. A. (1997) Glucose transporter proteins in brain: delivery of glucose to neurons and glia. *Glia* **21,** 2–21.
4. Bradbury, M. W. (1985) The blood–brain barrier. Transport across the cerebral endothelium. *Circ. Res.* **57,** 213–222.
5. de Boer, A. G., van der Sandt, I. C., and Gaillard, P. J. (2003) The role of drug transporters at the blood-brain barrier. *Annu. Rev. Pharmacol. Toxicol.* **43,** 629–656.
6. Zhang, Y., Schuetz, J. D., Elmquist, W. F., and Miller, D. W. (2004) Plasma membrane localization of multidrug resistance-associated protein homologs in brain capillary endothelial cells. *J. Pharmacol. Exp. Ther.* **311,** 449–455.
7. Seetharaman, S., Barrand, M. A., Maskell, L., and Scheper, R. J. (1998) Multidrug resistance-related transport proteins in isolated human brain microvessels and in cells cultured from these isolates. *J. Neurochem.* **70,** 1151–1159.
8. Regina, A., Koman, A., Piciotti, M., et al. (1998) Mrp1 multidrug resistance-associated protein and P-glycoprotein expression in rat brain microvessel endothelial cells. *J. Neurochem.* **71,** 705–715.
9. Doyle, L. A. and Ross, D. D. (2003) Multidrug resistance mediated by the breast cancer resistance protein BCRP (ABCG2). *Oncogene* **22,** 7340–7358.
10. Suzuki, H. and Sugiyama, Y. (2000) Role of metabolic enzymes and efflux transporters in the absorption of drugs from the small intestine. *Eur. J. Pharm. Sci.* **12,** 3–12.
11. Borst, P., Evers, R., Kool, M., and Wijnholds, J. (1999) The multidrug resistance protein family. *Biochim. Biophys. Acta* **1461,** 347–357.
12. Wolburg, H., Neuhaus, J., Kniesel, U., et al. (1994) Modulation of tight junction structure in blood-brain barrier endothelial cells. Effects of tissue culture, second messengers and cocultured astrocytes. *J. Cell Sci.* **107,** 1347–1357.
13. Wolburg, H., Neuhaus, J., Kniesel, U., et al. (1994) Modulation of tight junction structure in blood-brain barrier endothelial cells. Effects of tissue culture, second messengers and cocultured astrocytes. *J. Cell Sci.* **107,** 1347–1357.
14. Gaillard, P. J., Voorwinden, L. H., Nielsen, J. L., et al. (2001) Establishment and functional characterization of an in vitro model of the blood–brain barrier, comprising a co-culture of brain capillary endothelial cells and astrocytes. *Eur. J. Pharm. Sci.* **12,** 215–222.
15. Parran, D. K., Magnin, G., Li, W., Jortner, B. S., and Ehrich, M. (2005) Chlorpyrifos alters functional integrity and structure of an in vitro BBB model: co-cultures of bovine endothelial cells and neonatal rat astrocytes. *Neurotoxicology* **26,** 77–88.

16. Roux, F., Durieu-Trautmann, O., Chaverot, N., et al. (1994) Regulation of gamma-glutamyl transpeptidase and alkaline phosphatase activities in immortalized rat brain microvessel endothelial cells. *J. Cell Physiol.* **159,** 101–113.
17. Durieu-Trautmann, O., Bourdoulous, S., Roux, F., Bourre, J. M., Strosberg, A. D., and Couraud, P. O. (1993) Immortalized rat brain microvessel endothelial cells: II–Pharmacological characterization. *Adv. Exp. Med. Biol.* **331,** 205–210.
18. Roux, F., Durieu-Trautmann, O., Bourre, J. M., Strosberg, A. D., and Couraud, P. O. (1993) Immortalized rat brain microvessel endothelial cells: I–Expression of blood-brain barrier markers during angiogenesis. *Adv. Exp. Med. Biol.* **331,** 201–204.
19. Begley, D. J., Lechardeur, D., Chen, Z. D., et al. (1996) Functional expression of P-glycoprotein in an immortalised cell line of rat brain endothelial cells, RBE4. *J. Neurochem.* **67,** 988–995.
20. Brust, P., Friedrich, A., Krizbai, I. A., et al. (2000) Functional expression of the serotonin transporter in immortalized rat brain microvessel endothelial cells. *J. Neurochem.* **74,** 1241–1248.
21. Sampaio-Maia, B., Serrao, M. P., and Soares-da-Silva, P. (2001) Regulatory pathways and uptake of L-DOPA by capillary cerebral endothelial cells, astrocytes, and neuronal cells. *Am. J. Physiol.* **280,** C333–C342.
22. Friedrich, A., Prasad, P. D., Freyer, D., Ganapathy, V., and Brust, P. (2003) Molecular cloning and functional characterization of the OCTN2 transporter at the RBE4 cells, an in vitro model of the blood-brain barrier. *Brain Res.* **968,** 69–79.
23. Regina, A., Roux, F., and Revest, P. A. (1997) Glucose transport in immortalized rat brain capillary endothelial cells in vitro: transport activity and GLUT1 expression. *Biochim. Biophys. Acta* **1335,** 135–143.
24. el Hafny, B., Bourre, J. M., and Roux, F. (1996) Synergistic stimulation of gamma-glutamyl transpeptidase and alkaline phosphatase activities by retinoic acid and astroglial factors in immortalized rat brain microvessel endothelial cells. *J. Cell Physiol.* **167,** 451–460.
25. Yang, J., Mutkus, L. A., Sumner, D., et al. (2001) Transendothelial permeability of chlorpyrifos in RBE4 monolayers is modulated by astrocyte-conditioned medium. *Brain Res. Mol. Brain Res.* **97,** 43–50.
26. Pachot, J. I., Botham, R. P., Haegele, K. D., and Hwang, K. (2003) Experimental estimation of the role of P-Glycoprotein in the pharmacokinetic behaviour of telithromycin, a novel ketolide, in comparison with roxithromycin and other macrolides using the Caco-2 cell model. *J. Pharm. Pharm. Sci.* **6,** 1–12.
27. Brown, R. C., Mark, K. S., Egleton, R. D., and Davis, T. P. (2004) Protection against hypoxia-induced blood-brain barrier disruption: changes in intracellular calcium. *Am. J. Physiol. Cell Physiol.* **286,** C1045–C1052.
28. Ambudkar, S. V. (1998) Drug-stimulatable ATPase activity in crude membranes of human MDR1-transfected mammalian cells. *Methods Enzymol.* **292,** 504–514.
29. Senior, A. E., al-Shawi, M. K., and Urbatsch, I. L. (1998) ATPase activity of Chinese hamster P-glycoprotein. *Methods Enzymol.* **292,** 514–523.

30. Robey, R. W., Honjo, Y., van de Laar, A., et al. (2001) A functional assay for detection of the mitoxantrone resistance protein, MXR (ABCG2) *Biochim. Biophys. Acta* **1512,** 171–182.
31. Robey, R. W., Steadman, K., Polgar, O., et al. (2004) Pheophorbide a is a specific probe for ABCG2 function and inhibition. *Cancer Res.* **64,** 1242–1246.
32. Lee, Y. J., Kusuhara, H., Jonker, J. W., Schinkel, A. H., and Sugiyama, Y. (2005) Investigation of efflux transport of dehydroepiandrosterone sulfate and mitoxantrone at the mouse blood-brain barrier: a minor role of breast cancer resistance protein. *J. Pharmacol. Exp. Ther.* **312,** 44–52.
33. Shapiro, A. B., Fox, K., Lam, P., and Ling, V. (1999) Stimulation of P-glycoprotein-mediated drug transport by prazosin and progesterone. Evidence for a third drug-binding site. *Eur. J. Biochem.* **259,** 841–850.
34. Daoud, R., Kast, C., Gros, P., and Georges, E. (2000) Rhodamine 123 binds to multiple sites in the multidrug resistance protein (MRP1) *Biochemistry* **39,** 15,344–15,352.
35. Robey, R. W., Honjo, Y., Morisaki, K., Nadjem, T. A., Runge, S., Risbood, M., et al. (2003) Mutations at amino-acid 482 in the ABCG2 gene affect substrate and antagonist specificity. *Br. J. Cancer* **89,** 1971–1978.
36. Cornwell, M. M., Safa, A. R., Felsted, R. L., Gottesman, M. M., and Pastan, I. (1986) Membrane vesicles from multidrug-resistant human cancer cells contain a specific 150- to 170-kDa protein detected by photoaffinity labeling. *Proc. Natl. Acad. Sci. USA* **83,** 3847–3850.
37. Rabindran, S. K., He, H., Singh, M., Brown, E., Collins, K. I., Annable, T., and Greenberger, L. M. (1998) Reversal of a novel multidrug resistance mechanism in human colon carcinoma cells by fumitremorgin C. *Cancer Res.* **58,** 5850–5858.
38. Jodoin, J., Demeule, M., Fenart, L., Cecchelli, R., Farmer, S., Linton, K. J., et al. (2003) P-glycoprotein in blood-brain barrier endothelial cells: interaction and oligomerization with caveolins. *J. Neurochem.* **87,** 1010–1023.
39. Algers, G., Karlsson, B., and Sellstrom, A. (1986) On the composition and characteristics of microvessels isolated from the rabbit and bovine brain. *Neurochem. Res.* **11,** 661–670.
40. Wolff, J. E. A., Belloni-Olivi, L., Bressler, J. P., and Goldstein, G. W. (1992) Gamma-glutamyl transpeptidase activity in brain microvessels exhibits regional heterogeneity. *J. Neurochem.* **58,** 909–915.
41. Lasbennes, F. and Gayet, J. (1984) Capacity for energy metabolism in microvessels isolated from rat brain. *Neurochem. Res.* **9,** 1–10.
42. Betz, A. L. and Goldstein, G. W. (1986) Specialized properties and solute transport in brain capillaries. *Annu. Rev. Physiol.* **48,** 241–250.
43. Miller, D. S., Nobmann, S. N., Gutmann, H., Toeroek, M., Drewe, J., and Fricker, G. (2000) Xenobiotic transport across isolated brain microvessels studied by confocal microscopy. *Mol. Pharmacol.* **58,** 1357–1367.
44. Goldstein, G. W., Wolinsky, J. S., Csejtey, J., and Diamond, I. (1975) Isolation of metabolically active capillaries from rat brain. *J. Neurochem.* **25,** 715–717.

2

Interactions of Intestinal Epithelial Cells With Bacteria and Immune Cells

Methods to Characterize Microflora and Functional Consequences

Geraldine Canny, Alexander Swidsinski, and Beth A. McCormick

Summary

Epithelial cells at all mucosal surfaces are potentially apposed to bacteria, particularly in the intestine. It is established that intestinal epithelial cells (IECs) represent an important barrier between lamina propria cells and the potentially harmful lumenal contents. In addition, IECs are important immunoeffector cells with the capacity to release cytokines, chemokines, and other molecules involved in antigen presentation and immune defense. The interaction of IECs with intestinal bacteria can result in a decrease in barrier function and the development of inflammation, which is known to be an important factor in the development of intestinal pathology. The potential role of such crosstalk between bacteria and other intestinal cell types in normal physiology and/or pathophysiology is therefore a topic of intense investigation. In this chapter, we provide protocols for the identification of bacteria that are associated with the epithelium and mucosa in addition to functional assays examining the interactions of neutrophils with epithelial cells and epithelial cell-mediated killing of bacteria.

Key Words: Structural organization of intestinal microbiota; fluorescence *in situ* hybridization (FISH); bacterial–epithelial interactions; neutrophil transepithelial migration; bacterial killing by epithelial cells.

1. Introduction

Intestinal epithelial cells (IECs) represent the first line of defense against potentially harmful bacteria present in the lumen. The human intestine is colonized by a huge number of bacteria, with approx 10^{12} bacteria contained in a gram of colonic content, for example *(1)*.

From: *Methods in Molecular Biology, vol. 341: Cell–Cell Interactions: Methods and Protocols*
Edited by: S. P. Colgan © Humana Press Inc., Totowa, NJ

In addition, IECs are important immunoeffector cells with the capacity to release cytokines, chemokines, and other molecules involved in antigen presentation and immune defense ***(2,3)***. Although bacteria are clearly necessary for the development of normal immune function in the intestine, they also are involved in the pathogenesis of inflammatory bowel disease (IBD). Under normal circumstances, bacterial–epithelial interactions may result in the generation of tolerogenic signals, and no obvious inflammation ensues ***(2)***. The underlying mechanisms involved in normal individuals that are abrogated in IBD remain largely unknown. However, a significant number of both clinical and laboratory findings have provided evidence for the contribution of luminal bacteria to the pathogenesis of IBD ***(4)***, with considerable data from practically all animal models implicating bacteria in this process ***(5)***. The potential role of such crosstalk between bacteria and other intestinal cell types in normal physiology and/or pathophysiology is therefore a topic of intense investigation. In this chapter, we provide protocols for the identification of bacteria that are associated with the epithelium and mucosa in addition to functional assays determining the interactions of neutrophils with epithelial cells and the killing of bacteria by enterocytes. A large influx of neutrophils into the mucosa from the underlying vasculature occurs during intestinal bacterial infections ***(6,7)***. A marked infiltration of activated lymphocytes, macrophages, and granulocytes into the gut mucosa is also a salient feature in the pathology of IBD ***(8,9)***. This chronic influx and activation of immunocytes result in the sustained overproduction of reactive metabolites of oxygen and nitrogen, and it is thought that some of the intestinal injury and dysfunction observed in IBD results from the elaboration of these reactive species ***(10)***. Neutrophils traversing the intestinal epithelium can lead to the formation of crypt abscesses. Moreover, large-scale transepithelial polymorphonuclear (PMN) leukocyte migration causes decreased barrier function ***(11)***. Bacteria can use secretion systems to insert proteins into enterocytes and thus activate inflammatory signaling pathways. For example, SipA, a protein secreted by *Salmonella typhimurium*, is necessary and sufficient to drive PMN transmigration across model intestinal epithelia ***(12)***. The transepithelial migration protocol described in this chapter is of great use in understanding the dynamics and functional consequences of neutrophil interaction with enterocytes, the mediators produced by both cell types, how this process contributes to pathology, and how it might be prevented therapeutically ***(13)***.

The gastrointestinal expression of antimicrobial peptides is evolutionarily conserved ***(14)*** but it is a relatively recent discovery that epithelial cells have the capacity to actively kill bacteria. Indeed, antimicrobial peptides and proteins are thought to constitute an important facet of innate immunity in the intestine. Most antimicrobial peptides and proteins expressed by mammalian epithelial cells are members of protein families that mediate nonoxidative microbial cell killing by phagocytes *(15)*. Indeed enterocytes express members of the defensin family *(3)* and other antimicrobial proteins, such as BPI *(16)*, which are thought to be involved in the maintainance of homeostasis toward intestinal microflora and their products such as lipopolysaccharide (LPS). It is noteworthy that a deficiency in the production of certain defensins may play a role in the aetiopathogenesis of Crohn's disease *(17)*. The killing assay provided here can be used to determine the contribution of IEC to mucosal defense and whether this is subject to regulation.

Formation of sessile communities and their inherent resistance to antibiotics and host immune attack are increasingly identified as a source of many recalcitrant bacterial infections. These include periodontal disease, endocarditis, chronic obstructive lung disease, and foreign body-related infections *(18,19)*. With the exception of *Helicobacter pylori*, the impact of biofilms on the pathogenesis of intestinal diseases is unknown. We describe an easy practicable fluorescence *in situ* hybridization (FISH) methodology achieving high-quality pictures by accurate resolution of spatial structure and the composition of the microbiota associated with intestinal tissues. Probes are inexpensive and can be purchased from many oligonucleotide manufacturers (such as MWG Biotech, Ebersberg, Germany) using different fluorochromes. Cy3, Cy5, or fluorescein isothiocyanate (FITC) fluorochromes do not bleach quickly, have little auto fluorescence background, and allow high-quality micrographs (**Fig. 1**). Cy3 provides better resolution than Cy5, and Cy5 is better than FITC. The results with other fluorochromes are less encouraging. Presently, the sequences of more than 200 FISH probes targeting the bacterial rRNA at domain group and species levels are described in the literature. Some of the probes frequently used for evaluation of fecal communities are listed in **Table 1** *(20–23)*.

Contrary to expectations, none of the FISH probes that we have tested are absolutely specific. Depending on the microbial community investigated (e.g., human or animal intestine, pancreatic duct, gallstones, biliary stents) all FISH probes demonstrated some kind of cross-hybridization at the conditions of optimal stringency. Many probes delivering specific results in human intestine were widely cross hybridizing in murine material. Cross-hybridization of probes observed when characterizing biofilms within billiary stents and gallstones did not occur while investigating colonic microbiota even within the same patient. The choice of FISH probes must therefore be adjusted to the specific requirements of the investigation. FISH is an excellent tool for the assessment of spatial structure. However, care must be taken when interpreting the results, and the presence of specific bacterial groups must be confirmed using alternative methods such as culturing, polymerase chain reaction with subsequent cloning, and sequencing. Taken together, the methods

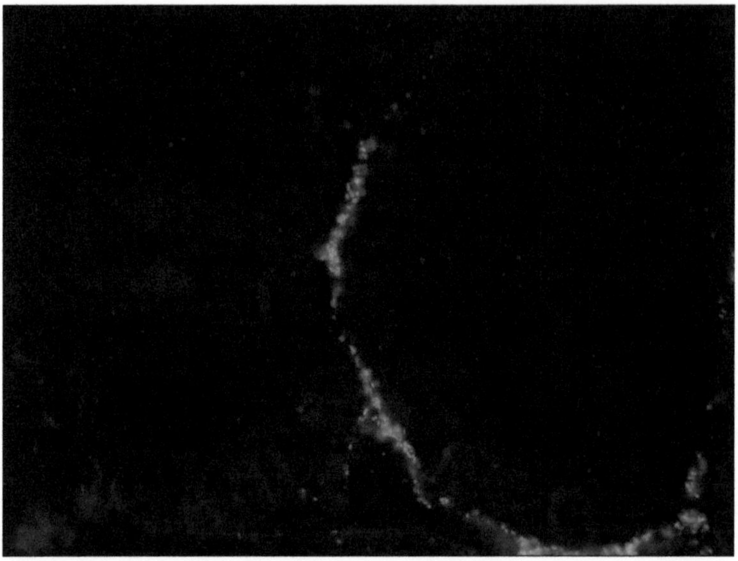

Fig. 1. Bacterial flora attached to the ileal mucosa in a patient with Crohn's disease (at a magnification of ×1000). The triangle of lumen is at the bottom right. Bacteria are tightly attached to the intact mucosal surface. *Bacteroides* is detected using an orange-labeled Cy3 probe, Erec (*Eubacterium rectale Clostridium coccoides* group), is stained red (Cy5). No bacteria are found in normal controls (data not shown). (Please *see* color insert following p. 50 for a color version of this figure.)

Table 1
FISH Probes

Name	Target	Reference
Eub338	Virtually all bacteria, Kingdom (*Eu*)*Bacteria*	***20***
Ebac	*Enterobacteriaceae*	***21***
Erec482	*Clostridium coccoides-Eubacterium rectale* group	***22***
Chis150	*Clostridium histolyticum* group	***22***
Clit135	*Clostridium lituseburense* group (incl. *Clostridium difficile*)	***22***
Strc493	*Streptococcus* group	***22***
Ecyl	*Eubacterium cylindroides* and other	***23***
Phasco	*Phascolarctobacterium faecium* group	***23***
Veil	*Veillonella* group	***23***
Rbro, Rfla	*Ruminococcus bromii, Ruminococcus flavefaciens* and other	***23***

provided here are of considerable use in characterizing intestinal bacterial flora and examining their interactions with other cell types in this environment to elucidate functional implications in health and disease.

2. Materials

2.1. Fluorescence In Situ *Hybridization*

1. Carnoy-solution (ethanol 6:glacial acetic acid 3:chloroform 1 from Sigma, St. Louis, MO) stored in the refrigerator.
2. Clean, filtered paraffin wax held at 2–4°C above its melting point (Kendall, Mansfield, MA).
3. A cold plate to rapidly cool the wax.
4. A supply of molds in which to embed the tissues.
5. SuperFrost® glass slides (Fisher Scientific, Pittsburgh, PA).
6. PAP-PEN (Birmingham, UK).
7. Forceps.
8. Xylene (Sigma).
9. Absolute alcohol.
10. FISH Probes. For commonly used probes, *see* **Table 1**.
11. Hybridization buffer for (2 mL): 360 μL of 5 *M* NaCl; 40 μL of 1 *M* Tris-HCl, pH 7.4, x mL (*see* **Table 2A**) formamide, deionized, 10 μL of 10% sodium dodecyl sulfate (the amount will vary depending on the probe). With the exception of formamide, which must be deionized, all stock solutions can be stored at room temperature (RT) for 1 mo.
12. Posthybridization wash buffer (for 50 mL): 1 mL of 1 *M* Tris-HCl pH 7.4, y mL of 5 *M* NaCl (depending on the formamide concentration in the hybridization buffer, *see* **Table 2B**), 500 μL of 0.5 *M* EDTA (if formamide is used), 50 μL of 10% sodium dodecyl sulfate. This buffer should be made up fresh. Buffers can be made according to **Table 2B**.
13. DAPI (Sigma), store at –20°C. Dilute to a working solution of 0.5 or 1 μg/mL.
14. For mounting SlowFade (Molecular Probes), ProLong anti-fade (Molecular Probes), or CitiFluor (CitiFluor, Ltd.) can be used.

2.1.1. Cell Culture

1. Cell culture medium: T84 intestinal epithelial cells are grown as monolayers in a 1:1 mixture of Dulbecco's Modified Eagle's Medium and Ham's F12 Nutrient mixture (Invitrogen) supplemented with 14 m*M* $NaHCO_3$; 15 m*M* HEPES buffer, pH 7.5, 40 mg/L penicillin, 8 mg/L ampicillin, 90 mg/L streptomycin (all Sigma); and 5% newborn calf serum (Sigma).
2. Solution of trypsin (0.25%) and EDTA (1 m*M*) from Invitrogen.
3. Hank's Balanced Salts Solution (HBSS) buffers. HBSS with Ca^{2+} and Mg^{2+} is termed HBSS+ (for 1 liter: 990 mL of deionized-distilled H_2O, 10 mL of 1 *M* HEPES, 9.75 g of HBSS+ powder [Sigma], 0.05 mL 10N NaOH; pH 7.4). HBSS without Ca^{2+} and Mg^{2+} is termed HBSS– (for 1 liter: 990 mL of deionized-distilled

Table 2A
Hybridization Buffer Composition

Formamide (%)	Formamide (µL) (=x)	Distilled H_2O (µL)
0	0	1600
5	100	1500
10	200	1400
15	300	1300
20	400	1200
25	500	1100
30	600	1000
35	700	900
40	800	800
45	900	700
50	1000	600
55	1100	500
60	1200	400
65	1300	300

Table 2B
Posthybridization Wash Buffer Composition

Formamide (%)	NaCl m*M* (µL)	5 *M* NaCl (µL) (=y)	0.5 *M* EDTA (µL)
0	900	9000	0
5	636	6300	0
10	450	4500	0
15	318	3180	0
20	225	2150	500
25	159	1490	500
30	112	1020	500
35	80	700	500
40	56	460	500
45	40	300	500
50	28	180	500
55	20	100	500

H_2O, 10 mL of 1 *M* HEPES, 9.5 g of HBSS– powder [Sigma], 0.05 mL 10 *N* NaOH; pH 7.4). All chemicals are obtained from Sigma.

4. Polystyrene T162 Tissue Culture Flasks with Phenolic Cap supplied by Fisher Scientific (Costar).

5. 0.33-cm^2 (5 μm pore size) collagen-coated polycarbonate filters (Costar Corp., Cambridge, MA).
6. Triton X-100 (10% solution; Sigma).
7. *N*-formyl-Met-Leu-Phe (1 μ*M* stock solution; Sigma).

2.1.2. Solution of Rat Tail Collagen

1. Rat tails.
2. 1% acetic acid (Sigma).
3. Spectropor no. 1 dialysis tubing (Fisher Scientific).
4. 70% ethanol.

2.1.3. Bacterial Growth Medium

1. Luria-Bertani broth (LB), per liter: 10 g tryptone/5 g yeast extract/10 g NaCl (Difco Laboratories Inc.).
2. Luria agar is LB containing 12 g of Bacto agar (Difco Laboratories Inc.) per liter.
3. MacConkey agar (Difco Laboratories) is prepared according to package instructions (50 g powder per liter of deionized-distilled H_2O).
4. Gentamicin (Sigma): 50 μg/mL prepared in deionized-distilled H_2O.

2.1.4. Neutrophil Isolation

1. 2% Gelatin (Sigma) prepared in HBSS. Make up 2% gelatin fresh per each use by preparing 3 g gelatin in 150 mL of HBSS– as follows: Preheat HBSS– to no more than 45°C (gelatin has collagen in it that could degrade). Then, add gelatin while stirring continually. Visually confirm that all of the gelatin is dissolved and let the solution cool to 37°C before using (do not over cool). Prepare 50 mL of 2% gelatin per 60 mL of blood drawn.
2. 13% Acid-citrate-dextrose per liter: 13.7 g of citric acid, 25 g of sodium citrate, 20 g of dextrose (all chemicals from Sigma).
3. Lysis buffer (keep at 4°C), per liter: 8.29 g of NH_4Cl, 1 g of $NaHCO_3$, 0.038 g of EDTA (all chemicals Sigma).
4. Hemocytometer.
5. Siliconized Pasteur pipets (Sigmacote; Sigma).
6. 2% Acetic acid (Sigma).

2.1.5. Myeloperoxidase Assay

1. 2,2′- azino-*bis*-(3-ethylbenzthiazoline-6-sulfonic acid) (ABTS) substrate solution. ABTS is made fresh before every use: 28 mg ABTS, 45 mL H_2O, 5 mL citrate buffer.
2. Citrate buffer (store at 4°C), for 500 mL: 400 mL of deionized-distilled H_2O, 73.5 g of sodium citrate, 52.5 g of citric acid, pH 4.2.
3. 30% H_2O_2. All chemicals obtained from Sigma.
4. 96-well plates.

2.2. Epithelial Cell-Killing Assay

1. Cell culture: Caco2 cells are cultured in Dulbecco's Modified Eagle's Medium supplemented with 10% fetal bovine serum, 100 U/mL of penicillin, and 100 μg/mL Streptomycin sulfate. All reagents were obtained from Sigma. Other necessary cell culture materials are listed in **Subheading 2.1.1**.
2. Sterile HBSS– (Sigma). Store at 4°C. Killing assays are conducted in the presence of HBSS– as calcium and magnesium inhibit BPI-mediated activity *(24)*; *see* **Note 1**.
3. Sterile water.
4. Sterilized Eppendorf tubes (Fisher Scientific).
5. Cell scrapers (Fisher Scientific).
6. Luria Bertani broth and agar plates, prepared as detailed in **Subheading 2.1.3**.

3. Methods

3.1. Fluorescence In Situ Hybridization

For Probe fixation, *see* **Note 2**.

1. Place the biopsy in an Eppendorf tube with 1.5 mL of cold Carnoy solution. Leave large probes longer in Carnoy (on average 1 h for each millimeter of tissue thickness).
2. Transfer the biopsy from Carnoy to the chilled absolute ethanol twice and from ethanol to xylene twice. Incubate the biopsy for 15 min in each. Embed the biopsy in liquid paraffin.
3. Paraffin embedding (*see* **Note 3**). Select the mold; there should be sufficient room for the tissue allowing for at least a 2-mm surrounding margin of wax.
4. Fill the mold with paraffin wax.
5. Using warm forceps, select the tissue, taking care that it does not cool.
6. Chill the mold on the cold plate, orienting the tissue and pressing it into the wax with warmed forceps. This step ensures that the correct orientation is maintained and the tissue surface to be sectioned is kept flat.
7. Insert the identifying label or place the labeled embedding ring or cassette base onto the mold.
8. Cool the block on the cold plate, or carefully submerge it under water when a thin skin has formed over the wax surface.
9. Remove the block from the mold.
10. Cross check block, label, and worksheet.
11. Cut the paraffin blocks to 2- to 10-μm thick slices and put them on super frost glass slices (two slices for each appointed FISH probe). It is important to have a properly fixed and embedded block and sharp knife or many biases can be introduced in the sectioning. Common biases include tearing, ripping, "venetian blinds," holes, and folding.
12. Once sections are cut, they are floated on a warm water bath, which helps remove wrinkles. They are then picked up on a glass microscopic slide.

13. Place the glass slides in a warm oven at 50°C for approx 60 min to help the section adhere to the slide.
14. Deparaffinize the slides by running them through xylene to alcohol shortly before hybridization: 4X 3-min incubations in xylene and 4X 3-min incubations in absolute ethanol; all steps can be carried out at RT. Change the ethanol and xylene solutions every 40 slices containing intestinal contents to prevent fecal bacteria floating in the fluid and being found on the slices.
15. Heat the slices once more in a warm oven at 50°C for approx 25 min.
16. Demark the area of the hybridization with a PAP-PEN and air-dry.
17. Prepare the hybridization chamber using a Tupperware container that seals well. Darken the chamber to avoid bleaching of the fluorescence. A platform can be created for the slides, that must by absolutely horizontal over the hybridization time. Soak several paper towels with hybridization buffer to keep the chamber humidified so the hybridizations do not dry out. Individual slides can also be hybridized in humidified, 50-mL plastic screw cap tubes.
18. Mix 16 μL of hybridization buffer with 2 μL of the FISH probes (~10 ng of each probe) and distribute it over the biopsy surface. Be careful not to scratch the probe while distributing it. Do not use cover slips; they can mechanically disturb the mucus layer and spatial structures.
19. Place slides in box and incubate for 60 to 90 min in a moisture chamber at probe-specific temperature (*see* **Note 4**).
20. Posthybridization washes; remove the slices from the hybridization chamber. They should not be dry! Rinse the slides with MQ H_2O and transfer them immediately into a 50-mL tube with washing buffer prewarmed to the required hybridization temperature (two slides in each tube). Wash for 5 min in a water bath at the required hybridization temperature.
21. After 5 min of washing, rinse the probes with MQ H_2O, air-dry.
22. DAPI counter staining; apply 20 μL of DAPI to each section, stain for 5 min at RT, wash with MQ H_2O, and air-dry.
23. Mounting; apply anti-fading mountant to the section. Add cover slip and view under an epifluorescent microscope and take pictures as desired (*see* **Note 5**). Because the slides fade quickly, they should be viewed within 2 d and stored in the dark.

3.2. Enteric Pathogen Interactions With Model IEC-Polarized Monolayers

3.2.1. Preparation of Rat Tail Collagen

The isolation of rat tail collagen is performed by the method of Cereijido et al. *(25)*.

1. Strip the skin from the rat tail; the tendons are almost pure collagen.
2. Soak 0.75 g of collagen in 70% ethanol for 20 min to dehydrate. Next, solubilize the tendons in 100 mL of 1 % acetic acid (v/v) precooled to 4°C. Stir the mixture overnight at 4°C.

At this point, the solution should be viscous and some undissolved material will remain.

3. Spin the collagen solution at 25,000*g* for 30 min at 4°C. Discard only the pellet and put the collagen preparation into dialysis tubing (Spectropor no. 1), which has been presoaked in distilled-deionized H_2O for 1 h.
4. Place the tubing in a 2 L beaker, add cold distilled-deionized H_2O, and dialyze for 24 h at 4°C against two to three changes of H_2O.
5. Remove the collagen from the dialysis tubing and dilute the solution if necessary. The collagen solution should be a little thinner than the consistency of honey.
6. Store at 4°C for 2 mo.

3.2.2. Preparation of T84 Cells Grown on Collagen-Coated Polycarbonate Filters

1. Conventional inserts are prepared for the use in bacterial invasion assays ***(26)***. A 50-µL aliquot of the rat tail collagen mixture is placed in each well. Care should be taken to make sure collagen is evenly distributed across the filter and that the collagen stock solution is maintained at 4°C.
2. Allow plates to completely dry (3–24 h).
3. Next, the T84 cells, which are grown on 162-cm^2 flasks are trypsinized, and 7.5×10^5 cells are then suspended in 75 µL of media and are placed in each transwell™ (Costar). A total 900 µL of cell culture media is placed in the outside chamber. Polarized monolayers prepared in this fashion can be used 6 to 14 d after plating. A steady-state resistance is reached 4 to 6 d after plating, with variability largely related to the cell passage number.
4. Inverted monolayers are prepared for use in neutrophil transmigration assays ***(26)***. The transwells (Costar) are removed from their respective wells and are placed upside down in a large gridded Petri dish; one invert to each petri dish grid square (each Petri dish holds 24 inverts).
5. A 50-µL aliquot of rail tail collagen is placed on the filter. When adding the collagen to the filter it is important to keep the solution at 4°C. Let the collagen dry on the filter overnight in the covered Petri dish.
6. Carefully add 75 µL of a 7.5×10^5 T84 cell suspension to each invert and incubate at 37°C overnight. The next day, use forceps to aseptically flip the inverts into a 24-well plate. Add 1 mL of cell culture medium to the outside well and 200 µL to the inside of the invert. Polarized monolayers prepared in this fashion can be used 6 to 14 d after plating. A steady-state resistance is reached 4 to 6 d after plating, with variability largely related to the cell passage number.

3.2.3. Growth of Bacteria

Nonagitated microaeophilic cultures of *S. typhimurium* are prepared by inoculating 10 mL of LB broth with 0.01 mL of a stationary-phase culture, followed by overnight incubation (18 h) at 37°C. Bacteria from such cultures are in late logarithmic phase of growth and represent approx 5–7 × 10^8 colony-forming units (CFUs)/mL. CFUs are determined by serially diluting the bacteria and plating onto MacConkey agar medium.

3.2.4. Neutrophil Isolation

The neutrophil isolation procedure is based on the method of Henson and Oades ***(27)***.

1. From a healthy human volunteer, draw 60 mL of blood directly into a 50-mL syringe that contains 6 mL of 13% acid-citrate-dextrose (pH 4.0).
2. Spin for the buffy coat (400*g*, 20 min at RT).
3. Using prepared, siliconized pasteur pipets, remove top yellow serum layer down through the white interface by gentle aspiration. The interface contains lymphocytes, platelets, and monocytes, while the neutrophils are in the red layer, and just below the white interface.
4. Once the buffy-coat is removed, dispense 20 mL of the blood per 50 mL Falcon™ centrifuge tube and then add the 2% gelatin bringing the final volume to 50 mL (i.e., should have three tubes/60-mL blood draw). Let stand for 25 min at 37°C, during which time the red blood cells will settle to the bottom of the tube.
5. Using a 10-mL pipet, combine as much of the clear supernatant as possible in new tubes and do not contaminate with the bottom red blood cell layer.
6. Spin the clear supernatant at RT in a tabletop centrifuge for 10 min at 400*g*.
7. Remove the supernatant by aspiration and gently resuspend the white pellet in 3 mL cold (4°C) lysis buffer per 50-mL tube.
8. Next, bring the volume of cold lysis buffer to 50 mL, and centrifuge at 4°C for 10 min at 1200*g*. Aspirate off the supernatant and discard.
9. Gently resuspend the pellet in 5 mL cold HBSS– and wash with 50 mL of cold HBSS–. Resuspend the pellet in 3 mL cold HBSS– and carefully break up the pellet by gentle pipetting.
10. To count the neutrophils, remove 5 µL of the suspension and place this in 500 µL of 2% acetic acid. Acetic acid prevents swelling of the neutrophils and lyses the red blood cells, permitting one to view the morphology of cells. Add 12 µL of the neutrophil suspension to each side of the hemocytometer and count the number of cells in each of the chambers and average the counts.
11. Resuspend the neutrophils at a final concentration of 5 × 10^7/mL in HBSS–. PMN are kept on ice in HBSS– and are used within 1 hr of being isolated. It is also important to keep the PMN in HBBS– buffer since the addition of divalent cations will cause them to become activated.

3.2.5. Invasion of S. typhimurium Into T84 Intestinal Epithelial Monolayers

Infection of T84 monolayers is performed by the method of Lee and Falkow ***(28)*** with slight modification ***(26)***.

1. Inserts with attached monolayers are lifted from the wells, drained of media by inverting, and gently washed by immersion in a beaker containing HBSS+, warmed to 37°C. Cells are incubated in HBBS+ buffer because the absence of divalent cations will cause the opening of intestinal epithelial tight junctions.
2. Next, the monolayers are placed in a new 24-well tissue culture plate bathed with 1.0 mL HBSS+ in the lower (outer) well and 0.05 mL HBSS+ added to the upper (inner well) and are then equilibrated for 30 min at 37°C.
3. 7 µL of each HBSS+ washed, bacterial sample (representing an inoculation ratio of 10:20 bacteria:epithelial cell) is added per monolayer for 1 h. If necessary, the bacterial invasion can be followed throughout a time-course of 1 h. Two populations of bacteria generally are assessed: cell associated and internalized. Cell-associated bacteria represent the population of bacteria adherent to and/or internalized into the T84 monolayers and are released by incubation with 0.1 mL of 1% Triton X-100 for 20 min at 4°C. Internalized bacteria are those obtained from the lysis of epithelial cells with 1% Triton X-100, 90 min after the addition of gentamicin (50 µg/mL). Gentamicin, an aminoglycoside antibiotic, does not permeate eukaryotic plasma membranes and is therefore cytolytic only to extracellular populations of bacteria while intracellular bacterial populations remain viable ***(29)***.
4. For both cell associated and internalized bacteria, 0.9 mL of LB is then added and each sample is vigorously mixed and quantified by plating for CFUs on MacConkey agar. To determine the number of adherent organisms the internalized population is subtracted from the cell-associated population (i.e., cell associated – internalized = adherent).

3.2.6. PMN Transepithelial Migration

Evaluation of neutrophil migration across monolayers of polarized intestinal epithelia stimulated by *S. typhimurium* is performed by the method of McCormick et al. ***(26)*** and is illustrated in **Fig. 2**.

1. T84 inverted monolayers are extensively rinsed in HBSS+ to remove residual serum components.
2. *S. typhimurium* is prepared by washing twice in HBSS+ and then are suspended to a final concentration of 5×10^9/mL. Then, 100-µL aliquots of the bacterial suspension are microcentrifuged at 13,000*g* for 3 min and resuspended in HBSS+ a final volume of 25 µL.
3. Inverted cell monolayers are removed from each well and placed in a moist chamber such that the epithelial apical membrane surface is oriented upward.
4. The bacterial suspension is gently distributed onto the apical surface and incubated for 45 min at 37°C.
5. Nonadherent bacteria are then removed by washing three times in HBSS+ buffer. The monolayers are then transferred, by inverting, back into the 24-well tissue culture tray containing 1.0 mL of HBSS+ buffer, in the lower (apical membrane now colonized with *Salmonella*) reservoir and 160 µL in the upper (basolateral interface) reservoir.

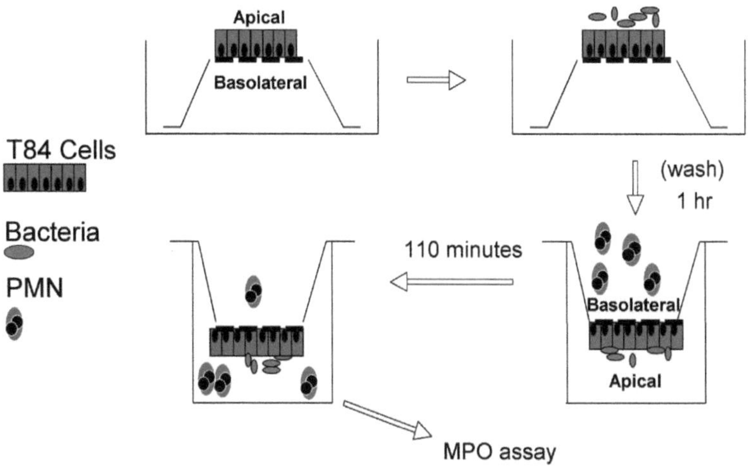

Fig. 2. Schematic for the protocol of *S. typhimurium*-induced neutrophil transepithelial migration. Inverted polarized, T84 cells are grown on the under-surface of, Transwell permeable supports. In this orientation, the apical surface of T84 cells was first colonized by *S. typhimurium* for 1 h. After the establishment of *S. typhimurium* infection, purified human peripheral blood neutrophils are then placed in the upper reservoir, interfacing with the epithelial basolateral membrane for 2 h. Such a maneuver is performed to achieve appropriate physiological orientation (i.e., neutrophils interacting basolaterally/*Salmonella* interacting apically).

6. To the basolateral bath, 40 mL (10^6) of isolated neutrophils are added to each monolayer and incubated for 110 min at 37°C. Positive control transmigration assays are performed by the addition of chemoattractant (1 μ*M* fMLP) to the opposing apical reservoir. All experiments are performed at 37°C.
7. Transmigration is quantified by a standardized assay for the neutrophil azurophilic granule marker myeloperoxidase. After each transmigration assay, nonadherent PMN are washed from the surface of the monolayer and neutrophil cell equivalents, estimated from a standard curve, are assessed as the number of neutrophils that had completely transversed the monolayer (i.e., into the apical reservoir).

3.2.7. Myeloperoxidase Assay

This assay is used to determine neutrophil transmigration across epithelial monolayers ***(26)***.

1. To assess the transmigrated neutrophil population, 50 μL of 10% Triton X-100 stock solution is added to the bottom reservoir. Rotate the plate carefully (setting ~150 rpm) at 4°C for at least 20 min.
2. During this time period, make up ABTS substrate solution and just before use add 1 μL of 30% H_2O_2 per mL of ABTS solution.

3. Prepare neutrophil standards by serially diluting 1×10^6 neutrophils down to 5×10^3.
4. Add 50 µL of citrate buffer to each well.
5. Remove 100 µL of sample from each well and add it into a 96-well plate in duplicate.
6. Add 100 µL of the ABTS solution (supplemented with H_2O_2) to the prepared 96-well plates.
7. Place at 37°C and let develop for 15 to 20 min.
8. Read the plate in an ELISA plate reader at a wavelength of 405 nm.

3.3. Epithelial Cell-Killing Assay

1. Culture *S. typhimurium* in LB broth at 37°C until the culture reaches an absorbance of 0.8 when spaced at 600 nm. This corresponds to approx 1×10^9 CFU/mL. If the bacteria have grown further, dilute the culture to this absorbance in HBSS–. These bacteria are in log phase.
2. Centrifuge the bacterial culture at 3000*g* for 10 min. Aspirate off the broth and resuspend the pellet in 5 mL of prewarmed HBSS–. Repeat this procedure once to wash the bacteria. Resuspend the bacteria in 5 mL HBSS–.
3. Fill and label Eppendorfs with 1 mL of HBSS–, these are used to dilute the samples. For purposes of accuracy it is important to pipet exact volumes.
4. Fill a 15-mL sterilin for each well of cells with 5 mL ice-cold sterile water and keep the sterilins on ice.
5. Take the plate of epithelial cells, aspirate off the medium and wash gently with 1 mL of warm HBSS–.
6. Reaspirate and add 1 mL of HBSS– to the cells.
7. Then incubate the cells with 100 µL of the bacterial suspension. We generally use 6 well plates of 7- to 10-d-old Caco2 cells but this method can be modified for cells cultured on inserts by adding HBSS– apically and basolaterally and incubating the cells with apically applied bacteria, thereby mimicking the physiological scenario. Incubate HBSS– and bacteria in an additional plate to serve as a bacterial control. This bacterial standard allows for control of bacterial death and growth independent of epithelial cells and will be used to calculate percentage killing.
8. Incubate the cell culture plates at 37°C on a rotating platform set to 60 rpm.
9. At desired time-points, scrape the cells using a cell scraper along with the supernatants into 5 mL ice-cold water to hypotonically lyse the epithelial cells and prevent further killing or growth of bacteria.
10. Take samples from epithelial lysates and the bacterial standard and serially dilute them by transferring 10 µL of lysate into 1 mL HBSS– and 100 µL of this dilution into 1 mL HBSS–. To reduce variability this volume should be added without touching the liquid in the Eppendorf. Plate triplicate 50-µL drops of 1/1000 and 1/10,000 dilutions on agar (this allows for replicates to be plated) and culture

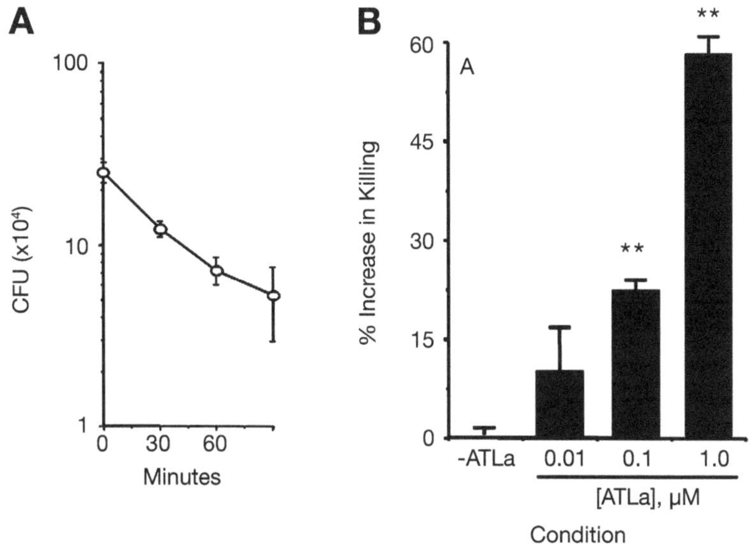

Fig. 3. **(A)** A time course depicting the killing of a smooth strain of *Salmonella typhimurium* by Caco2 cells. **(B)** Epithelial pre-exposure to the stable Lipoxin A_4 analog ATLa enhances this killing in a concentration-dependant fashion. Results are expressed as mean ± SEM.

the plates at 37°C overnight. Determine the percentage bacterial killing using the following equation:

Mean standard CFU – mean sample CFU/Mean standard CFU × 100 = % killing

where mean sample CFU was the mean of triplet drops for each dilution of the cellular lysate and standard was that of the bacteria alone. Results can also be expressed as CFU or CFU/mL. Alternatively, the plating may be performed using the "pour plate method" if desired. For Caco2 cells, significant killing has occurred by 15 min and, in our hands, it frequently is maximal by approx 60 min. For results of a typical experiment using a smooth strain of *Salmonella, see* **Fig. 3**. Care must be taken to ensure that the bacteria are not dying throughout the course of the experiment, thus negating the results. In this regard, *Salmonella* are viable for several hours in HBSS.

To prove that the killing is mediated by a specific peptide or protein, an inhibitory antibody or antisera may be used. In this case, prepare the antibody or antisera at the appropriate dilution and after washing the epithelial monolayer add 1 mL of this solution to the cells. Incubate for 30 min at 4°C. This is done to allow attachment of the antibody and prevent non-specific binding.

Thereafter, allow the plate to warm to 37°C, add the bacterial suspension and continue with the procedure above. Additional controls should be included to rule out the possibility of the antibody affecting bacterial growth.

4. Notes

1. Initial growth curves should be conducted to relate absorbance to bacterial number. We have used this protocol to assess the contribution of BPI to bacterial killing mediated by intestinal epithelial cells but the assay can be adapted for other mucosal epithelial cells using a relevant strain of bacteria. Anionic sites in the Lipid A region of LPS are normally occupied by divalent cations (Ca^{2+} and/or Mg^{2+}) which result in tight LPS packing. Bacterial/permeability increasing protein at antibacterial concentrations effectively competes for these sites in the LPS molecule and conversely, bound bacterial/permeability increasing protein can be displaced by these cations ***(24)***.
2. Before the spatial organization of intestinal microbiota can be investigated, it must be preserved. Shock frosting of the probes allows the preparation of native tissue slices with bacteria attached to the mucosa and intestinal contents. However, the fluorescent *in situ* hybridization takes place in an aqueous environment at temperatures of 46°C and higher. Hybridization occurs during a period of at least 90 min, which is necessary to achieve sufficient signals, but during this time bacteria are being progressively detached from the tissue and the spatial structure is lost. Formalin disintegrates spatial structures even before hybridization and tends to swell tissues rendering loose bacterial conglomerates within mucus more vulnerable to disintegration. The use of nonaqueous fixatives to preserve the mucus layer is crucial. We achieved excellent results preserving the material of human biopsies, surgically removed tissues and whole animal intestines with Carnoy solution. Once the structure of microbial communities is stabilized FISH proves to be a simple and reliable method.
3. The mechanical treatment of tissues while obtaining specimens leads to biases caused by dragging, crushing, and disrupting of normal anatomy. Fecal bacteria may be integrated into the distorted tissues and found at atypical anatomic sites and even within single cells. The optical integrity of the epithelial layer or of the tissue at the site of investigation is extremely important. Normally the autofluorescence background of human tissues allows good visualization and orientation within histological structures and the mucus layer. However, at least two additional stains (Alcian blue, hematoxylin and eosin, or other) for each biopsy or tissue specimens should be performed and are extremely helpful especially during optimization. No evaluation should be attempted when signs of tissue damage are obvious.
4. When optimizing probes for specificity, it is necessary to vary the hybridization temperature or formamide concentrations (increased temperature or formamide concentration means increased specificity). If possible do not use temperatures higher than 50°C or formamide concentrations higher than 55%. High temperatures result in mucus detachment and formamide introduces biases in eucaryotic tissues leading to bizarre fluorescent signals. Whenever possible, short hybridiza-

tion times of 90 min are preferable, and the hybridization should not exceed 6 h. In most cases, 90 min provides excellent resolution when Cy3, Cy5, or FITC fluorochromes are used.

5. The quantification of bacteria within spatial structures cannot be directly verified by stepwise dilution of bacteria. We use the approximation based on the following model. A 10-μL suspension of bacteria of 10^7 cells per mL applied to a glass surface in a circle of 1 cm results in 40 cells per average microscopic field of 200 μm diameter at a magnification of ×1000. One bacterium thus occupies an area of 785 μm^2. One bacterium averages 0.8 μm in size. Under these assumptions, a surface of 100 μm^2 containing one (or 100) bacterial cells will correspond to a concentration of 1×10^9 (or 1×10^{11}) cells per microliter in tissue slides that are 10-μm thick. 250 bacterial cells per 100 μm^2 does not allow the visual distinction of single bacteria but spaces between bacteria can still be seen; cases such as this were therefore assigned a concentration of 10^{11} cells per milliliter. A homogeneous carpet without any empty spaces between them occurs as bacteria increase to 2500 cells/100 μm^2; these cases were assigned a concentration of 10^{12} cells per milliliter. In cases of extremely high bacterial concentrations sections of 4 or even 2 μm in size can be performed. The spatial organization and the composition of the biofilm can be evaluated using a cumulative multi-step extension analysis which includes detection of bacteria using single FISH probes (Cy3) and DAPI DNA counter-stain. For each sample, probes that positively hybridize with more than 1% of the bacteria that are visualized by DAPI are further combined with each other in pairs and triplets and applied simultaneously using different fluorochromes in a single hybridization (Cy3/FITC/Cy5/ DAPI corresponds to orange/green/dark red/blue). This permits a three- or four-color analysis of the population structure within the same microscopic field. The hybridizations with different probe combinations are cumulatively extended until the position of all relevant bacterial groups to each other and their relative concentrations are clarified. When probes for unrelated bacterial groups hybridize with the same bacteria, it is necessary to adjust the hybridization stringency (*see* **Note 4**) until a clear differentiation of the bacterial groups is possible. Probes cross-hybridizing even under high stringency conditions should be excluded from the evaluation.

References

1. Berg, R. D. (1996) The indigenous gastrointestinal microflora. *Trends Microbiol.* **4,** 430–435.
2. Sansonetti, P. J. (2004) War and peace at mucosal surfaces. *Nat. Rev. Immunol.* **4,** 953–964.
3. Ouellette, A. J. (2004) Defensin-mediated innate immunity in the small intestine. *Best Pract. Res. Clin. Gastroenterol.* **18,** 405–419.
4. Swidsinski, A., Ladhoff, A., Pernthaler, S., et al. (2002) Mucosal flora in inflammatory bowel disease. *Gastroenterology* **122,** 44–54.
5. Fiocchi, C. (1998) Inflammatory bowel disease; etiology and pathogenesis. *Gastroenterology* **115,** 182–205.

6. Turnbull, P. C. and Richmond, J. E. (1978) A model of salmonella enteritis: the behaviour of, Salmonella enteritidis in chick intestine studies by light and electron microscopy. *Br. J. Exp. Pathol.* **59,** 64–75.
7. Wallis, T. S., Hawker, R. J., Candy, D. C., et al. (1989) Quantification of the leucocyte influx into rabbit ileal loops induced by strains of, Salmonella typhimurium of different virulence. *J. Med. Microbiol.* **30,** 149–156.
8. MacDermott, R. P. (1999) Chemokines in the inflammatory bowel diseases. *J. Clin. Immunol.* **19,** 266–272.
9. Ajuebor, M. N. and Swain, M. G. (2002) Role of chemokines and chemokine receptors in the gastrointestinal tract. *Immunology* **105,** 137–143.
10. Pavlick, K. P., Laroux, F. S., Fuseler, J., et al. (2002) Role of reactive metabolites of oxygen and nitrogen in inflammatory bowel disease (1,2). *Free Radic. Biol. Med.* **33,** 311–322.
11. Nash, S., Stafford, J., and Madara, J. L. (1987) Effects of polymorphonuclear leukocyte transmigration on the barrier function of cultured intestinal epithelial monolayers. *J. Clin. Invest.* **80,** 1104–1113.
12. Lee, C. A., Silva, M., Siber, A. M., Kelly, A. J., Galyov, E., and McCormick, B. A. (2000) A secreted *Salmonella* protein induces a proinflammatory response in epithelial cells, which promotes neutrophil migration. *Proc. Natl. Acad. Sci. USA* **97,** 12,283–12,288.
13. Mrsny, R. J., Gewirtz, A. T., Siccardi, D., et al. (2004) Identification of hepoxilin, A3 in inflammatory events: a required role in neutrophil migration across intestinal epithelia. *Proc. Natl. Acad. Sci. USA* **101,** 7421–7426.
14. Zasloff, M. (1992) Antibiotic peptides as mediators of innate immunity. *Curr. Opin. Immunol.* **4,** 3–7.
15. Nissen-Meyer, J. and Nes, I. F. (1997) Ribosomally synthesized antimicrobial peptides: their function, structure, biogenesis, and mechanism of action. *Arch. Microbiol.* **167,** 67–77.
16. Canny, G., Levy, O., Furuta, G. T., et al. (2002) Lipid mediator induced expression of bactericidal/permeability-increasing protein (BPI) in human mucosal epithelia. *Proc. Natl. Acad. Sci. USA* **99,** 3902–3907.
17. Fellermann, K., Wehkamp, J., Herrlinger, K. R., and Stange, E. F. (2003) Crohn's disease: a defensin deficiency syndrome? *Eur. J. Gastroenterol. Hepatol.* **15,** 627–634.
18. Costerton, J. W., Veeh, R., Shirtliff, M., Pasmore, M., Post, C., and Ehrlich, G. (2003) The application of biofilm science to the study and control of chronic bacterial infections. *J. Clin. Invest.* **112,** 1466–1477.
19. Wilson, M. (2001) Bacterial biofilms and human disease. *Sci. Prog.* **84,** 235–254.
20. Amann, R., Krumholz, L., and Stahl, D. A. (1990) Fluorescent-oligonucleotide probing of whole cells for determinative, phylogenetic, and environmental studies in microbiology. *J. Bacteriol.* **172,** 762–770.
21. Bohnert, J., Hübner, B., and Botzenhart, K. (2002) Rapid identification of Enterobacteriaceae using a novel 23S rR. N.A-targeted oligonucleotide probe. *Int. J. Hyg. Environ. Health* **203,** 77–82.

22. Franks, A. H., Harmsen, H. J., Raangs, G. C., Jansen, G. J., Schut, F., and Welling, G. W. (1998) Variations of bacterial populations in human feces measured by fluorescent in situ hybridization with group-specific 16S rR. N.A-targeted oligonucleotide probes. *Appl. Environ. Microbiol.* **64,** 3336–3345.
23. Harmsen, H. J., Raangs, G. C., He, T., Degener, J. E., and Welling, G. W. (2002) Extensive set of 16S rRNA-based probes for detection of bacteria in human feces. *Appl. Environ. Microbiol.* **68,** 2982–2990.
24. Elsbach, P. and Weiss, J. (1992) Oxygen-independent antimicrobial systems of phagocytes, in *Inflammation: Basic principles and Clinical Correlates, Second Edition* (Gallin, J. I., Goldstein, I. M., and Snyderman, R., eds.), Raven, New York, pp. 603–636.
25. Cereijido, M., Robbins, E. S,. Dolan, W. J., Rotunno, C. A., and Sabitini, D. D. (1978) Polarized monolayers formed by epithelial cells on a permeable and translucent support. *J. Cell Biol.* **77,** 853–880.
26. McCormick, B.A., Colgan, S. P., Delp-Archer, C., Miller, S. I., and Madara, J. L. (1993) *Salmonella typhimurium* attachment to human intestinal epithelial monolayers: transcellular signaling to subepithelial neutrophils. *J. Cell Biol.* **123,** 895–907.
27. Henson, P. M. and Oades, Z. G. (1975) Stimulation of human neutrophils by soluble and insoluble immunoglobulin aggregates: secretion of granule constituents and increased oxidation of glucose. *J. Clin. Invest.* **56,** 1053–1061.
28. Lee, C. A. and Falkow, S. (1990) The ability of *Salmonella* to enter mammalian cells is affected by bacterial growth state. *Proc. Natl. Acad. Sci. USA* **87,** 4304–4308.
29. Lissner, C. R., Swanson, R. N., and O'Brien, A. D. (1983) Genetic control of the innate resistance of mice to *Salmonella typhimurium* expression of the Ity gene in peritoneal and splenic macrophages is located in vitro. *J. Immunol.* **131,** 3006–3013.

3

The Role of Junctional Adhesion Molecules in Interactions Between Vascular Cells

Triantafyllos Chavakis and Valeria Orlova

Summary

Adhesive interactions between cells regulate tissue integrity as well as the process of inflammatory cell recruitment. Such intercellular interactions are regulated by adhesion receptors and can be homotypic, that is, between cells of the same type, for example, between adjacent endothelial cells in the vasculature, as well as heterotypic, that is, between different cells, such as the leukocyte endothelial interactions that take place during leukocyte extravasation. Emerging evidence points to the importance of the family of junctional adhesion molecules (JAMs), which are localized in interendothelial contacts and are implicated in the regulation of leukocyte extravasation. JAMs are members of the immunoglobulin superfamily and can undergo both homophilic and heterophilic interactions. This chapter deals with the role of JAMs in the regulation of adhesive interactions between vascular cells.

Key Words: Cell adhesion; transmigration; junction; JAM, inflammation; homophilic; heterophilic; co-immunoprecipitation; leukocyte; endothelial.

1. Introduction

Junctional adhesion molecules (JAMs) belong to the CD2 subgroup of the immunoglobulin superfamily and consist of two extracellular Ig-like domains, followed by a single transmembrane region and a short cytoplasmic tail. The genes for human JAM-A, JAM-B, and JAM-C are localized on the chromosomes 1, 21, and 11, respectively, and the encoded mature glycoproteins have molecular masses of 32, 40, and 43 kDa, respectively ***(1–3)***. JAM-A is found on different circulating blood cells, including granulocytes, monocytes, lymphocytes, platelets, and erythrocytes, as well as on endothelial and epithelial cells ***(4,5)***. JAM-B particularly localizes on vascular and lymphatic endothelium and especially at high endothelial venules ***(6)***. JAM-C is expressed on endothelial

From: *Methods in Molecular Biology, vol. 341: Cell–Cell Interactions: Methods and Protocols*
Edited by: S. P. Colgan © Humana Press Inc., Totowa, NJ

cells, and platelets ***(7–9)*** and on epithelial cells ***(10)***. All three JAMs have a class-II PDZ domain-binding motif at their final carboxy-terminus (three C-terminal amino acids). This motif predisposes the molecules to interact with PDZ-domain-containing molecules, such as the ones found in tight junctions (TJs ***[2]***). Such PDZ-domain molecules include zonula occludens (ZO)-1 ***(11)***, AF6/afadin ***(12)***, MUPP 1 (multi-PDZ-domain-protein 1 ***[13]***), and PAR-3 (partitioning defective), which forms a complex with atypical protein kinase C and PAR-6 ***(14,15)***. The association of JAMs with TJ-marker proteins such as ZO-1 can be studied with coimmunoprecipitation experiments. The codistribution of these molecules can also be confirmed by immunofluorescence studies. Thus, JAMs may be involved in a multiprotein complex located at TJs and, because ZO-1 and AF-6 are linked directly to the actin cytoskeleton ***(16)***, JAMs may be involved in the dynamic regulation of junction assembly. Here, a particular role may be assigned to the homophilic interaction between JAMs on two adjacent cells.

As opposed to the homophilic interaction, JAMs also can undergo heterophilic interactions and several lines of evidence point to the fact that JAMs also are engaged as counter-receptors for leukocyte integrins: in particular, JAM-A binds the leukocyte integrin $\alpha L\beta 2$ (LFA-1 ***[17]***), JAM-B associates with $\alpha 4\beta 1$ (VLA-4 ***[18]***), and JAM-C interacts with $\alpha M\beta 2$ (Mac-1 ***[9]***). The latter interaction also mediates leukocyte-platelet interactions. Moreover, the cellular localization of JAMs and these heterophilic binding capacities pointed to a potential role of JAMs in leukocyte extravasation. These heterophilic interactions of JAMs with leukocyte integrins and their potential applications in leukocyte transendothelial migration can be investigated by specific assays, including the binding of isolated proteins in vitro, the adhesion of leukocytes, as well as their transendothelial migration in vitro.

2. Materials

2.1. Cell Culture and Lysis for the Preparation of Samples

1. Endothelial cell growth medium (PromoCell, Heidelberg, Germany) supplemented with growth factors and 5% FCS (PromoCell).
2. Alpha-Minimal Essential Media growth medium (PAA, Linz, Austria) supplemented with 10% heat-inactivated fetal calf serum (FCS; Biowest, Nuaillé, France; *see* **Note 1**), 1% penicillin/streptomycin/glutamine, 1% Na-Pyruvate (GIBCO/BRL, Bethesda, MD).
3. Zeocin (Invitrogen, Carlsbad, CA); Accutase (PAA); Trypsin solution (10X) (GIBCO/BRL).
4. To prepare 0.2% gelatin solution, dilute 2% gelatin stock-solution (Sigma-Aldrich, Munich, Germany) in Hank's Balanced Salt Solution.
5. NP-40 lysis buffer: 50 m*M* Tris-HCl, pH 7.4, 150 m*M* NaCl, and 0.5% NP-40. Buffer should be stored at 4°C. Before use, add protease inhibitor cocktail tablets (Complete Mini, EDTA-free, Roche, Mannheim, Germany), phosphatase inhibitor

cocktail 1 (Sigma), 2.5 m*M* Na-ortovanadate (Stock 200 m*M*, Sigma; *see* **Note 2**), and 1 m*M* NaF (Sigma).
6. Cell scrapers (Greiner Bio-One, Frickenhausen, Germany).

2.2. Immunoprecipitation

1. Mouse monoclonal antibodies against human JAM-C (Gi11) were provided by Dr. S. Santoso (Giessen, Germany *[9]*).
2. UltraLink Immobilized Protein A/G (PIERCE, Rockford, IL).
3. Sample buffer (4X): 240 mM Tris-HCl, pH 6.8, 40% (w/v) glycerol, 8% (w/v) sodium dodecyl sulfate (SDS), 0.1% (w/v) bromophenol blue.

2.3. SDS-PAGE

1. Separating buffer (4X): 1.5 *M* Tris-HCl, pH 8.8, 0.4% SDS. Store at 4°C.
2. Stacking buffer (4X): 0.5 *M* Tris-HCl, pH 6.8, 0.4% SDS. Store at 4°C.
3. 10% (w/v) Ammonium persulfate solution (APS), dissolve in water, and store at room temperature (RT; up to 1 wk) or at 4°C.
4. Thirty percent acrylamide/*bis* solution (37.5:1) and TEMED (Sigma).
5. Running buffer (5X): 125 m*M* Tris-HCl, 1.25 *M* glycine, 0.5% (w/v) SGS. Store at RT.
6. Prestained protein molecular weight marker (Fermentas, Hanover, MD).

2.4. Western Blot

1. Transfer buffer: 20 m*M* Tris, 150 m*M* glycine, 20% (v/v) methanol.
2. Gel-blotting paper (Schleicher & Schuell, Dassel Germany) and Nybond-Enhanced Chemiluminescence (ECL) Nitrocellulose membrane (Amersham Biosciences, Freiburg, Germany).
3. PBS-T20: Prepare 10X phosphate-buffered saline (PBS) stock solution of 1.37 *M* NaCl, 27 m*M* KCl, 81 m*M* $Na_2HPO_4x2H_2O$, and 18 m*M* KH_2PO_4. Dilute PBS to 1X and add 0.05% (v/v) Tween-20.
4. Blocking buffer: 5% (w/v) nonfat dry milk in PBS-T20.
5. Primary antibody dilution buffer: 1% (w/v) BSA in PBS-T20.
6. Primary antibody: Rabbit anti-ZO-1 (Zymed, South San Francisco, CA).
7. Secondary antibody: Pork anti-rabbit IgG-HRP (Dako Cytomation, Hamburg, Germany).
8. ECL Western blotting detection reagent and Hyperfilm ECL (Amersham Bioscience).

2.5. Immunofluorescence

1. Eight-well chamber slides (Permanox; Nunc, Roskilde, Denmark).
2. Human umbilical vein endothelial cells (HUVECs) grown on eight-well chamber slides precoated with 0.2% gelatin for 36 to 48 h at 37°C.
3. The methanol used to fix cells should be ice cold.
4. Primary antibodies: MAb Gi11 against JAM-C (*see* **Subheading 2.2.**) and rabbit polyclonal antibody against ZO-1 (Zymed).

5. Antibody-dilution buffer: 5% goat serum + 1% bovine serum albumin (BSA) in PBS.
6. Secondary antibodies: rhodamine-coupled anti-mouse IgG (Dianova, Hamburg, Germany) and fluorescein-coupled anti-rabbit IgG.
7. Vectashield to mount slides (Vector Laboratories, Burlingame, CA).

2.6. Enzyme-Linked Immunosorbent Assay for Ligand–Receptor Interactions

1. Coating buffer (HEPES buffer): 20 m*M* HEPES, 150 m*M* NaCl, 1 m*M* Mn^{2+}, pH 7.4.
2. Blocking buffer: HEPES buffer supplemented with 3% BSA, pH 7.4.
3. Washing buffer: HEPES buffer supplemented with 0.05% Tween-20.
4. Binding buffer: HEPES buffer supplemented with 0.05% Tween-20 and 0.5% BSA (*see* **Note 3**).
5. Maxisorp plates (Greiner).
6. Purified Mac-1, LFA-1 were provided by Dr. C. Gahmberg (Helsinki, Finland). Soluble recombinant human JAM-C was produced as previously described ***(19)***. Purified recombinant intercellular adhesion molecule (ICAM)-1 was provided by Dr. S. Bodary (Genentech, CA; *see* **Note 4**).
7. Primary antibodies: mouse monoclonal antibody (MAb) to human JAM-C, Gi11 ***(9)***, and mouse MAb to human ICAM-1 (DAKO Cytomation). Secondary antibody: peroxidase-conjugated antibody to mouse immunoglobulins (DAKO Cytomation).
8. Substrate for quantification of bound secondary antibody: 2,2′-azino-*bis*-(3-ethylbenzthiazoline-6-sulfonic acid) (ABTS; Boehringer Mannheim, Germany): ABTS-substrate solution: 0.022 g/mL in double-distilled (dd) H_2O; ABTS-Buffer: 0.1 *M* Na-acetate, 0.05 *M* NaH_2PO_4, pH 5.5; Hydrogen peroxide solution: 30% H_2O_2 (w/w). Mix ABTS-substrate solution/ABTS–Buffer/Hydrogen peroxide solution 4.76/95.17/0.07 (v/v/v), e.g., 0.75 mL of ABTS-substrate solution + 15 mL of ABTS-Buffer + 0.01 mL of hydrogen peroxide solution.
9. Microplate reader: BIO-TEK (Neufahrn, Germany).

2.7. THP-1 Cell Adhesion to Immobilized Soluble Human JAM-C

1. THP-1 cells were cultivated in RPMI supplemented with 10% FCS (Gibco, Bethesda, MD).
2. Coating buffer: 15 m*M* Na_2CO_3, 35 m*M* $NaHCO_3$, pH 9.6.
3. Blocking buffer: PBS supplemented with 3% BSA.
4. Washing buffer: PBS.
5. Adhesion buffer: Serum-free RPMI (*see* **Note 5**).
6. Phorbol 12-myristate 13-acetate (PMA; Sigma) is dissolved at 1 mg/mL in dimethyl sulfoxide and stored in single use aliquots at –80°C.
7. Antibodies that were used in the adhesion assay: MAb IB4 against β2 integrin subunit (CD18; Alexis, Gruenberg, Germany); MAb L15 against LFA-1 (kindly provided by Dr. C. Figdor, Nijmegen, The Netherlands); MAb LPM19c against

Mac-1 (kindly provided by Dr. A. May, Munich, Germany); MAb 6S6 against β1 integrin subunit (CD29) (Chemicon, Temecula, CA).
8. Soluble recombinant human JAM-C (*see* **Subheading 2.6.**).
9. Ice-cold methanol-acetone 1:1 (v/v) for fixing cells.
10. Staining solution: 0.5% (w/v) Crystal violet (Sigma), 20% methanol, 80% dd H_2O.
11. Destaining solution: acetic acid/methanol/dd H_2O (10/30/60, v/v/v).
12. Tissue culture 96-well plates (Greiner).
13. Microplate reader (BIO-TEK, Neufahrn, Germany).

2.8. THP-1 Cell Transendothelial Migration

1. 6.5-mm Transwells with a 8-µm pore size (Costar, Bodenheim, Germany).
2. HUVECs cultivated in endothelial cell growth medium (PromoCell) supplemented with growth factors and 5% FCS on 0.2% gelatin-precoated transwells.
3. Monocyte chemoattractant protein (MCP)-1 (R&D Systems, Wiesbaden, Germany) is dissolved at 10 µg/mL in PBS containing 0.1% BSA.
4. Antibodies that were used in the adhesion assay: MAb LPM19c against Mac-1; MAb L15 against LFA-1; MAb Gi18 against platelet endothelial cell adhesion molecule-1 (PECAM-1) (kindly provided by Dr. S. Santoso, Giessen, Germany).
5. Soluble recombinant human JAM-C (*see* **Subheading 2.6.**).

3. Methods

3.1. Cell Culture and Lysis

1. Primary HUVECs were grown on gelatin-coated 100-mm dishes until confluence. When confluent, they were passaged with trypsin/EDTA to provide new cultures on 100-mm tissue dishes (for maintenance) or on eight-well chamber slides (for immunofluorescence), or on transwell filters (for transmigration). For maintenance, cells were split 1:4 to approach confluence after 72 h (*see* **Note 6**).
2. Chinese hamster ovary (CHO) cells transfected with JAM-C *(9)* were grown on 100-mm dishes in alpha-Minimal Essential Media media supplemented with zeocin 200 ng/mL. Cells were split 1:10 every 4 d with accutase treatment.
3. For each IP, approx 2×10^7 JAM-C-transfected CHO cells were used. Cells were washed twice with ice-cold PBS and 500 µL of NP-40 lysis buffer were added. After scraping, lysates were incubated for additional 30 min on ice and then precleaned by centrifugation at 16,000*g* for 15 min at 4°C. The supernatant was transferred in to a fresh vial and then processed for immunoprecipitation. A 100-µL aliquot was kept for further Western blot analysis and frozen at –20°C.

3.2. Immunoprecipitation

1. Extracts were precleaned by rotating at 4°C with Protein A/G (50 µL per 1 mL) for 30 min. After centrifugation, supernatants were transferred to a new tube.
2. MAb Gi11 against JAM-C or an irrelevant antibody (each at 4 µg/mL) was added to the cell lysate. The lysate-antibody mixture was vortexed and incubated over night with gentle rotation at 4°C.

3. The next morning, 50 μL of UltraLink Protein A/G are added to the cell extract and incubated rotating for additional 2 h at 4°C.
4. The beads are then washed three times with NP-40 lysis buffer, resuspended in 60 μL of 2X sample buffer, and boiled for 5 min at 95°C. Thereafter, samples were separated by sodium dodecyl sulfate polyacrylamide gel electrophoresis (SDS-PAGE).

3.3. SDS-PAGE

1. For SDS-PAGE, we used Mini Protean II Gel System (Bio-Rad, Munich, Germany). Glass plates were rinsed with detergent and dried with paper towel before the procedure.
2. A 1.5-mm thick 6% gel was prepared by mixing 3 mL of separation buffer (4X), 2.4 mL acrylamide solution, 6.5 mL of water, 120 mL of APS solution, and 12 mL of TEMED. Pour 7.5 mL of the gel solution and overlay with 0.5 mL of water. After the gel is polymerized, remove water with Whatman paper. The stacking gel is prepared by mixing 1 mL of stacking buffer (4X), 0.4 mL of acrylamide, 2.56 mL of water, 100 mL of APS, and 9 mL of TEMED. Pour the gel and insert the comb.
3. Prepare 1X running buffer by mixing four parts of water with one part of the 4X running buffer.
4. After the stacking gel is polymerized, remove the comb carefully and wash the wells with running buffer.
5. Add running buffer to the top and bottom chamber. Boil samples for 5 min at 95°C and load into the wells.
6. Run gel at 100 V for 2.5 h.

3.4. Western Blotting

1. Transfer is performed using Mini Trans-Blot System (Bio-Rad).
2. Prepare blotting buffer. Cut membrane and two pieces of blotting paper (they should be larger than the gel). Soak membrane with water and than equilibrate membrane and blotting paper in blotting buffer for 15 min.
3. Assemble blotting sandwich on transfer cassette as follows: sponge prewetted in blotting buffer, then blotting paper, membrane, and the gel. There should be no bubbles between blotting paper, membrane, and the gel. Cover gel with another sheet of blotting paper and sponge. Close transfer cassette and put it to the transfer tank filled with blotting buffer. Perform transfer for 3 h 100 mA at RT.
4. When transfer is finished, take out membrane and place it to the chamber with blocking buffer. Incubate membrane in blocking buffer for 1 h at RT with shaking and thereafter wash once with PBS-T20.
5. Add primary antibody against ZO-1, diluted 1:2500 in PBS-T20/1% BSA and incubate overnight at 4°C with shaking.
6. Discard antibody solution (*see* **Note 7**). Wash membrane three times for 5 min each in PBS-T20.
7. Add freshly prepared secondary antibody diluted 1:10,000 in PBS-T20/1% BSA and incubate with shaking for 45 min at RT.
8. Wash membrane in PBS-T20 three times for 5 min each.

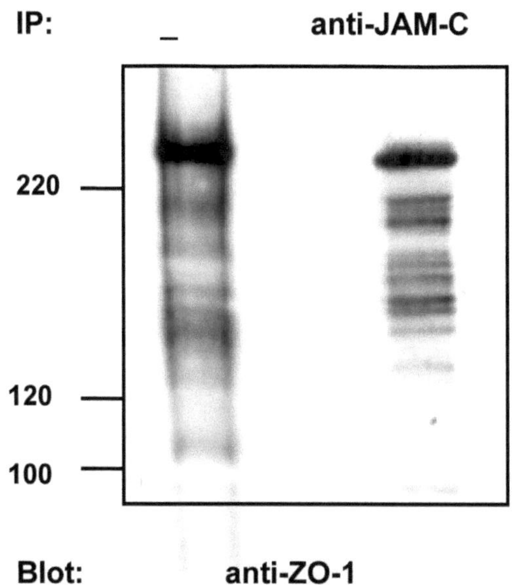

Fig. 1. Coimmunoprecipitation of JAM-C with ZO-1 in chinese hamster ovary (CHO) cells transfected with JAM-C. Detection of ZO-1 by Western blot in JAM-C transfected CHO cell lysates (— indicates no immunoprecipitation) or in CHO-JAM-C cell lysates that were subjected to immunoprecipitation with MAb against JAM-C, as indicated.

9. Discard secondary antibody and place membranes on saran wrap protein side up. Prepare ECL reagent by mixing equal parts of detection reagent 1 with detection reagent 2. Add ECL reagent to the membrane and incubate for 1 min. Discard ECL reagent and rinse membrane with 1X PBS. Then, cover membrane with new piece of saran wrap, put to the X-ray cassette and expose to film (**Fig. 1**).

3.5. Immunofluorescence

1. HUVECs were passaged and 30,000 cells/well plated onto eight-well chamber slides that were precoated with 0.2% gelatin (2 h at 37°C) and were allowed to adhere and grow to confluence for 36–48 h at 37°C.
2. After washing the cells three times with PBS, the cells were fixed with ice-cold methanol for 15 min.
3. Methanol is discarded, the cells are washed twice with PBS, and then blocked in antibody dilution buffer for 1 h at RT.
4. The blocking solution is removed and replaced with MAb Gi11 against JAM-C (10 μg/mL) and rabbit polyclonal antibody against ZO-1 (5 μg/mL) in antibody dilution buffer for 2 h at 22°C.
5. The primary antibody is removed and the samples washed three times for 2 min each with PBS. Then, rhodamine-coupled anti-mouse IgG or fluorescein-coupled anti-rabbit IgG is added (each at 1:100) for 1 h at 22°C.

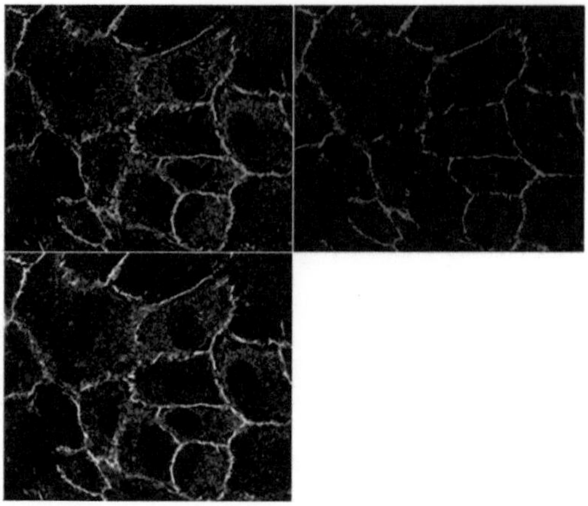

Fig. 2. JAM-C localization at interendothelial borders and colocalization with ZO-1. Representative immunofluorescence of HUVECs showing JAM-C distribution (incubation with MAb Gi11 followed by secondary rhodamine-coupled anti-mouse IgG) together with ZO-1 distribution (incubation with rabbit anti-ZO-1 followed by fluorescein isothiocyanate-coupled anti-rabbit IgG). Double stained images were merged to analyze colocalization of JAM-C with ZO-1. (Please *see* color insert following p. 50 for a color version of this figure.)

6. The secondary antibody is removed, the cells are washed three times with PBS, and DAPI is added for 10 min at RT to stain the DNA and identify the nuclei.
7. The samples are then washed five times in PBS. The gasket is then removed from the chamber slide, and then mounting medium (Vectashield) and a cover slip are added.
8. The slides are viewed and double-stained images are merged to analyze colocalization (**Fig. 2**).

3.6. Enzyme-Linked Immunosorbent Assay for Ligand–Receptor Interactions

1. The buffers used are described in **Subheading 2.6**.
2. Maxisorp plates are coated with Mac-1, LFA-1, or BSA (5 μg/mL) dissolved in coating buffer overnight at 4°C (*see* **Note 8**).
3. Discard the integrins and BSA and wash the wells twice in washing buffer. Wells are then blocked with blocking buffer for 1 h at RT.
4. The wells are then washed twice in washing buffer. Binding of soluble human JAM-C (0–20 μg/mL) or soluble human intercellular adhesion molecule-1 to the immobilized integrins or to BSA-coated wells is performed in a final volume of 50 μL in binding buffer for 2 h at 22°C. In **Fig. 3**, binding of 10 μg/mL JAM-C and 10 μg/mL ICAM-1 is shown.
5. After the incubation period the ligands are discarded and wells are washed three times in washing buffer. To detect bound JAM-C or ICAM-1, the appropriate

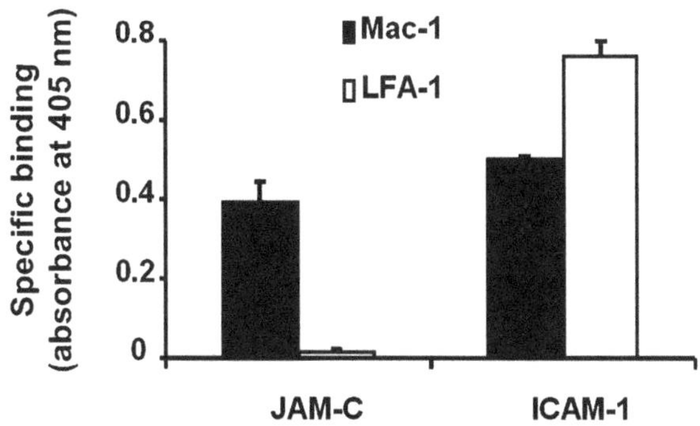

Fig. 3. Interaction between purified JAM-C and Mac-1 proteins. The binding of JAM-C or ICAM-1 to immobilized Mac-1 **(filled bars)** or to immobilized LFA-1 **(open bars)** is presented. Binding of ICAM-1 or JAM-C to BSA was subtracted to obtain specific binding to the integrins. Specific binding is expressed as absorbance at 405 nm. Data are mean + SD (n = 3) of a typical experiment.

primary antibodies (*see* **Subheading 2.6.**) in binding buffer are added to the wells (50 μL/well) at a dilution of 1:1000 for 2 h at 22°C.

6. After the incubation period, the primary antibodies are discarded and the wells are washed three times in washing buffer. Then, the secondary antibody in binding buffer is added (50 μL/well) at a dilution of 1:2000 for 1 h at 22°C. The plate should be kept in a dark place after addition of the secondary antibody.
7. The secondary antibody is discarded and wells are washed four times with washing buffer.
8. Bound secondary antibody is quantified by using the substrate containing ABTS (*see* **Subheading 2.6.**; 150 μL/well). The plate should be kept in a dark place at RT and incubated for 5 to 120 min so that the change of color takes place.
9. The colorimetric reaction is quantified by measuring the absorbance at 405 nm in microplate reader. Nonspecific binding of JAM-C or ICAM-1 to BSA-coated wells is used as blank and is subtracted to calculate specific binding of these ligands to the integrins **(Fig. 3)**.

3.7. THP-1 Cell Adhesion to Immobilized Soluble Human JAM-C

1. The buffers used are described in **Subheading 2.7.**
2. 96-well tissue culture plates are coated with JAM-C or BSA (each 10 μg/mL) dissolved in coating buffer overnight at 4°C.
3. Discard the JAM-C and BSA and wash the wells once with PBS. Wells are then blocked with blocking buffer for 1 h at RT.
4. During the 1-h incubation: THP-1 cells (cells grow in suspension) are collected from the tissue culture plate and washed in serum-free RPMI medium. For the washing, cells are centrifuged for 5 min at 1200 rpm, the pellet is resuspended in

10 mL serum-free RPMI (keep a small aliquot, e.g., 10 or 20 μL to measure cell number), and then cells are centrifuged once more for 5 min at 200*g*. During this centrifugation step the cell number should be estimated (*see* **Note 9**). The cells should then be resuspended in serum-free RPMI at a density of 2×10^6/mL.

5. After the 1-h incubation, the wells are washed twice in washing buffer. 50 μL of serum-free RPMI is added to each well without or with 50 ng/mL PMA (*see* **Notes 5** and **10**) and in the absence or presence of competitors (competitors are described in **Subheading 2.7.**; *see* **Note 11**). Immediately thereafter, 50 μL of the THP-1 cell solution should be added per well (i.e., 100,000 cells/well). Incubate cells at 37°C for 40 to 60 min.
6. After the incubation period, the wells are washed twice with PBS, to remove non-adherent cells. Adhesion buffer or PBS should be aspirated at this washing step.
7. Adherent cells are then fixed with methanol aceton (1:1; 100 μL/well) at 4°C for 30 min.
8. Discard methanol-aceton and stain adherent cells with staining solution (50 μL/well) for 30 min at RT.
9. Remove the staining solution and wash the plates in a bucket with tap water for several times. Let the plate dry. Add the destaining solution (100 μL/well) to the stained adherent cells. Wait for 5 min and then measure absorbance at 595 nm in a microplate reader. Nonspecific adhesion of THP-1 cells to BSA-coated wells is used as blank and is subtracted to calculate specific cell adhesion to JAM-C (*see* **Note 12**; **Fig. 4A**).

3.8. THP-1 Cell Transendothelial Migration

1. HUVECs are passaged and are plated on 0.2% gelatin-precoated 6.5-mm transwells with a 8-μm pore size in endothelial cell growth medium supplemented with growth factors and 5% FCS at a density of 30,000 cells per well. At this density the wells are confluent after 36–48 h incubation in a humidified atmosphere (37°C, 5% CO_2; *see* **Note 13**).
2. After this incubation period, chambers are examined microscopically for integrity and uniformity of endothelial monolayers. Monolayers are washed once with PBS and medium is switched to serum-free conditions (endothelial cell growth medium without FCS). If needed, stimuli are added (e.g., tumor necrosis factor-α, 10 ng/mL for 6 h at 37°C) to the endothelial cells, which will upregulate the expression of adhesion molecules on HUVECs.
3. At the beginning of the transmigration assay, THP-1 cells are collected from the tissue culture plate and washed in serum-free RPMI medium. For the washing, cells are centrifuged for 5 min at 1200 rpm, the pellet is resuspended in 10 mL serum-free RPMI (keep a small aliquot, for example, 10 or 20 μL to measure cell number), and then cells are centrifuged once more for 5 min at 1200 rpm. During this centrifugation step, the cell number should be estimated. The cells should then be resuspended in serum-free RPMI at a density of 4×10^6/mL.
4. The transwell filters (inserts) are removed from their wells and 600 μL migration assay medium (serum-free RPMI in the absence or presence of 50 ng/mL MCP-1)

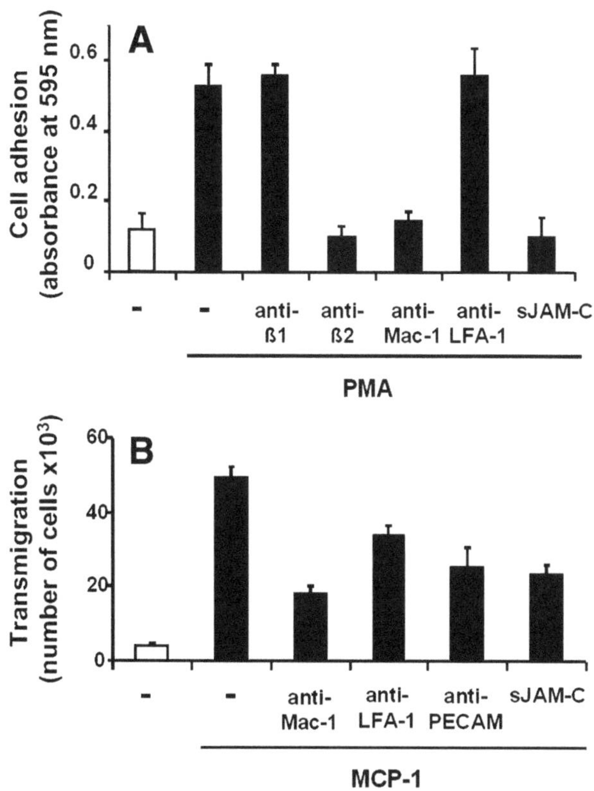

Fig. 4. Interaction between leukocyte Mac-1 and endothelial JAM-C. (**A**) Adhesion of THP-1 cells to immobilized human JAM-C is shown without (**open bar**) or with PMA (**filled bars**) and in the absence (-) or presence of blocking MAb to β1-integrin, MAb to β2-integrin, MAb to Mac-1, MAb to LFA-1 (each at 20 μg/mL), or soluble human JAM-C (20 μg/mL). Cell adhesion is represented as absorbance at 590 nm. Adhesion of THP-1 cells to BSA was subtracted to obtain specific adhesion. Data are mean ± SD (n = 3) of a typical experiment. (**B**) The *trans*-endothelial migration of THP-1 cells toward serum-free medium without (**open bar**) or with 50 ng/mL MCP-1 (**filled bars**) is shown in the absence (-) or presence of MAb to Mac-1, MAb to LFA-1, MAb to PECAM-1 (each at 20 μg/mL), or soluble human JAM-C (20 μg/mL). Data are mean ± SD (n = 3) of a typical experiment.

is added to the lower compartment (lower well). HUVECs are then washed twice with serum-free medium.

5. The transwell filters are brought back to their original wells, which now contain 600 μL of migration assay medium without or with MCP-1 in the lower compartment. The medium is then aspirated from the transwells and an aliquot of 100 μL of THP-1 cells (4×10^5) is then added to the upper compartment on top of the

endothelial monolayer. The competitors (different MAb or soluble proteins, as described under 2.8.) are immediately added (*see* **Note 11**) to the transwell filters.

6. After incubation for 4 h at 37°C the transwell filters (inserts) are removed from their wells to stop the experiment. The inserts are washed twice in PBS and fixed with methanol/acetone (1 : 1), stained with crystal violet and mounted on glass slides in order to confirm the confluency of the endothelial monolayer of the filters after the assay (for details on staining procedure please *see* **Subheading 3.7.**).
7. The number of transmigrated cells in the lower compartment is estimated by measuring cell number of an aliquot of the lower well with a cell counter (CASY-Counter, Schärfe-System, Reutlingen, Germany; *see* **Note 12**; **Fig. 4B**).

4. Notes

1. Heat inactivation is performed by incubating FCS for 30 min at 56°C and thereafter serum is stored in single-use aliquots at –20°C.
2. Activated Na-ortovanadate solution is prepared by dissolving an appropriated amount of reagent into ddH_2O, adjusting pH to 10.0 and boiling for approx 10 min (until it turns colorless). This procedure is repeated one more time and Na-ortovanadate is stored in 1-mL aliquots at –20°C.
3. To keep the immobilized integrins in an active conformation, the supplementation of 1 m*M* Mn^{2+} to the coating, washing, blocking and binding buffers is extremely important. Without Mn^{2+} binding of integrins to their ligands will be very weak.
4. Recombinant ICAM-1 or ICAM-1-Fc chimera can also be commercially purchased (e.g., by R&D Systems).
5. In cell adhesion experiments instead of using serum-free RPMI, a buffer containing 20 m*M* HEPES, pH 7.5; 140 mM NaCl; and 2 mg/mL glucose may be used as the adhesion buffer. In this case, 1 m*M* MnCI should be used as the stimulus for cell adhesion instead of PMA.
6. We recommend that early-passage HUVECs (up to fourth passage) are used.
7. It is possible to reuse primary antibody for several times. For storage 0.02% sodium azide should be added and solution might be kept at 4°C.
8. When the interaction between β2-integrins and their ligands is investigated, it is strongly recommended that the integrin is immobilized and the ligand is in solution.
9. 10 or 20 μL of the cells should be diluted in 10 mL of NaCl and the number of cells should be estimated in a cell counter. We use CASY-Counter (Schaerfe-System). Alternatively, the number of the cells can also be measured in a Neubauer chamber.
10. Besides PMA and MnCl, another way to specifically stimulate β2-integrin-dependent adhesion is by using stimulating mAb, such as Kim127 or Kim185 (use at 20 μg/mL; kindly provided by Dr. M. Robinson, Slough Berkshire, UK).
11. In some cases an additional preincubation of the competitors with the THP-1 cells is recommended rather than the single coincubation of the competitors during the adhesion or transmigration assay in order to obtain higher degrees of inhibition. This can be performed by preincubating the antibodies or soluble proteins with the THP-1 cells for 30 min on ice. In this way the inhibitory antibodies such as the blocking MAb against Mac-1 (*see* **Subheading 2.7.**) or soluble JAM-C can already

bind to and block Mac-1. When cell–cell adhesion, e.g., THP-1 cell adhesion to HUVECs or THP-1 transendothelial migration is assessed, such a preincubation can be performed on both the THP-1 cells or the HUVECs as appropriate.

12. We have used the same protocols for adhesion and transmigration to test the HUVEC adhesion and transendothelial migration of freshly isolated neutrophils from peripheral blood ***(19,20)***. The pattern observed was comparable to results with THP-1 cells.
13. It is very important that no medium is added to the lower compartment before the beginning of the transmigration assay. In particular, serum-containing medium in the lower compartment might prevent HUVECs from growing to confluence on the transwell filter.

References

1. Bazzoni, G. (2003) The JAM family of junctional adhesion molecules. *Curr. Opin. Cell. Biol.* **15,** 525–530.
2. Ebnet, K., Suzuki, A., Ohno, A., and Vestweber, D. (2004) Junctional Adhesion Molecules (JAMs): more molecules with dual functions? *J. Cell. Sci.* **117,** 19–29.
3. Keiper, T., Santoso, S., Nawroth, P. P., Orlova, V., and Chavakis, T. (2005) The role of junctional adhesion molecules in cell–cell interactions. *Histol. Histopathol.* **20,** 197–203.
4. Kornecki, E., Walkowiak, B., Naik, U. P., and Ehrlich, Y. H. (1990) Activation of human platelets by a stimulatory monoclonal antibody. *J. Biol. Chem.* **265,** 10,042–10,048.
5. Sobocka, M. B., Sobocki, T., Banerjee, P., et al. (2000) Cloning of the human platelet F11-receptor: a cell adhesion molecule member of the immunoglobulin superfamily involved in platelet aggregation. *Blood* **95,** 2600–2609.
6. Palmeri, D., van Zante, A., Huang, C. C., Hemmerich, S., and Rosen, S. D. (2000) Vascular endothelial junction-associated molecule, a novel member of the immunoglobulin superfamily, is localized to intercellular boundaries of endothelial cells. *J. Biol. Chem.* **275,** 19,319–19,345.
7. Aurrand-Lions, M. A., Duncan, L., Ballestrem, C., and Imhof, B. A. (2001) JAM-2, a novel immunoglobulin superfamily molecule, expressed by endothelial and lymphatic cells. *J. Biol. Chem.* **276,** 2733–2741.
8. Aurrand-Lions, M. A., Johnson-Leger, C., Wong, C., DuPasquier, L., and Imhof, B. A. (2001) Heterogeneity of endothelial junctions is reflected by differential and specific subcellular localization of three JAM family members. *Blood* **98,** 3699–3707.
9. Santoso, S., Sachs, U. J. H., Kroll, H., et al. (2002) The junctional adhesion molecule 3 (JAM-3) on human platelets is a counterreceptor for the leukocyte integrin Mac-1. *J. Exp. Med.* **196,** 679–691.
10. Zen, K., Babbin, B. A., Liu, Y., Whelan, J. B., Nusrat, A., and Parkos, C. A. (2004) JAM-C is a component of desmosomes and a ligand for CD11b/CD18-mediated neutrophil transepithelial migration. *Mol. Biol. Cell.* **15,** 3926–3937.
11. Bazzoni, G., Martinez-Estrada, O. M., Orsenigo, F. M., Cordenonsi, M., Citi, S., and Dejana, E. (2000). Interaction of junctional adhesion molecule with the

tight junction components ZO-1, cingulin, and occludin. *J. Biol. Chem.* **275,** 20,520–20,526.

12. Ebnet, K., Schulz, C. U., Meyer Zu Brickwedde, M. K., Pendl, G. G., and Vestweber, D. (2000) Junctional adhesion molecule interacts with the PDZ domain-containing proteins AF-6 and ZO-1. *J. Biol. Chem.* **275,** 27,979–27,988.
13. Hamazaki, Y., Itoh, M., Sasaki, H., Furuse, M., and Tsukita, S. (2002) Multi-PDZ domain protein 1 (MUPP1) is concentrated at tight junctions through its possible interaction with claudin-1 and junctional adhesion molecule. *J. Biol. Chem.* **277,** 455–461.
14. Ebnet, K., Suzuki, A., Horikoshi, Y., et al. (2001) The cell polarity protein ASIP/PAR-3 directly associates with junctional adhesion molecule (JAM). *EMBO J.* **20,** 3738–3748.
15. Ebnet, K., Aurrand-Lions, M., Kuhn, A., et al. (2003) The junctional adhesion molecule (JAM) family members JAM-2 and JAM-3 associate with the cell polarity protein PAR-3: A possible role for JAMs in endothelial cell polarity. *J. Cell. Sci.* **116,** 3879–3891.
16. Fanning, A. S., Jameson, B. J., Jesaitis, L. A., and Anderson, J. M. (1998) The tight junction protein ZO-1 establishes a link between the transmembrane protein occludin and the actin cytoskeleton. *J. Biol. Chem.* **273,** 29,745–29,753.
17. Ostermann, G., Weber, K. S., Zernecke, A., Schroeder, A., and Weber, C. (2002) JAM-1 is a ligand of the beta(2) integrin LFA-1 involved in transendothelial migration of leukocytes. *Nat. Immunol.* **3,** 151–158.
18. Cunningham, S. A., Rodriguez, J. M., Arrate, P. M., Tran, T. M., and Brock, T. A. (2002) JAM2 interacts with alpha4beta1. Facilitation by JAM3. *J. Biol. Chem.* **277,** 27,589–27,592.
19. Chavakis, T., Keiper, T., Matz-Westphal, R., et al. (2004) The junctional adhesion molecule-C promotes neutrophil transendothelial migration in vitro and in vivo. *J. Biol. Chem.* **279,** 55,602–55,608.
20. Chavakis, T., Athanasopoulos, A., Rhee, J. S., et al. (2005) Angiostatin is a novel anti-inflammatory factor by inhibiting leukocyte recruitment. *Blood* **105,** 1036–1043.

4

Cell–Cell Interactions on Solid Matrices

Nancy A. Louis, Dionne Daniels, and Sean P. Colgan

Summary

Models to study molecular, biochemical, and functional responses in vitro generally incorporate an individual cell type or group of cells organized in a random fashion. Normal physiological responses in vivo require that individual cell types be oriented in an organized fashion with three-dimensional architecture and appropriately positioned cellular interfaces. Much recent progress has been made in the development and implementation of models to study cell–cell contact using substrate grown cells. Here, we summarize the use of membrane permeable supports to study functional responses in appropriately positioned cell types. These models incorporate two or more different cells cultured in physiologically positioned locales on solid substrates. Models incorporating nonadherent cells (e.g., leukocytes) in co-culture with such models also are discussed. Such models have been used extensively to discovery both cell-bound as well as soluble mediators of physiological and pathophysiological processes.

Key Words: Cell–cell interactions; substrate; endothelia; epithelia; leukocyte; co-culture.

1. Introduction

In intact tissues, matrix-bound cells must communicate not only with outside environment but also with other cells in the local microenvironment. For example, mucosal epithelial cells provide a barrier that separates lumenal and vascular compartments. This monolayer of cells provides barrier function and serves as a conduit for vectorial ion movement, the transport event responsible for mucosal hydration *(1)*. By secreting solutes and transporting fluid across the epithelium, epithelia are able to coordinate compositional changes of the lumenal compartment. A number of stimuli, including hormones, neurotransmitters, and cytokines, have been shown to directly regulate both epithelial barrier function and ion transport *(2)*.

In intact mucosal tissues, including the lung and intestine, epithelial cells are anatomically positioned in close proximity to a number of subepithelial cell

From: *Methods in Molecular Biology, vol. 341: Cell–Cell Interactions: Methods and Protocols*
Edited by: S. P. Colgan © Humana Press Inc., Totowa, NJ

types, including lymphocytes, fibroblasts, smooth muscle cells, and endothelia. These subepithelial cell populations contribute to epithelial function through paracrine pathways *(3)*. Under these circumstances, for example, the vascular endothelium functions as more than a passive conduit for blood components and synthesizes many compounds that precisely regulate blood vessel tone, vascular composition, and leukocyte movement *(4,5)*. Under such conditions, endothelial cells themselves respond to a variety of proinflammatory stimuli that, in turn, can release inflammatory mediators such as cytokines and bioactive lipids *(4)*. The vital role of the endothelium in coordinating inflammation and the proximity of the vasculature to the epithelium provides a potential paracrine crosstalk pathway between these two cell types.

Here, we discuss the design and implementation of models to study interactions between two or more cell types on solid matrices. These culture models provide ideal opportunities to identify paracrine mediators of physiological and pathophysiological importance and allow flexibility in the use of multiple functional responses. For the purposes of this chapter, we summarize models implementing mucosal epithelial cells as at least one component, but it should be noted that nearly any matrix-bound cell can be integrated into these models.

2. Materials

2.1. Epithelial and Endothelial Cell Culture

1. T84 cells: 1:1 ratio of Dulbecco's Modified Eagle's Medium (Gibco/BRL, Bethesda, MD) and Hams F-112 Medium supplemented with 10% fetal bovine serum (Invitrogen, Carlsbad, CA), 15 m*M* HEPES buffer (pH 7.5; Sigma, St. Louis, MO), 14 m*M* Na HCO_3 (Sigma), 40 mg/L penicillin (Sigma), 8 mg/L ampicillin (Sigma), and 9 mg/L streptomycin (Sigma).
2. Human microvascular endothelial cells (HMECs): Dulbecco's Modified Eagle's Medium supplemented with 10% fetal bovine serum; 15 mM HEPES buffer, pH 7.5; 14 mM Na HCO_3; 40 mg/L penicillin; 8 mg/L ampicillin; 9 mg/L streptomycin; 10 ng/mL epidermal growth factor (Collaborative Biomedical Products, Bedford, MA); and 1 μg/mL hydrocortisone (Sigma).
3. Membrane permeable supports in various formats and pore sizes are available from Corning-Costar (Cambridge, MA).
4. Endotoxin (List Biologicals, Campbell, CA).
5. Forskolin (Sigma): 10 m*M* stock solution made in DMSO (Sigma) and stored at –20°C.
6. Mineral oil (Sigma).
7. Hank's Buffered Salt Solution (HBSS) (Sigma) supplemented with 10 m*M* HEPES.
8. Modified HBSS (without Ca^{2+} or Mg^{2+}; Sigma) supplemented with 10 m*M* HEPES.
9. Voltage clamp (Iowa Dual Voltage Clamps, Bioengineering, University of Iowa) interfaced with an equilibrated pair of calomel electrodes and a pair of Ag–AgCl electrodes.
10. Evohm (World Precision Instruments Inc.).

2.2. Neutrophil Transmigration

1. Histopaque 1077 (Sigma).
2. Gelatin (100 bloom, Fisher Scientific).
3. Ammonium chloride lysis buffer: mix 8.29 g of NH_4Cl (Sigma), 1 g of $NaCO_3$ (Sigma), and 0.038 g of EDTA (disodium salt, Sigma) in 1 L of distilled and deionized water.
4. Acid citrate dextrose is prepared by mixing 13.7 g of citric acid (Sigma), 25 g of sodium citrate (Sigma) and 20 g of dextrose (Sigma) in 1 L of distilled and deionized water.
5. Citrate buffer: titrate 1 *M* sodium citrate (Sigma) with 1 *M* citric acid (Sigma) to final pH of 4.2.
6. 2,2′-azino-*bis*-(3-ethylbenzothiazoline-6-sulfonic acid) (ABTS) reagent is made by adding 50 mg ABT to 100 m*M* citrate buffer and a final concentration of 0.03% H_2O_2 (Sigma).
7. *N*-formyl-methionyl-leucyl-phenylalanine (fMLP; Sigma) prepared as a 10 m*M* stock in dimethyl sulfoxide and stored at –20°C.

3. Methods

Cell culture models on solid matrices have proven important in the modeling of normal biochemical and molecular signaling in vitro, which is particularly important for cells that grow as polarized monolayers (e.g., epithelial cells). This principle has been defined to test and implement a number of cell–cell interaction models, including endothelial–epithelial interactions ***(3)***, fibroblast–epithelial interactions ***(6)***, neutrophil–epithelial interactions ***(7,8)***, mononuclear cell–epithelial interactions ***(9,10)***, and neutrophil interactions with residual matrix proteins ***(11)***. Examples of each of these cultures, as well as sample sets of data, are provided in this chapter.

3.1. Preparation of Epithelial–Endothelial Co-Cultures

1. T84 cells are trypsinized using standard methods of trypsin and EDTA solution (*see* **Note 1**).
2. T84 cells are washed in phosphate-buffered saline, diluted in media to a final concentration of × 10^5/mL, and plated on the underside of polycarbonate inserts (termed *inverts*; *see* **Notes 2** and **3**), using no more that 82 µL of cell suspension. For these purposes, cells are carefully beaded onto the underside of the insert. Alternatively, cloning rings (6.5 mm diameter) can be glued to the underside of supports to provide a reservoir for plating cells.
3. After plating, inverts are carefully transferred to a tissue culture incubator, and incubated at 37°C for 12 to 18 h. Monolayers are re-oriented (**Fig. 1A**) into T84 cell media.
4. HMECs are trypsinized using standard methods of trypsin and EDTA solution.

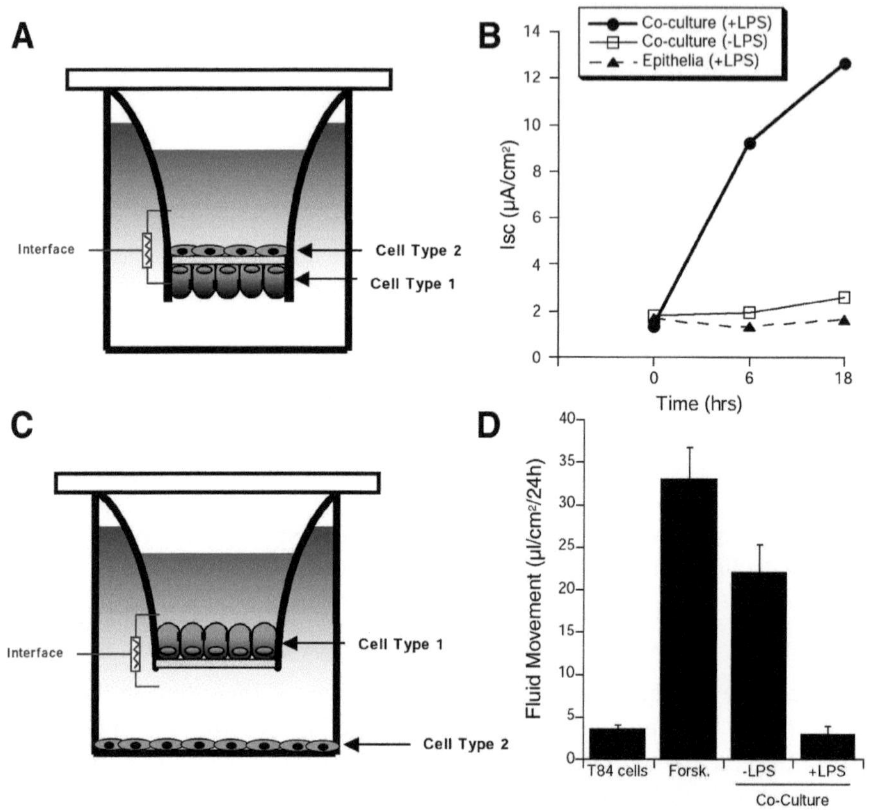

Fig. 1. Models of noncontact substrate grown co-cultures. **(A)** Cell type 1 (e.g., epithelia) and cell type 2 (e.g., endothelia) are plated on opposing surfaces of membrane permeable supports (0.4-µm pores). Cells are allowed to grow to confluence, and functionally assessed. **(B)** Example of results derived from such a co-culture model. In this example, cultures of T84 intestinal epithelial cells (cell type 1) in co-culture with human microvascular endothelia (cell type 2) were incubated in the absence (□) and presence of lipopolysaccharide (LPS) (100 ng/mL, ●) for indicated times and examined for generation of epithelial electrogenic Cl^- secretion (measured as a short circuit current, Isc). T84 monolayers alone (▲) were incubated with LPS as a control. Results represent data pooled from four monolayers in each condition. **(C)** Cell type 1 (e.g., epithelia) were plated on membrane permeable supports, whereas cell type 2 (e.g., endothelia) were plated in the adjacent well to the permeable support. **(D)** Provides an example of results derived from such a co-culture model. In this example, co-cultures were incubated in the presence and absence of LPS (100 ng/mL) for 24 h and examined for vectorial basolateral-to-apical fluid transport. T84 cells alone served as a negative control and T84 cells exposed to forskolin (10 µ*M*) in the presence of 3-isobutyl-1-methylxanthine (5 m*M*) served as the positive control for induction of fluid transport. Data are pooled from three monolayers in each condition, and results are expressed as the mean ± standard error of the mean (SEM) fluid movement over the course of 24 h. (Please *see* color insert following p. 50 for a color version of this figure.)

5. HMECs are washed in phosphate-buffered saline, diluted in media to a final concentration of 3.5×10^5/mL, and plated on the inner well of polycarbonate inserts using 100 µL of cell suspension. HMECs are then fed 100 µL of media on the apical surface and 1 mL of media on the basolateral surface. Plates are then incubated at 37°C for 3 to 5 d.
6. Typical experimental results assessing functional epithelial endpoints of endothelial activation is shown in **Fig. 1B**. For example, the addition of lipopolysaccharide (100 ng/mL) to endothelial–epithelial co-cultures (model shown in **Fig. 1B**) resulted in a time-dependent increase in epithelial chloride secretion (measured as a change in short circuit current. or ΔIsc; *see* **Note 4**).
7. Alternatively, endothelial cells can be plated opposing epithelial cells in the adjacent well of a 24-well plate (**Fig. 1C**). For these purposes, epithelial cells are plated in the inner well of polycarbonate supports.
8. Such co-cultures can be used to assess fluid transport in vitro *(3)*. The apical solution of confluent T84 cell monolayers grown on 0.33-cm^2 permeable supports was replaced with 30 µL of media and layered with 60 µL of warm, sterile mineral oil to minimize evaporation. In some monolayers, the cyclic adenosine monophosphate agonist forskolin (50 µ*M*) was added to the basolateral solution to promote fluid movement and serves as a positive control. After 24 h, the apical solution was collected, centrifuged at 10,000*g* and quantified with a calibrated pipet and weighed on a balance.

3.2. Neutrophil Transmigration Assays

1. T84 cells plated as inverts on the underside of polycarbonate inserts (*see* **Subheading 3.1.**). For these purposes, it is necessary to plate the cells on substrates with 5-µm pores to allow neutrophils to passage through the substrate. Such studies can also be performed on co-cultures (*see* **Fig. 2A** and **Subheading 3.1.**).
2. Neutrophils are purified from whole venous blood by density centrifugation. Anticoagulated whole blood. Tubes are centrifuged at 400*g* for 20 min at 25°C. Plasma and mononuclear cells are removed by aspiration. Red blood cells were sedimented using 2% gelatin, and residual red blood cells were removed by lysis in ice-cold NH_4Cl buffer.
3. Neutrophils are resuspended to a final concentration of 5×10^7 in modified HBSS.
4. Neutrophils (1×10^6) are added to the upper chambers of transwell inverts (**Fig. 2**). A chemotactic gradient is established by adding fMLP to the lower chambers (1 µ*M* final concentration in HBSS).
5. Transmigration is carried out at 37°C for 60 to 120 min.
6. Transmigrated polymorphonuclear leukocytes are quantified by assaying for the neutrophil azurophilic marker myeloperoxidase. Neutrophils are lysed by the addition of Triton X-100 to a final concentration 0.5%. Samples are incubated at 4°C for 25 min on a rotating platform at 80 rpm.
7. The samples are acidified with citrate buffer (final concentration 100 m*M*, pH 4.2).

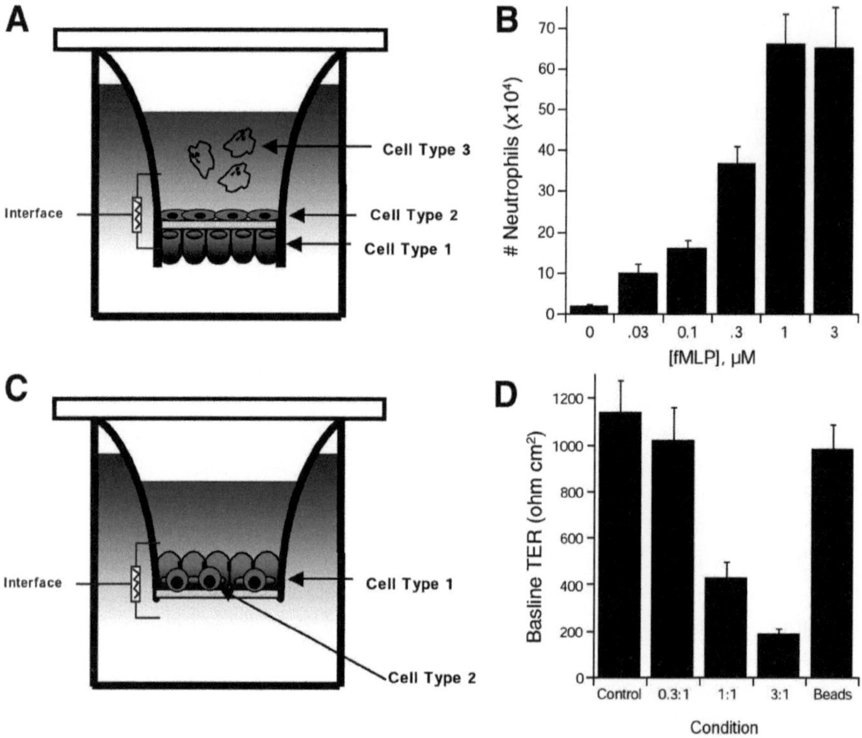

Fig. 2. Models of contact substrate grown co-cultures. **(A)** Cell type 1 (e.g., epithelia) and cell type 2 (e.g., endothelia) are plated on opposing surfaces of membrane permeable supports (5 μm pores). Cells are allowed to grow to confluence, and functionally was assessed as a substrate for migration of human neutrophils (cell type 3). **(B)** An example of results derived from such a model. In this example, cultures of T84 intestinal epithelial cells (cell type 1) in co-culture with human microvascular endothelia (cell type 2) were washed, and a gradient of indicated concentrations of the neutrophil chemoattractant fMLP was placed in the lower chamber. Purified human neutrophils (1×10^6) were placed added to the inner well and transmigration over the course of 90 min was determined by quantification of the detergent extractable neutrophil marker myeloperoxidase. Data are pooled from three monolayers in each condition, and results are expressed as the mean ± SEM number of transmigrated neutrophils. **(C)** Cell type 1 (e.g., epithelia) were plated on membrane permeable supports (0.4 μm) in direct co-culture with cell type 2 (e.g., mononuclear cells). **(D)** An example of results derived from such a co-culture model. In this example, co-cultures of T84 epithelial cells and human mononuclear cells were plated at indicated mononuclear cell:epithelial ratios (where control condition is epithelia alone and beads condition represents three latex beads per epithelial cell). Co-cultures were grown for 6 d and assessed for barrier function by measurement of transepithelial resistance. Data are pooled from three monolayers in each condition and results are expressed as the mean ± standard error of the mean (SEM) transepithelial resistance (ohm · cm^2). (Please *see* color insert following p. 50 for a color version of this figure.)

8. An aliquot of sample (70 μL) is added to an equal volume of ABTS solution in a 96-well plate. The resulting absorbance is determined on a plate reader at 405 nm.
9. Typical results of a transmigration are shown in **Fig. 2B**.

3.3. Preparation of Epithelial–Mononuclear Cell Co-Cultures

1. Mononuclear cells are isolated from the interface of density gradients (*see* **Subheading 3.2.**) and quantified by counting on a hemocytometer.
2. T84 cells are trypsinized using standard methods of trypsin and EDTA solution.
3. After trypsinization, epithelial cells are mixed with mononuclear cells in various ratios; for example, 0.3 mononuclear cells per epithelial cell (0.3 : 1), 1 mononuclear cell per epithelial cell (1 : 1), and 3 mononuclear cells per epithelial cell (3 : 1; **Fig. 2C**). Latex beads (11.9 ± 1.9 μm diameter, neutral charge, Seradyn Inc., Indianapolis, IN) were mixed with epithelial cells at the time of plating (three beads per epithelial cell).
4. Epithelial/mononuclear cell mixtures are plated on 0.33-cm^2 ring supported polycarbonate filters (0.4-μm pore size).
5. Co-cultures are grown in T84 cell media to which human recombinant interleukin-2 is added at a final concentration of 2 n*M*. These epithelial–lymphocyte co-cultures are used 6 to 10 d after plating.
6. Typical experimental results assessing functional epithelial endpoints (e.g., barrier function) are shown in **Fig. 2D**.

3.4. Preparation of Residual Matrix Cultures

1. Endothelial or epithelial cultures are established on 0.33-cm^2 ring-supported polycarbonate filters (5-μm pore size; *see* **Subheading 3.1.**).
2. Removal of cells from the polycarbonate filters is accomplished by incubation in 5 m*M* EDTA in modified HBSS at 37°C (*see* **Note 5**).
3. Filters are vigorously washed in HBSS and used to assess migration of neutrophils (*see* **Subheading 3.2.**).
4. Typical experimental results comparing neutrophil transendothelial and transmatrix migration are shown in **Fig. 3B**.

4. Notes

1. Trypsinizing T84 cells can take anywhere between 5 and 20 min. Trypsin–EDTA becomes less active with repeated freeze–thaw cycles. To reduce the time it takes to trypsinize T84 cells, use trypsin that has been aliquoted to avoid repeated freeze–thaw cycles.
2. Generally, cells grow better if the inserts are collagen coated the night before plating. The surfaces are coated with 50 to 150 μL (depending on the diameter of the surface) of type IV collagen and allowed to dry for 12 to 18 h in the hood under ultraviolet light. The type IV collagen is dissolved in 0.2% acetic acid and then diluted in 60% EtOH. Dissolving the collagen can take up to 3 d. Allow the inserts to air dry, and plate cells directly over the collagen coating. No washing is necessary.

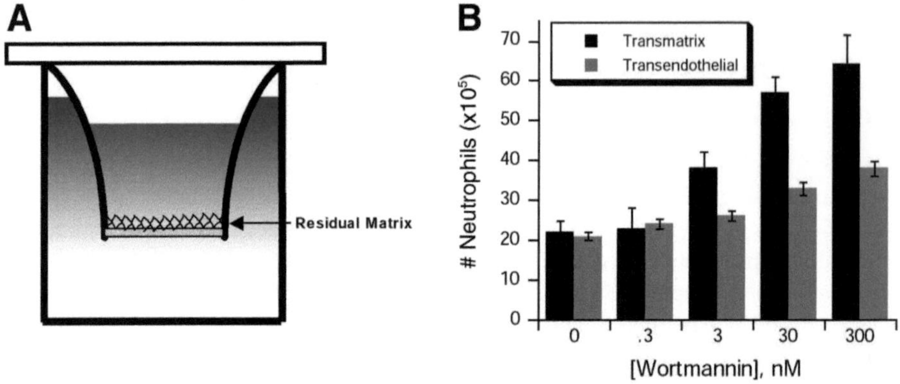

Fig. 3. Models of residual matrix co-cultures. **(A)** Cells (e.g., endothelia) are plated on membrane permeable supports (5-µm pores) and allowed to grow to confluence. Cells are then removed by vigorous washing in EDTA (2 m*M*), resulting in residual matrix proteins attached to the membrane support. **(B)** An example of results derived from such a model. In this example, neutrophils were pre-exposed to indicated concentrations of the phosphoinositide-3-kinase inhibitor wortmannin and assessed for fMLP ($10^{-8}M$)-stimulated transmigration across confluent endothelia (■) or across matrix-coated substrates (■). Transmigrated neutrophils were quantified by determination of myeloperoxidase content after termination of the assay. Data are derived from four monolayers in each condition with results expressed as mean ± standard error of the mean (SEM) number transmigrating neutrophils.

3. Individual inserts are kept sterile and flipped upside down into a large Petri dish.
4. To measure agonist-stimulated short circuit currents (Isc), transepithelial potentials and resistance, a commercially available voltage clamp (Iowa Dual Voltage Clamps, Bioengineering, University of Iowa) interfaced with an equilibrated pair of calomel electrodes and a pair of Ag-AgCl electrodes is utilized, as described in detail elsewhere ***(12)***. Using these values and Ohm's law, transepithelial resistance and transepithelial current were calculated. Fluid resistance within the system accounts for less than 5% of total transepithelial resistance. HBSS was used in both apical and basolateral baths during all experiments, unless otherwise noted. Cl^- secretory responses are expressed as a change in short circuit current (ΔIsc) necessary to maintain zero potential difference across the monolayer.
5. An alternative method for generating a general matrix has also been used ***(11)***. Acellular, matrix-coated permeable support substrates can be generated as follows: bare polycarbonate permeable supports (Corning-Costar, Cambridge, MA, 5-µm pore size) are precoated for 30 min with 50 µL of 0.1% gelatin (derived from porcine skin, 175 Bloom, Attachment Factor, Cascade Biologics, Portland, OR) followed by the addition of media (containing 10% newborn calf serum, Gibco,

Grand Island, NY) for an additional 30 min. Inserts are rinsed in HBBS and polymorphonuclear leukocyte transmigration assessed using similar conditions as described in **Subheading 3.2.**

Acknowledgments

This work was supported by National Institutes of Health grants DK50189 and HL60569 (to S. P. C.) and by DK62007 (to N. A. L.).

References

1. Barrett, K. E. and Keely, S. J. (2000) Chloride secretion by the intestinal epithelium: molecular basis and regulatory aspects. *Annu. Rev. Physiol.* **62**, 535–572.
2. Colgan, S. P., Furuta, G. T., and Taylor, C. T. Cytokines and epithelial function, in Microbial Pathogeneis and the Intestinal Epithelial Cell (Hecht, G., ed.), ASM Press, Washington, DC, pp. 61–78.
3. Blume, E. D., Taylor, C. T., Lennon, P. F., Stahl, G. L., and Colgan, S. P. (1998) Activated endothelial cells elicit paracrine induction of epithelial chloride secretion: 6-keto-$PGF_{1\alpha}$ is an epithelial secretagogue. *J. Clin. Invest.* **102,** 1161–1172.
4. Pober, J. S. and Cotran, R. S. (1990) Overview: the role of endothelial cells in inflammation. *Transplantation* **50,** 537–541.
5. Vane, J. R., Anggard, E. E., and Botting, R. M. (1990) Regulatory functions of the vascular endothelium. *N. Eng. J. Med.* **323,** 27–36.
6. Berschneider, H. M. and Powell, D. W. (1992) Fibroblasts modulate intestinal secretory responses to inflammatory mediators. *J. Clin. Invest.* **89,** 484–489.
7. Parkos, C. A., Colgan, S. P., Delp, C., Arnaout, M. A., and Madara, J. L. (1992) Neutrophil migration across a cultured epithelial monolayer elicits a biphasic resistance response representing sequential effects on transcellular and paracellular pathways. *J. Cell. Biol.* **117,** 757–764.
8. Parkos, C. A., Delp, C., Arnaout, M. A., and Madara, J. L. (1991) Neutrophil migration across a cultured intestinal epithelium: dependence on a CD11b/CD18-mediated event and enhanced efficiency in the physiologic direction. *J. Clin. Invest.* **88,** 1605–1612.
9. Kaoutzani, P., Colgan, S. P., Cepek, K. L., et al. (1994) Reconstitution of cultured intestinal epithelial monolayers with a mucosal-derived T lymphocyte cell line: Modulation of epithelial phenotype dependent on lymphocyte-basolateral membrane apposition. *J. Clin. Invest.* **94,** 788–796.
10. McKay, D. M. and Singh, P. K. (1997) Superantigen activation of immune cells evokes epithelial (T84) transport and barrier abnormalities via, I. F.N-gamma and, T. N.F-alpha: inhibition of increased permeability, but not diminished secretory responses by TGF-beta2. *J. Immunol.* **159,** 2382–2390.
11. Bruyninckx, W. J., Comerford, K. M., Lawrence, D. W., and Colgan, S. P. (2001) Phosphoinositide 3-kinase modulation of beta(3)-integrin represents an endogenous "braking" mechanism during neutrophil transmatrix migration. *Blood* **97,** 3251–3258.
12. Dharmsathaphorn, K. and Madara, J. L. (1990) Established intestinal cell lines as model systems for electrolyte transport studies. *Methods Enzymol.* **192,** 354–389.

5

Cell–Cell Interactions in the Kidney

Inducible Expression of Mutant G Protein α-Subunits in Madin–Darby Canine Kidney Cells for Studies of Epithelial Cell Junction Structure and Function

Ernesto Sabath and Bradley M. Denker

Summary

Disrupted epithelial cell junctions are a hallmark of numerous disease processes. The signaling mechanisms regulating barrier function and re-establishment of intact junctions after injury and during development are complex and tightly regulated. We have shown that heterotrimeric G proteins regulate assembly and maintenance of epithelial cell barrier in cultured Madin–Darby canine kidney (MDCK) cells. The inducible expression of mutant signaling molecules (constitutively active or dominant negative) in polarized cells is a useful strategy for elucidating the role(s) for specific proteins. Using tetracycline-off inducible expression of wild-type and constitutively active Gα12, we have demonstrated a fundamental role for Gα12 in regulating the junction of MDCK cells. Inducible expression permits the comparison of the identical cell line in the presence and absence of the protein of interest and minimizes variation arising from distinct clones. The methods described here are applicable to virtually any protein that may regulate maintenance or assembly of the epithelial cell junctional complex.

Key Words: G proteins; Madin–Darby canine kidney cells; tight junctions; signaling; epithelial cells; kidney; ZO-1.

1. Introduction

G proteins (guanine nucleotide binding proteins) are molecular switches relaying information from upstream molecules (receptor or equivalent) to downstream targets (effectors). The fundamental "switch" in these proteins occurs after an upstream interaction (i.e., receptor) induces a change in conformation, resulting in the dissociation of bound guanosine diphosphate (GDP). Guanosine triphosphate (GTP) normally is present in cells at higher concentrations than GDP, and

From: *Methods in Molecular Biology, vol. 341: Cell–Cell Interactions: Methods and Protocols*
Edited by: S. P. Colgan © Humana Press Inc., Totowa, NJ

GTP binds to the G protein. The G protein remains activated and capable of downstream interactions until the intrinsic GTPase activity hydrolyzes bound GTP to GDP (reviewed in **refs.** ***1*** and ***2***). There are two broad families of G proteins: monomeric (single subunit) GTPases of the Ras Superfamily, and heterotrimeric (composed of Gα GTPase and Gβγ). Both families of G proteins regulate epithelial cell junctions ***(3,4)***. Regulating these signaling pathways holds promise for novel therapies to prevent or treat disorders resulting in disruption of junctional integrity. Madin–Darby canine kidney (MDCK) cells (derived from canine kidney, distal tubule, or collecting duct) are used widely for studies of epithelial cell polarity and cell junctions. MDCK cells have proven to be a useful model for studies of tight junction assembly through the "calcium switch" ***(5)***. When cultured in low calcium (μ*M*), cells lose polarity, junctional proteins become intracellular, and monolayer integrity is lost. With switching to normal calcium medium ("calcium switch"), there is coordinated assembly of the tight junctions thought to recapitulate fundamental processes involved in junctional biogenesis during development and after injury ***(6)***. Synchronized tight junction assembly without new protein synthesis allows the process to be studied biochemically. Functionally, the integrity of the barrier is monitored through the measurement of transepithelial resistance (TER) at specific time points during the switch process. Recently, this model has been modified to permit the effects of an induced protein on the development of the epithelial barrier to be studied. An example of an inducible system that will be described in detail in this chapter is the "doxycycline switch." In the most common MDCK cell system (tetracycline-off [Tet-off]), the removal of doxycycline (dox) induces protein expression, permitting the effects of the transfected protein on the cell to be studied. We have used this methodology to elucidate the role of the Gα12-subunit on the epithelial cell barrier ***(7,8)***.

2. Materials

2.1. Plasmids and DNA Preparation

1. pTRE (Clontech, Palo Alto, CA).
2. Complimentary DNA (cDNA) of interest and control.
3. Restriction enzymes.
4. Gel extraction kit (QIAquick; Qiagen, Valencia, CA).
5. Calf intestinal alkaline phosphatase (New England Biolabs, Beverly, MA).
6. T4 DNA Ligase 400,000 U/mL (New England Biolabs).
7. DH5α competent cells (Invitrogen, Carlsbad, CA).
8. Luria-Bertani (LB) broth powder (Sigma, St. Louis, MO).
9. Bacteriological agar (Sigma).
10. Ampicillin sodium salt powder (Sigma).

11. Petri dishes 100 mm (BD Falcon).
12. Qiagen Plasmid Mini or Midi-Kit (Qiagen).

2.2. Cell Culture

1. MDCK type II cells stably expressing the tetracycline (Tc)-controlled transactivator for use in the Tet-off gene expression system (Clontech, Palo Alto, CA).
2. Dulbecco's Modified Eagle's Medium (500 mL) with 4.5 g/L glucose and L-glutamine (Cellgro, Herndon, VA).
3. Fetal bovine serum Tet System Approved (Clontech). Heat inactivate at 55°C for 30 min. Use at 5% (25 mL) final concentration.
4. Solution of 10,000 U/mL penicillin G sodium and 10,000 μg/mL streptomycin sulfate (Gibco-BRL, Carlsbad, CA). Add 5 mL of stock solution to medium.
5. G418 Powder, 1 g (Cellgro, Herndon, VA). Prepare stock solution at a concentration of 50 mg/mL. Add 1 mL to 500 mL of medium.
6. Hygromycin B from *Streptomyces hygroscopicus*. Stock solution (50 mg/mL; Roche Diagnostics, Mannheim, Germany.) Add 1 mL to medium.
7. Dox (Sigma) or Tc (*see* **Note 1**). Store in aliquots (concentration 1 mg/mL) at –20°C, store in the dark and protect from light with foil. Add 20 μL for a final concentration of 40 ng/mL.
8. Phosphate-buffered saline (PBS) tablets (Sigma). To prepare 1X solution, dissolve one tablet/200 mL H_2O and sterilize.
9. Trypsin 0.25% and EDTA (Sigma).
10. Tissue culture microscope.

2.3. Stable Transfection and Selection of Antibiotic-Resistant Clones

1. Lipofectamine 2000 1 mL (Invitrogen).
2. OptiMEM I Reduced Serum Medium for transfections (Gibco, Carlsbad, CA).
3. pTRE Vector-Gene X.
4. pTK-Hyg (*see* **Note 2**; Clontech).
5. Cloning cylinders (Scienceware, Pequannock, NJ). An alternative is to cut 200-μL pipet tips approx 0.5 cm from wide end and autoclave.
6. High vacuum grease (Dow Corning, Fisher Scientific). Autoclave.
7. Tissue culture plates 24-, 12-, and 6-well plates, and 60-mm plates (BD Falcon).

2.4. Sodium Dodecyl Sulfate-Polyacrylamide Gel Electrophoresis

1. Stacking buffer (4X): Tris-HCl 0.5 *M*, pH 6.8, 0.4% sodium dodecyl sulfate (SDS).
2. Separating buffer (4X): Tris-HCl 1.5 *M*, pH 8.8, 0.4% SDS.
3. 30% Acrylamide Solution (**Caution:** When acrylamide is unpolymerized, it is a neurotoxin that may be harmful if inhaled, swallowed, or exposed directly to skin).
4. TEMED (Sigma).
5. Ammonium persulfate (APS). Prepare 10% solution in water and store at –20°C (Sigma).
6. Isobutanol, water saturated.

7. Electrode (running) buffer (10X): 0.25 *M* Tris-HCl, 1.9 *M* glycine; add SDS 0.1% before use.
8. Sample buffer (3X): 187.5 m*M* Tris-HCl, pH 6.8, 6% SDS, 30% glycerol, 0.003% bromophenol blue. Add β-mercaptoethanol 5% before use.
9. Prestained molecular weight markers (Bio-Rad, Hercules, CA).

2.5. Cell Lysis

1. PBS (*see* **Subheading 2.2., item 8**).
2. Tissue Buffer: 10 m*M* Na–HEPES, pH 7.5, 1% NP-40, 0.25% Na–deoxycholate, 0.1% SDS, 150 m*M* NaCl, 1 m*M* EDTA, 1 m*M* phenylmethyl sulfonyl fluoride, protease inhibitor cocktail (Complete, Roche. One tablet in 1 mL of water makes a solution 50X); Na_3VO_4 (tyrosine-phosphatase inhibitor) 1 m*M*; NaF (serine-phosphatase inhibitor) 25 m*M*; and 1-mL tuberculin syringe with 27-gage needle.
3. Tabletop centrifuge in a cold room at 4°C.
4. Liquid nitrogen (*see* **Note 3**).
5. Water bath at 37°C.

2.6. Western Blot

1. Electrophoretic transfer cell (Bio-Rad).
2. Nitrocellulose Pure Transfer Membranes (GE Osmonics Labstore, Minnetonka, MN).
3. Transfer buffer (glycine 39 m*M*, Tris-base 48 m*M*, SDS 0.04%, methanol 20%).
4. Filter paper, Whatmann 3MM (Fisherbrand).
5. GB003 thick 0.8 filter paper (Schleicher & Schuell BioScience, Keene, NH).
6. Ponceau-S Solution (Sigma).
7. Tris-buffer saline with Tween-20 (TBST): 20 m*M* Tris-HCl, pH 7.6, 137 m*M* NaCl, 0.05% Tween-20.
8. Blocking-buffer: TBST containing 5% dry milk or 2% albumin powder.
9. Enhanced chemiluminescent reagents (Pierce, Rockford, IL).
10. Rabbit Gα12 antibody (Santa Cruz Biotechnologies, Santa Cruz, CA).
11. Goat anti-rabbit IgG, secondary antibody (Santa Cruz Biotechnologies).

2.7. TER Measurements

1. Polycarbonate filters (Transwell, Costar, Acton, MA): tissue culture-treated, 12-mm diameter, and 0.4-μm pore size.
2. MDCK II cells (stably transfected).
3. Millipore ERS Electrical Resistance System (Millipore, Billerica, MA).
4. Dulbecco's Modified Eagle's Medium with additives.

3. Methods

The Tet systems permit gene expression to be tightly regulated in response to varying concentrations of Tc or dox. In the Tet-off system, gene expression is turned on when Tc or dox is removed from the culture medium. In contrast, expression is turned on in the Tet-on system by the addition of dox (*see* **Note 1**

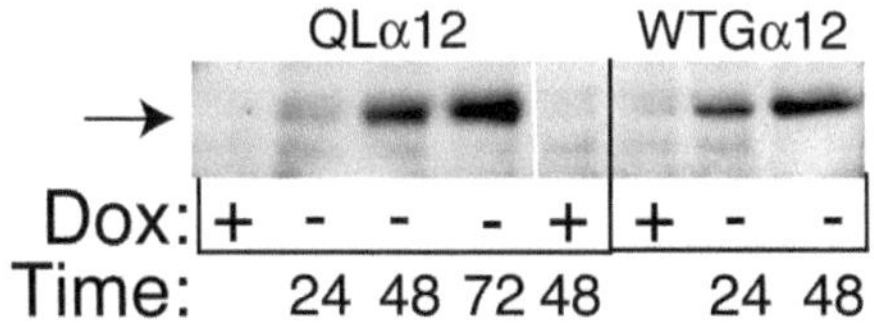

Fig. 1. Gα12 Western blot of inducible Gα12 expression in MDCK cells. Approximately 50 μg of cell lysates were analyzed by SDS-PAGE and Western blot as described. Monolayers were maintained in +dox for several days before switching to –dox for the indicated times. There is a small amount of endogenous Gα12 visible in the +dox lysates that is also seen in Tet-off MDCK cells (not shown). Both wild-type and activated (QL) α12 are rapidly induced with dox removal. The readdition of dox for 48 h (QLα12) demonstrates complete resuppression of protein expression.

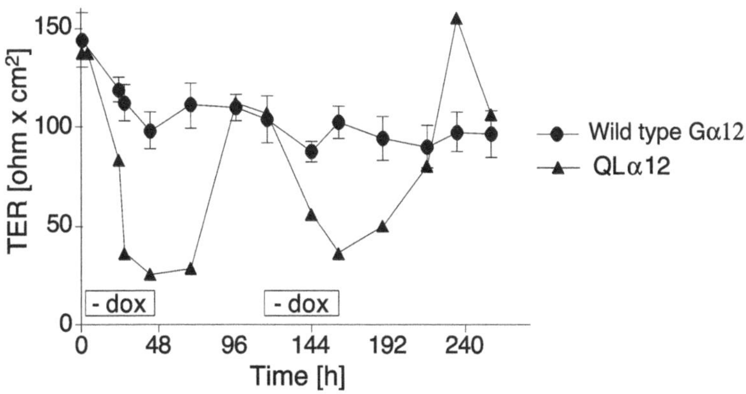

Fig. 2. QLα12 expression results in loss of barrier function. TER was monitored daily for 10 d in wild-type and QLα12-MDCK cells ± dox. The rectangles denote the time interval during which dox was removed. In wild-type Gα12-MDCK cells, there was little change in TER during the 10-d period regardless of Gα12 expression. With QLα12 expression, there was a significant decrease in TER correlating with QLα12 expression, and the effect on TER was completely reversible with readdition of dox.

and **Figs. 1** and **2**). Background or leaky expression of Protein X in the absence of induction is low with pTRE; however, it is important to add fresh dox every 2 d (*see* **Note 4**). A major advantage of this system is the comparison of the same clone in conditions ± dox (without or with *gene X* expression). In addition to ± *gene X* expression, it is also recommended to prepare a parallel transfection with a control gene. For example, in our analysis of Gα12, two different cell lines were established; one expressing the constitutively activated QLα12 gene and another expressing the wild-type protein (*see* **Figs. 1** and **2**).

3.1. Creating the Gene-Specific Expression Vector

1. Transform each of the plasmids into a DH5α *Escherichia coli* (or equivalent) to ensure a renewable source of DNA.
2. Linearize pTRE Tet-vector using the appropriate restriction enzymes contained in the multicloning site. Cut your X-gene cDNA with the same restriction enzymes.
3. Run an agarose gel, check the digestion of the plasmids, and excise the bands corresponding at the Tet-vector and X-gene cDNA with a new razor blade; avoid long exposure to the ultraviolet light because it can crosslink the DNA (*see* **Note 5**). Take a picture from the remaining gel to be sure that no DNA remains.
4. Purify the DNA using a gel-extraction kit or a polymerase chain reaction-purification kit.
5. Add calf intestinal alkaline phosphatase to the pTRE vector for 1 h at 37°C to decrease the probability for self-ligation.
6. To perform ligation, use a ratio of pTRE:X gene cDNA of 1:3, add 10X ligase buffer (*see* **Note 6**), 1 μL of T4 Ligase, and water (to complete 10–15μL/reaction). Incubate for 1 h at room temperature. For blunt-end ligation, incubate for 16 h at 16°C (*see* **Note 7**).
7. Perform transformation in DH5α cells.
8. Purify DNA from bacteria with Miniprep kit according to the manufacturer's instructions.
9. Corroborate your ligation product by restriction analysis and/or sequencing.
10. Produce 50–100 μg of DNA using the Midiprep kit according to manufacturer's instructions.

3.2. Cell Culture

1. Culture the MDCK II cells in Dulbecco's Modified Eagle's Medium with 4.5 g/L glucose and L-glutamine. Add fetal calf serum 5%, penicillin–streptomycin (5 mL), and G418 (1 mL) to the medium.
2. Grow the cells at 37°C in a humidified chamber with 5% CO_2. Split (trypsinize) the cells when 90 to 100% confluent. Remove the Dulbecco's medium, wash twice with PBS (prewarmed to 37°C), add trypsin–EDTA (2 mL/10 cm plate). Incubate at 37°C for 5 to 15 min. Check under microscope for detachment of cells. Add Dulbecco's Medium (8 mL) to quench the trypsin and split into new plates at 1:5–10 (*see* **Note 8**).

3.3. Preparing Frozen Stocks

1. Trypsinize the desired number of plates or flasks.
2. Centrifuge cells at 100*g* for 3 to 5 min and aspirate the supernatant.
3. Resuspend the pellet in freezing medium. We resuspend cells from a confluent 10-cm plate in 1 mL and divide into 2 × 0.5 mL aliquots for long-term storage. Freezing medium can be purchased from Sigma (cat. no. C6164) or cells can be frozen in 90% fetal bovine serum and 10% dimethyl sulfoxide.
4. Dispense 0.5-1-mL aliquots into sterile cryovials.

5. Freeze gradually. Start with –20°C for 1 to 2 h. Transfer to –80°C overnight and then store in liquid nitrogen.

3.4. Titrating Antibiotics (Kill Curve)

Before using hygromycin or puromycin to establish your stable cell lines, it is recommended to titrate the selection agent to determine the optimal concentration for selection.

1. Plate cells at normal dilutions (usually 1:10) in a six-well plate. Allow them to attach for 24 h, then add complete medium plus varying amounts (0, 50, 100, 200, 400, 800 µg/mL) of hygromycin.
2. Incubate the cells for 10 to 14 d, replacing the selective medium every 2 to 3 d. Examine the dishes for viable cells every 2 d.
3. For selecting stable transformants, use the lowest concentration that demonstrates significant cell death at approx 5 d and 100% mortality at 10 to 14 d. For Tet-off MDCK cells, we have found 200 µg/mL hygromycin to be optimal but it may vary from lot to lot.

3.5. Stable Transfection and Colony Selection

The goal with the stable transfection is to generate a cell line that gives low background and high expression of Protein X. Both expression levels and induction of Protein X can be significantly affected by the site of plasmid integration. Insertion near an enhancer may result in high basal expression of Protein X, whereas other insertion sites may result in suboptimal induction. To find the clone with the highest induction and lowest background, we recommend that you grow and analyze as many clones as feasible. We usually aim for 30 to 50 clones initially; 20 to 50% will be lost on subsequent processing (to mininize the loss of clones, it is important to avoid long exposure times to trypsin).

1. Grow cells to 80% confluency in complete medium.
2. Transfect pTRE-*Gene X* (25 µg/100 cm^2 dish) and your Selection Marker (pTK-Hyg, or pPUR; 2.5–5 µg/100 cm^2 dish). The plasmids can be linearized before transfection and this may increase transfection efficiency.
3. To transfect MDCK II cells, use a lipid agent, such as Lipofectamine 2000.
4. We recommend adding dox (40 ng/mL) to the transfection medium within 10 to 12 h of the transfection. This step is particularly important if your protein product may be toxic to the cells.
5. Split the cells (1:10) 24 h after the transfection. Usually, a 10-cm plate will be split into ten 10-cm plates.
6. Allow cells to grow 24 h, then add hygromycin (or puromycin) at the optimal concentration (previously determined by killing curve). Sometimes we have added hygromycin immediately after the first split (**step 5**) and have found comparable results.

7. Replace medium containing hygromycin (or puromycin) every 2 d. Fresh dox must be added every 2 d for Tet-off cells (*see* **Note 4**).
8. After about 5 d, clear patches of plastic devoid of cells should begin to be apparent at low magnification under the tissue culture microscope. Ideally, cells at this stage should not be trypsinized and split (however, if your cells become confluent, you should trypsinize and split again).
9. After 2 to 4 wk, hyg- (or puro-) resistant colonies will begin to become visible to the naked eye and can be viewed by low power on the tissue culture scope. View and circle with a marker on the bottom of the plate to know where to place the cloning cylinder or pipet ring. Avoid waiting until your colonies get confluent, but ideally they should be greater than 50 cells.
10. Viewing the plate from the bottom, you may distinguish small white spots that are colonies. Mark them and confirm under the microscope. Expect 0 to more than 30 distinct colonies per 10-cm plate. It is important that you look carefully at plate periphery because sometimes the colonies prefer this location.
11. Wash entire plate twice with PBS (prewarmed to 37°C).
12. According to the size of the colony (usually 10–100 cells) choose a cloning cylinder, place vacuum grease on lower lip, and place over the margins of your colony. Add trypsin–EDTA 50 to 100 μL per ring. Incubate at 37°C for 5 min and look for cell detachment under the microscope. To avoid excessive trypsinization, check cells every 2 min. After cells detach, add complete medium and transfer the cells to a 24 well-plate.
13. Grow the cells with complete medium plus hygromycin and dox. When confluent, transfer to a 12-well plate. When clones nearly confluent, repeat this process for a six-well plate.
14. When the cells are confluent in a six-well plate, we recommend splitting the entire well into a 10-cm plate. When this plate is confluent, we trypsinize and freeze 90% of the cells to insure a continued lineage. The remaining 10% of cells are placed in a six-well plate for analysis.
15. When a clone in the six-well plate is confluent, the media is changed to –dox for 72 h to induce protein expression. Make a cell lysate and determine protein expression by SDS/Western blot. Controls include one +dox cell lysate and one Tet-off cell lysate on the Western (*see* **Note 9**).

3.6. Cell Lysis

1. Wash monolayer with PBS three times.
2. Scrape with rubber policeman in 0.5 to 5 mL of tissue buffer plus protease inhibitors and place in a 1.5-mL Eppendorf tube.
3. Freeze in liquid nitrogen; thaw in 37°C water bath. Be careful not to let sample warm. Repeat two more times.
4. Triturate by passing 10 to 20 times through a 27-gage needle.
5. Centrifuge samples at 3000 rpm for 5 min to pellet nuclei and debris.
6. Determine protein concentration and store samples at –80°C.

3.7. Sodium Dodecyl Sulfate-Polyacrylamide Gel Electrophoresis

Gα-subunits migrate at 39 to 43 kDa on SDS-polyacrylamide gel electrophoresis (PAGE). Therefore, we use 10 to 12% acrylamide gels.

1. The glass of your gel system should be cleaned with a rinsable detergent (Alconox, Alconox, NY). Rinse extensively with distilled water and do the last wash with ethanol 70%. Dry at room temperature.
2. Assemble your gel system. Make sure that your system is not leaky.
3. Prepare a 0.75-mm thick, 10% gel using 30% acrylamide, 1.25 mL of separating buffer, 25 μL of 10% APS, and 5 μL of TEMED. Pour the gel, leaving space for the stacking gel, and overlay with water-saturated isobutanol. The gel should polymerize in approx 25 to 30 min.
4. Pour off the isobutanol and rinse with distilled water until alcohol odor is gone.
5. Prepare the stacking gel. Pour the stacking gel solution on to the top of the separating gel, and insert the comb. Once your gel has polymerized, remove the comb and fill the spaces with running (electrode) buffer.
6. Add sample-buffer to your cell lysate (usually 10–60 μg of protein); heat your samples for 4 to 5 min at 95°C.
7. With an electrophoresis tip, fill the wells of the stacking gel with your samples and add your molecular weight marker (Bio-Rad) to one or more lanes.

Connect your device to the power supply. The gel can be run overnight at 40–50 V (constant voltage) or at 15–20 mA/gel until the sample has exited the stacking gel and entered the running gel. Current can be increased to 30–40 mA/gel until the blue dye reaches the bottom (usually ~2–3 h). Gels run in this manner should be cooled.

3.8. Western Blot

1. After SDS-PAGE, transfer is done with the Bio-Rad transfer system.
2. Disconnect the gel unit from the power supply and disassemble. Discard the stacking gel.
3. To assembly your transfer system, you need two foam sheets, two thick filter paper sheets, two thin filter paper sheets, and one nitrocellulose membrane. We usually use squares of 14 cm^2; try to get all pieces to be the same size. Pour transfer buffer in a big tray and assemble components immersed in transfer buffer. Assemble a transfer cassette (be sure to assemble in order): place two foam sheets, one thick filter paper sheet, one thin filter paper sheet, and then nitrocellulose. The gel is laid on the nitrocellulose and the sandwich completed by placing thin filter paper followed by thick filter paper and the foam sheet.
4. Place your cassette in the transfer tank; your nitrocellulose membrane should be oriented to the anode (red color) and the gel to the cathode (black color). To avoid overheating your system, place your transfer tank on ice and in a cold room. You can transfer your proteins overnight at 0.25 mA or in 2 h at 0.5 to 0.75 mA.

5. Once the transfer is complete, remove the cassette and disassemble. Mark the location of the prestained markers. To assess the quality of transfer, place membrane in Ponceau-S solution for 3 min, wash first with water and then with TBST and look for protein bands. Optional: take a photocopy for later comparison.
6. Incubate nitrocellulose membrane in blocking solution (TBST + 5% dry milk) for 1 h at room temperature or overnight at 4°C on a rocking platform.
7. Discard blocking solution and rinse membrane once with TBST.
8. Place membrane inside a plastic sealable bag, add Gα12 primary antibody (diluted 1:1000 in TBST), and seal. Incubate overnight at 4°C on a rocking platform.
9. Remove the primary antibody (*see* **Note 10**) and wash the membrane three times, 10 min each with TBST.
10. Add goat anti-rabbit IgG secondary antibody (diluted 1:10,000 in TBST) and incubate for 1 h at room temperature in a rocking platform.
11. Remove the secondary antibody and wash six times for 5 min each with TBST.
12. Prepare enhanced chemiluminescent solution mixing 2 mL of each component. Add this solution to membrane, taking care to cover all the surface of your membrane. Incubate for 5 min.
13. Remove membrane and place into film cassette with intensifying screens and expose to X-ray film for various times to optimize visualization of the protein.

3.9. TER Measurements

1. Split cells with trypsin–EDTA.
2. Plate the cells on polycarbonate filters (Transwell, Costar) at confluent density (3×10^5 cells/cm^2). From a 10-cm plate in 10 mL of medium, place 0.5 mL of cell suspension in the top chamber. Add 1 mL of medium to the lower reservoir. Leave one filter blank with no cells.
3. Allow 24 to 48 h for attachment.
4. TER stabilizes over the course of 2 to 4 d after plating; once TER is stabilized you can switch your cells to –dox.
5. To measure TER, rinse the Millipore ERS electrodes in 70% ethanol, allow them to air-dry, and then rinse them with sterile PBS. When obtaining measurements, it is important to always place the electrode in the same orientation (*see* **Note 11**). One prong of the chopstick electrode (shorter) is placed inside the suspended filter and the other in the outer well. Press the measure button, wait approx 1 s for the value to stabilize and record. Subtract the value of your blank (this number should be around 100 Ω/cm^2). (**Note:** we find our Tet-off cells with TER ~75–90 Ω/cm^2.)

4. Notes

1. Dox has some advantages over Tc, including a lower concentration for inactivation and a longer half-life in culture.
2. There are other vectors with the Tet-responsive cassette. The pTRE2pur (Clontech, Palo Alto, CA) confers resistance to puromycin instead of hygromycin. There is no clear advantage for either of the antibiotics. The pTRE-Tight (Clontech) is reported to minimize the basal expression of most proteins. Other TRE vectors

permit placement of an epitope tag to aid in protein detection (i.e., pTRE-myc; Clontech).

3. There are other alternatives to extract proteins from the cell lysate, such as sonication on ice.
4. This is the most important procedure to avoid leaky protein expression. Store Dox in the dark at –20°C for no more than 6 mo. Add fresh Dox to medium every 2–3 d, and replace the medium of cells every 2–3 d.
5. This step is important for success in ligation. Exposure of DNA to ultraviolet light can result in damage to the ends and lower ligation efficiency.
6. Use fresh ligase buffer and avoid many thaw–freeze cycles because this may inactivate the ATP.
7. For ligations, we have obtained the same results incubating the ligation mixture for 1 h at room temperature in comparison with 16 h at 16°C.
8. When trypsin is thawed at 37°C, it can lose activity. Aliquot your trypsin to avoid freeze–thaw cycles.
9. On occasion, well-characterized stable cell lines can lose their responsiveness to Tc or Dox. This can occur after changing lots serum and may be the result of contamination with antibiotics. If a sudden loss of responsiveness is observed, check serum by performing a dose–response curve. It is recommended to subclone and freeze stocks of your cells at various stages.
10. The primary antibody can be saved for subsequent experiments. Add 0.02% final concentration of sodium azide (prepare a 10% stock solution), and storage at 4°C.
11. TER measurements are inherently slightly variable. Measure TER with all the filters in the same orientation, angle and depth of electrodes, media volume and duration of "pulse." Filters that have had TER recently measured (within last 15–20 min) may have TER values that are noticeably higher if remeasured within this time frame.

References

1. Li, G. and Zhang, X. C. (2004) GTP hydrolysis mechanism of Ras-like GTPases. *J. Mol. Biol.* **340,** 921–932.
2. Robishaw, J. D. and Berlot, C. H. (2004) Translating G protein subunit diversity into functional specificity. *Curr. Opin. Cell Biol.* **16,** 206–209.
3. Jou, T. S., Schneeberger, E. E., and Nelson, W. J. (1998) Structural and functional regulation of tight junctions by RhoA and Rac1 small GTPases. *J. Cell Biol.* **142,** 101–115.
4. Denker, B. M., Saha, C., Khawaja, S., and Nigam, S. K. (1996) Involvement of a heterotrimeric G protein α subunit in tight junction biogenesis. *J. Biol. Chem.* **271,** 25,750–25,753.
5. Gonzalez-Mariscal, L., Contreras, R. G., Bolivar, J. J., Ponce, A., Chavez De Ramirez, B., and Cereijido, M. (1990) Role of calcium in tight junction formation between epithelial cells. *Am. J. Physiol.* **259,** 978–986.
6. Rodriguez-Boulan, E. and Nelson, W. J. (1989) Morphogenesis of the polarized epithelial cell phenotype. *Science* **245,** 718–725.

7. Meyer, T. N., Hunt, J., Schwesinger, C., and Denker, B. M. (2003) Gα12 regulates epithelial cell junctions through Src tyrosine kinases. *Am. J. Physiol.* **285,** 1281–1293.
8. Meyer, T. N., Schwesinger, C., and Denker, B. M. (2002) Zonula occludens-1 is a scaffolding protein for signaling molecules. Gα12 directly binds to the Src homology 3 domain and regulates paracellular permeability in epithelial cells. *J. Biol. Chem.* **277,** 24,855–24,858.

6

Nucleotide Metabolism and Cell–Cell Interactions

Holger K. Eltzschig, Thomas Weissmüller, Alice Mager, and Tobias Eckle

Summary

Interactions between the vascular endothelium and polymorphonuclear leukocytes (PMNs) are central to PMN emigration into inflamed tissues, and to neutrophil–endothelial crosstalk pathways that modulate inflammatory responses and vascular barrier function. For example, during episodes of inflammation, the transendothelial migration (TEM) of PMNs potentially disturbs vascular barrier and gives rise to intravascular fluid extravasation and edema. However, because of the close special relationship between PMNs and the vascular endothelium, TEM creates an ideal situation for neutrophil–endothelial crosstalk. While investigating innate mechanisms to dampen intravascular fluid loss and edema occurring during TEM, we observed that PMNs release adenine nucleotides after activation (adenosine triphosphate [ATP] and adenosine monophosphate [AMP]). ATP and AMP are metabolized by endothelial cell-surface enzymes, the ecto-apyrase (CD39, metabolizes ATP to AMP) and the 5′-ecto-nucleotidase (CD73, metabolizes AMP to adenosine). Adenosine generated in this fashion can activate endothelial adenosine receptors, leading to increases in intracellular cyclic AMP and resealing of the endothelial junctions, thereby promoting vascular barrier function. This crosstalk pathway provides an endogenous mechanism to dampen vascular leak syndrome during neutrophil–endothelial interaction. In other words, during TEM, neutrophils close the door behind them.

Key Words: Adenosine; adenosine monophosphate; adenosine triphosphate, ectonucleotidase; endothelium; neutrophil; CD73; CD39; inflammation.

1. Introduction

Neutrophil-mediated tissue injury is a prominent component in organ damage during inflammation *(1,2)*. Endothelial cells, which line the inner lumen of blood vessels and largely determine overall vascular permeability *(3)*, are primary targets for polymorphonuclear leukocyte (PMN) during uncontrolled inflammatory responses *(4)*. Endothelial injury may result in increased endothelial paracellular permeability and decreased endothelial barrier function. The resultant local tissue

From: *Methods in Molecular Biology, vol. 341: Cell–Cell Interactions: Methods and Protocols*
Edited by: S. P. Colgan © Humana Press Inc., Totowa, NJ

edema and intravascular loss of fluid and solutes may further exacerbate tissue injury via interference with cellular function and limitation of regional blood flow *(5)*.

To understand such interactions between PMNs and the vascular endothelium and potentially identify therapeutic targets for limiting vascular leak syndrome associated with inflammation and hypoxia, the first step is to identify biological mechanisms that control endothelial barrier function and regulate vascular leak. For example, macromolecule transit across blood vessels has evolved to be tightly controlled. Relatively low macromolecular permeability of blood vessels is essential for maintenance of a physiologically optimal equilibrium between intravascular and extravascular compartments *(6,7)*. Endothelial cells comprise the inner cell layer of blood vessels and are the predominant cell type determining vascular permeability *(3)*. Disturbance of endothelial barrier during disease states can lead to deleterious loss of fluids and plasma protein into the extravascular compartment. Such disturbances in endothelial barrier function are prominent in disorders such as shock and ischemia-reperfusion and contribute significantly to organ dysfunction *(1,2,8)*. However, the direct relationship between PMN activation and increased endothelial permeability is not clear. Although PMN accumulation and increased vascular permeability often are coincidental, particularly at sites of inflammation/hypoxia, PMN activation can occur with limited or no net changes in endothelial permeability *(9–12)*. Moreover, although PMN depletion has been demonstrated to decrease organ injury in some models of ischemia and reperfusion *(13)*, PMNs have been demonstrated to exert a protective effect in other models *(14–16)*. Therefore, further information regarding PMN–endothelial cell interactions and their influence on endothelial permeability may provide a better understanding of the regulation of endothelial permeability.

Approximately 70 million PMNs exit the vasculature per minute *(17)*. These inflammatory cells move into underlying tissue by initially passing between endothelial cells that line the inner surface of blood vessels. This process, referred to as transendothelial migration (TEM), is particularly prevalent in inflamed tissues. Understanding the biochemical details of leukocyte–endothelial interactions is currently an area of concentrated investigation, and recent studies of genetically modified animals have suggested that specific molecules may establish "bottlenecks" to the control of the inflammatory response *(18)*. For example, detailed studies have revealed that the process of leukocyte TEM entails a concerted series of events involving intimate interactions of a series of leukocyte and endothelial glycoproteins that include selectins, integrins, and members of the immunoglobulin supergene family (e.g., intercellular adhesion molecule-1 *[19–21]*). Moreover, histological studies of TEM reveal that PMNs initially adhere to endothelium, move to nearby interendothelial junctions via diapedesis, and insert pseudopodia into the interendothelial paracellular space *(22)*. Successful TEM is accomplished by temporary PMN self-deformation with localized widening of the interendothelial junction. After TEM, adjacent endothelial cells appear to "reseal," leaving no residual interendothelial gaps *(22)*. These histological studies are consistent with the observation that leukocyte TEM may result in little or no change in endothelial permeability to macromolecules *(9–12,23)*. In the absence of this tight and dynamic control of endothelial morphology and permeability, interendothelial gap formation during leukocyte TEM could lead to marked increases in endothelial permeability. However, only limited information exists regarding the biochemical events which maintain and dynamically regulate endothelial permeability in the setting of either PMN activation or TEM *(22,24)*.

Despite the severe course in which inflammatory diseases can proceed, most inflammation is self-limiting. One important factor may be increased production of endogenous adenosine, a naturally occurring anti-inflammatory agent *(25–27)*. Several lines of evidence support this assertion. First, adenosine receptors are widely expressed on target cell types as diverse as leukocytes, vascular endothelia, and mucosal epithelia and have been studied for their capacity to modulate inflammation *(27)*. Second, murine models of inflammation provide evidence for adenosine receptor signaling as a mechanism for regulating inflammatory responses in vivo in endogenous anti-inflammation. For example, mice deficient in A_{2A}-adenosine receptor ($AdoRA_{2A}$) show increased inflammation-associated tissue damage *(28)*. Third, hypoxia is a common feature of inflamed tissues *(29)* and is accompanied by significantly increased levels of adenosine *(30–33)*. At present, the exact source of adenosine is not well-defined but likely results from a combination of increased intracellular metabolism and amplified extracellular phosphohydrolysis of adenine nucleotides via surface ecto-nucleotidases.

Previous studies have indicated that activated PMN release a number of soluble mediators that regulate vascular permeability during transmigration. For - example, PMNs have been demonstrated to actively release adenine nucleotides (in the form of adenosine triphosphate [ATP] or adenosine monophosphate [AMP]). On the endothelial surface, ATP can be converted by the ecto-apyrase (CD39) to AMP, which in turn is converted by the 5′-ectonucleotidase (CD73) to adenosine *(14,15,34,35)*. Adenosine generated in this fashion is then available to activate adenosine receptors either on the endothelial surface or on neutrophils to modulate inflammatory responses.

To pursue studies of neutrophil–endothelial interaction, we first describe an isolation technique of fresh PMNs from human blood and a high-performance liquid chromatography (HPLC)-based technique to measure nucleotide/nucleoside release from activated neutrophils. Then, we introduce an in vitro model of endothelial permeability that allows the in vitro measurement of endothelial barrier function. Finally, we describe techniques to measure induction of CD73 or CD39 protein and function on the endothelial surface.

2. Materials

2.1. Isolation of Human Neutrophils

1. Citrate blood: 10 mL of Monovets containing Na–citrate, 1 mL (Sarstedt, Nümbrecht, Germany).
2. Percoll (Amersham Biosciences, Uppsala, Sweden).
3. NH_4Cl lysis buffer: 420 mg of NH_4Cl, 4 g of $NaHCO_3$, and 18.5 mg of EDTA; bring with H_2O to 500 mL.
4. Hank's Balanced Salts Solution (HBSS)– (Gibco/Invitrogen) containing no calcium and no magnesium, buffered with HEPES (Gibco/Invitrogen) 10 m*M* to pH 7.4 (10 mL of the 1 *M* stock solution is added to 1 L of HBSS).

2.2. Preparation of Activated PMN Supernatants and Measurement of Nucleotide Contents

1. *N*-formyl-Met-Leu-Phe (fMLP; Sigma-Aldrich, Munich, Germany).
2. HBSS– (Gibco/Invitrogen) containing no calcium and no magnesium, buffered with HEPES (Gibco/Invitrogen) 10 m*M* to pH 7.4 (10 mL of the 1 *M* stock solution is added to 1 L of HBSS).
3. HBSS+ (Gibco/Invitrogen) containing calcium and magnesium, buffered with HEPES (Gibco/Invitrogen) 10 m*M* to pH 7.4 (10 mL of the 1 *M* stock solution is added to 1 L of HBSS).
4. CHRONO-LUME reagent (Crono-log Corp, Haverton, PA).
5. For activation of murine PMN, use Leukotriene B4 (Calbiochem, San Diego, CA).

2.3. Cell Isolation and Cell Culture

1. HBSS– (Gibco/Invitrogen) containing 1% antibiotic–antimycotic solution (Sigma-Aldrich).
2. Lactated Ringer's Solution (Fresenius, Bad Homburg, Germany).
3. Collagenase A solution (Roche, Mannheim, Germany) in HBSS– to 1 mg/mL.
4. Modified Medium 199, Medium 199 (Cambrex, Verbier, Belgium) and fetal calf serum (FCS; PAA, Cölbe, Germany) + 1% Na Pyruvate (Sigma-Aldrich) + 1% antibiotic/antimycotic solution (Sigma-Aldrich).
5. Antibiotic solution: antibiotic–antimycotic solution (Sigma-Aldrich).
6. Endothelial cell basal medium (Promo Cell, Heidelberg, Germany) supplemented with endothelial cell growth media (Promo Cell).
7. Accutase (PAA).
8. Trypsin–EDTA (Cambrex).
9. Trypsin neutrolizing solution (Cambrex).
10. Human microvascular endothelial cells (HMEC) media: modified MCDB 131 (Gibco/Invitrogen) with L-glutamin 5 m*M* (Gibco), 10% fetal calf serum (PAA),

1 μg/mL hydrocortisone (Sigma), and epidermal growth factor (BD Biosciences, Bradford, MA).

2.4. Endothelial Permeability Assay

1. HBSS+ (Gibco/Invitrogen) containing calcium and magnesium, buffered with 10 m*M* HEPES (Gibco/Invitrogen) pH 7.4 (10 mL of the 1 *M* stock solution is added to 1 L of HBSS).
2. HBSS– (Gibco/Invitrogen) containing no calcium and no magnesium, buffered with 10 m*M* HEPES (Gibco/Invitrogen) pH 7.4 (10 mL of the 1 *M* stock solution is added to 1 L of HBSS).
3. 70-kDa fluorescein isothiocyanate (FITC)-labeled dextran (Molecular Probes, Eugene, OR).

2.5. Immunoprecipitation

1. Biotin (Sigma-Aldrich, 1 m*M* final concentration).
2. NH_4Cl buffer (150 m*M*): for 50 mL add 0.4 g of NH_4Cl to 50 mL of H_2O.
3. Radio-immunoprecipitation (RIPA) buffer: 250 m*M* NaCl, 10 m*M* Tris-HCl, pH 8.1, 1% Triton X-100 (Sigma-Adrich) 1%; NP 40 (Igepal/CA630, Sigma-Aldrich); and 1 m*M* EDTA. Add protease/phosphatase inhibitors immediately before use: 1 m*M* phenylmethylsulfonyl fluoride, 1 μg/mL leupeptin, and 100 μ*M* sodium orthovanadate.
4. Protein G–Sepharose (Amersham Bioscience).
5. Reducing sample buffer: 2.5% sodium dodecyl sulfate, 0.38 *M* Tris-HCl, pH 6.8, 20% glycerol, and 0.1% bromophenol blue.
6. Wash buffer: 30 g of NaCl, 2 mL of Tween-20 (Sigma-Aldrich), and 10 mL of 1 *M* HEPES to 1000 mL with H_2O.
7. Blocking buffer: use wash buffer with 3% bovine serum albumin.
8. Streptavidin-peroxidase (Sigma-Aldrich).
9. Enhanced chemiluminescence (Amersham Pharmacia Biotechnology).

2.6. Measurement of Endothelial Surface Enzyme Activity of the Ecto-Apyrase (CD39) and the 5′-Ectonucleotidase (CD73)

1. Etheno-ATP (E-ATP; Molecular Probes).
2. Etheno-AMP (E-AMP; Sigma-Aldrich).
3. Etheno-adenosine (Sigma-Aldrich).

Avoid frequent thawing of the etheno-compounds because they will lose their activity. Thaw once, and aliquot them as needed for each experiment.

3. Methods

3.1. Isolation of Human Neutrophils (PMNs)

PMNs have to be freshly isolated from whole blood obtained by venipuncture from human volunteers and anticoagulated with acid citrate. We use 10-mL

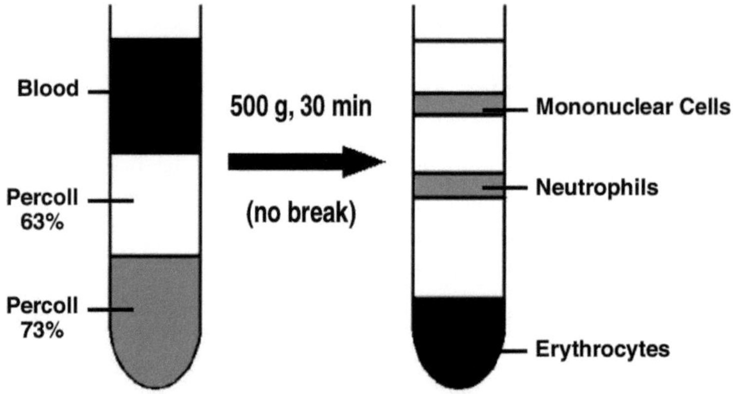

Fig. 1. Isolation of PMNs using double-density centrifugation of whole blood. Whole blood is obtained by venipuncture from human volunteers and anticoagulated with acid citrate. To remove platelets, plasma, mononuclear cells and erythrocytes, a double-density centrifugation technique is applied using the Percoll system. For this purpose 4 mL of Percoll, prepared to a density of 63% (*see* manufacturer's instructions), is placed into a clear 15-mL tube, and 4 mL of Percoll, density 73%, is carefully placed into the bottom. Then, 4 mL of whole blood is placed on top **(left)**. The tubes are centrifuged at 500*g* for 30 min at room temperature, no break. After the centrifugation, three bands can be visualized in the tube: below are the erythrocytes, the second band consists of the PMNs, and the third band contains the mononuclear cells **(right)**.

monovets containing Na–citrate (Sarstedt, Nümbrecht, Germany) to draw the blood. To remove platelets, plasma, mononuclear cells, and most of the erythrocytes, a double-density centrifugation technique is applied using the Percoll system (*see* **Note 1**).

1. In short, 4 mL of Percoll, prepared to a density of 63% (*see* manufacturer's instructions) is placed into a clear 15-mL tube, and 4 mL of Percoll, density 73%, is carefully placed below.
2. Then, 4 mL of the anticoagulated whole blood is placed carefully on top (**Fig. 1**).
3. The tubes are centrifuged at 500*g* for 30 min at room temperature, no break.
4. After the centrifugation, three bands can be visualized in the tube. At the bottom are the erythrocytes, the second band represents the PMNs, and the third band contains the mononuclear cells (**Fig. 1**). To isolate the neutrophils, the plasma and the band containing the mononuclear cells are removed to gain easier access to the PMN-containing band.
5. Next, the band with the neutrophils is transferred into a 50-mL tube. From this point forward, PMNs should be kept strictly at 4°C (*see* **Note 2**).

Residual erythrocytes are removed by lysis in cold NH_4Cl buffer. For this purpose, PMN are resuspended into 30 mL of ice-cold NH_4Cl buffer, incubated at

4°C for 10 min, and centrifuged at 170*g* (10 min, 4°C, break on). The supernatant is discarded, and the cells are washed with HBSS– twice (400*g*, 10 min, 4°C, break on). Remaining cells should consist of 97% or greater PMNs, as can be assessed by microscopic evaluation. PMNs should be studied within 2 h of their isolation. In general, approx 0.5 to 1 × 10^8 PMNs can be isolated from 50 mL of fresh blood.

3.2. Preparation of Activated PMN Supernatants and Measurement of Nucleotide Contents

To measure nucleotide release from activated PMN, freshly isolated human neutrophils are activated with fMLP, and samples from the supernatant are taken at different time-points after activation and analyzed by HPLC.

1. In short, freshly isolated human PMNs (10^7 cells in 1 mL of HBSS+ with 10^{-6} *M* fMLP) are incubated end-over-end at 37°C.
2. At 1, 5, 10, and 15 min, 200-µL samples are transferred into an ice-cold Eppendorf cup and immediately pelleted (1000*g* for 20 s at 4°C).
3. The resultant cell-free supernatants are filtered (0.45-µm Phenomenex, Torrance, CA) and kept strictly on ice until assessed by HPLC. Alternatively, supernatants can be stored at –80°C until HPLC measurements are performed. To determine nucleotide content, the cell-free supernatants are resolved by HPLC (Hewlett-Packard, Palo Alto, CA; model 1050) with an HP 1100 diode array detector by reverse-phase on an HPLC column (Luna 5-µm C18, 150 × 4.60 mm; Phenomenex, Torrance, California, CA) with 100% H_2O mobile-phase. ATP and adenine nucleotides can thus be identified by their chromatographic behavior (retention time, ultraviolet absorption spectra, and co-elution with standards).

In experiments when only the ATP content of the supernatant is of interest, a luciferase-based technique can be applied. This technique has a higher resolution because ATP concentrations in the picco-molar range can easily be detected. To do this, PMN are activated as outlined previously, supernatants are collected, and the ATP content can be quantified using CHRONO-LUME reagent (Crono-log Corp, Haverton, PA). Luciferase activity can be assessed on a luminometer (Turner Designs Inc., Sunnyvale, CA) and compared with internal ATP standards (*see* **Notes 3** and **4**).

3.3. Endothelial Cell Isolation and Culture

For assessment of endothelial barrier function in vitro, human umbilical vein endothelial cells (HUVECs) or immortalized HMEC-1 were used. HUVECs were isolated from a 10- to 15-cm long "fresh" piece of umbilical cord (minimum 10 cm). The umbilical cord was transported on ice from the labor floor to the laboratory in a sterile bottle without any solution. The residual blood within the umbilical cord provides the best media for the cells until isolation.

To isolate the HUVECs from the cord, the umbilical cord is first flushed with HBSS– (containing no calcium and no magnesium). The vein is identified by its larger size (compared with the arteries) and lack of a muscular tissue surrounding the lumen, and filled all the way up with collagenase A (1 mg/mL). Both ends of the umbilical cord are sealed with strings, and the collagenase A filled umbilical cord is incubated in 37°C in warm lactated Ringer's solution for exactly 5 min. The collagenase A solution (containing the umbilical vein endothelial cells) is then flushed out of the vein using modified Medium 199 (*see* **Subheading 2.1.**) into a 50-mL Falcon tube. The cells are centrifuged at 110*g* for 8 min, no break at room temperature. The supernatant is discarded and to wash the cells, the pellet is carefully resuspended into 20 mL of modified Medium 199 and centrifuged again (110*g*, 8 min, room temperature, no break). After discarding the supernatant, the pellet is resuspended into 20 mL of endothelial cell basal medium and transferred into a collagen-coated cell culture bottle. Cells are cultured at 37°C containing 5% CO_2. The media has to be changed the day after the primary cell preparation and thereafter every other day. The time until the first passage of the cells differs considerably; therefore, cells have to be observed carefully before the first passage. Full confluency usually is reached within 1 to 2 wk. At that time cells should be split using trypsin–EDTA and trypsin-neutralizing solution (*see* **Subheading 2.1.**) and cultured as described previously. HUVECs can be used up to four passages; then, the cells should be discarded and new cells should be freshly prepared from an umbilical cord (*see* **Note 5**).

HMEC-1 are harvested with Accutase and incubated at 37°C in 95% air/5% CO_2. Modified MCDB 131 is used for cell culture. Endothelial cell purity can be assessed by phase microscopic "cobblestone" appearance and uptake of fluorescent acetylated low-density lipoprotein.

3.4. Endothelial Permeability Assay

To study details of neutrophil endothelial interaction with regard to endothelial barrier function, we culture endothelial cells (HMEC-1 or HUVECs) on a permeable support system with a separate apical and basal portion (so-called "insert"). As soon as the cultured endothelium forms a confluent monolayers, the system can be used to investigate endothelial barrier function. For example, activated neutrophils can be added in different concentration to the apical portion of the insert. By using a tracer substance that is applied to the apical compartment, the flux of the tracer substance can be measured at different time points from the basal portion, thereby determining transcellular flux rates and measuring endothelial barrier function. We have previously used this technique to measure endothelial barrier function in the presence of different concentrations of activated neutrophils or the supernatant of activated neutrophils. Interestingly, activated PMNs or their supernatant decreased endothelial flux

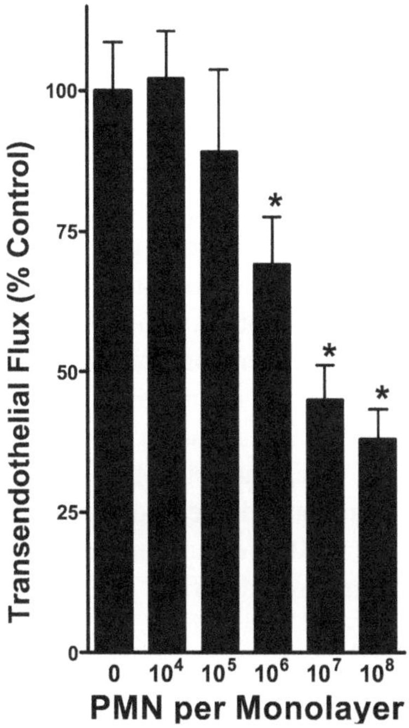

Fig. 2. Activated PMNs promote endothelial barrier function. PMNs activated with fMLP (10^{-6} *M*) and FITC-dextran 70 kDa were added to HMEC-1 monolayers. Transendothelial flux was calculated by linear regression (six samples over 60 min) and normalized as percent of control (HBSS+ only). Data are derived from six monolayers in each condition; results are expressed as mean of percent control flux ± standard deviation. *$p < 0.05$.

of 70 kDa dextran, thereby promoting barrier function in the below described model (**Fig. 2**).

For preparation of experimental endothelial monolayers used to measure endothelial barrier function, confluent endothelial cells are seeded at approx 1×10^5 cells/cm^2 onto either permeable polycarbonate inserts or 100-mm Petri dishes. In short, HMEC-1 are seated on polycarbonate permeable inserts (0.4-μm pore, 6.5-mm diameter; Costar Corp., Cambridge, MA). To do this, HMEC-1 from a cell culture bottle (area 75 cm^2) of fully confluent HMEC-1 are detached with trypsin and suspended in 20 mL of HMEC media. As next, 1 mL of HMEC media (without cells) is placed on the basal portion of the insert and 80 μL of the cell suspension is carefully placed on the apical portion of the insert. A few drops of media are then added to the apical portion. Endothelial barrier function can be studied when controls seated on Petri dishes reach full confluency (usually 5–7 d after seeding; *see* **Note 6**). To study endothelial barrier function, inserts are placed in HBSS+-containing wells (0.9 mL), and HBSS+ (alone or with PMN, PMN supernatant, or ATP) is added to the apical portion (100 μL). At the start of the assay (t = 0), FITC-labeled dextran 70 kDa (concentration 3.5 μ*M*) is added to fluid within the apical portion of the insert. The size of FITC–dextran, 70 kDa, approximates that of human albumin, both of which have been used in similar endothelial paracellular permeability models ***(36,37)***. Fluid from the opposing well (basal portion) are sampled (50 μL) over the course of 60 min (t = 20, 40, and 60 min). Fluorescence intensity of each sample can then be measured (excitation, 485 nm; emission, 530 nm; Cytofluor 2300; Millipore Corp., Waters Chromatography, Bedford, MA) and FITC–dextran concentrations are determined from standard curves generated by serial dilution of FITC–dextran. Paracellular flux is calculated by linear regression of sample fluorescence ***(37)***. Consistent with observations of other investigators, control experiments demonstrate decreased paracellular permeability with forskolin or 8-bromo-cyclic AMP ***(38)*** and increased paracellular permeability with thrombin and hydrogen peroxide ***(39)***.

3.5. Immunoprecipitation

To show expression or induction of a specific cellular protein on the endothelial surface, surface proteins can be labeled with biotin. Using an avidin blot, these proteins can be quantified. We have previously used this technique to show expression and induction of the CD39 and CD73 by hypoxia ***(14,34)***.

To pursue this, HMEC-1 or HUVECs are grown to full confluency in 100-mm Petri dishes. The labeling has to be performed in the cold (4°C) to stabilize surface proteins (*see* **Note 7**). Therefore, Petri dishes with cells are taken to a cold room (4°C) and placed on a cold surface (4°C). Cells are washed twice with ice-cold HBSS+ and are labeled with biotin (final concentration 1 m*M*). For this purpose, cover cells completely with 1 m*M* Biotin in HBSS+, enough to cover all cells (minimum 3 mL), and place cells on a rotator (slow settings) for 20 min at 4°C. As next, cells are rinsed twice with HBSS+. Then, endothelia are incubated with 150 m*M* NH_4CL buffer (10 mL/plate, 30 min at 4°C), rinsed twice with HBSS+, and lysed with RIPA buffer. For this purpose, 1.25 mL of RIPA buffer is given to each 100-mm Petri dish, cells are scraped with a cell-scraper or a pipet tip and transferred into an Eppendorf cup (avoid bubbles). The Eppendorf cups are rotated end-over-end for 30 min, and cell debris is removed by centrifugation (1000*g* at 4°C for 5 min). Immunoprecipitation is then performed with mono or polyclonal antibodies (e.g., for human CD39 [Research Diagnostics, Inc. Flanders, NJ; 5 μg/mL], CD73 with MAb 1E9 [5 μg/mL, a kind gift from Dr. Linda Thompson, Oklahoma Medical Research Foundation]) followed by addition of 100 μL of pre-equilibrated protein

G-Sepharose (use RIPA buffer for pre-equilibration) and overnight incubation (end over end at 4°C; *see* **Note 8**). Washed immunoprecipitates are boiled in reducing sample buffer, separated by sodium dodecyl sulfate polyacrylamide gel electrophoresis, transferred to nitrocellulose, and blocked overnight in blocking buffer. Biotinylated proteins are labeled with streptavidin–peroxidase and visualized by enhanced chemiluminescence (*see* **Note 9**).

3.6. Measurement of Endothelial Surface Enzyme Activity of the Ecto-Apyrase (CD39) and the 5′-Ectonucleotidase (CD73)

To assess enzyme activity of CD39 and CD73 on the surface on endothelia, the etheno-compounds E-ATP (Molecular Probes) and E-AMP (Sigma-Aldrich) can be used, respectively (*see* **Note 10**). Because of their etheno-groups, these compounds exhibit different elution characteristics on the HPLC compared with the native compounds (ATP or AMP) and, therefore, endogenous nucleotide concentrations will not affect the measurements. To measure CD39 activity, HMEC-1 or HUVECs are grown to full confluency on a 60-mm Petri dish. HBSS+ (with addition of the CD73 inhibitor αβ-methylene-adenosine 5′-diphosphate [ADP; 50 μ*M*] to prevent further metabolism of E-AMP to E-adenosine) is added to the endothelial monolayers. After 10 min, E-ATP (final concentration 5 μ*M*) is added. Samples are taken at 0, 5, 10, and 30 min, removed, acidified to pH 3.5 with HCl, spun (10,000*g* for 20 s at 4°C), filtered (0.45 μm), and frozen (–80°C) until analysis via reverse-phase HPLC using a Reprosil-Pur C18-AQ 5 μ*M* column (Maisch, Ammerbuch, Germany). The ratio of E-ATP and E-AMP is measured with a mobile phase (1 mL/min over the course of 15 min) of 15% methanol/NH_4-acetate (0.1 molar) buffer adjusted to pH 5.0. Absorbance is measured with a fluorescence detector at 410 nm with excitation at 315 nm. Retention time for E-ATP is approx 3 min, for E-AMP approx 6 min. CD39 activity is expressed as percent conversion of E-ATP to E-AMP in this time frame (*see* **Note 11**).

To measure CD73 activity, cells are washed with HBSS+, and HBSS+ with or without αβ-methylene-ADP is added to the endothelial monolayers (grown on 60-mm Petri dishes). After 10 min, E-AMP (final concentration 5 μ*M*) is added. Samples are taken at 0, 5, 10, and 30 min, removed, acidified to pH 3.5 with HCl, spun (10,000*g* for 20 s, 4°C), filtered (0.45 μm), and frozen (–80°C) until analysis via reverse-phase HPLC using a Reprosil-Pur C18-AQ 5 μ*M* column (Maisch, Ammerbuch, Germany). Ratio of E-AMP and E-adenosine are measured with a mobile phase (1 mL/min over the course of 8 min) of 25% methanol/NH_4-acetate (0.1 *M*) buffer adjusted to pH 5.0. Absorbance is measured with a fluorescence detector at 410 nm after excitation at 315 nm. Retention time for E-AMP is approx 3 min; for E-adenosine, it is approx 5 min.

Retention time for E-adenosine is confirmed using commercially available E-adenosine (Sigma-Aldrich). CD73 activity is expressed as percent conversion of αβ-methylene-ADP inhibited E-AMP conversion to E-adenosine in this time frame.

4. Notes

1. For neutrophil isolation, it is critical that the freshly drawn blood and both Percoll solutions have the same temperature. Otherwise, the double-density centrifugation cannot be performed successfully, and the different bands cannot be distinguished clearly.
2. To be able to measure release of nucleotides upon PMN activation, it is critical to avoid activation of PMNs in the process of isolation and washing. Therefore, all solutions used during this process should not contain calcium (e.g., HBSS–). Also, after the initial isolation, PMNs should be kept on ice all the time, and centrifuges should be used at 4°C. Do not let keep PMNs in the pellet; instead, keep them in solution in HBSS–.
3. To activate PMNs, neutrophils in HBSS– are pelleted by centrifugation (400*g* for 10 min at 4°C) and transferred into warm HBSS+ (37°C), containing fMLP (100 n*M*).
4. In case that murine PMNs are studied, the neutrophils cannot be activated by fMLP because murine PMNs do not express fMLP receptors. Instead of fMLP, murine PMN can be activated using leukotriene B4 (i.e., leukotriene B_4 ***[40]***).
5. Don't use PBS to wash or flush HUVECs.
6. When changing the media or the HBSS solution of inserts used to measure endothelial barrier function in vitro, one has to be extremely careful not to touch the cell layer with a pipet. If that happens, the barrier is lost, and the tracer substance will immediately equilibrate between apical and basal compartment.
7. For surface labeling of endothelial proteins with biotin, it is critical to perform the whole experiment at 4°C to stabilize the confirmation and position of surface proteins.
8. When performing end-over-end incubations for immunopreciptation, the Eppendorf cups should be locked tightly and sealed with plastic foil to avoid leakage and loss of the lysates.
9. Do not perform an avidin blot with milk powder. Instead, use 3% bovine serum albumin resolved in wash buffer.
10. The etheno-compounds in the supernatant of HMEC-1 or HUVECs are unstable, unless acidified and cooled to 4°C immediately.
11. It is important to filter the samples thoroughly before injection into the HPLC. Otherwise, HPLC filter systems may be blocked or increased pressure readings will occur.

Acknowledgments

This work was supported by Fortune grant 1319-0-0 and DFG grant EL274/2-2 to HKE.

References

1. Waxman, K. (1996) Shock: ischemia, reperfusion, and inflammation. *New Horiz.* **4,** 153–160.
2. Kloner, R.A., Ellis, S. G., Lange, R., and Braunwald, E. (1983) Studies of experimental coronary artery reperfusion. Effects on infarct size, myocardial function, biochemistry, ultrastructure and microvascular damage. *Circulation* **68,** I8–I15.
3. Ramirez, C., Colton, C., Smith, K., Stemerman, M., and Lees, R. (1984) Transport of 125I-albumin across normal and deendothelialized rabbit thoracic aorta in vivo. *Arterioscler. Thromb. Vasc. Biol.* **4,** 283–291.
4. Takano, T., Clish, C. B., Gronert, K., Petasis, N., and Serhan, C. N. (1998) Neutrophil-mediated changes in vascular permeability are inhibited by topical application of aspirin-triggered 15-epi-lipoxin a4 and novel lipoxin b4 stable analogues. *J. Clin. Invest.* **101,** 819–826.
5. Sunnergren, K. P. and Rovetto, M. J. (1987) Myocyte and endothelial injury with ischemia reperfusion in isolated rat hearts. *Am. J. Physiol.* **252,** H1211–H1217.
6. Rippe, B. and Haraldsson, B. (1994) Transport of macromolecules across microvascular walls: the two-pore theory. *Physiol. Rev.* **74,** 163–219.
7. Lum, H. and Malik, A. B. (1994) Regulation of vascular endothelial barrier function. *Am. J. Physiol.* **267,** L223–L241.
8. Kloner, R.A., Ganote, C. E., and Jennings, R. B. (1974) The "no-reflow" phenomenon after temporary coronary occlusion in the dog. *J. Clin. Invest.* **54,** 1496–1508.
9. Dejana, E., Corada, M., and Lampugnani, M. G. (1995) Endothelial cell-to-cell junctions. *FASEB J.* **9,** 910–918.
10. Huang, A. J., Furie, M. B., Nicholson, S. C., Fischbarg, J., Liebovitch, L. S., and Silverstein, S. C. (1988) Effects of human neutrophil chemotaxis across human endothelial cell monolayers on the permeability of these monolayers to ions and macromolecules. *J. Cell Physiol.* **135,** 355–366.
11. Lewis, R. E. and Granger, H. J. (1988) Diapedesis and the permeability of venous microvessels to protein macromolecules: the impact of leukotriene B4 (LTB4). *Microvasc. Res.* **35,** 27–47.
12. Meyrick, B., Hoffman, L. H., and Brigham, K. L. (1984) Chemotaxis of granulocytes across bovine pulmonary artery intimal explants without endothelial cell injury. *Tissue Cell* **16,** 1–16.
13. Welbourn, C. R., Goldman, G., Paterson, I. S., Valeri, C. R., Shepro, D., and Hechtman, H. B. (1991) Pathophysiology of ischaemia reperfusion injury: central role of the neutrophil. *Br. J. Surg.* **78,** 651–655.
14. Eltzschig, H. K., Ibla, J. C., Furuta, F. T., et al. (2003) Coordinated adenine nucleotide phosphohydrolysis and nucleoside signaling in posthypoxic endothelium: role of ectonucleotidases and adenosine A2B receptors. *J. Exp. Med.* **198,** 783–796.
15. Eltzschig, H. K., Thompson, L. F., Karhausen, J., et al. (2004) Endogenous adenosine produced during hypoxia attenuates neutrophil accumulation: coordination by extracellular nucleotide metabolism. *Blood* **104,** 3986–3992.

16. Lennon, P. F., Taylor, C. T., Stahl, G. L., and Colgan, S. P. (1998) Neutrophil-derived 5′-adenosine monophosphate promotes endothelial barrier function via CD73-mediated conversion to adenosine and endothelial A2B receptor activation. *J. Exp. Med.* **188,** 1433–1443.
17. Dancey, J. T., Deubelbeiss, K. A., Harker, L. A., and Finch, C. A. (1976) Neutrophil kinetics in man. *J. Clin. Invest.* **58,** 705–715.
18. Ley, K. (2001) Pathways and bottlenecks in the web of inflammatory adhesion molecules and chemoattractants. *Immunol. Res.* **24,** 87–95.
19. Gonzalez-Amaro, R. and Sanchez-Madrid, F. (1999) Cell adhesion molecules: selectins and integrins. *Crit. Rev. Immunol.* **19,** 389–429.
20. Springer, T. A. (1994) Traffic signals for lymphocyte recirculation and leukocyte emigration: the multistep paradigm. *Cell* **76,** 301–314.
21. Shimizu, Y., Rose, D. M., and Ginsberg, M. H. (1999) Integrins in the immune system. *Adv. Immunol.* **72,** 325–380.
22. Dejana, E., Spagnuolo, R., and Bazzoni, G. (2001) Interendothelial junctions and their role in the control of angiogenesis, vascular permeability and leukocyte transmigration. *Thromb. Haemost.* **86,** 308–315.
23. Shaw, J. O. and Henson, P. M. (1982) Pulmonary intravascular sequestration of activated neutrophils: failure to induce light-microscopic evidence of lung injury in rabbits. *Am. J. Pathol.* **108,** 17–23.
24. Ley, K. (2002) Integration of inflammatory signals by rolling neutrophils. *Immunol. Rev.* **186,** 8–18.
25. Hasko, G., Sitkovsky, M. V., and Szabo, C. (2004) Immunomodulatory and neuroprotective effects of inosine. *Trends Pharmacol. Sci.* **25,** 152–157.
26. Linden, J. (2001) Molecular approach to adenosine receptors: receptor-mediated mechanisms of tissue protection. *Ann. Rev. Pharmacol. Toxicol.* **41,** 775–787.
27. Sitkovsky, M. V., Lukashev, D., Apasov, S., et al. (2004) Physiological control of immune response and inflammatory tissue damage by hypoxia-inducible factors and adenosine A2A receptors. *Ann. Rev. Immunol.* **22,** 657–682.
28. Ohta, A. and Sitkovsky, M. (2001) Role of G-protein-coupled adenosine receptors in downregulation of inflammation and protection from tissue damage. *Nature* **414,** 916–920.
29. Cramer, T., Yamanishi, Y., Clausen, B. E., et al. (2003) HIF-1alpha is essential for myeloid cell-mediated inflammation. *Cell* **112,** 645–657.
30. MacLean, D. A., Sinoway, L. I., and Leuenberger, U. (1998) Systemic hypoxia elevates skeletal muscle interstitial adenosine levels in humans. *Circulation* **98,** 1990–1992.
31. Mo, F. M. and Ballard, H. J. (2001) The effect of systemic hypoxia on interstitial and blood adenosine, AMP, ADP and ATP in dog skeletal muscle. *J. Physiol. (Lond.)* **536,** 593–603.
32. Phillis, J. W., O'Regan, M. H., and Perkins, L. M. (1992) Measurement of rat plasma adenosine levels during normoxia and hypoxia. *Life Sci.* **51,** PL149–PL152.

33. Saito, H., Nishimura, M., Shinano, H., et al. (1999) Plasma concentration of adenosine during normoxia and moderate hypoxia in humans. *Am. J. Respir. Crit. Care Med.* **159,** 1014–1018.
34. Synnestvedt, K., Furuta, G. T., Comerford, K. M., Louis, N., Karhausen, J., Eltzschig, H. K., et al. (2002) Ecto-5′-nucleotidase (CD73) regulation by hypoxia-inducible factor-1 mediates permeability changes in intestinal epithelia. *J. Clin. Invest.* **110,** 993–1002.
35. Thompson, L. F., Eltzschig, H. K., Ibla, J. C., et al. (2004) Crucial Role for Ecto-5′-Nucleotidase (CD73) in Vascular Leakage during Hypoxia. *J. Exp. Med.* **200,** 1395–1405.
36. Mizuno-Yagyu, Y., Hashida, R., Mineo, C., Ikegami, S., Ohkuma, S., and Takano, T. (1987) Effect of PGI2 on transcellular transport of fluorescein dextran through an arterial endothelial monolayer. *Biochem. Pharmacol.* **36,** 3809–3813.
37. Sanders, S. E., Madara, J. L., McGuirk, D. K., Gelman, D. S., and Colgan, S. P. (1995) Assessment of inflammatory events in epithelial permeability: a rapid screening method using fluorescein dextrans. *Epithelial Cell Biol.* **4,** 25–34.
38. Casnocha, S. A., Eskin, S. G., Hall, E. R., and McIntire, L. V. (1989) Permeability of human endothelial monolayers: effect of vasoactive agonists and cAMP. *J. Appl. Physiol.* **67,** 1997–2005.
39. Garcia, J. G., Siflinger-Birnboim, A., Bizios, R., Del Vecchio, P. L., Fenton, 2nd, J. W., and Malik, A. B. (1986) Thrombin-induced increase in albumin permeability across the endothelium. *J. Cell Physiol.* **128,** 96–104.
40. Haribabu, B., Verghese, M. W., Steeber, D. A., Sellars, D. D., Bock, C. B., and Snyderman, R. (2000) Targeted disruption of the leukotriene B4 receptor in mice reveals its role in inflammation and platelet-activating factor-induced anaphylaxis. *J. Exp. Med.* **192,** 433–438.

7

Analysis of Mammalian Sperm–Egg Membrane Interactions During In Vitro Fertilization

Genevieve B. Wortzman, Allison J. Gardner, and Janice P. Evans

Summary

The interactions between egg and sperm are among the most fascinating in cell biology. These interactions include cell–cell adhesion and then membrane fusion between the two gametes. This chapter details the experimental methods used to for gamete culture and in vitro fertilization using mouse sperm and eggs.

Key Words: In vitro fertilization; cell adhesion; membrane fusion; cell–cell fusion.

1. Introduction

This chapter focuses on the cell–cell interactions between mouse sperm and egg, namely the interactions between sperm and egg plasma membranes, which include sperm–egg adhesion (or binding) and sperm–egg fusion. General aspects of the cell and molecular biology of fertilization have been addressed in recent review articles *(1,2)*. It should be noted that normally the sperm interacts with the egg on three different levels. The first two interactions are with the egg's two extracellular coats, the cumulus layer and the *zona pellucida* (ZP). After penetrating these two layers, the sperm gains access to the perivitelline space, between the inner surface of the ZP and the egg plasma membrane; it is in this space that sperm–egg membrane interactions occur. For most experiments for the study of sperm–egg adhesion and fusion, the cumulus layers and the ZP are removed from the eggs, as addressed in the chapter, with relevant aspects of gamete cell biology discussed in the appropriate portions of the Notes section. Mammalian eggs have not yet completed meiosis (in contrast to the sperm, which complete meiosis during spermatogenesis in the testis). Instead, most mammalian eggs are arrested in metaphase of meiosis II, with

From: *Methods in Molecular Biology, vol. 341: Cell–Cell Interactions: Methods and Protocols*
Edited by: S. P. Colgan © Humana Press Inc., Totowa, NJ

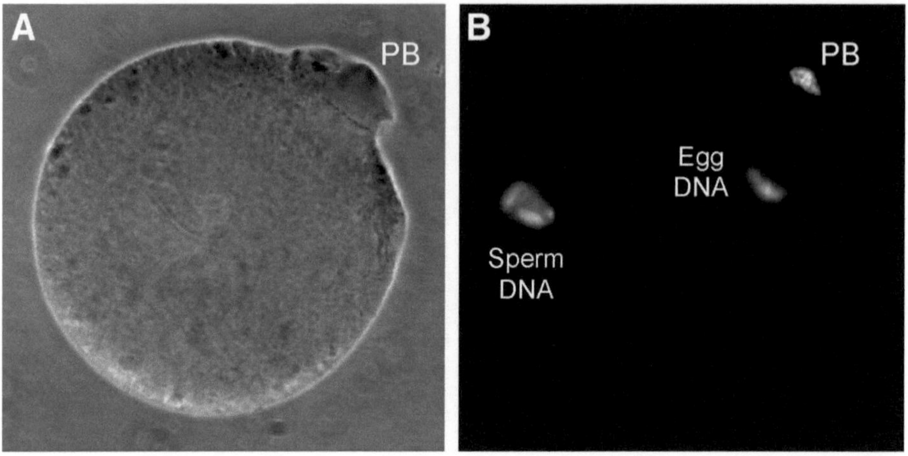

Fig. 1. DAPI-stained fertilized egg, approx 60 min after insemination. **(A)** Phase contrast image. **(B)** Fluorescent image of the DAPI-stained DNA. This egg has nearly completed meiosis, as indicated by the emission of the polar body (PB). The sperm nucleus is present in the egg cytoplasm; this will ultimately form a male pronucleus which will fuse with the female pronucleus to create the new diploid organism.

fertilization by the sperm providing the trigger to complete meiosis *(3)*. This activation by the sperm results in cytokinesis that produces a small cell, called the polar body, containing half of the egg's DNA (**Fig. 1**). This leaves a haploid set of DNA in the egg that will ultimately fuse with the DNA delivered by the sperm, at approx 7 to 8 h after fertilization.

The processes of fertilization are highly conserved between mammalian species. For experimental studies, the mouse is most commonly used because gamete collection and in vitro fertilization (IVF) procedures have been very well refined (**Fig. 2**). IVF is more difficult in other species, for example, low fertilization rates are common with rat gametes, and polyspermy (fertilization by more than one sperm) is common with pig gametes. It should be noted that the main barrier for species specificity of fertilization is the ZP, whereas sperm–egg membrane fusion is much less species-specific than sperm–ZP interactions. Mouse sperm can fuse with the eggs of rat, guinea pig, and rabbit *(4)*. ZP-free eggs from golden hamsters can be fertilized by sperm from a very wide variety of species (including human, bat, and dolphin *[4]*). Because of logistical and ethical problems associated with doing experiments with human sperm and eggs, it is common for studies of human sperm to use ZP-free hamster eggs. However, for general interests in cell–cell interactions, it is more appropriate to use sperm and eggs from the same species because potentially misleading results can come from studies of heterologous gametes *(5)*. This chapter

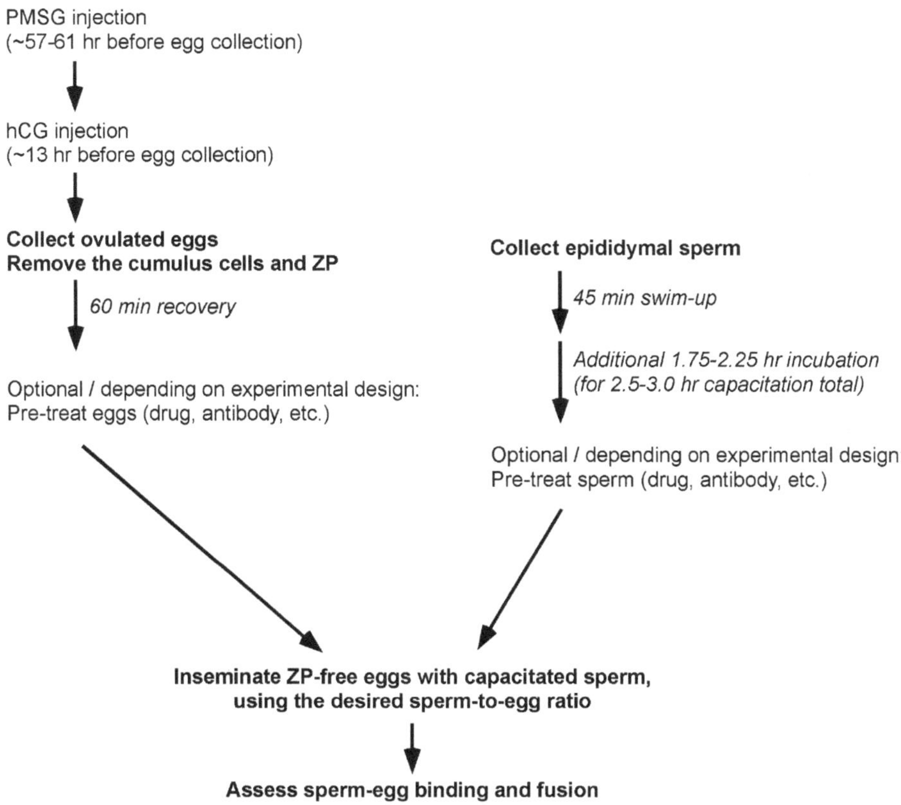

Fig. 2. Flow chart of the experimental design for an IVF study.

presents protocols for gamete collection and IVF using the mouse as an experimental model system. The reader also is advised that there are a number of other issues related to the experimental procedures used to assess sperm–egg interactions that are beyond the scope of this chapter (the interested reader is referred to Section 3 in **ref. *6***).

2. Materials

1. Mice: the females most commonly used for superovulation and egg collection are CF-1 (non-Swiss albino) or CD-1 (ICR) from Harlan or Charles River. An inexpensive choice for males is CD-1 retired breeders; sperm quality from these males generally is good, although there will be occasional mice with suboptimal sperm (e.g., low count, self-aggregating and clumpy). An alternative source of sperm are F1 hybrid males, such as from Jackson Laboratories, although these tend to be significantly more expensive. Insemination conditions (optimal sperm concentrations, insemination times) may vary for gametes from mice of different strains.

2. Gonadotropins are the hormones used to induce females to ovulate. Pregnant mare's serum gonadotropin (PMSG) has activity similar to the naturally occurring pituitary hormone follicle-stimulating hormone and is used to prime the females for ovulation. Human chorionic gonadotropin (hCG) has activity similar to the naturally occurring pituitary hormone luteinizing hormone, and is used to induce ovulation.

   PMSG (Calbiochem, San Diego, CA; cat. no. 367222, 5000 IU) is dissolved in 100 mL of phosphate-buffered saline (PBS) for a final concentration of 5 U/0.1 mL. hCG from human pregnancy urine (Sigma, St. Louis, MO; cat. no. CG-10, 10,000 U, or cat. no. C1063, 2500 IU) is dissolved in PBS for a final concentration of 5 U/0.1 mL. Store aliquots of 600 µL (enough for three to six mice) at –20 or –80°C. (PMSG in particular may lose effectiveness when stored below –20°C.)
3. Whitten's medium stock *(7)*: begin with high-quality water (such as sterile surgical irrigation water; Abbott Laboratories, North Chicago, IL or water for embryo transfer; Sigma; cat. no. W1503). Add 109.5 m*M* NaCl, 4.7 m*M* KCl, 1.2 m*M* KH_2PO_4, 1.2 m*M* $MgSO_4$, 5.5 m*M* glucose, 0.23 m*M* pyruvic acid, 4.8 m*M* lactic acid, 0.01% gentamycin, and 0.0001% Phenol Red. For Whitten's-HEPES medium (W-H), add 7 m*M* $NaHCO_3$ and 15 m*M* HEPES and adjust pH to 7.4. For Whitten's bicarbonate medium (W-B), add 22 m*M* $NaHCO_3$. Filter through a surfactant-free cellulose acetate 0.2-µm filter, and store at 4°C (*see* **Note 1**).
4. W-B medium with 15 mg/mL bovine serum albumin (W-B/15). This medium is made the night before or the day of the IVF experiment by adding 15 mg/mL of bovine serum albumin (BSA; Albumax I, Gibco-BRL, Gaithersburg, MD) to the desired volume of W-B. Typically 5 to 10 mL is made, depending on the scale of the experiment. Invert gently to dissolve the BSA into solution (do not vortex). Add apprpox 5 µL of 1 *N* NaOH per 1 mL of medium; the medium color will change from orange-ish to pinkish-purple. Gently filter through a 0.2-µm surfactant-free cellulose acetate syringe filter, and then incubate at 37°C, 5% CO_2 in air (*see* **Note 2**).
5. Whitten's bicarbonate medium with 0.01–0.1% polyvinyl alcohol (W-B/PVA) (PVA, Sigma; cat. no. P-8136). We make this in batches of approx 50 mL, and our preferred concentration of PVA is 0.05% (*see* **Note 3**).
6. Pipets for handling eggs: Pasteur pipets made of borosilicate glass (Fisher; cat. no. 13-678-20D; do not use not soda lime glass) are heated in a flame and pulled thin. Using a Bunsen burner, hold both ends of the pipet so the flame touches one spot on the narrow region of the pipet. As the glass softens, in one motion quickly lift the pipet out of the flame and pull the ends of the pipet apart. Done correctly, the pipet is intact with a very narrow section in the middle. After the glass has cooled, slide the thinned section of glass between your index finger and thumb with light pressure until the pipet breaks. This will result in a pipet with a narrow tip and a clean edge (*see* **Note 4**).

   Egg handling is managed by mouth pipetting. Homemade mouth pipets are made of mouthpieces (HPI Hospital Products, Altamonte Springs, FL) and conventional rubber tubing (Latex amber thick wall tubing, 1/8 in. inner diameter, 1/16 in. wall thickness, 1/4 in. outer diameter; Fisher; cat. no. 14-178-5B).

7. Culture dishes: sterile polystyrene, not tissue culture treated, 35 × 10 mm (Falcon; cat. no. 351008) and 60 × 15 mm (Falcon; cat. no. 351007).
8. Mineral oil: Sigma cat. no. M-3516, stored at 37°C, 5% CO_2.
9. Hyaluronidase 0.025% in W-H with 3 mg/mL BSA (W-H/HY): measure 2.5 mg of hyaluronidase (type IV-S from bovine testis, Sigma cat. no. H3884) and 30 mg BSA (Gibco-BRL Albumax I) into a 15-mL polystyrene culture tube. Add 10 mL of W-H; invert or vortex to mix.
10. Acidic Minimal Essential Medium (MEM)-compatible (MEMCO) for ZP removal *(8)*: in high-quality water (*see* **Subheading 2.3.**), dissolve 10 m*M* HEPES, 1 m*M* NaH_2PO_4, 0.8 m*M* $MgSO_4$, 5.4 m*M* KCl, and 116.4 m*M* NaCl. Adjust pH to 1.5. Store aliquots of 10 mL at –20°C, heat to thaw. Working stock can be kept at room temperature.
11. Chymotrypsin for ZP removal: stock of α-chymotrypsin (type II from bovine pancreas; Sigma) is dissolved at 10 mg/mL in PBS and stored at –20°C. Working solutions are prepared immediately before use by diluting to 10 µg/mL final concentration and/or 100 µg/mL final concentration in W-B with 3 mg/mL BSA added (W-B/3, as described for W-B/15 previously; *see also* **Note 5**).
12. MitoTracker for labeling of sperm: MitoTracker Green (Molecular Probes, Eugene, OR; cat. no. M-7514) is dissolved in dimethyl sulfoxide to a stock concentration of 1 mM and stored at –20°C protected from light. Working solutions are prepared by diluting to 100 n*M* in Whitten's medium containing 15 mg/mL BSA (*see* **Note 6**).
13. DAPI for labeling of live eggs: 1–10 µg/mL DAPI (Sigma), diluted in W-B/15.

3. Methods

3.1. Superovulation of Mice

Three days before the experiment, female mice are given 100 to 200 µL of PMSG by intraperitoneal injection, approx 57 to 61 h before egg collection and 44 to 48 h before the hCG injection. PMSG will stimulate maturation ovarian follicles. The evening before the experiment, each mouse is given an injection 100 to 200 µL of hCG. This hCG injection is done 44 to 48 h after injection of PMSG and 12 to 13 h before the planned collection of eggs (*see* **Note 7**).

3.2. Egg Collection

1. Prepare a 35-mm dish with five 35 µL drops of W-B/15 covered with warm mineral oil.
2. Sacrifice females with CO_2 followed by cervical dislocation.
3. Open the body cavity by making a U-shaped incision in the lower abdomen up to the liver; push aside the skin flap and intestines to the side to gain access to the reproductive tract.
4. Remove the oviducts (just below the ovary) and place in a 200 to 300 µL of W-H to rinse off blood and other fluid.

5. In a 60-mm culture dish, put a drop of W-H/HY for every two oviducts and two drops of W-B for washing the eggs.
6. Transfer the oviducts to the W-H/HY drops and puncture the swollen ampulla region (in which a "cloud" of eggs should be visible) with a 27-gage needle to release the cumulus-encased eggs. The eggs can be gently pipetted through a wide-bore pipet to help break up the complexes and give the hyaluronidase access to the cumulus cells' extracellular matrix. As the cumulus cells disperse from the eggs in the W-H/HY, transfer the cumulus cell-free eggs to a wash drop of W-B (*see* **Note 8**).
7. Wash the cumulus-free eggs through the second W-B drop and then a fresh W-B/15 drop.
8. Transfer the eggs to the previously prepared 35-mm culture dish with five 35-μL drops of W-B/15, covered with warm oil. Eggs should be cultured in W-B/15 at 37°C, 5% CO_2 until the next step.

3.3. Sperm Preparation (see Note 9)

1. Sacrifice one male mouse with CO_2 followed by cervical dislocation. Clip away the skin/muscle layer of the lower abdomen in an inverted "U" shape; pull flap up. Gently clip the cauda epididymis and upper part (~1 cm) of vas deferens free from surrounding tissue.
2. Add the two caudae epididymides to 250 μL of W-B/15 covered in warm mineral oil in a 35-mm Petri dish. Clean scissors and gently mince the tissue. Incubate the drop at 37°C, 5% CO_2 for 10 min to allow sperm to swim out into the medium. Remove the tissue from the sperm suspension. Gently draw up the 250-μL sperm suspension into a pipet and carefully transfer the sperm suspension to the bottom of a 1.5 mL of W-B/15 in a 12 × 75-mm polystyrene culture tube. After 45 min, remove the top 440 μL from the tube and place this sperm suspension in a fresh culture dish and cover with warm mineral oil (*see* **Note 10**).
3. To determine the concentration of sperm (sperm/milliliter), prepare a 1:10 dilution of swim-up sperm in water and transfer 10 μL of this to a hemacytometer. Count the number of sperm in the 4 × 4 double-line-bordered square. Repeat for two other squares and average these values. To determine the number of sperm/mL, multiply this average sperm count value by the dilution factor (which is 10, for the 1:10 dilution in water) and also by 10^4 (a correction factor for the hemacytometer). This will give you the number of sperm per mL in the swim-up stock suspension (*see* **Note 11**).

3.4. Two Methods of ZP Removal From Eggs (see Note 12)

3.4.1. Acid Solubilization of the ZP

1. Set up a Petri dish with one drop of acidic MEMCO and several drops of W-B/15.
2. Work in batches of 5 to 20 eggs, depending on skill level. Transfer ZP-intact eggs to the drop of Acidic MEMCO and pipet the eggs up and down for approx 15 s, watching closely as the ZP dissolve.

3. As soon as the ZP have dissolved, wash the eggs by transferring them through adjacent drops of W-B/15. Transfer ZP-free eggs to fresh drops of W-B/15 covered in warm mineral oil; culture at 37°C, 5% CO_2 in air for 60 min to allow eggs to recover from ZP removal.

3.4.2. Protease Treatment and Mechanical Removal of the ZP

1. Set up a Petri dish with one drop of 10 μg/mL chymotrypsin in W-B/3, one drop of 100 μg/mL chymotrypsin in W-B/3, three drops of W-B/3, and three drops of W-B/15, covered in warm mineral oil. (Each drop can be ~50–100 μL.)
2. Transfer eggs to the drop of 10 μg/mL chymotrypsin in W-B/3 at 37°C and check for ZP swelling after approx 2 min. If the ZP appears swollen, use a thin-bore pipet to shear off the ZP from the egg. If the ZP is not swollen, incubate an additional approx 2 min. If the ZP is not swollen after this second incubation, transfer to the drop containing 100 μg/mL chymotrypsin in W-B/3 for approx 1 min (*see* **Note 5**).
3. After shearing off the ZP, immediately wash eggs through the W-B/3 drops then the W-B/15 drops.

3.5. Optional Additional Step: Labeling of Live Sperm With MitoTracker Green (see Note 13)

1. Collect epididymal sperm as described in **Subheading 3.3.** (sperm collection) and allow the sperm to capacitate for 2.5 to 3 h before use.
2. Approximately 15 min before the start of IVF, prepare the labeling solution for the sperm. Dilute the 1 m*M* MitoTracker Green stock to 200 n*M* with W-B/15 (i.e., double the final working concentration of MitoTracker Green).
3. Dilute the 200 n*M* MitoTracker Green solution to the working concentration of 100 n*M* by adding an equal volume of sperm to be labeled (*see* **Note 14**).
4. Incubate the sperm in the MitoTracker Green-containing medium for 10 min at 37°C in a dark, humidified atmosphere of 5% CO_2.
5. Dilute the labeled sperm to the desired concentration for IVF with W-B/15 (*see* **Note 6**).

3.6. Optional Supplement: Loading Live Eggs With DAPI (see Note 15)

1. Dilute DAPI in W-B/15 as noted in **Subheading 2.11.** (1–10 μg/mL final concentration). Prepare approx 25-μL drops of this medium and cover with warm mineral oil.
2. Approximately 25 min before the start of the insemination, transfer eggs to the W-B/15 containing DAPI and incubate for 15 min, then wash through 3 to 4 50- to 100-μL drops of W-B/15 before insemination (*see* **Note 16**).

3.7. IVF (see Notes 17 and 18)

1. Prepare a suspension of sperm at the desired concentration in W-B/15, typically 25,000 to 100,000 sperm/mL (*see* **Note 17**). Make 10-μL drops of this sperm suspension and cover with warm mineral oil. (If you have 20 eggs per experimental group, make two 10-μL drops of sperm suspension.)

2. Add 10 eggs per 10-µL drop of sperm suspension. Co-culture the eggs with sperm in these insemination drops at 37°C, 5% CO_2 in air for desired length of time (*see* **Note 17**).
3. To remove loosely attached sperm from the eggs, wash the eggs using a pipet approx 1.5X the diameter of the eggs. Transfer the eggs to the top of an approx 75-µL drop of W-B/15, blowing gently (as if blowing a whistle quietly); allow the eggs to settle to the bottom. Repeat two more times (*see* **Notes 19** and **20**).
4. To prepare the eggs for viewing on a fluorescent microscope to assess sperm–egg interactions, fix the eggs in 3.7% to 4.0% paraformaldehyde in PBS for 15 to 60 min. Fixed eggs are then permeabilized in PBS containing 0.05% Triton X-100 (3–15 min), washed briefly in PBS two to three times, and then mounted in mounting medium (e.g., VectaShield, Vector Laboratories) containing 1.5 µg/mL DAPI to stain the DNA in the eggs and sperm. Eggs also can be viewed live (particularly if DAPI-loaded eggs were used to assess sperm–egg fusion).

4. Notes

1. The culture medium should be prepared using glassware designated only for media preparation. Do not use metal spatulas to weigh out the dry reagents. W-H and W-B can be prepared together. To prepare 500 mL of each, add all the ingredients up to the phenol red and bring up the volume to 900 mL; then divide into two batches of 450 mL and add the remaining reagents for either W-H or W-B.
2. We have had good success with Whitten's medium for IVF. There are also commercially available media for IVF (e.g., Human Tubal Fluid), although making your own Whitten's medium is more economical. W-H medium is used for collection and other work in regular atmosphere, whereas W-B medium, containing bicarbonate only (no HEPES), is used for culture in 5% CO_2. Care should be taken to maintain appropriate pH of the W-B medium, as shifts in the pH may affect the health of eggs and sperm and fertilization success.
3. The addition of BSA or PVA keeps eggs from sticking to the culture dish. If making W-B with 0.1% PVA, it is likely that not quite all the PVA will go into solution; it is recommended that the medium be filtered before use.
4. Pipet pulling is a process of trial and error; you pull a couple dozen pipets, and then check them once you are at the dissecting microscope working with the eggs. The size of the pipet opening will vary, but a standard size for moving eggs will have a diameter slightly larger than the egg. It is advisable to have separate pipets for separate experimental groups (e.g., a control antibody and an experimental antibody), to avoid contamination between groups.
5. The times stated are approximate, but it is recommended that the eggs not be incubated for an extended period of time with chymotrypsin. As noted in **Note 9**, with the chymotrypsin treatment some eggs may never undergo ZP swelling; these eggs should be discarded.
6. Because the concentration of MitoTracker Green in the medium needs to be diluted before fertilization to reduce levels of background staining (i.e., dye uptake by the mitochondria of the eggs), label a sperm concentration that is about five fold higher

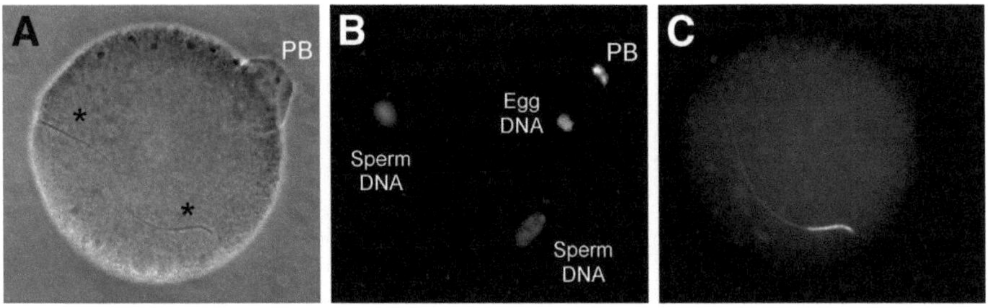

Fig. 3. Fluorescent images of an egg inseminated with unlabeled and MitoTracker Green-labeled sperm. **(A)** Phase contrast image with asterisks indicating the location of two sperm tails. **(B)** Fluorescent image showing DAPI staining of the egg DNA, polar body DNA (PB), and two sperm nuclei. **(C)** Image photographed with the FITC filter, to show MitoTracker Green staining. This image shows that one of the two sperm that fertilized the egg was labeled with MitoTracker Green.

than what will be used for insemination. Immediately before IVF, dilute the sperm to the desired concentration with W-B/15. *See* **Fig. 3** for images of MitoTracker-labeled sperm fertilizing an egg.

7. The number of female mice used will depend on the planned scope of an experiment. It is preferable to have 20 to 30 eggs per experimental group for a given day's experiment (with a given experimental series being repeated at least three times). Average egg yield is approx 20 eggs per mouse but can vary widely. Healthy eggs are round with clear cytoplasm; the first polar body (the product of the first meiotic division) may or may not be present. Discard eggs that have a brown, granular cytoplasm, eggs that are fragmented, eggs that still have a germinal vesicle, or eggs that have extruded a second polar body. Eggs should be collected at approx 12 to 13 h after hCG injection for best quality and egg health ***(9)***.
8. Cumulus cells are somatic cells that are derived from the ovarian follicle. Cumulus cell detachment from the eggs happens somewhat asynchronously; some eggs are clean in approx 1 min and some require 5 min. Pipetting through a wide-bore pipet can facilitate cumulus removal and exposure of the hyaluronic acid (the extracellular matrix in which the cumulus cells are embedded) to the hyalurdonidase. Transfer an egg to W-B as soon as it is cumulus cell-free. Try to minimize the time the eggs are in hyaluronidase (excessive exposure to hyaluronidase has been associated with spontaneous egg activation). It is not necessary to remove 100% of the cumulus cells, as a few residual cumulus cells will detach with ZP removal.
9. Sperm must undergo a maturational change known as capacitation in order to be able to fertilize eggs. Capacitation occurs in vitro in the appropriate culture medium (such as Whitten's medium) containing energy substrates, a cholesterol acceptor (e.g., BSA), $NaHCO_3$, Ca^{2+}, and low K^+ and physiological Na^+ concentrations ***(10)***. During capacitation, a subset of sperm will undergo spontaneous acrosome exocytosis; this is crucial for fertilization of ZP-free eggs as only acrosome-reacted

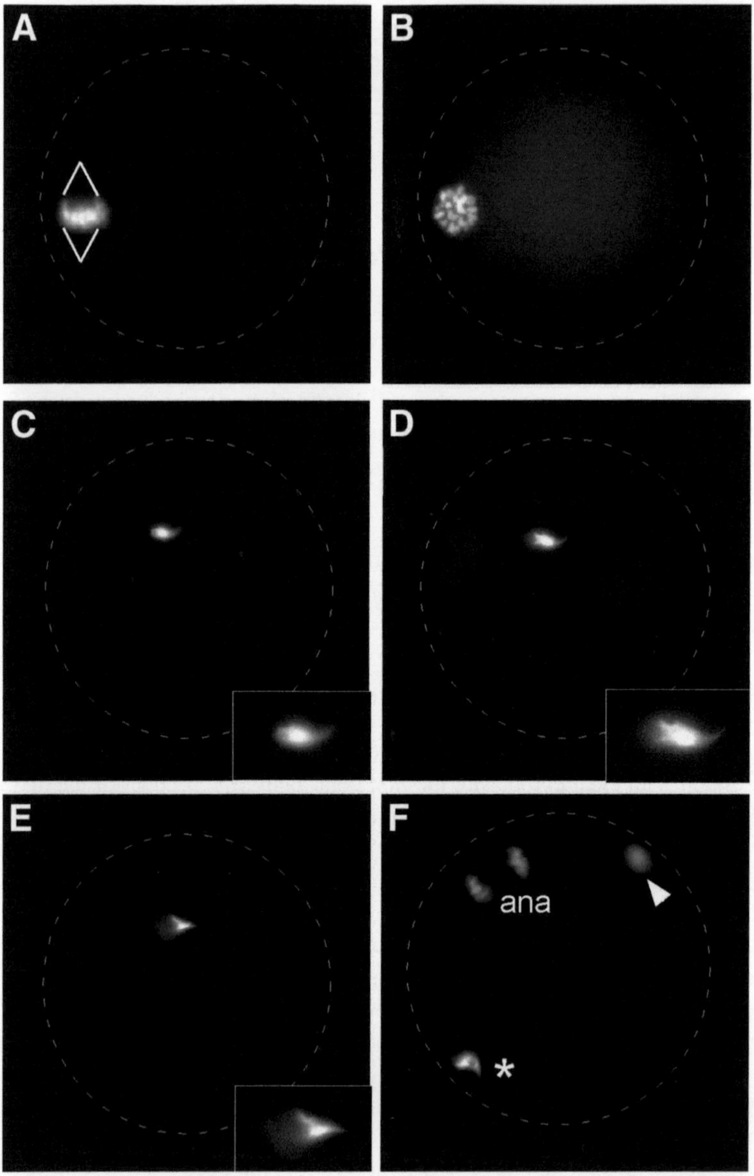

Fig. 4. DAPI-stained eggs, showing morphology of egg DNA and sperm DNA at different stages of fertilization. Dotted lines show the approximate outline of the egg; mouse eggs are approx 80 μm in diameter. **(A)** Unfertilized metaphase II egg, with a side view of the chromosomes on the metaphase plate. White lines indicate the directions of the microtubules holding the chromosomes in place. **(B)** Unfertilized metaphase II egg, with an en face view of the chromosomes on the metaphase plate. Microtubules of the meiotic spindle are extending backward and forward from the

sperm are capable of binding and fusing with the egg plasma membrane ***(11)***. It is advisable to allow 2.5 to 3 h of capacitation time for these IVF experiments. Therefore, this is a key factor when planning the timing of sperm collection.

10. The "swim-up" preparation for sperm selects for the most motile sperm and reduces the risk of including any dead or immotile sperm in the preparation. A typical sperm concentration for the sperm swim-up preparation is $1–5 \times 10^6$ sperm/mL.
11. For example, if you counted 20, 19, and 30 sperm in three separate squares on the hemacytometer, the calculation would be to determine the sperm concentration would be:

$$(20 + 19 + 30)/3 \times 10 \times 10^4 = 2.3 \times 10^6 \text{ sperm/mL}$$

Please note that it is important to prepare the sperm dilution in water (not medium) so that the sperm stop swimming to facilitate counting.

12. There are two commonly used methods for ZP removal: low pH to solubilize the ZP and chymotrypsin treatment to induce ZP swelling and thus facilitate mechanical removal. (It is possible to remove the ZP by mechanical manipulation alone, but this is extremely difficult and often results in low yields of ZP-free eggs because so many eggs lyse during the procedure.) We primarily use the acid solubilization method, although there are advantages and disadvantages to each method. The chymotrypsin-mechanical method has the disadvantage of resulting in lower yields of eggs than the acid solubilization method, as some ZP will never swell sufficiently for removal. The exposure of surface proteins to a protease like chymotrypsin can also be disadvantageous, although the exposure of surface molecules to low pH has the potential to alter them (we have observed the reversible modification of one monoclonal antibody epitope on eggs exposed to low pH *[12]*).
13. MitoTracker Green-labeling of sperm can be useful for a variety of experimental designs. We have used labeled sperm to inseminate early zygotes, that is, eggs that have been previously fertilized. Sperm labeling also can be used to distinguish between two different experimental groups, such as control versus drug-treated or wild-type versus knockout.
14. The MitoTracker Green signal will fade if left exposed to light for a prolonged period of time. Protect MitoTracker aliquots and labeled sperm suspensions from

---

Fig. 4. (*continued*) plane of the image. **(C)** An egg with a bound sperm, with inset showing a close-up of the sperm DNA. (The egg DNA is out of the plane of focus.) **(D)** An egg with a recently fused sperm, with inset showing a close-up of the sperm DNA. (The egg DNA is out of the plane of focus.) Note the difference in appearance (slight flaring) at the posterior end of the sperm nucleus as compared to the bound sperm in **(C)**. **(E)** Eggs with sperm at a later stage of decondensation, with inset showing a close-up of the sperm DNA. (The egg DNA is out of the plane of focus.) The flare at the posterior region of the sperm nucleus has become more pronounced. **(F)** An egg with two fused sperm. The egg has exited from metaphase II arrest and is progressing in to anaphase (labeled "ana"). The sperm heads are marked with an asterisk and an arrowhead; the one marked with the arrowhead is somewhat beyond the plane of focus.

light and view labeled sperm by fluorescence microscopy as soon as possible (preferably on the same day the experiment is performed).

15. Labeling of eggs with DAPI (or Hoechst *[13]*) allows for the detection of sperm–egg fusion by observing the transfer of the DNA-staining dye from the egg cytoplasm to the sperm nucleus once the sperm has fused with the egg. Alternatively, sperm-egg fusion can be detected simply by staining the fertilized eggs with a DNA-specific dye, as sperm head morphology changes upon sperm-egg fusion and with the initiation of sperm nucleus decondensation. This method is easier, although recognizing the earliest stages of sperm decondensation requires some practice; stages of sperm nucleus decondensation are illustrated in **Fig. 4**.
16. It is important to wash the eggs thoroughly so that no DAPI is left in the culture medium to stain the sperm.
17. Insemination conditions are selected based on the fertilization end point of interest, the desired rate of fertilization for control eggs, and the tolerance for polyspermy. A 15-min insemination protocol has been used to assess sperm–egg adhesion before any decreases in egg membrane receptivity to sperm that can occur approx 60 min after fertilization ***(14–16)***. Inseminations of 45 to 60 min routinely are used to examine sperm–egg adhesion and fusion. Typical sperm concentrations used to inseminate ZP-free eggs are in the range of 10,000 to 250,000 sperm/mL. In our experience (with eggs from CF-1 females and sperm from CD-1 males), a 60 min insemination with 25,000 sperm/mL (i.e., a sperm:egg ratio of 25:1) will result in 70 to 80% of the egg fertilized with low levels of polyspermy. A 60-min insemination with 100,000 sperm/mL (i.e., a sperm:egg ratio of 100:1) will result in 90 to 100% of the eggs fertilized with 10 to 50% of the eggs being polyspermic (fertilized by more than one sperm). The emission of the second polar body, indicative of exit of the egg from meiotic arrest, will be apparent at approx 60 to 90 min after insemination.
18. Experimental manipulations of sperm or eggs can be performed at several points in this protocol. The gametes can be pretreated with antibodies to surface molecules or with pharmacological reagents before and/or during insemination (**Fig. 1**). If a drug has reversible effects, then it must be present during the insemination; this then should be taken into account when interpreting results, as both the eggs and sperm would have been exposed during this time period.
19. It is critical that these washes are kept consistent between experimental groups for accurate assessment of sperm adhesion to the egg membrane. All washes of all experimental groups should be performed by the same person using the same pipet, preferably by an individual who does not know which experimental group is which to avoid bias.
20. For experiments in which is it desired to aggressively remove the majority of bound sperm, eggs can be washed more forcefully in W-B containing 0.05% PVA.

Acknowledgment

Research in our laboratory is supported by the National Institute of Child Health and Human Development and by the March of Dimes.

References

1. Evans, J. P. and Florman, H. M. (2002) The state of the union: the cell biology of fertilization. *Nat. Cell Biol.* **4(Suppl.),** s57–s63.
2. Primakoff, P. and Myles, D. G. (2002) Penetration, adhesion, and fusion in mammalian sperm-egg interaction. *Science* **296,** 2183–2185.
3. Runft, L. L., Jaffe, L. A., and Mehlmann, L. M. (2002) Egg activation at fertilization: where it all begins. *Dev. Biol.* **245,** 237–254.
4. Yanagimachi, R. (1988) Sperm-egg fusion. *Curr. Top. Memb. Transport.* **32,** 3–43.
5. Primakoff, P. and Hyatt, H. (1986) An antisperm monoclonal antibody inhibits sperm fusion with zona-free hamster eggs but not homologous eggs. *Fertil. Steril.* **46,** 489–493.
6. Evans, J. P. (1999) Sperm disintegrins, egg integrins, and other cell adhesion molecules of mammalian gamete plasma membrane interactions. *Front. Biosci.* **4,** D114–D131.
7. Whitten, W. K. (1971) Nutrient requirements for the culture of preimplantation embryos in vitro. *Adv. Bio. Sci.* **6,** 129–139.
8. Evans, J. P., Schultz, R. M., and Kopf, G. S. (1995) Identification and localization of integrin subunits in oocytes and eggs of the mouse. *Mol. Reprod. Dev.* **40,** 211–220.
9. Xu, Z., Abbott, A., Kopf, G. S., Schultz, R. M., and Ducibella, T. (1997) Spontaneous activation of ovulated mouse eggs: time-dependent effects on M-phase exit, cortical granule exocytosis, maternal messenger ribonucleic acid recruitment, and inositol 1,4,5-trisphosphate sensitivity. *Biol. Reprod.* **57,** 743–750.
10. Visconti, P. E., Westbrook, V. A., Chertihin, O., Demarco, I., Sleight, S., and Diekman, A. B. (2002) Novel signaling pathways involved in sperm acquisition of fertilizing capacity. *J. Reprod. Immunol.* **53,** 133–150.
11. Yanagimachi, R. (1994) Mammalian fertilization, in *The Physiology of Reproduction,* 2nd ed. (Knobil, E. and Neill, J. D., eds.), Raven Press, Ltd., New York, pp. 189–317.
12. Evans, J. P., Kopf, G. S., and Schultz, R. M. (1997) Characterization of the binding of recombinant mouse sperm fertilin β subunit to mouse eggs: Evidence for adhesive activity via an egg β_1 integrin-mediated interaction. *Dev. Biol.* **187,** 79–93.
13. Conover, J. C. and Gwatkin, R. B. L. (1988) Pre-loading of mouse oocytes with DNA-specific fluorochrome (Hoechst 33342) permits rapid detection of sperm oocyte-fusion. *J. Reprod. Fert.* **82,** 681–690.
14. Redkar, A. A. and Olds-Clarke, P. J. (1999) An improved mouse sperm-oocyte plasmalemma binding assay: Studies on characteristics of sperm binding in medium with or without glucose. *J. Androl.* **20,** 500–508.
15. Zhu, X., Bansal, N. P., and Evans, J. P. (2000) Identification of key functional amino acids of the mouse fertilin β (ADAM2) disintegrin loop for cell-cell adhesion during fertilization. *J. Biol. Chem.* **275,** 7677–7683.
16. McAvey, B. A., Wortzman, G. B., Williams, C. J., and Evans, J. P. (2002) Involvement of calcium signaling and the actin cytoskeleton in the membrane block to polyspermy in mouse eggs. *Biol. Reprod.* **67,** 1342–1352.

8

Collagen Gel Contraction Assay

Peter Ngo, Punitha Ramalingam, Jonathan A. Phillips, and Glenn T. Furuta

Summary

Mucosal tissues undergo contraction and relaxation on a continuous basis. In its normal state, the pliable intestinal tract is characterized by a rhythmic pattern of contractions controlled by its intrinsic neuronal innvervation. In chronic inflammatory diseases such as Crohn's disease, the intestine can become stiff and fibrotic, losing much of its normal motility. Although muscle fiber contraction accounts for much of this activity, contraction of nonmuscle tissue is constantly occurring in events associated with chronic inflammation, such as wound healing, scar formation, and tissue remodeling. However, the physiological and pathological mechanisms defining these events are not well defined. Tissue contraction is a dynamic event characterized by both intracellular and extracellular events. A number of cells, such as fibroblasts, epithelial cells, lymphocytes, and eosinophils, normally reside within the gastrointestinal tract. Additionally, the extracellular matrix is composed of a complex infrastructure that includes collagen and other molecules. The manner in which these two components interact is not certain, but the use of recent model systems has provided insights into these processes. The collagen lattice contraction assay provides a model for tissue contraction that takes advantage of the finding that cell-populated collagen hydrogels contract over time in a predictable, consistent manner. This model allows for investigation of the influence of specific agonists on the rate and extent of matrix contraction.

Key Words: Contraction; collagen hydrogel; collagen lattice; collagen gel.

1. Introduction

Mucosal tissues are composed of several molecular and cellular components. Extracellular matrices can be remodeled by resident cells, such as fibroblasts, to form scar tissue and to contract. In vitro models of tissue wound healing, remodeling, and contraction offer investigators a highly controlled and flexible environment to study the influences of a variety of factors contributing to matrix contraction. Because collagen is the principle extracellular matrix molecule

From: *Methods in Molecular Biology, vol. 341: Cell–Cell Interactions: Methods and Protocols*
Edited by: S. P. Colgan © Humana Press Inc., Totowa, NJ

found in injured tissue, various models have employed collagen as a matrix for imitating the wound environment.

In 1979, Bell et al. ***(1)*** took advantage of the solubility of collagen type 1 in acidic solutions to create cell-populated collagen hydrogels. Measuring the dimensional changes of these collagen gels allows investigators to evaluate the influence of certain variables on the rate and extent of matrix contraction and remodeling in a three-dimensional system ***(2–4)***.

2. Materials

2.1. Cell Culture and Counting

1. Dulbecco's Modified Eagle's Medium (DMEM; Gibco, Invitrogen Co., Grand Island, NY).
2. Fetal bovine serum (Gibco).
3. Solution of penicillin–streptomycin (10,000 U/mL penicillin G sodium and 10,000 μg/mL streptomycin sulfate in 0.85% saline; Gibco).
4. Solution of trypsin (0.25%) and EDTA from Gibco.
5. Trypan Blue stain (Gibco).
6. Hy-Lite Counting Chamber (Fisher, Pittsburgh, PA).

2.2. Collagen Lattices

1. Rat tail collagen type 1 (Sigma, St. Louis, MO); store at 2 to 8°C.
2. NaOH (1 *M*; Sigma).
3. Acetic acid, Glacial 99.9% (J.T. Baker, Phillipsburg, NJ).
4. 0.22-μm syringe-driven filter units (Millipore Corp. Bedford, MA).
5. 200-μL pipet tips (ESP 200; Continental Lab Products, San Diego, CA).
6. 24-well cell culture cluster plates (Corning, Corning, NY).

2.3. Image Acquisition/Analysis

1. Canon A70 Digital Camera 3.2 megapixel (or equivalent).
2. ImageJ Software v. 1.32 (National Institutes of Health, Bethesda, MD). This software is available for download in the public domain at http://rsb.info.nih.gov/ij/.
3. Portable Lightbox (Apollo, Court, NY).

3. Methods

This method defines preparation of collagen solutions, optimization of gel solidification conditions, quantification of gel contraction, and interpretation of data. This model system permits the study of specific agonists or inhibitors on contraction of different cell populations in a three dimensional system.

3.1. Preparation of Collagen for Use in Collagen Gels

1. Acetic acid solution (0.2%) is made from 100% glacial acetic acid and water (*see* **Note 1**).
2. Filter sterilize the solution with a 0.2-μm filter and cool to 4°C.

3. Type 1 collagen is mixed with 0.2% acetic acid solution under sterile conditions to make 6 mg/mL of collagen solution (100 mg collagen dissolved into 16.7 mL acetic acid solution in 50-mL conical tube).
4. Gently agitate the collagen solution at 4°C. This process takes 2 to 5 d and collagen must be completely in solution before proceeding (*see* **Note 2**).
5. After the collagen is entirely dissolved into solution, dilute with equal volume of filter-sterilized water to make a 3 mg/mL collagen solution in 0.1% acetic acid.
6. Equilibrate with gentle shaking at 4°C for 1 d.
7. Collagen solution should be stored at 4°C.

3.2. NaOH Titration of Collagen

Perform this assay whenever using a new batch of collagen to optimize solidification. After identifying the optimal amount of NaOH to add to the collagen/media mixture, use the same quantity for all subsequent gels.

1. Use eight clear-colored 1.6-mL Eppendorf tubes for pH titration of gels.
2. Add 0.4 mL of cell specific media or DMEM to each Eppendorf.
3. Add 0.2 mL of collagen solution (3 mg/mL in 0.1% acetic acid) to one tube containing media. Immediately add 1 μL of 1 *M* NaOH and pipet mixture up and down three times with 1-mL pipet. DO NOT DELAY (*see* **Note 3**).
4. Repeat the above step with increasing amounts of 1 *M* NaOH solution (1–8 μL) to determine the amount of NaOH needed to produce a well-solidified gel with neutral pH (*see* **Note 4**).
5. Allow the mixture to solidify for 20 min. The final collagen concentration should be 1 mg/mL. Compare rigidity and color of gels titrated with different volumes of NaOH to determine which volume of NaOH produces a well-solidified gel with neutral pH. The least amount of NaOH needed to turn the phenol red media indicator a light pink color will produce the most rigid collagen gels.

3.3. Preparing Cells Before Suspension in Collagen Gels

Populating collagen gels with cells requires careful attention to cell concentrations. Determining optimal cell concentration requires performance of a cell concentration curve. Below is a protocol that provides uniform distribution for a final population of 1×10^5 cells/mL of Caco2 epithelial cells. During this portion of the procedure cells can be resuspended in media containing agonist/antagonists to determine effects on collagen gel contraction.

1. A confluent layer of Caco2 cells should be detached from the culture vessel using warmed Trypsin-EDTA solution (0.25% trypsin + EDTA-4Na in Hank's Balanced Salts Solution; *see* **Note 5**).
2. Suspend cells into complete culture media (~10mL media per 75-cm² vessel). DMEM with 10% fetal calf serum and 1% penicillin–streptomycin was used for Caco2 cells (*see* **Note 6**).
3. Count cells using Trypan Blue stain and counting chamber.

4. Centrifuge the remaining cell suspension at 500*g* for 10 min.
5. Remove supernatant and add fresh warm culture media (± experimental variable) to achieve a concentration of 1.5×10^5 cells/mL (*see* **Note 7**). Cells should be resuspended at 1.5 times the final desired cell concentration as cells are further diluted with addition of collagen solution.
6. Mix cells gently into suspension.

3.4. Pouring Cell Populated 1 mg/mL Collagen Gels

Creating consistent, homogeneous, cell populated collagen gels requires practice. Because of the rapid solidification of the collagen gels, it is important to calculate the necessary volumes of all reagents and ensure that adequate quantities are readily available before starting the protocol. The following is a protocol using 24-well plates as molds for 500-µL collagen gels.

1. Collagen gels should be prepared under sterile conditions.
2. Add 0.66 volumes of well-mixed cell suspension (± experimental variable) to a sterile tube.
3. Add 0.33 volumes of 3 mg/mL collagen solution to the cell suspension (*see* **Note 8**).
4. Quickly add the appropriate volume of 1 *M* NaOH (*see* **Subheading 3.2.**) and mix the solution up and down three times.
5. Immediately transfer 500 µL of the mixture to a 1.9-cm^2 well (*see* **Note 9**).
6. Allow gels to solidify, covered at room temperature for 20 min.
7. Gently add a minimum of an equal volume (500 µL) of culture media (± experimental variable) to each well.
8. Dissociate the gel from its mold by gently running the tip of a 200-µL pipet tip along gel edges being careful not to shear or tear gels.
9. Resuspend gels by gently pulling the edges of the gel away from the mold using the pipet tip (*see* **Note 10**). Gently swirl plate to make sure that gel is free from the plate.
10. Replace 24-well plate into incubator at 37°C, humidified 5% CO_2.

3.5. Documenting Contraction

Contraction occurs in a time dependent manner. The rate and degree of contraction depends on the cell type and concentration, and the presence of any contraction agonists or inhibitors in the media. The concentration of serum in the media also affects contraction. **Figure 1** illustrates contraction of Caco2 populated collagen gels at 24 h.

1. At predetermined time-points, remove the 24-well plate from the incubator for image acquisition (*see* **Note 11**).
2. Place the 24-well plate on top of a lightbox.
3. Using a digital camera at a fixed distance above the gels, obtain an image at each of the time-points (*see* **Note 12**).
4. Return the gels to the incubator.

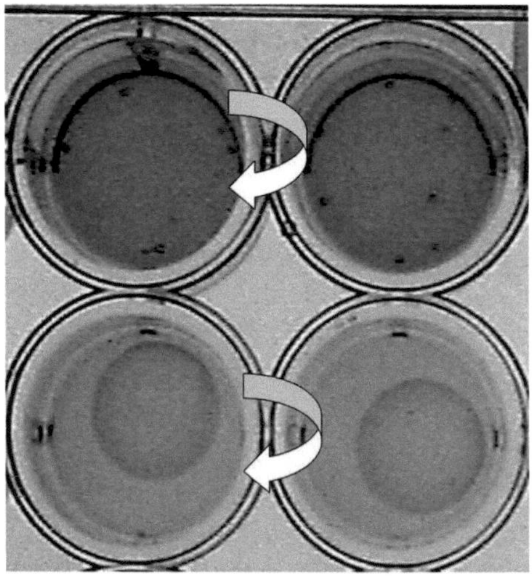

Fig. 1. Photograph of collagen gels taken at 24 h. The top two wells contain acellular collagen gels showing no contraction. The bottom two wells contain Caco2-populated gels displaying uniform, symmetric contraction.

3.6. Interpretation of Data

The degree and rate of contraction of gels exposed to various conditions can be evaluated by calculating the gel surface area from acquired images. To standardize comparison, results can be reported as a percent of initial surface area for each given time point. **Figure 2** shows results at 24 h from a collagen gel contraction assay with 3T3 fibroblast cell concentration curve. Increasing cell concentrations result in increased collagen lattice contraction.

1. Images can be analyzed with National Institutes of Health ImageJ software.
2. Trace the outline of each collagen gel and calculate the surface area according to ImageJ software instructions (*see* **Note 13**).
3. Report the surface area at each time point as a percentage of initial gel surface area.

4. Notes

1. Unless stated otherwise, all solutions should be prepared in water that has resistivity of 18.2 MΩ-cm and total organic content of less than five parts per billion. This standard is referred to as "water" in this text.
2. Gentle agitation with a shaker may be used to aid in dissolving collagen, but forceful agitation or repeated inversion may disassociate collagen fibrils, leading to poorly solidifying gels.
3. A collagen lattice will begin to form immediately when collagen, media, and NaOH are mixed together. Reagents and equipment (NaOH, collagen solution, and 1-mL

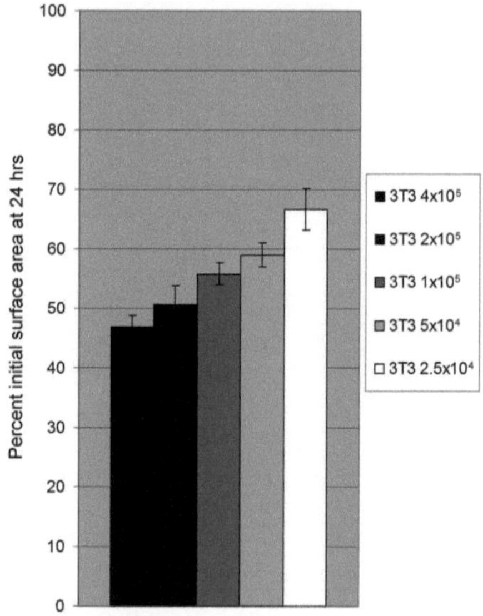

Fig. 2. Results from 3T3 fibroblast collagen gel contraction assay with 3T3 cell concentration curve. Increasing 3T3 concentrations result in increased collagen gel contraction. Results represent means ± standard deviation of mean of four lattices.

pipet set to 500 μL of volume) should be readily available for immediate use. Repeated pipetting or agitation after the gel has begun to solidify will shear collagen and produce a poorly solidified gel.

4. After mixing, gels should be a homogeneous color. A nonhomogeneous gel indicates inadequate mixing.
5. Trypsin not only aids in lifting cells from culture vessels but also breaks up aggregates of cells to produce a more even distribution of cells within gels.
6. Contraction of collagen gels is dependent on the presence of serum in media and gels. Gel contraction increases in a concentration dependent manner with increasing serum so consistent serum concentrations should be used for all conditions within a particular assay ***(4)***. Media containing 10% fetal bovine serum yields consistent results.
7. This step allows resuspension of cells at desired concentrations. In addition, cells can be resuspended in solutions containing contraction agonists or inhibitors. This step should be performed for all cell populations as it also removes trypsin from the suspension which may interfere with collagen gel contraction.
8. When making each 500-μL collagen gel, it is adviseable to mix 400 μL of cell suspension and 200 μL of collagen solution; the additional volume allows for easier transfer of mixtures into molds without incorporating bubbles in gels. Multiples of these volumes can be used to create experimental replicates.

9. Be careful not to introduce bubbles into the mixture as these will solidify in the gel. For replicates, a 5-mL stripette works well to both mix and quickly dispense 500-μL of mixture into each well. Drawing up more than the needed volume into the stripette will prevent air bubbles from incorporating into the last poured gel.
10. All gels should be freed from well walls in the same manner. Running the pipet tip around edges of detached gels should cause the gel to turn easily within the well, confirming that it is not adherent to the well bottom.
11. Depending on the rate and degree of contraction with each specific cell type and concentration, photographs should be obtained at different time points, that is, 0, 6, 12, 24, and 48 h after gels are poured typically yields an adequate initial time course.
12. Using a Canon Powershot A70 3.2 megapixel digital camera, a distance of 40 cm with 3X optical zoom yields good quality images of the 24-well plate and collagen gels. It is best to obtain several photos at each time point to ensure an image with good resolution is obtained. With increasing cell concentrations and with increasing contraction the gel edge is easier to visualize allowing for easier analysis of contraction. Alternatively, a flatbed digital scanner can be used to obtain images for analysis.
13. Adjusting the contrast or brightness of each image prior to analysis with ImageJ software may allow easier surface area calculation. To increase accuracy in outlining gels, images can be magnified using ImageJ software without affecting surface area calculations. The outline of wells in the 24-well plate provides an optimal reference for initial surface area of each gel.

References

1. Bell, E., Ivarsson, B., and Merrill, C. (1979) Production of a tissue-like structure by contraction of collagen lattices by human fibroblasts of different proliferative potential in vitro. *Proc. Natl. Acad. Sci. USA* **76,** 1274–1278.
2. Levi-Schaffer, F., Garbuzenko, E., Rubin, A., et al. (1999) Human eosinophils regulate human lung- and skin-derived fibroblast properties in vitro: a role for transforming growth factor beta (TGF-beta). *Proc. Natl. Acad. Sci. USA* **96,** 9660–9665.
3. Zagai, U., Skold, C. M., Trulson, A., Venge, P., and Lundahl, J. (2004) The effect of eosinophils on collagen gel contraction and implications for tissue remodelling. *Clin. Exp. Immunol.* **135,** 427–433.
4. Gillery, P., Maquart, F. X., and Borel, J. P. (1986) Fibronectin dependence of the contraction of collagen lattices by human skin fibroblasts. *Exp. Cell Res.* **167,** 29–37.

9

Methods to Assess Tissue Permeability

Juan C. Ibla and Joseph Khoury

Summary

An essential requirement for adequate organ performance is the formation of permeability barriers that separate and maintain compartments of distinctive structure. The endothelial cell lining of the vasculature defines a semipermeable barrier between the blood and the interstitial spaces of all organs. Disruption of the endothelial cell barrier can result in increased permeability and vascular leak. These effects are associated with multiple systemic disease states. The mechanisms that control barrier function are complex and their full understanding requires a multidisciplinary approach. In vivo permeability data often complement molecular findings and add power to the studies. The interaction of multiple cell types and tissues present only on mammalian models allow for testing of hypothesis and to establish the physiological significance of the results. In this chapter, we describe methods that can be used systematically to measure the permeability characteristics of several organs.

Key Words: Permeability assay; vasculature; edema; inflammation; Evan's blue; fluorescence; bioparticles; water content; wet/dry ratio.

1. Introduction

Mucosal surfaces provide a physical barrier between the environment and the internal layers of many organs. The dual role of maintaining close contact, yet separated biological compartments requires multiple interactions between endothelia, epithelia, and *trans*-membrane adhesion domains. This highly adapted cellular hierarchy denotes a specific anatomical location within an organ (e.g., polarized epithelia) and assigns individual cell types vital tasks for organ function. The endothelial lining of all organs is charged to interact with the systemic circulation, transduce soluble messages, and control regional blood flow and organ development *(1,2)*. Insight into the mechanisms that regulate these natural barriers is necessary for the understanding or normal physiology and disease progression. It is well established that abnormal barrier function is

From: *Methods in Molecular Biology, vol. 341: Cell–Cell Interactions: Methods and Protocols*
Edited by: S. P. Colgan © Humana Press Inc., Totowa, NJ

present in a multitude of disease states and that available methods can be used to estimate the severity of diseases and response to treatment *(3)*. Several methods have been developed to study endothelial permeability in the vasculature, both in vivo and in vitro *(4)*. These include blood clearance and diffusion methods that can be performed both invasively and noninvasively in animal and human subjects *(5)*. Experimentally, tissue permeability methods can be accomplished with equipment available in standard laboratory settings and the results obtained are easy to interpret and highly reproducible.

2. Materials

2.1. Miles Assay for Vascular Permeability

1. Mice.
2. Phosphate-buffered saline (PBS; Gibco BRL, Invitrogen, Carlsbad, CA).
3. Evan's blue (Sigma, St. Louis, MO).
4. Ligand, cytokine, growth factor, or condition being tested.
5. Syringe with needles.
6. Surgical dissecting equipment.
7. Photographic equipment.
8. Formamide (Sigma).
9. Water bath at 55°C.
10. Spectrophotometer at 450 nm and 590 to 620 nm (SpecraMax 250, Molecular Devices, Sunnyvale, CA).
11. 96-well plate or cuvette for spectrophotometer (Corning Costar, Acton, MA).

2.2. Fluorescent-Labeled Molecules

1. Mice.
2. PBS.
3. Fluorescein isothiocyanate (FITC)–dextran, molecular weight 4 to 40 kDa (Sigma), or fluorescent-labeled microspheres (Molecular Probes, Eugene, OR).
4. 1-mL syringe and 23-gage needles.
5. 1.5-mL microcentrifuge tubes (Eppendorf, Westbury, NY).
6. Fluorescent plate reader (Cytofluor 2300, Millipore, Framingham, MA).
7. Fluorescent microscope (Olympus, B.H2, Melville, NY).
8. Plates for fluorescence (Corning Costar).

2.3. Labeled Bioparticles

1. Fluorescent bacteria commercially labeled with flourescein, Alexa.
2. Fluor (488/594) and Texas Red dye among others (Molecular Probes).
3. Sterile PBS or normal saline.
4. 2 m*M* sodium azide (Sigma).
5. Vortex (Bio-Rad, Hercules, CA).
6. Hemocytometer (Fisher Scientific, Pittsburgh, PA).

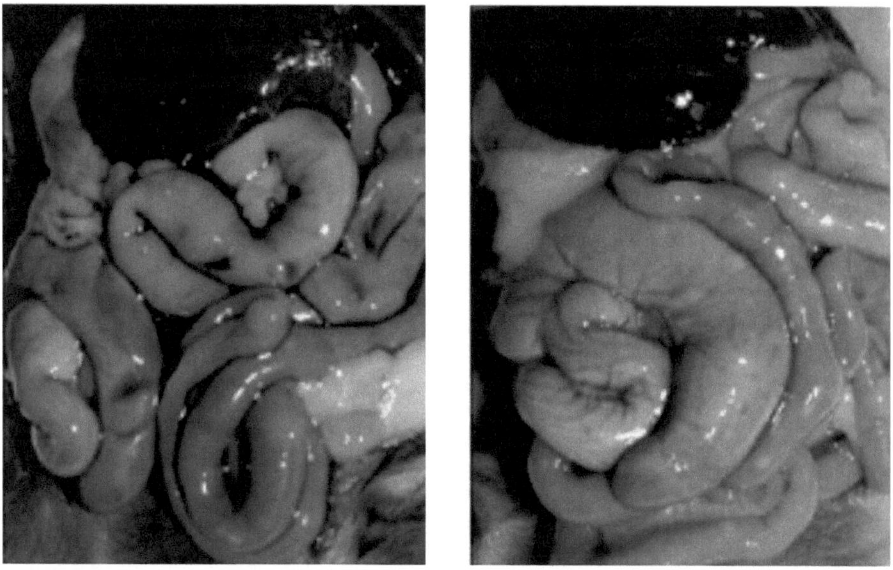

Fig. 1. Photographs representing normal and increased vascular permeability in mice. C57Bl6 male mice were injected with 0.2 mL of Evan's blue via tail vein. Animals were then subjected to 4 h of normobaric hypoxia (8% O_2, 92% N_2) or room air conditions. Postmortem macroscopic examination of intra-abdominal viscera displayed significant differences between experimental and control groups. (Please *see* color insert following p. 50 for a color version of this figure.)

7. Microscope (Olympus BH2).
8. Mice.

2.4. Water Content in Tissues

1. Microcentrifuge tubes (Eppendorf).
2. High-fidelity electronic balance (Mettler-Toledo, P.M 400, Columbus, OH).
3. Speed-Vac (Vaccufuge, Eppendorf).
4. Tissue.

3. Methods

3.1. Miles Assay for Vascular Permeability

The classical method in which vascular permeability can be measured in vivo is by the Miles Assay for Vascular Permeability, also known as the Evan's blue dye method ***(6)***. Evan's blue is a marker that specifically binds to albumin, allowing for a quantification of vascular leakage into the extravascular tissues. The accumulated dye can be quantified by the use of a spectrophotometer. Results are often evident at the macroscopic level, as depicted in **Fig. 1**.

1. Reconstitute, Evan's blue to a final concentration of 0.5% in PBS.
2. 0.2 mL of Evan's blue is administered by tail vein injection per mouse.
3. The stimulation of choice is then administered with a subset of mice not receiving the stimulation as controls.
4. The mice are then sacrificed.
5. Dissection of the organs of interest are performed and photographed as necessary.
6. Evan's blue concentration is quantified by eluting 50 mg of individual organs in 0.5 mL of formamide at 55°C from 2 h to overnight.
7. Aliquots of the eluted dye are then measured by absorbance at 610 nm with subtraction of reference absorbance at 450 nm.

3.2. Fluorescent-Labeled Molecules

Fluorescent dyes are linked to either dextran or polysterene microspheres that could be used as tracers for permeability. These tracers may then be visualized either directly in vivo by aid of intravital microscopy or in tissue sections. Several dyes and molecule sizes may be used to assay both epithelial and endothelial permeability. Although the initial method employed by, Rudolph and Heymann *(7)* used radiolabeled microspheres, growing concerns over environmental and health issues as well as cost of handling, disposal and half-life of the radiolabel have brought to the development of nonisotope fluorescent or colored molecules. The method we describe here uses FITC–dextran *(8)*, but it may easily be adapted for various fluorescent-labeled microspheres (*see* **Note 1**).

1. Reconstitute FITC–dextran in, P.BS to a concentration of 80 mg/mL.
2. Administer FITC–dextran to each animal 60 mg/100 g body weight (*see* **Note 2**).
3. Expose animals to condition of interest for appropriate amount of time (2–6 h).
4. Collect whole blood by cardiac puncture in anesthetized animals using the syringe and needle.
5. Transfer blood to a microcentrifuge tube, let sit on ice for 1 h, spin at 77,000*g* for 5 min, and collect supernatant (serum).
6. On a fluorescence plate, make serial dilutions of FITC–dextran in PBS (range from 10 to 0.015 μg/mL), to unknown wells, add 60 μL 1:3 diluted serum in PBS per well (triplicate).
7. Read plate at excitation wavelength 485 nm and emission 530 nm.
8. Measurements are recorded as nanograms of FITC–dextran per microliter of serum.

3.3. Labeled Bioparticles

The dynamic interaction of cellular components within a tissue can be analyzed in vivo by the kinetics of labeled bioparticles during exposure to critical physiological conditions. This technique utilizes a series of fluorescent labeled, heat or chemically killed bacteria of a variety of sizes and antigenic characteristics. These methods have been employed to study a diverse group of parameters including epithelial permeability, phagocytosis and opsonization. Results can be obtained by fluorescent microscopy, quantitative spectrophotometry and flow cytometry ***(9,10)***. We have used these methods to study bacterial translocation in transgenic mice exposed to mucosal barrier disruptive conditions (hypoxia).

1. Reconstitute labeled bacteria (100 mg, 3×10^8 *Escherichia coli*) to 20 mg/mL in sterile PBS or normal saline (*see* **Note 3**).
2. Add 2 m*M* sodium azide to bacterial suspension and store at 4°C. Without the addition of sodium azide, bacterial suspension should be used within 1 d.
3. Vigorously vortex the solution 3×15 s at the highest setting.
4. Count the number of bioparticles per millititer using a hemocytometer in a phase-contrast or fluorescence microscope.
5. Dilute bioparticles to 4×10^8 per 0.5 mL.
6. Vortex aliquots of bioparticles 3×15 s at the highest setting.
7. Using a 1-mL syringe, inject bioparticle suspension into mice by oral/esophageal injection.
8. Expose animals to desired experimental conditions.
9. Harvest target organs and serum.
10. Fluorescence can be are quantified directly in serum or by eluting organ aliquots in formamide at 55°C from 2 h to overnight.
11. Dissected organs can be fixed in formalin embedded and cut for microscopy.

3.4. Water Content in Tissues

Tissue inflammation is often characterized by cellular infiltration and accumulation of extracellular water (edema). Quantifying water content in tissues can be a useful measurement to determine the integrity of vascular and epithelial barriers. This parameter can often be combined with other permeability assays to corroborate the effect of experimental conditions or pharmacological interventions ***(11,12)***. We describe a simple and highly accurate method to measure water content in tissues using dry to wet weight ratios. Multiple organs can be assayed simultaneously as shown in **Fig. 2**.

1. After the experimental conditions have been completed, collect 20 to 50 mg of tissue of interest from a representative portion of the organ (*see* **Note 4**).
2. Obtain wet weight on an electronic high fidelity balance and record (tissue alone and tube containing tissue).
3. Place samples on labeled microcentrifuge tubes (make sure that sample is placed at the bottom of the tube).
4. Place tubes in speed-vac. Set up for vacuum mode at 65°C for 12 to 18 h (*see* **Note 5**).
5. Obtain dry weight as described previously.
6. Subtract dry value from wet weight.

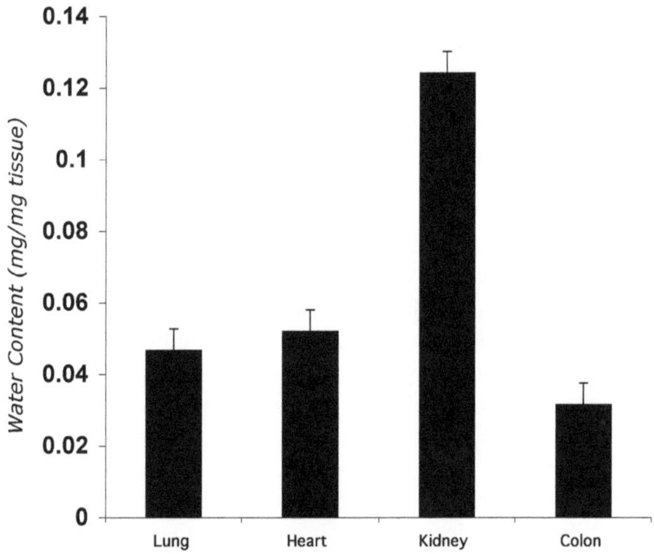

Fig. 2. Vascular permeability profile of different organs in mice. All organs were harvested and water content quantified by wet/dry ratios. Multiple organs were profiled simultaneously. Data are expressed in mg/mg of tissue.

4. Notes

1. If using fluorescent microspheres, fluorescence may be measured directly in serum or in frozen sections of tissues using fluorescence microscopy.
2. For intestinal epithelial permeability, FITC–dextran solution is given by oral gavage. For pulmonary epithelial permeability, FITC–dextran is injected directly into the trachea of anesthetized animals.
3. Bioparticles are provided as lyophilized powders. Upon receipt, they should be stored frozen at –20°C. In the case of fluorescent-labeled products, these should be protected from light exposure.
4. Smaller amounts of tissue (10–30 mg) from organs with high fat content (intestine) or high cellular density (kidney) can be used, since these tissues will take longer to dry.
5. If tissues from different organs are processed simultaneously, check the vacuum run periodically (every 6–8 h). Lung tissue depending on the amount will typically require less time to dry compared with other organs. Over exposure of the samples may result in protein evaporation, tissue destruction and inaccurate readings.

Acknowledgment

This work was supported by a fellowship award from the Foundation for Anesthesia, Education and Research (to J.C.I.).

References

1. Beck, K. F., Eberhardt, W., Frank, S., et al. (1999) Inducible NO synthase: role in cellular signaling. *J. Exp. Biol.* **202,** 645–653.
2. Bertuglia, S. and Giusti, A. (2005) Role of nitric oxide in capillary perfusion and oxygen delivery regulation during systemic hypoxia. *Am. J. Physiol. Heart Circ. Physiol.* **288,** H525–H531.
3. Hofmann, T., Stutts, M. J., Ziersch, A., et al. (1998) Effects of topically delivered benzamil and amilorideon nasal potential difference in cystic fibrosis. *Am. J. Resp. Crit. Care Med.* **157,** 1844–1849.
4. Karhausen, J., Ibla, J. C., and Colgan, S. P. (2003). Implications of hypoxia on mucosal barrier function. *Cell. Mol. Biol.* **49,** 77–87.
5. Bleeker-Rovers, C. P., Boerman, O. C., Rennen, H. J., et al. (2004) Radiolabeled compounds in diagnosis of infectious and inflammatory disease. *Curr. Pharm. Design.* **10,** 2935–2950.
6. Miles, A. A. and Miles E. M. (1952) Vascular reactions to histamine, histamine-liberator, and leukotaxine in the skin of guinea pigs. *J. Physiol. (London)* **118,** 228–257.
7. Rudolph, A. M. and Heymann, M. A. (1967) The circulation of the fetus in utero. Methods for studying distribution of blood flow, cardiac output and organ blood flow. *Circ. Res.* **21,** 163–184.
8. Synnestvedt, K., Furuta, G. T., Comeford, K. M., et al. (2002) Ecto-5′-nucleotidase (CD 73) regulation by hypoxia-inducible factor-1 (HIF-1) mediates permeability changes in intestinal epithelia. *J. Clin. Invest.* **110,** 993–1002.
9. Samel, S., Keese, M., Kleczka, M., et al. (2002) Microscopy of bacterial translocation during small bowel obstruction and ischemia in vivo—a new animal model. *BMC Surg.* **2,** 6.
10. Sorrells, D. L., Friend, C., Koltuksuz, U., et al. (1996) Inhibition of nitric oxide with aminoguanidine reduces bacterial translocation after endotoxin challenge in vivo. *Arch. Surg.* **131,** 1155–1163.
11. Thompson, L. F., Eltzschig, H. K., Ibla, J. C., et al. (2004) Critical role for the ecto-5′-nucleotidase (CD73) in vascular leak during hypoxia. *J. Exp. Med.* **200,** 1395–1405.
12. Toung, T. J., Chang, Y., Lin, J., and Bhardwaj, A. (2005) Increases in lung and brain water following experimental stroke: effect of mannitol and hypertonic saline. *Crit. Care Med.* **33,** 203–208; discussion 259–260.

10

Bacterial–Bacterial Cell Interactions in Biofilms

Detection of Polysaccharide Intercellular Adhesins by Blotting and Confocal Microscopy

Kimberly K. Jefferson and Nuno Cerca

Summary

Adhesive interactions between bacterial cells coupled with adherence to a solid surface can lead to the formation of a biofilm. The important role of biofilm formation in the pathogenesis of certain types of infection, especially those involving indwelling medical devices, is becoming increasingly apparent. Critical to the development of a biofilm is the elaboration of exo-polysaccharide that contributes to substrate and intercellular adhesion. The synthesis and secretion of large exo-polysaccharides is a metabolically expensive process and is therefore often suppressed under conditions that favor the planktonic mode of growth. One way to identify the environmental cues that cause a given bacterial species to switch to the biofilm mode of growth is to monitor exo-polysaccharide elaboration in vitro. The exo-polysaccharide involved in biofilm formation in a number of bacterial species is a polymer of *N*-acetyl-glucosamine. In this chapter, we outline two methods that use wheat germ agglutinin, a lectin that binds to *N*-acetyl-glucosamine, to evaluate extracellular polysaccharide production by a variety of bacterial species.

Key Words: Biofilms; exo-polysaccharide; polysaccharide matrix; bacteria; lectins; wheat germ agglutinin; confocal microscopy.

1. Introduction

"United we stand, divided we fall" is one of the most famous quotes of the legendary Greek fabulist Aesop. The truism is as befitting to bacterial consortia as it is to human societal interactions. Interactions between bacterial cells and adherence to a solid surface, mediated in large part by exo-polysaccharide (EPS) intercellular adhesins, result in cohesive, structured communities called biofilms, which are able to withstand many adverse environmental conditions that would destroy individual free-floating bacteria. For reasons that are not completely

From: *Methods in Molecular Biology, vol. 341: Cell–Cell Interactions: Methods and Protocols*
Edited by: S. P. Colgan © Humana Press Inc., Totowa, NJ

understood, many bacterial and fungal species switch from a planktonic to a biofilm mode of growth under certain environmental conditions or during infection. Many types of infections, including osteomyelitis, endocarditis, and catheter or other medical device-related infections, involve biofilm formation, which is problematic for a number of reasons. Biofilms are tenacious and resilient. Physical forces such as the shear force of blood flow or urine flow across the surface of intravenous or urinary catheters do not wash them away. Biofilms also are resistant to phagocytosis and can tolerate levels of antibiotics 10- to 1000-fold higher than the levels required to inhibit or kill free-floating bacteria. Thus, biofilm infections often are chronic or relapsing, frequently are refractory to antimicrobial chemotherapy, and often necessitate the removal of the colonized tissue or device.

The biofilm EPS matrix is an important intercellular adhesin and is therefore fundamental to biofilm development and structure. EPS also plays a role in protecting the bacteria from adverse conditions and has been implicated in resistance of *Staphylococcus epidermidis* to phagocytosis ***(1)***. Elaboration of EPS is a key factor in the switch from planktonic to biofilm growth ***(2)***. Antibodies that react specifically to the EPS of certain bacterial species have been produced and used to monitor the effect of different mutations or environmental conditions on EPS production; however, EPS-specific antibodies are not always readily available ***(3,4)***. The biofilm EPS elaborated by a number of bacterial species, including *Staphylococcus aureus* and *Staphylococcus epidermidis*, *Escherichia coli*, and *Acinetobacter*, is a polymer of *N*-acetyl-glucosamine referred to as PNAG, PGA, or polysaccharide intercellular adhesin ***(4–7)***. The sugar residue *N*-acetyl-glucosamine is a ligand for the lectin wheat germ agglutinin (WGA). Horseradish peroxidase-labeled and fluorescently labeled WGA is commercially available and can be used to detect EPS semiquantitatively by immunoblotting and by confocal microscopy, respectively ***(8,9)***. Thus, we have developed a technique that can be used to detect EPS production by a number of bacterial species for which a specific anti-polysaccharide antibody is not readily available.

2. Materials

2.1. Cell Culture

1. Culture medium is chosen depending on the environmental conditions being tested. A rich medium such as tryptic soy broth + 1% glucose is recommended when it is desirable to maximize biofilm formation.
2. For EPS blots, use 50-mL polystyrene conical tubes.
3. For confocal microscopy, use either 35-mm collagen-coated glass-bottom microwell dishes (Mat-Tek, Ashland, MA) or 35-mm tissue culture-treated polystyrene wells (Corning, Corning, NY).

2.2. Preparation of EPS Extract

1. dH_2O (distilled water) and ddH_2O (deionized, distilled water).
2. 0.5 *M* EDTA (American Bioanalytical, Natick, MA), pH 8.0. Prepare by adding 186.12 mg of EDTA to 900 mL of dH_2O, add 5 *N* NaOH dropwise, and mix continuously until the pH reaches 8.0 and the EDTA is completely dissolved.
3. Proteinase K 20 mg/mL (Qiagen, Valencia, CA).
4. 60 and 80°C heat-blocks.

2.3. Blotting and Detection of EPS

1. Vacuum manifold slot-blotter or dot-blotter.
2. Nitrocellulose (Amersham, Piscataway, NJ).
3. Tris-buffered saline (TBS). Prepare a stock solution of 1 *M* Tris-HCl pH 7.5. For 1 L of TBS, combine 100 mL of 1 *M* Tris-HCl, pH 7.5., 9 g NaCl, and 900 mL of dH_2O and store at room temperature (RT).
4. TBS + 0.05% Tween-20 (TBST).
5. Blocking solution. Prepare fresh by dissolving 1% w/v bovine serum albumin (Sigma, St. Louis, MO) in TBST.
6. WGA–horseradish peroxidase-labeled lectin from *Triticum vulgaris* (HRP; Sigma). Prepare a stock solution by dissolving 1 mg WGA–HRP in 1 mL of ddH_2O. Store 10-μL aliquots at –20°C.
7. Chemiluminescent HRP substrate (ECL kit, Amersham).
8. Plastic wrap.
9. Film cassette.
10. Kodak BioMax MS X-ray film (Amersham).

2.4. Confocal Microscopy

1. Phosphate-buffered saline (PBS). To prepare a 10X stock solution, dissolve 1.44 g KH_2PO_4, 90 g NaCl, and 7.95 g Na_2HPO_4 in 1 L of dH_2O and adjust the pH to 7.5.
2. WGA-Oregon green conjugate (WGA-Green; Molecular Probes, Eugene, OR). Prepare a stock solution by dissolving 10 mg WGA–Green in 1 mL of H_2O. Store in aliquots of 25 μL at –80°C.

3. Methods

3.1. Preparation of EPS Samples

1. Grow bacteria to stationary phase in 5 mL of medium.
2. Approximately 5×10^9 bacterial cells are collected by centrifugation at 8000*g*. The culture supernatant may be discarded (*see* **Note 1**) or saved for analysis.
3. Resuspend bacterial cells in 180 μL of 0.5 *M* EDTA by pipetting up and down or by gentle sonication (*see* **Note 2**) using a probe-type sonicator. Transfer the suspension to a microfuge tube and boil for 5 min to kill the bacteria and to aid in release of EPS from the cell surface.

4. Centrifuge the samples at 18,000*g* to remove the cells, and transfer 150 µL of the EPS-containing supernatant to a fresh microfuge tube.
5. Add 15 µL of proteinase K (2 mg/mL final concentration) and incubate the samples at 60°C for 1 h. After protein digestion, heat-inactivate the proteinase K at 80°C for 30 min.

3.2. Blotting and Detection of EPS

1. Cut a piece of nitrocellulose to fit into the slot-blot apparatus, and pre-wet with TBS.
2. Sandwich the nitrocellulose between the gasket and top plate of the blotting apparatus and the turn on the clamp vacuum. Fill the wells with TBS and turn on the sample vacuum until the liquid has been drawn through the membrane. Turn off the sample vacuum.
3. A series of 2- to 10-fold dilutions of the EPS samples should be made in TBS. Add the samples and dilutions to the wells of the vacuum manifold and turn on the vacuum until the samples have been completely drawn through the nitrocellulose membrane.
4. Disassemble the vacuum manifold and allow the nitrocellulose membrane to dry thoroughly at RT.
5. Wet the membrane with TBS and block in 25 mL of blocking solution (*see* **Note 3**) for 1 h at RT with rocking.
6. Add 15 mL of 50 ng of WGA–HRP + 10 mg bovine serum albumin/mL TBST and incubate for 30 min at RT with rocking.
7. Wash the membrane 3 times for 5 min with at least 25 mL of TBST.
8. Rinse the membrane briefly with dH_2O. Lift the membrane with forceps or gloved fingers and gently shake off excess water. Lay the membrane on a flat surface such as a clean benchtop.
9. Work very quickly from this step forward. Thoroughly mix 2 mL of ECL reagent 1 and 2 mL of reagent 2 and immediately pipet onto the membrane. Surface tension will hold the solution on the blot. Leave the solution on the blot for exactly 1 min.
10. Lift the blot and shake off excess solution. Wrap the blot in plastic wrap and tape to the inside of a film cassette. Take the film cassette to a darkroom and lay a piece of film on the blot under safelight conditions. Close the cassette and expose the film for 1 min. Remove and develop the first film, add a second film to the blot and start a stopwatch to monitor the exposure-time of the second film.
11. Gage the exposure time for the second film after the first film has been developed. A sample blot illustrating results obtained with EPS-positive and EPS-negative strains of staphylococci and *E. coli* is shown in **Fig. 1**.

3.3. Prepare Biofilms for Confocal Microscopy

1. Inoculate 2.5 mL of the medium from liquid cultures and grow biofilms at 37°C with rocking for 18 to 72 h (*see* **Note 4**) in 35-mm dishes (*see* **Note 5**).
2. Remove medium from biofilm and rinse with PBS (*see* **Note 6**).

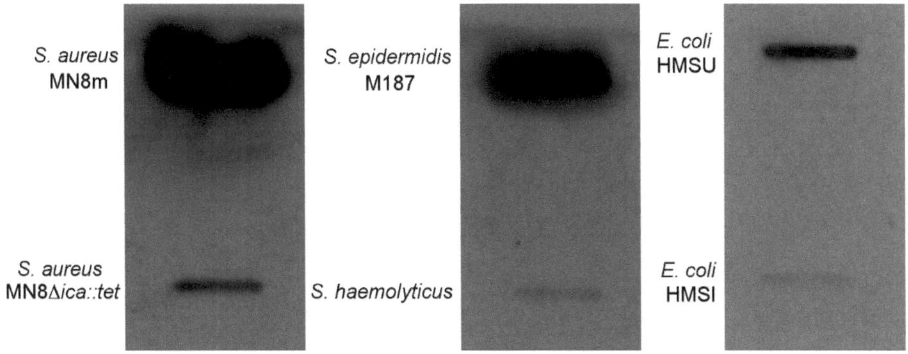

Fig. 1. Analysis of EPS elaboration by blotting. Surface extracts from staphylococci and *E. coli* were blotted onto nitrocellulose and probed with WGA–HRP. Strains are indicated to the left of each blot: PNAG-hyper-producing *S. aureus* strain MN8m, the PNAG-negative *S. aureus* strain MN8Δ*ica::tet*, PNAG-producing *S. epidermidis* strain M187, PNAG-negative staphylococcal species *S. haemolyticus* (strain M176), *pga*-positive, *E. coli* clinical isolate (strain HMS-U) from a urinary tract infection, and a *pga*-negative, *E. coli* urinary tract infection isolate (strain HMS-I).

3. Dilute WGA–Green in PBS (final concentration 0.1 mg/mL) and add 1 mL to the biofilm. Incubate in the dark for 15 min (*see* **Note 7**).
4. Pipet off WGA–Green solution and rinse the biofilm with 2 mL of PBS.

3.4. Detection of EPS by Confocal Microscopy

1. Biofilms should be observed by confocal microscopy using an immersion objective (Achroplan ×63/0.95W).
2. The bacterial cells are detected by the refraction of light in the red spectrum (*see* **Note 8**) and WGA–Green is detected by fluorescence in the green spectrum, using a single channel analysis (*see* **Note 9**).
3. Excitation wavelengths are set at 633 nm with a output power of 70% for the detection of bacterial cells and 488 nm with a output power of 10% for the detection of WGA–Green. The excitation beam splitter is HFT UV/488/543/633.
4. The filter used to detect the light refracted by bacterial cells is LP650, and the filter for detection of light emitted by WGA–Green is BP 505-530. The beam splitter is NFT 545.
5. The pinholes of both channels are adjusted to obtain equivalent stack sizes.
6. A fast scan is performed to adjust the detector gain and detector offset of both channels. The color palette is set to Range Indicator, and the detector gain is adjusted to minimize background color intensity (an indication of minimum saturation of the detector gain). The detector offset should also be adjusted to minimize color intensity if background pixels.
7. A fast scan is performed to determine the top and bottom of the biofilm.

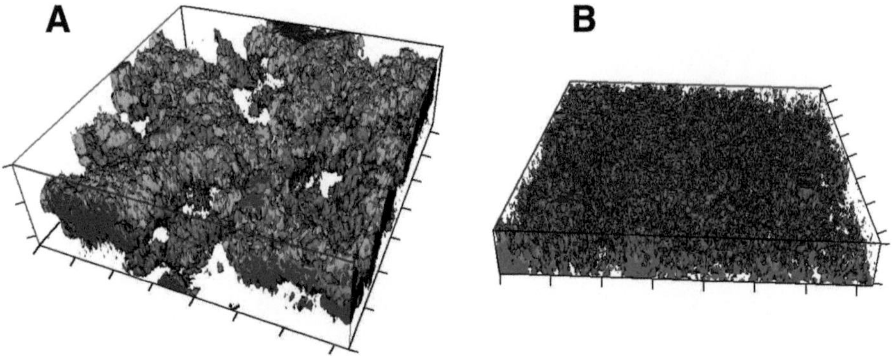

Fig. 2. Use of confocal microscopy to analyze EPS elaboration. The intercellular adhesin (*ica*) locus in *S. aureus* encodes the proteins necessary for PNAG synthesis so *ica*-positive and *ica*-negative mutants were used as positive and negative controls. Biofilms produced by (**A**) the strong PNAG-producing *S. aureus* strain MN8m and (**B**) the PNAG-negative *S. aureus* strain MN8Δ*ica::tet* were stained with WGA–Green and imaged by confocal microscopy. Biofilm cells are presented in red and PNAG in green. Each bar segment represents 20 μm. (Please *see* color insert following p. 50 for a color version of this figure.)

8. Once the z-settings have been set, a full resolution scan is done. Sample images of EPS-positive and EPS-negative strains of *S. aureus* are shown in **Fig. 2**.

4. Notes

1. EPS may be secreted into the culture fluid; therefore, the investigator may wish to reserve the supernatant and blot it onto the nitrocellulose membrane. However, noncell-associated EPS does not appear to contribute to biofilm development ***(10)***.
2. EPS-producing bacteria may be mucoid and difficult to resuspend. Sonication may be used to disrupt cell clusters but should be at low power to prevent release of cell wall components that may contain *N*-acetyl-glucosamine and could cross-react with WGA–HRP.
3. Skim milk contains sugars that can interact with WGA, and should not be used as a blocking agent.
4. In some bacterial species, biofilm maturation requires several days of growth ***(11)***. If bacteria are allowed to grow for more than 18 h, then the medium should be replaced every 12 h.
5. Because of variations in surface proteins and hydrophobicity among different bacterial species, glass or tissue culture-treated polystyrene may provide a better substrate for adherence and biofilm formation. This must be determined experimentally for each species. Most bacteria will not, however, adhere to polystyrene that has not been tissue culture-treated.

6. Loosely adherent biofilms should be washed very gently or the initial washing step may be skipped.
7. The 15-min incubation time should be adhered to as longer incubation times can actually reduce the level of fluorescence.
8. Fluorescent labeling of bacterial cells can produce artifactual results and certain markers can interfere with the binding or detection of WGA–Green. The detection of unlabeled bacterial cells by refracted light overcomes this problem.
9. If there is a chance that emission or reflected light spectra could overlap, then multi track analysis should be used, instead of single track.

Acknowledgments

We thank Michelle Ocana and the Center for Brain Imaging, Harvard Center for Neurodegeneration and Repair (Harvard University, Boston, MA) for use of their confocal microscope. This work was funded by National Institutes of Health grant R21AI61590-01 from the National Institute of Allergy and Infectious Diseases and by the FCT and Portuguese Fulbright Commission.

References

1. Vuong, C., Voyich, J. M., Fischer, E. R., et al. (2004) Polysaccharide intercellular adhesin (PIA) protects *Staphylococcus epidermidis* against major components of the human innate immune system. *Cell Microbiol.* **6,** 269–275.
2. Jefferson, K. K. (2004) What drives bacteria to produce a biofilm? *FEMS Microbiol. Lett.* **236,** 163–173.
3. Cramton, S. E., Ulrich, M., Götz, F., and Doring, G. (2001) Anaerobic conditions induce expression of polysaccharide intercellular adhesin in *Staphylococcus aureus* and *Staphylococcus epidermidis. Infect. Immun.* **69,** 4079–4085.
4. Kaplan, J. B., Velliyagounder, K., Ragunath, C., Rohde, H., Mack, D., Knobloch, J. K., and Ramasubbu, N. (2004) Genes involved in the synthesis and degradation of matrix polysaccharide in *Actinobacillus actinomycetemcomitans* and *Actinobacillus pleuropneumoniae* biofilms. *J. Bacteriol.* **186,** 8213–8220.
5. Mack, D., Fischer, W., Krokotsch, A., et al. (1996) The intercellular adhesin involved in biofilm accumulation of *Staphylococcus epidermidis* is a linear beta-1, 6-linked glucosaminoglycan: purification and structural analysis, *J. Bacteriol.* **178,** 175–183.
6. Maira-Litran, T., Kropec, A., Abeygunawardana, C., Joyce, J., Mark, G., 3rd, Goldmann, D. A., and Pier, G. B. (2002) Immunochemical properties of the staphylococcal poly-*N*-acetylglucosamine surface polysaccharide. *Infect. Immun.* **70,** 4433–4440.
7. Wang, X., Preston, J. F., 3rd, and Romeo, T. (2004) The *pgaABCD* locus of *Escherichia coli* promotes the synthesis of a polysaccharide adhesin required for biofilm formation, *J. Bacteriol.* **186,** 2724–2734.
8. Leriche, V., Sibille, P., and Carpentier, B. (2000) Use of an enzyme-linked lectinsorbent assay to monitor the shift in polysaccharide composition in bacterial biofilms, *Appl. Environ. Microbiol.* **66,** 1851–1856.

9. Thomas, V. L., Sanford, B. A., Moreno, R., and Ramsay, M. A. (1997) Enzyme-linked lectinsorbent assay measures N-acetyl-D-glucosamine in matrix of biofilm produced by *Staphylococcus epidermidis*. *Curr. Microbiol.* **43,** 249–254.
10. Vuong, C., Kocianova, S., Voyich, J. M., et al. (2004) A crucial role for exopolysaccharide modification in bacterial biofilm formation, immune evasion, and virulence. *J. Biol. Chem.* **279,** 54,881–54,886.
11. Costerton, J. W., Stewart, P. S., and Greenberg, E. P. (1999) Bacterial biofilms: a common cause of persistent infections. *Science* **284,** 1318–1322.

11

A Biochemical Method for Tracking Cholera Toxin Transport From Plasma Membrane to Golgi and Endoplasmic Reticulum

Heidi E. De Luca and Wayne I. Lencer

Summary

Asiatic cholera is a rapidly progressing disease resulting in extreme diarrhea and even death. The causative agent, cholera toxin, is an AB_5-subunit enterotoxin produced by the bacterium *Vibrio cholera*. The toxin must enter the intestinal cell to cause disease. Entry is achieved by the B-subunit binding to a membrane lipid that carries the toxin all the way from the plasma membrane through the *trans*-Golgi to the endoplasmic reticulum (ER). Once in the ER, a portion of the A-subunit, the A1 chain, unfolds and separates from the B-subunit to retro-translocate to the cytosol. The A1 chain then activates adenylyl cyclase to cause disease. To study this pathway in intact cells, we used a mutant toxin with C-terminal extension of the B-subunit that contains *N*-glycosylation and tyrosine-sulfation motifs (CT-GS). This provides a biochemical readout for toxin entry into the trans Golgi (by ^{35}S-sulfation) and ER (by *N*-glycosylation). In this chapter, we describe the methods we developed to study this trafficking pathway.

Key Words: Retrograde transport; ganglioside; lipid trafficking; cholera toxin; retro-translocation; *N*-glycosylation; tyrosine sulfation; endoplasmic reticulum; *trans*-Golgi network.

1. Introduction

Cholera toxin (CT) typifies the structure and function of the AB_5 toxins, all of which must enter the cytosol of host cells by traveling backwards from the plasma membrane (PM) to the endoplasmic reticulum (ER) and retro-translocating to the cytosol *(1,2)*. This pathway is unusual because membrane proteins rarely move backwards from the PM to the ER *(3,4)* and because it is dependent on toxin binding to a membrane lipid, ganglioside GM1 for CT, and not a membrane protein *(5)*.

From: *Methods in Molecular Biology, vol. 341: Cell–Cell Interactions: Methods and Protocols*
Edited by: S. P. Colgan © Humana Press Inc., Totowa, NJ

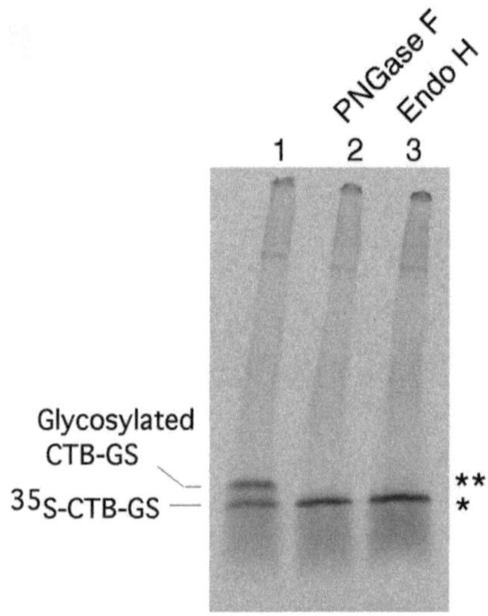

Fig. 1. Phosphorimage showing sulfation (marked by *) and *N*-glycosylation (marked by **) of CT-GS B-subunit 3 h after apical and basolateral application to T84 cells (*see* **Subheading 3.1.**). Immunoprecipitates were divided into three equal samples and treated with PNGaseF **(lane 2)**, EndoH **(lane 3)**, or mock treated **(lane 1)** before analysis by SDS-PAGE. (Reprinted with permission from **ref. *5***.)

To study the toxin pathway, we modified an approach originally developed by Rapak and Johannes ***(6,7)***. It uses a mutant toxin tagged with tyrosine-sulfation and *N*-glycosylation motifs and exploits the subcellular localization of the enzymes that catalyze these reactions in native cells. Post-translational tyrosine-sulfation of nascent proteins occurs in the *trans*-Golgi network (TGN) and *N*-glycosylation occurs in the ER ***(6,8,9)***. As such, measurement of these modifications on CT gives a biochemical measure of toxin transport from PM through TGN to ER (**Fig. 1**). In addition, because the ER-localized mutant CT is modified by *N*-glycosylation, the ER fraction of toxin can be isolated by affinity purification using lectins that bind the oligosaccharide (**Fig. 2**; ***[5]***).

Two methods are discussed in this chapter. One describes the measurement of toxin transport from PM to TGN to ER, as reported by glycosylation and sulfation, and the other describes the affinity purification of the toxin fraction localized in ER. The sulfation–glycosylation assay depends on immunoprecipitation of the B-subunit and subsequent detection of the ^{35}S-sulfated fraction of this B-subunit by sodium dodecyl sulfate polyacrylamide gel electrophoresis (SDS-PAGE) and autoradiography. *N*-glycosylation is detected by an *N*-glycanase-sensitive shift in apparent molecular weight as observed by SDS-PAGE. In some cells, only a very small fraction of toxin moves through

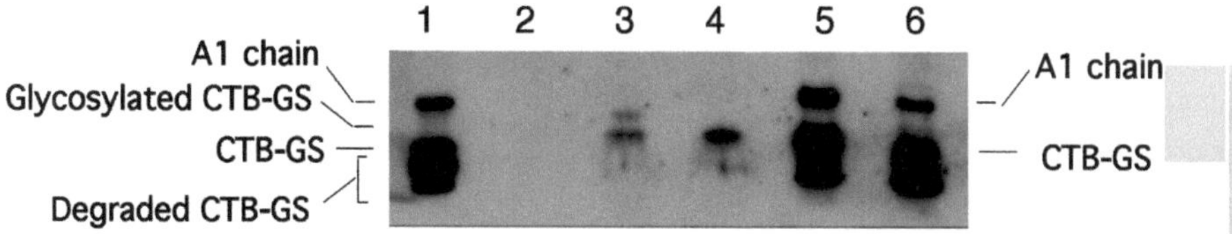

CT-GS	-	-	-	+	+	Single Pull Down
	+	+	+	-	-	Double Pull Down
	4	37	37	4	37	Incubation (°C)
	-	-	+	-	-	PNGaseF

Fig. 2. Immunoblot for A- and B-subunits after double affinity precipitation (*see* **Subheading 3.2.**, **lanes 2–4**) or single precipitation by GM1 beads alone (**lanes 5** and **6**) in Vero cells. **Lane 4** shows the same sample as in lane 3 treated with PNGaseF. **Lane 1** shows purified CT-GS as control. (Reprinted with permission from **ref. 5**.)

the Golgi and into the ER. In this method, because only the fraction of toxin that traveled through the Golgi is radiolabeled, the assay is extremely sensitive and specific for Golgi/ER transport.

The second method, isolation of the ER glycosylated toxin, describes a technique for affinity isolation of the small fraction of the glycosylated B-subunit. This method depends on two sequential affinity precipitations. The first one isolates all the *N*-glycosylated proteins with a glucose/mannose specific lectin, and the second one isolates the subfraction of toxin B-subunit from this collection of *N*-glycosylated proteins. The double-affinity precipitation method proved to be highly specific and allowed us to determine the conformation of toxin in the ER. Since protein disulfide isomerase can interact only with the proteolytically activated CT A-subunit to unfold and dissociate the A1 chain from the B-subunit, the CT-GS toxin used in this assay is activated by trypsin incubation before use.

2. Materials

See **Table 1** for a summary of buffers used in these assays.

2.1. Sulfation and Glycosylation Assay

1. Hank's Balanced Salt Solution (HBSS) without sulfate: Dissolve dry HBSS without calcium chloride, magnesium sulfate, phenol red, and sodium bicarbonate (Sigma, St Louis, MO) in 900 mL of water. Adjust to 1 m*M* magnesium chloride, 1 m*M* calcium chloride, and 10 m*M* HEPES final concentration. Adjust the pH to 7.4, bring volume to 1 L with water, and store at 4°C for up to 4 d (*see* **Note 1**).
2. HBSS with sulfate: dissolve dry HBSS without phenol red and sodium bicarbonate (Sigma) in 900 mL of water. Adjust to 10 m*M* HEPES final concentration, adjust the pH to 7.4, bring volume to 1 L with water, and store at 4°C for up to 4 d.

Table 1
Summary of Buffers Used in the Assays

Buffer	Buffer name	Composition	Protocol steps
A	Lysis stock buffer	20 m*M* Tris-base, 150 m*M* NaCl, 5 m*M* EDTA, 20 m*M* Triethanolamine, pH to 8.0 with HCl	
B	SDS-based lysis buffer	0.5% (w/v) SDS, 0.2% (w/v) bovine serum albumin in lysis stock buffer (Buffer A)	**Subheading 3.1.2., step 1**
C	Triton dilution buffer	2.5% (v/v) Triton X-100 (Sigma) in lysis stock buffer (Buffer A)	**Subheading 3.1.2., step 3**
D	Mixed micelle buffer	1% (v/v) Triton X-100, 0.2% (w/v) SDS, 8% (w/v) sucrose in lysis stock buffer (Buffer A)	**Subheading 3.1.2., step 10**
E	Sucrose final wash buffer	8% (w/v) sucrose in lysis stock buffer (Buffer A)	**Subheading 3.1.2., steps 10** and **11**
F	TN lysis buffer	10 m*M* Tris-base, pH 7.4, 150 m*M* NaCl, 5 m*M* EDTA, 1% (v/v) Triton X-100, 1.75% (w/v) NOG, 0.2% (w/v) gelatin, 0.1% sodium azide	**Subheading 3.2.2., steps 5** and **6**
G	ConA-binding buffer	10 m*M* Tris-base, pH 7.4, 6 m*M* $CaCl_2$, 6 m*M* $MnCl_2$, 6 m*M* $MgCl_2$, 0.5 *M* NaCl, 0.2% (w/v) gelatin	**Subheading 3.2.3., step 1**
H	TN wash buffer for ConA beads	10 m*M* Tris-base, pH 7.4, 0.5 *M* NaCl, 1% (v/v) Triton X-100, 1.75% (w/v) NOG, 1 m*M* $CaCl_2$, 1 m*M* $MnCl_2$, 1 m*M* $MgCl_2$, 0.1% (w/v) sodium azide in water	**Subheading 3.2.3., steps 2–4**
I	TN wash buffer for GM1 beads	10 m*M* Tris-base, pH 7.4, 0.15 *M* NaCl, 1% (v/v) Triton X-100, 1.75% (w/v) NOG, 1 m*M* $CaCl_2$, 1 m*M* $MnCl_2$, 1 m*M* $MgCl_2$, 0.1% (w/v) sodium azide in water	**Subheading 3.2.3., step 6**

SDS, sodium dodecyl sulfate; NOG, octyl β-D-glucopyranoside; TN, Triton-NOG; ConA, Concanavalin A.

3. ^{35}S-Sodium Sulfate (Sulfur-35 Radionuclide ~1000 Ci/mMole, Perkin Elmer, Boston, MA; *see* **Note 2**). All radioactive precautions must be taken when working with any kind of radioactivity, including ^{35}S. We routinely use a radioactive beta shield when working with this nuclide, in addition to gloves, goggles, and a lab coat. All radioactive waste must be disposed of properly and according to each institution's guidelines.
4. Sepharose CL-4B (Sigma) is washed three times in phosphate-buffered saline (PBS) and resuspended to 50% slurry in PBS/0.02% sodium azide before use. Wash sepharose by spinning at 3000*g* to pellet, remove supernatant, add wash buffer to fill tube, and rotate 1 to 2 min at 4°C. Spin again to pellet beads and repeat wash twice more. After the final wash, remove as much wash buffer as possible with a gel loading tip or small gage needle, and add wash buffer to create 50% slurry of the sepharose (*see* **Note 3**).
5. Protein A sepharose (Amersham Biosciences, Piscataway, NJ) is washed three times in PBS and resuspended to 50% slurry in PBS before use. Wash sepharose by spinning at 3000*g* to pellet, remove supernatant, add wash buffer to fill tube and rotate 1 to 2 min at 4°C. Spin again to pellet beads, and repeat wash twice more. After the final wash, remove as much wash buffer as possible with a gel loading tip or small gauge needle, and add wash buffer to create 50% slurry of the sepharose.
6. Lysis stock buffer (Buffer A): 20 m*M* Tris-base 150 m*M* NaCl, 5 m*M* EDTA, 20 m*M* triethanolamine, pH 8.0. The stock buffer can be stored for 3 mo at 4°C.
7. SDS lysis buffer (Buffer B): 0.5% (w/v) SDS, 0.2% (w/v) bovine serum albumin in Lysis stock buffer (Buffer A). Buffer B must be kept at room temperature; otherwise, the SDS will precipitate out of solution. It can be stored at 4°C but must be warmed to room temperature to redissolve SDS crystals. Add protease inhibitors just before use (*see* **Note 4**).
8. Triton dilution buffer (Buffer C): 2.5% (v/v) Triton X-100 in Buffer A.
9. Mixed micelle buffer (Buffer D): 1% (v/v) Triton X-100, 0.2% (w/v) SDS, and 8% (w/v) sucrose in Buffer A.
10. Sucrose final wash buffer (Buffer E): 8% (w/v) sucrose in Buffer A.
11. Rubber cell scraper (Fisher Scientific, Pittsburgh, PA).
12. 2X Laemmli sample buffer (Bio-Rad, Hercules, CA): add dithiothreitol or 5% (v/v) β-mercaptoethanol to sample buffer immediately before use. β-mercaptoethanol should be handled in a fume hood with gloves because of its toxicity and strong odor.
13. 10 to 20% Tris-HCl ready gel and gel running and transfer apparatus including cells, electrodes, and power supply (Bio-Rad). Precautions should be taken to avoid any electric shock stemming from the power supply unit. Carefully read the instructions of the power supply before use. Always ensure the power supply is off before connecting or disconnecting the electrodes to reduce the risk of electric shock.
14. Gel fixation buffer: 30% (v/v) methanol, 7% (v/v) acetic acid in water.
15. Rabbit anti-CT B-subunit antibody (3–5 mg/mL *[10,11]*) can be aliquoted and stored at –80°C.
16. CTB-GS and CT-GS protein (0.1–3 mg/mL *[5]*) aliquoted and stored at –80°C.

17. Radioactive containment box (*see* **Note 5**).
18. Gel-drying apparatus.
19. PhosphoImager.
20. Six-well tissue culture plate, with 35 mm diameter wells (Fisher Scientific).

2.2. Capture and Isolation of ER Glycosylated Toxin

1. *N*-*p*-Tosyl-L-phenylalanine chloromethyl ketone (TPCK)-treated trypsin (Sigma).
2. Soybean trypsin inhibitor (Sigma): We prepare a stock solution of 0.7 mg/25 μL in water just before use.
3. Triton-NOG lysis buffer (Buffer F): 10 m*M* Tris, pH 7.4, 150 m*M* NaCl, 5 m*M* EDTA; 1% (v/v) Triton X-100, 1.75% (w/v) octyl β-D-glucopyranoside (NOG), 0.2% (w/v) gelatin. Store at 4°C and use within 2 d. Add protease inhibitors just before use.
4. Concanavalin A (ConA)-binding buffer (Buffer G): 10 m*M*, Tris-base pH 7.4, 6 m*M* $CaCl_2$, 6 m*M* $MnCl_2$, 6 m*M* $MgCl_2$, 0.5 *M* NaCl, and 0.2% (w/v) gelatin. Store at 4°C and use within 2 d. Add protease inhibitors just before use.
5. ConA sepharose beads (Amersham Biosciences, Piscataway, NJ) are prewashed three times in a 1:1 solution of Buffer F:Buffer G. Wash sepharose by spinning at 3000*g* to pellet, remove supernatant, add wash buffer to fill tube, and rotate 1 to 2 min at 4°C. Spin again to pellet beads, and repeat wash twice more. After the final wash, remove as much wash buffer as possible with a gel loading tip or small gage needle, and add wash buffer to create 50% slurry of the sepharose.
6. Triton-NOG wash buffers (Buffers H and I): Start with a stock buffer of 10 m*M* Tris-base, pH 7.4; 1% (v/v) Triton X-100, 1.75% (w/v) NOG, 1 m*M* $CaCl_2$, 1 m*M* $MnCl_2$, and 1 m*M* $MgCl_2$ in water. For ConA bead wash buffer (Buffer H), add 0.5 *M* NaCl. For GM1 bead wash buffer (Buffer I), add 0.15 *M* NaCl. Store solutions at 4°C and use within 2 d. Add protease inhibitors just before use.
7. 0.2 *M* methyl-α-D-mannopyranoside (Sigma) is dissolved in Buffer H.
8. PBS: Prepare 10X stock with 1.48 *M* NaCl, 14 m*M* KH_2PO_4, 81 m*M* Na_2HPO_4, 26.8 m*M* KCl, adjust pH to 7.4, and bring volume to 1 L. Filter sterilize into an autoclaved bottle, and store at room temperature. Prepare working 1X solution by diluting one part 10X stock with nine parts water.
9. GM1 beads are prepared with lyso-monosialoganglioside GM1 (Matreya, Pleasant Gap, PA) and Dynabeads® M-270 Carboxylic Acid (Dynal, Oslo, Norway) according to Dynal's manufacturer instructions for bead coupling. The coupled beads are then washed in PBS three times before they are added to the sample. The beads are easily washed using the capture magnet and successive removal and addition of 1 mL of PBS. After the washes, bring the beads up to the original volume in PBS.
10. Peptide: *N*-Glycosidase F (PNGase F, Glycerol free) and associated buffers (New England Biolabs, Beverly, MA).
11. SDS loading buffer (6X): Mix 7 mL of 4X Tris-HCl/SDS pH 6.8 (0.5 *M* Tris-HCl containing 0.4% (w/v) SDS), 3 mL of glycerol, 1 g of SDS, 0.93 g of dithiothreitol, and 1.2 mg of bromophenol blue. Bring the volume to 10 mL with water, divide into 0.5-mL aliquots, and freeze at –80°C.

12. SDS running buffer: Make a 10X stock of 250 m*M* Tris, 1.92 *M* glycine, and 1% (w/v) SDS in water. DO NOT adjust pH. For a working stock, mix one part 10X buffer with nine parts water. Store 10X and 1X buffers at room temperature.
13. Transfer buffer: 200 mL of methanol, 3 g of Tris-base, 14.4 g of glycine are added to 800 mL of water. Mix and store at 4°C to chill before use. This solution is made fresh for every gel, but a 10X stock can be made with the salts and then diluted with methanol and water before use. The solution should be cooled to 4°C before use to help minimize any temperature increase during the transfer.
14. Tris-buffered saline (TBS): Make a 10X stock of TBS with 500 m*M* Tris 2 *M* NaCl, pH 7.4 in water. Prepare a working stock of 1X TBS by adding one part 10X TBS to nine parts water. Both 10X and 1X stocks can be stored at 4°C for 4 mo.
15. TBS with Tween-20 (TTBS; Sigma): Prepare a working stock of 1X TTBS by adding 1% (v/v) Tween-20 to 1X TBS. Both 10X and 1X stocks can be stored at 4°C for 4 mo.
16. Blocking buffer: Add 5% (w/v) of nonfat dry milk to TTBS. Stir 30 min at room temperature and filter into a vacuum flask using a büchner funnel and a piece of filter paper. Store up to 3 d at 4°C.
17. Anti-rabbit IgG (whole molecule)–peroxidase antibody produced in goat (secondary antibody, Sigma).
18. Enhanced chemiluminescent detection kit (Pierce Biotechnology, Rockford, IL).
19. Round tissue culture dishes, 60-mm diameter. The actual growth area diameter of these dishes is approx 50 mm (Fisher Scientific, Pittsburgh, PA).

3. Methods

The mutant cholera toxin used in the following assays was developed by Jobling and Fujinaga ***(5)***. The C-terminal motifs were specifically designed to provide maximum opportunity for glycosylation and sulfation and to exert minimal effect on normal cholera toxin function in the cell. Theories regarding optimal placement of these motifs, linker lengths, and conformation were considered when designing the mutant toxin. Functional tests revealed a minimal change in toxicity of the mutant toxin compared to the wild type.

The following methods can be modified to fit other cell or biological systems. Depending on the binding affinity of the toxin to the cell receptor, the conditions and incubation times should be modified to achieve maximal saturation binding at steady state. Although there are no specific cleavage sites within the linker, the appended tag is subject to nonspecific proteolysis. In some cell or in vivo systems, the addition of protease inhibitors is crucial to maintain the full-length tag to allow affinity precipitation with the lectin. To control for the isolation of the ER glycosylated toxin, a single affinity precipitation with GM1 alone is conducted with a fraction of the lysate. This will reveal, by immunoblot, the entire B-subunit toxin that was taken up by the cell during the incubation and control for any problems that may have occurred during the toxin uptake steps.

3.1. Sulfation and Glycosylation Assay

3.1.1. Depletion of Cell Sulfate and Toxin Incubation

1. Seed T84, Vero, Cos, or Y1 cells into a six-well plate and grow to 90 to 100% confluency.
2. Wash each well three times with 4 mL of 37°C HBSS without sulfate. This is easily done by aspirating off the media with a 5-mL pipet, gently adding the HBSS wash, and then repeating this process twice.
3. Add 4 mL of HBSS without sulfate to each well and incubate at 37°C for 30 min in a CO_2 incubator.
4. Aspirate off the HBSS without sulfate and replace with 4 mL fresh HBSS without sulfate. Incubate for 1 h at 37°C in a CO_2 incubator.
5. Aspirate the HBSS without sulfate and replace with 1 mL of 0.5 mCi/mL ^{35}S-sodium sulfate in HBSS without sulfate. Incubate the cells for 30 min at 37°C in a CO_2 incubator using a radioactive containment box to prevent contamination.
6. Add CTB-GS or CT-GS to a final concentration of 20 n*M* (or appropriate) and incubate for desired time (30 min to overnight) at 37°C in CO_2 incubator, using a radioactive containment box.
7. Aspirate off and properly dispose of ^{35}S toxin solution.
8. Wash cells three times with 3 mL of 4°C HBSS with sulfate. Add wash buffer to cells by gently pipetting against the wall of the well and gently aspirate off wash buffer. This technique helps prevent the cells from being washed off the well bottom.

3.1.2. Immunoprecipitation With CT antibody

1. Remove the last wash buffer from the wells and add 600 µL of room temperature Buffer B.
2. Using a rubber cell scraper, scrape the cells off the well bottom and collect in a 2-mL screw cap tube (*see* **Note 6**).
3. Heat the tube to 100°C for 5 min, then transfer the tube to ice, and add 600 µL of ice-cold Buffer C. This is an appropriate stopping point for the assay. If desired, after the addition of Buffer C, the samples can be stored at –80°C for later processing.
4. To fracture the DNA, put the sample tubes in an automatic vortex at 4°C, and vortex at maximum speed for 20 min. Pass the sample through a 25-gage needle five to six times to disrupt the DNA, and transfer the sample to a new tube.
5. Clarify the DNA by adding 30 µL of 50% Sepharose CL-4B slurry. Tumble sample for 5 min, spin at maximum speed (16,000g) for 3 min, and transfer the supernatant to a new tube. Repeat this clarification step one more time.
6. Preadsorb the sample by adding 15 µL of 50% protein A bead slurry. Tumble sample for 15 min at 4°C, spin at maximum speed for 1 min, and transfer the supernatant to a new tube.
7. Dilute the anti-CTB stock antibody 1:30 in PBS and add 100 µL of this to each tube. Tumble at 4°C for 1 h.

8. Add 15 µL of 50% protein A bead slurry to each tube and tumble at 4°C for 1 h to overnight.
9. Spin the sample at maximum speed for 1 min. Carefully remove the supernatant and store at –80°C (in case of recapture).
10. Wash the protein A bead pellet in 1 mL of ice-cold Buffer D. Add 1 mL of buffer to the pellet, incubate on ice for 2 min, and spin at maximum speed for 1 min. Remove the supernatant and repeat the wash four times for a total of five washes, then repeat the wash step once more with Buffer E.
11. Remove as much Buffer E as possible with a gel-loading tip or small gauge needle and then add 20 µL of 2X Laemmli sample buffer to the bead pellet. Use a 30-gage needle to make a hole in the top of the tube (to avoid explosion), and boil for 5 min.
12. Spin at maximum speed for 5 min and carefully remove the supernatant. This is the final sample and can be stored at –80°C until further processing.

3.1.3. SDS-PAGE and Phosphoimaging

1. Run the samples on a 10 to 20% gradient SDS-PAGE gel with approx 6–12 µL per lane for a phosphoimage. (For a Western blot, load 2–3 µL per lane. For more detailed SDS-PAGE and Western blotting instructions, *see* **Subheading 3.2.5.**)
2. Incubate the gel in fixation solution at room temperature for 1 h with gentle shaking.
3. Soak the gel in water at room temperature for 30 min with gentle shaking. During the 30-min incubation, change the water four times.
4. Dry the gel down onto a piece of filter paper using a vacuum driven gel-drying apparatus (or other quick gel-drying method).
5. Expose the dried gel in a PhosphoImager plate for 3 d to 2 wk and then read the plate on a PhosphoImager (*see* **Note 7**).

3.2. Isolation of ER Glycosylated Toxin

3.2.1. Nicking Toxin With Trypsin Treatment

1. Incubate 12 µg of CT-GS with 6 µg of TPCK treated trypsin for 30 min at 37°C. We usually add 6 µL of a 1 mg/mL solution of TPCK trypsin to 12 µg of CT-GS in PBS to a final volume of 180 µL for the incubation.
2. After the incubation, add 0.56 mg of soybean trypsin inhibitor to stop the trypsin digest. Bring the final volume to 12 mL with HBSS with sulfate to yield a solution of 10 n*M* CT-GS.

3.2.2. Toxin Incubation and Cell Lysis

1. Seed cells on a 6 cm^2 dish and grow until 90 to 100% confluent. For Vero cells, we typically seed them in the plates and allow them to grow for 2 to 3 d to adhere to the plate.
2. Wash the adherent cells three times with cold HBSS with sulfate. This is easily done by gently aspirating off the media, adding 2–3 mL of HBSS, and then aspirating off HBSS. Take care to do these steps gently because any turbulence may lift the cells off the plate.

3. Add 10 n*M* of the nicked CT-GS to the plate. This is usually done in a 1-mL volume. Incubate the cells for the desired time, realizing that even if the uptake incubation is for a short time, the internalization incubation needs to be 3–5 h at 37°C for glycosylation to occur (*see* **Note 8**).
4. After the incubation, wash the adherent cells four times with HBSS with sulfate. The first wash should be done with 37°C HBSS, and the subsequent three with 4°C HBSS. This will help to release any CT-GS bound to membranes, thus giving a cleaner pulldown.
5. Add 1 mL of Buffer F to each dish and let sit 10 min on ice. Scrape off the cells with a rubber cell scraper, and collect in a 1.5-mL Eppendorf tube. Lyse cells by pipetting up and down with a 1-mL tip seven to eight times.
6. Spin the lysate at 6000*g* for 10 min at 4°C to pellet insoluble material. Remove the supernatant (~1 mL) and transfer it to a new tube. Reserve 20 µL of the supernatant to run as "lysate" on the SDS-PAGE gel.

3.2.3. Double Pulldown With ConA and GM1

1. Add an equal volume of Buffer G to the lysate. Add 100 µL of ConA beads to the tube. Rotate for 1 h at 4°C (*see* **Note 9**).
2. Spin the samples at 3000*g* for 5 min at 4°C. Remove and save the supernatant (in case of recapture). Wash the ConA bead pellet with Buffer H. Add 1 mL of Buffer H to the bead pellet, resuspend the pellet for 1 to 2 min on a mixer or by inverting, then spin at 3000*g* for 5 min to pellet the beads. Repeat the wash three more times.
3. Use a small-gage needle or gel loading tip to remove as much buffer as possible from the bead pellet.
4. Elute the glycosylated toxin by adding 50 µL of 0.2 *M* methyl α-D-mannopyranoside in Buffer H to the beads. Rotate the sample for 1 h at 4°C for maximal elution. Spin the sample at 3000*g* for 5 min at 4°C. Collect the supernatant and save the beads in case of re-elution. If applicable, combine supernatants into one tube (*see* **Note 10**).
5. Bring the volume of the eluted sample to 1 mL with PBS (with protease inhibitors). Add 40 µL of GM1 beads. Rotate the samples overnight at 4°C (*see* **Note 11**).
6. Wash the GM1 beads four times with Buffer I. Each wash should allow the beads to rotate with Buffer I for 1 to 2 min. Using a small-gage needle or gel-loading tip, remove as much of the wash buffer as possible.

3.2.4. PNGase Treatment

1. To load sample directly onto an SDS-PAGE gel, add 20 µl 2X Laemmli sample buffer to GM1 beads and boil 10 min. Isolate the supernatant with the capture magnet. Load 3 µL of this sample onto an SDS-PAGE gel for a Western blot.
2. To proceed with a PNGase digest, add 20 µL of 2X denaturing buffer from the PNGase kit and boil 10 min. Using the capture magnet, remove the supernatant. Follow manufacturer's instructions for PNGase treatment. We typically mix 3 µL of eluted sample together with 2 µL of NP-40, 1 µL of 10X G7 buffer, 1 µL of

PNGase, 3 μL of water, and incubate at 37°C for 2 h. In parallel, prepare and incubate a control digest containing all the reagents of the digest with the exception of substituting water for PNGase. After digestion, add two microliters of 6X sample buffer to the digested samples, and boil 10 min. The entire 12 μL is loaded onto a reducing SDS-PAGE gel and Western blotted for CTB.

3.2.5. SDS-PAGE and Western Blot

1. The following SDS-PAGE instructions assume the use of a Bio-Rad Mini Protean Cell or Criterion Cell and Mini Trans Blot Cell or Criterion Blotter. If using other systems, follow the manufacturer's instructions. Load 3 μL of the undigested sample, 12 μL of the digested samples (± PNGase), and 2 μL of the reserved lysates into separate lanes of a 10 to 20% Tris-HCl ready gel (Bio-Rad). Assemble the gel apparatus and fill the gel box with SDS running buffer. Connect the gel box to a power supply and run at approx 30 to 35 mA per gel until the dye front has just reached the bottom of the gel.
2. After the gel is completed, disassemble the apparatus, remove the gel, and place it in a large (9 × 13 in) pan filled with transfer buffer. Open the transfer cassette and submerge the black side of the cassette in the pan. Place one pre-wet sponge onto the black side of the cassette (this should be the interior of the cassette), and layer successively one piece of filter paper, the gel, one piece of nitrocellulose, one piece of filter paper, and one sponge onto the cassette. Using the provided roller, very gently roll across the top of the transfer sandwich to remove any air bubbles. Close the cassette, assemble it the transfer cell with the black side of the cassette toward the black electrode, and fill the cell with transfer buffer. Connect to a power supply and transfer at 100 V for 45 min to 1 h (*see* **Note 12**).
3. After the transfer turn off the power, disassemble the transfer cell, remove the nitrocellulose blot, and properly dispose of the transfer buffer (*see* **Note 13**). The nitrocellulose should be placed in a small incubation dish with blocking buffer and gentle shaking for a minimum of 1 h at room temperature.
4. Remove the blocking buffer, add a 1:2000 dilution of CTB antibody in blocking buffer and incubate 1 h with gentle shaking at room temperature. After the primary incubation, wash the blot three times for 5 min each with TTBS. Remove the last wash buffer.
5. Add a 1:2500 dilution of Anti-Rabbit IgG (whole molecule)–Peroxidase secondary antibody in blocking buffer to the blot, and incubate for 1 h at room temperature. After the secondary incubation, wash the blot three times for 10 min each with TTBS, then two times for 10 min each with TBS.
6. Develop the blot with a chemiluminescent substrate for detection of horseradish peroxidase. We use the Pierce supersignal detection kit (Pierce, Rockford, IL).

4. Notes

1. When making solutions with water, always use milliQ water with a resistivity of 18 MΩ. This is assumed for all solutions that are made with "water."

2. When purchasing ^{35}S-sodium sulfate from Perkin Elmer, be sure to specify "high specific activity" and request the freshest lot. We always plan the experiment to coincide with the production of a fresh lot to ensure the highest activity.
3. To create 50% slurry, we typically determine the volume of beads by filling an identical tube to the same volume of water, measuring this volume, doubling it, and adding enough wash buffer to the tube with beads to reach the doubled volume.
4. We typically use an EDTA-free protease inhibitor tablet from Roche Diagnostics (Indianapolis, IN). The tablets are available in sizes to accommodate 10 or 50 mL. Various protease inhibitors can also be added individually to the solutions.
5. The radioactive containment box is used to prevent radio contamination of the CO_2 incubator. The box is made by affixing a piece of charcoal paper to a cut out section of the lid of a square (2.4 L) Tupperware® container. Charcoal pellets are added to the bottom of the container. The plate of cells with ^{35}S-sodium sulfate is placed on top of the pellet layer, and the lid is closed. This device helps to prevent ^{35}S contamination of the incubator.
6. It is necessary to use a screw cap tube because of the subsequent boiling step. The screw cap prevents explosion of the tube which could result in radioactive contamination of the surrounding workspace.
7. Do not use any type of plastic wrap to cover the gel because the ^{35}S signal is very weak. For T84 cells we find 3 d to 2 wk is necessary to see a signal; 1 to 3 d are enough for Vero, Cos, and Y1 cells.
8. We typically incubate with toxin for 45 min on ice, then discard the incubation buffer containing the toxin, add back 3 mL HBSS+, and incubate for an additional 4.5 h at 37°C. It is also a good idea to run a negative control in parallel, in which the entire toxin incubation is held at 4°C. Because the toxin is not taken up at this temperature, the pulldown should be negative for CT.
9. The volume of the lysate is usually 1 mL; therefore, 1 mL of ConA-binding buffer is added. If the volume of cell lysate differs, simply change the volume of ConA-binding buffer added to make a 1:1 dilution. The bed volume of 100 μL of 50% slurry of ConA beads is 50 μL.
10. A solution of up to 1 *M* methyl α-D-mannopyranoside can be used without difficulty. We have found that 0.2 *M* methyl α-D-mannopyranoside solution may not always elute the entire toxin sample off ConA beads. The 1-h incubation step can be reduced to 30 min if necessary; however, the best results and highest elution rates are observed with 1 hour incubation. It often is beneficial to repeat this elution step by adding another 50 μL of methyl α-D-mannopyranoside to the beads, rotating 1 h, spinning at 3000*g*, and combining this 50 μL with the first elution of 50 μL for a total of 100 μL.
11. An overnight incubation yields the best results for the GM1 pulldown. We suspect it takes some time for the CT to release from the membrane GM1 and bind to the GM1 beads.
12. It is imperative to follow the manufacturer's instructions regarding cooling of the tank. Typically sufficient cooling is achieved by either adding a frozen ice block to

the transfer cell, or using a cooling coil in the transfer cell. A rise in temperature will cause the gel to transfer poorly, and may also damage the transfer equipment.

13. The methanol in the transfer buffer may be considered hazardous waste in some institutions.

Acknowledgments

This work was supported by DK48106 to WIL and DK34854 to the Harvard Digestive Diseases Center.

References

1. Lencer, W. I. (2004) Retrograde transport of cholera toxin into the ER of host cells. *Int. J. Med. Microbiol.* **293,** 491–494.
2. Spangler, B. D. (1992) Structure and function of cholera toxin and the related *Escherichia coli* heat-labile enterotoxin. *Microb. Rev.* **56,** 622–647.
3. Volz, B., Orberger, G., Porwoll, S., Hauri, H. P., and Tauber, R. (1995) Selective reentry of recycling cell surface glycoproteins to the biosynthetic pathway in human hepatocarcinoma HepG2 cells. *J. Cell Biol.* **130,** 537–551.
4. Porwoll, S., Loch, N., Kannicht, C., et al. (1998) Cell surface glycoproteins undergo postbiosynthetic modification of their N-glycans by stepwise demannosylation. *J. Biol. Chem.* **273,** 1075–1085.
5. Fujinaga, Y., Wolf, A. A., Rodigherio, C., et al. (2003) Gangliosides that associate with lipid rafts mediate transport of cholera toxin from the plasma membrane to the ER. *Mol. Biol. Cell* **14,** 4783–4793.
6. Rapak, A., Falnes, P. Ø., and Olsnes, S. (1997) Retrograde transport of mutant ricin to the endoplasmic reticulum with subsequent translocation to the cytosol. *Proc. Natl. Acad. Sci. USA* **94,** 3783–3788.
7. Johannes, L., Tenza, D., Antony, C., and Goud, B. (1997) Retrograde transport of KDEL-bearing B-fragment of shiga toxin. *J. Biol. Chem.* **272,** 19,554–19,561.
8. Gavel, Y. and von Heijne, G. (1990) Sequence differences between glycosylated and non-glycosylated Asn-X-Thr/Ser acceptor sites: implications for protein engineering. *Protein Eng.* **3,** 433–442.
9. Bundgaard, J. R., Vuust, J., and Rehfeld, J. F. (1997) New consensus features for tyrosine O-sulfation determined by mutational analysis. *J. Biol. Chem.* **272,** 21,700–21,705.
10. Lencer, W. I., Constable, C., Moe, S., et al. (1995) Targeting of cholera toxin and *E. coli* heat labile toxin in polarized epithelia: role of C-terminal KDEL. *J. Cell Biol.* **131,** 951–962.
11. Lencer, W. I., Moe, S., Rufo, P. A., and Modara, J. L. (1995) Transcytosis of cholera toxin subunits across model human intestinal epithelia. *Proc. Natl. Acad. Sci. USA* **92,** 10,094–10,098.

12

Isolation and Culture of Murine Heart and Lung Endothelial Cells for In Vitro Model Systems

Yaw-Chyn Lim and Francis W. Luscinskas

Summary

The inflammatory response is a critical component of host defense. An important goal of our group has been to understand the endothelial-dependent mechanisms that mediate leukocyte recruitment during an inflammatory response. In this chapter, we present a detailed method for the isolation and in vitro culture of murine vascular endothelial cells from the lung and heart. The endothelial cells are of high purity (85–99%) and retain certain of their functional differences, including constitutive and cytokine inducible adhesion molecule expression and chemokine production. Such endothelial cells, therefore, can be used for various in vitro models including leukocyte adhesion assay or in depth biochemical analyses.

Key Words: Inflammation; immunity; leukocytes; in vitro culture; cell heterogeneity.

1. Introduction

The endothelial monolayer lining all blood vessels plays an active role in the recruitment of blood leukocytes to tissues and organs in response to infection, injury, and immune responses. It is well documented that during an inflammatory response, the endothelium becomes "activated" by inflammatory cytokines or bacterial endotoxins and regulates the expression of adhesion molecules and the production of chemokines that mediate blood leukocyte adhesion and transmigration *(1,2)*. Recent studies reveal that the endothelium is not homogenous because endothelial cells within various vascular beds and organs exhibit different cellular and biochemical properties that are specifically adapted to meet the requirements of their local environment *(3–6)*.

Leukocyte recruitment is a critical component of the inflammatory and immune response systems. However, when recruitment persists or becomes dysregulated, it can lead to numerous pathological conditions, including allergic reactions,

From: *Methods in Molecular Biology, vol. 341: Cell–Cell Interactions: Methods and Protocols*
Edited by: S. P. Colgan © Humana Press Inc., Totowa, NJ

reperfusion injury, organ rejection, and atherosclerosis and result in massive tissue and organ damage. Many studies have been devoted to the elucidation of the molecular and biochemical signals that mediate these interactions in attempt to formulate effective therapeutic interventions. Much of the knowledge on this complex phenomenon is generated from well-designed in vitro studies that use isolated human leukocytes and endothelial cells, such as human umbilical vein endothelial cells, in their model systems.

To study the mechanisms that initiate and propagate the immune and inflammatory responses, investigators increasingly have turned to gene knockout approaches to create mice deficient in one or more molecules implicated in these processes ***(7–9)***. However, the ability to set-up useful in vitro models using endothelium from these knockout mice remains limited because the methods to isolate and culture murine endothelial cells have met with varying success. This in part is caused by the amount of tissue available from a mouse as well as the difficulties isolating pure populations of endothelial cells. Nevertheless, the availability of a simple, reproducible, and well-tested method to isolate murine endothelial cells would be useful because it would enable us to culture cells from these genetically modified animals for use in various in vitro model systems ***(5,10–12)***.

In this chapter, we describe a detailed protocol for the isolation, culture, and characterization of murine lung and heart endothelial cells. This protocol has allowed us (and other groups) to generate endothelial cells from different mouse strains and backgrounds for in vitro models to study the role of adhesion molecules and chemokines in leukocyte–endothelial cell interactions under different biological conditions ***(5,13–15)***. A more detailed characterization of lung and heart endothelial cell phenotypes isolated using this protocol is available ***(5)***.

2. Materials

2.1. Endothelial Cell Isolation and Culture

1. Collagenase solution: Hank's Balanced Salt Solution (HBSS; Biowhittaker, cat. no. 10-543) containing 0.2 mg/mL (~180–200 U/mL) type 1 collagenase (Worthington Biochemical Corporation; cat no. 4197; *see* **Note 1**).
2. Base medium: Dulbecco's Modified Eagle's Medium (Invitrogen Corporation, cat. no. 11965-092) containing 25 m*M* HEPES (Sigma; cat. no. H0887), 20% fetal calf serum, 100 U/100mg/mL penicillin and streptomycin, and 2 m*M* L-glutamine (Invitrogen Corporation, cat. nos. 600-5140 and 320-5030, respectively; *see* **Note 2**).
3. Complete culture medium: 80 mL of base medium supplemented with heparin (Sigma, cat. no. H-3933, 100 µg/mL final concentration), endothelial cell growth stimulant (Biomedical Technologies, cat. no. BT-203, 100 µg/mL final concentration), nonessential amino acid (Invitrogen Corporation; cat. no. 11140050), and

sodium pyruvate (Invitrogen Corporation, cat. no. 11360070) at 1X final concentration, respectively.

4. Trypsin–versene mixture (0.5% trypsin–0.02% EDTA) from Invitrogen (cat. no. 610-5300) or Biowhittaker (cat. no. 17-161).
5. Cell dissociation buffer from Sigma (cat. no. C5789).
6. Precoat a 75-mm^2 flask with sterile filtered 0.1% gelatin solution (Sigma; cat. no. G1393) as follows: pipet 4 mL of 0.1% gelatin solution to generously cover the bottom of the flask, leave for 1 min, and remove any excess fluid. Air-dry flask completely (4 h under tissue culture hood) before use. Solution can be reused for coating additional flasks.
7. Dynal M-450 sheep anti-rat immunoglobulin (Ig)G magnetic beads (Dynal, cat. no. 110.07; concentration 4×10^8 beads/mL).
8. Purified rat anti-mouse CD31 (PECAM-1, clone MEC13.3, 0.5 mg/mL concentration, Pharmingen cat. no. 557355 or 553370) and rat-anti-mouse CD102 (intercellular adhesion molecule [ICAM]-2, clone 3C4, 0.5 mg/mL concentration, Pharmingen cat. no. 553326).
9. Dynal magnetic bead wash buffer (bead buffer): phosphate-buffered saline (PBS) containing 0.1% bovine serum albumin (Sigma, cat. no. A9647).

2.2. Endothelial Cell Characterization

2.2.1. Flow Cytometric Analysis (i.e., Fluorescence-Activated Cell Sorting)

1. Wash buffer/diluent–PBS containing 1% bovine serum albumin.
2. Anti-mouse CD31 conjugated to APC (Pharmingen, cat. no. 551262) and anti-mouse ICAM-2 (CD102) conjugated to fluorescein isothiocyanate (FITC; Pharmingen; cat. no. 557444). Use at 1:50 (v/v) in wash buffer/diluent.
3. Isotype-matched rat-IgG2a conjugated to FITC or APC (Caltag; cat. no. R2a01 and R2a05, respectively). Use at 1:50 (v/v) dilution.
4. Fixative: 1% formaldehyde (37% stock, Baker Chemical) diluted (v/v) in PBS, pH 7.4.

2.2.2. Fluorescent Immunoassay

1. Wash buffer/diluent–PBS containing 1% bovine serum albumin (*see* **Note 3**).
2. Rat antibodies against mouse ICAM-1 (clone YN1/1.7.4, ATCC no. CRL1878, ascites) and vascular cell adhesion molecule (VCAM)-1 (clone M/K2.7, ATCC no. CRL1909, culture supernatant) were used at 1:200 and 1:50 dilution, respectively. Purified mouse antibodies against mouse E-selectin (clone RME-1) and P-selectin (clone RMP-1) were kind gifts from Dr. Andrew Issekutz (Dalhousie University, Halifax, Canada) and were used at 10 µg/mL final concentration (*see* **Note 3**).
3. Negative control was rat IgG (Caltag Laboratories; cat. no. 10700; 10 µg/mL) as nonbinding MAb control. An irrelevant nonbinding mouse monoclonal antibody (clone K16/16, IgG1 nonbinding control; hybridoma culture supernatant *[16]*) was used undiluted.

4. FITC-conjugated anti-rat IgG (Caltag Laboratories, cat. no. M30101) or FITC-conjugated anti-mouse IgG (Caltag Laboratories, cat. no. M30101) was used at 1:80 dilution.

2.3. Mice

Optimal results were obtained using three mice of 7 to 9 d of age. We used mice from either C57/BL6 or the 129 background (Taconics Farm, Germantown, NY). The animals were sacrificed by carbon dioxide asphyxiation as approved by the panel on euthanasia at the American Veterinary Association (*see* **Note 4**).

2.4. Instruments and Disposable Supplies for Organ Isolation and Tissue Dissociation

All metal instruments were sterilized by autoclaving; sterile disposables were from B-D Falcon.

2.4.1. Animal Dissection

1. Two pairs of scissors and two pairs of forceps. Stand the instruments in 70% ethanol (to minimize contamination) when not in use.

2.4.2. Tissue Dissociation

1. Two pairs of scissors.
2. Two pairs of forceps.
3. 6-in. long 14-G metal cannula (Fisher Scientific, 6-in. popper needles; cat. no. 1482516N).
4. 70-μm pore size cell strainer (B-D Falcon, cat. no. 352350).
5. 50-mL screw cap centrifuge tubes (B-D Falcon, cat. no.352098).
6. 100-mm diameter cell culture dishes (B-D Falcon, cat. no.353003).
7. 5-mL round-bottom tube with snap-cap (B-D Falcon, cat. no. 352058).

3. Methods

3.1. Endothelial Cell Isolation and Culture

3.1.1. Precoating of Anti-Rat IgG Dynal Beads

In accordance with the manufacturer's recommendations, our standardized protocol for the Dynal bead coating used 1.25 μg of antibody per 10^7 beads. The bead concentration was maintained at 4×10^8 beads per milliliter.

1. Resuspend an aliquot of the Dynal magnetic beads in 1 mL of PBS + 0.1% bovine serum albumin (bead wash buffer), mix well, and place on Magnetic separator (Dynal MPC-S).
2. Remove supernatant and repeat wash three times.
3. Resuspend beads in original volume with bead wash buffer.
4. Add 5 μL (2.5 μg) of antibody for every 50 μL (2×10^7) of beads.

5. Incubate bead suspension on an end-over-end rotator overnight at 4°C or for 3 h at room temperature.
6. Repeat wash (**steps 1** and **2**) four times.
7. Resuspend immunobeads in the original volume with bead wash buffer (*see* **Note 5**). Store beads at 4°C and use within 1 to 2 wk.

3.1.2. Dissection

1. Sacrifice the mice and totally submerge the carcass into 70% ethanol, taking care that the skin and fur of the animal is soaked through.
2. Place two paper towels on a Styrofoam board and pin the mouse supine.
3. Pick up the skin above the pubis with a pair of forceps, make a small incision in the skin with a pair of scissors, and deglove the skin to above the chest.
4. Spray chest/abdomen with 70% ethanol and remove any loose fur from gloves and surgical field.
5. Cut open the anterior abdomen and take down the anterior diaphragm.
6. Grip the rib cage with the forceps and cut along the lateral walls of the chest.
7. Flip the "freed" rib cage back and pin away from the exposed thoracic cavity (*see* **Fig. 1A**).
8. Switch to a second set of dissection instruments.
9. Grasp the mediastinum and carefully cut the tissues and organs away from the vertebral column. Place the excised tissue (*see* **Fig. 1B**) into a 50-mL tube containing 30 mL of ice-cold base medium.
10. Repeat dissection procedure with the other two mice.

3.1.3. Tissue Dissociation

Transfer the tissues to a sterile hood for the remainder of the protocol.

1. Using fresh instruments, transfer the tissues to a sterile, disposable 100-mm diameter tissue culture dish.
2. Grasp the heart, cut it away from the mediastinal tissue and remove most of the left and right atria to reduce contamination of endothelial cells from large vessels (*see* **Fig. 1C**).
3. Make a central cut in the organ to open up the ventricles and put it into a fresh tube containing 30 mL of base medium.
4. Dissect out the lobes of the lung carefully; taking care to cut away any visible bronchi and mediastinal tissues (*see* **Fig. 1D**).
5. Cut the larger lobes of the lung into two and place them into another tube containing 30 mL of base medium.
6. Repeat with the mediastinum from the other two mice and pool the organs in their respective tubes. Label tubes to avoid cross contamination of organs.
7. Gently agitate the tubes (1 min) to wash out excess red blood cells from the organs.
8. Place the hearts on a new, dry 100-mm diameter dish and mince finely with scissors (*see* **Note 6**).

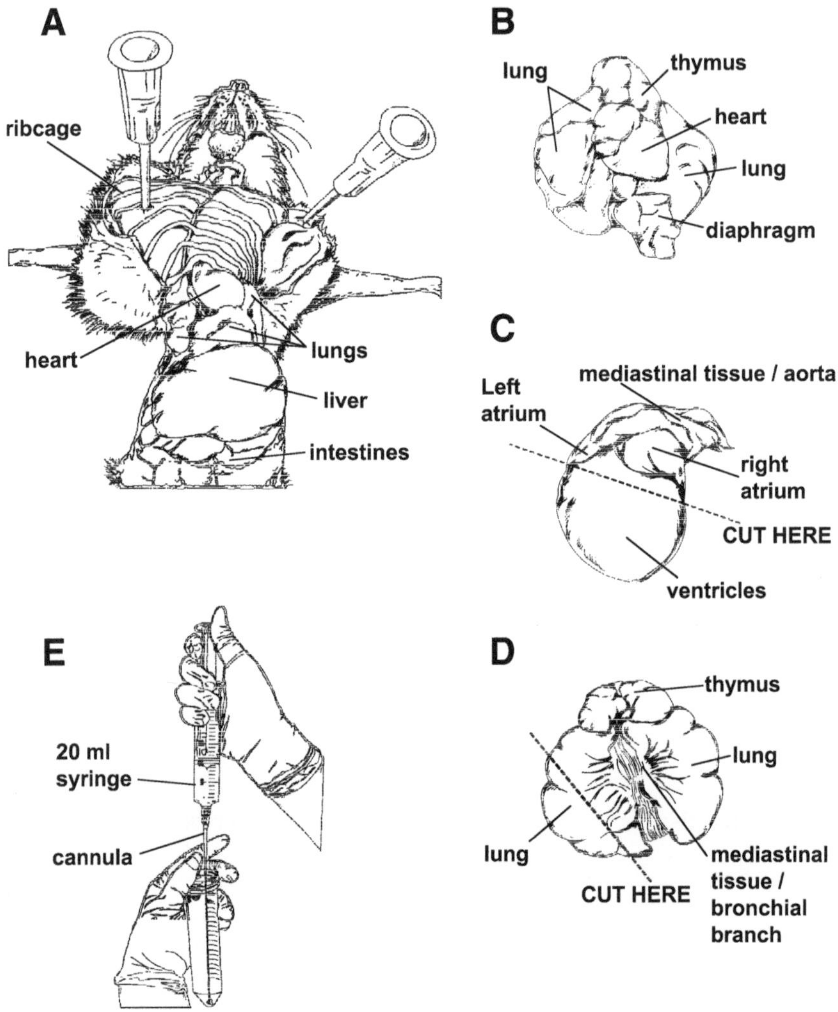

Fig. 1. Harvesting of organs and tissue dissociation. **(A)** The sacrificed mouse is pinned supine onto a styrofoam board and the rib cage is cut open and pinned away from the thoracic cavity; **(B)** the lungs, heart, and mediastinal tissues are carefully removed from the thoracic cavity and placed in a 50-mL tube containing cold base medium; **(C)** the ventricular heart tissue is carefully cut away from the atria and the large vessels; **(D)** the lung lobes are carefully dissected out to minimize contamination by the bronchial tree; and **(E)** the digested tissue is triturated multiple times to break up the cell clumps forming a single cell suspension.

9. Transfer the minced heart tissues to a 50-mL tube containing 15 mL of prewarmed collagenase solution. Wash all the tissue bits off the dish and into the tube of collagenase.
10. Repeat **steps 8** and **9** with the lung tissues.
11. Incubate at 37°C with gentle agitation for 45 min.
12. Attach a 20-mL syringe to a cannula; triturate the cellular suspension 12 to 20 times, taking care to avoid frothing (*see* **Fig. 1E**).
13. Filter the cell suspension through a 70-mm cell strainer into fresh 50-mL tube and rinse the sieve with 15 mL of base medium.
14. Use a new canula and sieve for each of the different tissues.
15. Centrifuge the cell suspension at 400*g* (~1300 rpm with a 15-cm diameter rotor) for 8 min at 4°C.
16. Discard supernatant and resuspend the heart endothelial cell preparation in 1 mL of cold base medium. Transfer to 5-mL tube.
17. Repeat **step 16** with the lung preparation using 2 mL of cold base medium.

3.1.4. Cell Sorting: Primary Immunobead Isolation

From this point onward, both the heart and lung preparations are treated identically.

1. Add 15 μL of anti-CD31 coated beads for every milliliter of cell suspension.
2. Incubate the tube on an end-over-end rotator for 10 min at room temperature.
3. Place the tubes on a magnetic separator and leave for 2 min. Cell-bound and unbound magnetic beads will be pulled to the magnet side of the tube.
4. Gently, without moving the tube, aspirate the supernatant using a 9-in. glass Pasteur pipet. The beads/cells pellet on the side of the tube should look somewhat foamy if there is a good yield of positively selected endothelial cells.
5. Remove the tube from the magnet and resuspend the pellet in 3 mL of base medium (*see* **Note 7**).
6. Replace the tube back on the magnetic separator for 2 min; aspirate the supernatant carefully.
7. Repeat washes (**steps 5** and **6**) four times (or more) until the supernatant appears clear.
8. After the last wash, resuspend beads/cell pellet in 3 mL of complete growth medium and transfer to a gelatin coated T75 culture flask. Add 7 mL of complete medium.
9. Culture in a standard CO_2 (5%) incubator.
10. After an overnight incubation, remove the non-adherent cells, wash the adherent cells twice with HBSS and add 10 mL of fresh complete medium.
11. Re-feed cultures every other day with 10 mL of fresh complete medium.

3.1.5. Second Immunobead Isolation

The primary cultures are sorted a second time (second immunobead isolation) at 70 to 80% confluence (usually 7–9 d after plating), to improve endothelial cell purity. These cultures should contain discrete colonies of endothelial cells with the

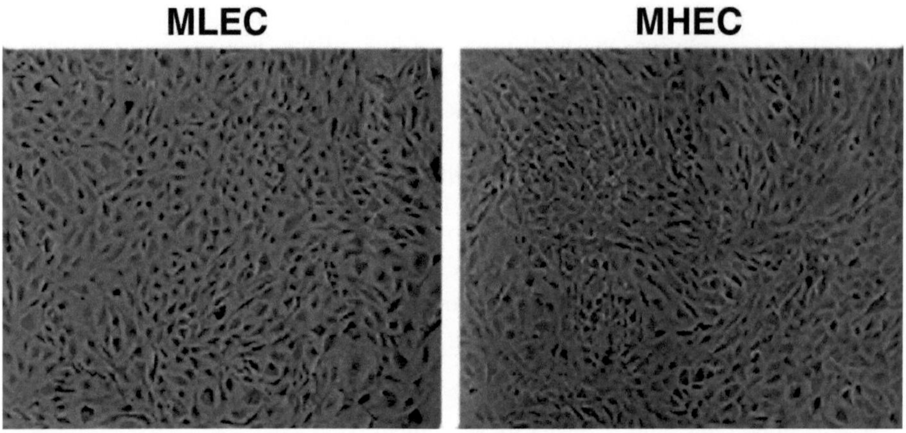

Fig. 2. Confluent mouse lung and heart endothelial cells in culture. Purified mouse endothelial cells were plated out in Dulbecco's Modified Eagle's Medium containing FCS, standard supplements, endothelial cell growth stimulant, heparin, sodium pyruvate, and nonessential amino acids on gelatin-coated 75-mm^2 flask. These cells grow as a monolayer in culture and exhibit the well-described cobblestone morphology.

classical "cobblestone" appearance when viewed by phase contrast microscopy (*see* **Fig. 2**). Routinely, we discard and do not second immunobead sorted cultures that contain less that 70% of "endothelial-like" cells as determined by cell morphology using phase contrast microscopy.

1. Remove culture supernatant and wash the endothelial monolayer twice with 7 mL of Ca/Mg-free HBSS.
2. Add 2mL of trypsin–EDTA, agitate gently, and incubate at 37°C for 3 min.
3. Examine the flask by phase contrast microscopy to confirm that endothelial cells have detached. Monitor this stage carefully to avoid over exposure to trypsin (*see* **Note 8**).
4. Neutralize the trypsin with 10 mL of base medium and transfer the detached cells to a 15-mL tube.
5. Centrifuge at 400*g* to pellet the cells.
6. Aspirate the supernatant; resuspend the cell pellet carefully in 2 mL of base medium and transfer to a round-bottom 5-mL tube.
7. Add 30 μL of anti-CD102 coated magnetic beads (prepared per **Subheading 3.1.1.**) and incubate for 10 min at room temperature on an end-over-end rotator (*see* **Note 9**).
8. Place the tube with the cell suspension on a magnetic separator for 2 min to retain cells; remove the supernatant.
9. Resuspend the positively selected cell pellet in 3 mL of base medium and replace on the magnetic separator. Repeat this wash step four times (**steps 8** and **9**).
10. After the last wash, resuspend the cells in complete medium and plate onto a gelatin-coated T75 culture flask (*see* **Note 10**).

3.1.6. In Vitro Cell Propagation

1. The endothelial cell cultures are fed on alternate days with 10 mL of complete culture medium.
2. At confluence, the cultures should be detached with trypsin–EDTA and subcultured at a split ratio of no more than 1:3.
3. The cells should be used in experiments from passage 2 to 4 as follows: The heart endothelial cells (MHECs) retain phenotypic markers (>80% positive for VE-cadherin, Tie2, and PECAM-1) at passage 4. In contrast, mouse lung endothelial cells (MLEC) retain these markers only up to passage 3 ***(5)***. The purity of cultured MLECs and MHECs will vary from preparation to preparation (*see* **Subheading 3.3.**).

3.2. Endothelial Lineage Characterization

3.2.1. Morphological Appearance

Both murine lung and heart endothelial cells (MLECs and MHECs, respectively) isolated by double-selection method are essentially of mixed vascular bed origin (microvessels, arterial, venous). At confluence, these monolayers exhibit classical cobblestone morphology typical of vascular endothelium (**Fig. 2**). However, overconfluent cultures tend to progressively acquire a more spindly appearance as the cells pack more closely together.

3.2.2. Detection of Endothelial Cell Markers by Two-Color Fluorescence-Activated Cell Sorting

The lineage and purity of the cell cultures can be confirmed by immunostaining the cells (at subculture 1 or 2) for surface expression of two endothelial cell markers used for the cell isolation, namely platelet endothelial cell adhesion molecule (PECAM)-1 and ICAM-2 or, alternatively, VE-cadherin.

1. Plate isolated endothelial cells in gelatin-coated 60-mm diameter tissue culture plates.
2. Wash the confluent monolayer of endothelial cells twice with HBSS to remove any traces of serum.
3. Add 5 mL of prewarmed nonenzymatic cell dissociation buffer and incubate at 37°C for 15 to 20 min. Monitor the cell detachment progress under a phase contrast microscope. When the cells have all rounded up, gently triturate the media to detach cells and to create single, unicellular suspension (*see* **Note 11**).
4. Add 8 mL of base medium and transfer the cell suspension to a fresh 15-mL centrifuge tube.
5. Centrifuge at 400*g* to pellet cells.
6. Resuspend cell pellet to 3 to 5 × 10^6 cells/mL in wash buffer/diluent and aliquot 100 μL of cell suspension into each tube (*see* **Note 12**).

7. Add 2 μL of anti-PECAM-1 APC and 2 μL of anti-ICAM-2 FITC monoclonal antibodies to tube no. 1.
8. Add appropriate isotype controls to tube no. 2.
9. Incubate on ice for 30 to 45 min (minimum is 30 min).
10. Add 1 mL of wash buffer/diluent to each tube, vortex gently, and centrifuge at 400*g*.
11. Repeat wash step with PBS.
12. Suspend stained endothelial cells in 200 to 300 μL of fixative (1% formaldehyde in PBS, pH 7.4).
13. Keep tubes containing stained cells tightly capped and covered with aluminum foil at 4°C until analysis.

3.2.3. Fluorescent Immunoassay of Constitutive and Inducible Adhesion Molecules

To characterize the ability of these cells to respond to proinflammatory cytokine stimulation, the expression of various constitutive and inducible adhesion molecules on unstimulated and tumor necrosis factor (TNF)-α-stimulated MLECs and MHECs were detected by fluorescent immunoassay. Confluent monolayers in 96-multiwell plates are stimulated with murine TNF-α (120 ng/mL in RPMI + 5% fetal calf serum (FCS) with L-glutamine and antibiotics, 100 μL/well) for 6 or 24 h. Monolayers incubated with medium alone serve as negative (unstimulated) control.

1. Wash wells with 200 μL of wash buffer/diluent twice (*see* **Note 13**).
2. Incubate triplicate wells with diluted primary antibody or with control rat IgG or murine nonbinding control K16/16 MAb (100 μL/well) for 45 min on ice.
3. Wash five times with wash buffer/diluent.
4. Incubate wells with diluted FITC-conjugated secondary antibody (diluted 100 μL/well) for 45 min on ice.
5. Wash five times with 200 μL wash buffer/diluent.
6. Wash once with PBS.
7. Read fluorescence at 485 nm in a fluorescence microtiter plate.

3.3. Quantification and Interpretation: General Conclusions

1. The endothelial cell purity varies from preparation to preparation, but generally twice-sorted isolated MLEC and MHEC cultures that are between 85 and 99% PECAM-1 and ICAM-2 double positive at sc1 or sc2 as determined by fluorescence-activated cell sorting. These cells also are negative for VEGF-R3, thus suggesting that the cultures were minimally contaminated by lymphatic endothelial cells *(5)*.
2. The mean fluorescence ± standard deviation (SD) for each adhesion molecule expression is calculated as follows:

   Step 1: Calculate the mean background fluorescence level from the mean (triplicate) negative control rat IgG, or K16/16 antibody wells.

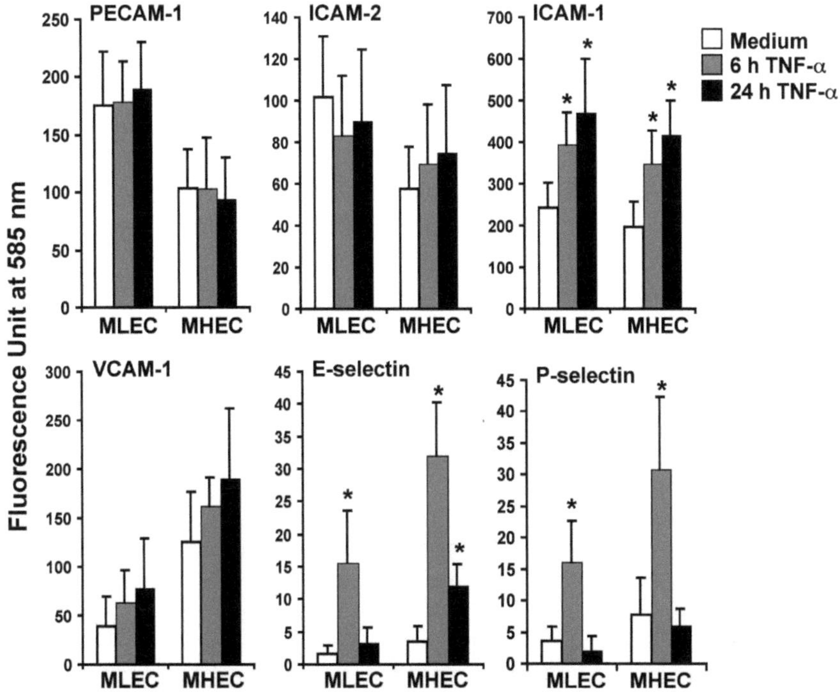

Fig. 3. Expression pattern of adhesion molecules on MLECs and MHECs. Constitutive expression of PECAM-1 and ICAM-2 is not influenced by TNF-α stimulation over the time period studied. In contrast, the expression of ICAM-1, E-selectin, and P-selectin is significantly elevated by continuous exposure to recombinant murine TNF-α. Surprisingly, expression of VCAM-1 on both MLEC and MHEC was not dramatically further increased by TNF-α. Data are mean ± SEM from three experiments. (From **ref. 5** with permission.)

Step 2: Deduct the mean background fluorescent level from each test well to obtain a corrected mean fluorescence intensity for each well.

Step 3: Determine the mean net fluorescence ± SD for each set of triplicate wells.

3. Treatment with TNF-α should not significantly increase the expression PECAM-1 and ICAM-2 on either MLEC or MHEC at any time point (**Fig. 3**, top panels) because these adhesion molecules are constitutively expressed (endothelial lineage markers) and not induced by TNF-α stimulation.
4. Both MLECs and MHECs show constitutive ICAM-1 expression in culture, which is significantly upregulated by TNF-α after 6 and 24 h of continuous treatment (**Fig. 3**, top panel).
5. Resting MLECs and MHECs did not express P- or E-selectin, but both molecules were induced after 6 h of treatment with TNF-α. The increased levels of both selectins returned to baseline by 24 h (**Fig. 3**, bottom panel).

6. VCAM-1 was constitutively expressed at high levels in MHECs, whereas the levels were significantly lower in MLECs. Treatment of MHECs with murine TNF-α for 6 or 24 h did not induce a further increase. MLEC expression of VCAM-1 was increased after both 6 and 24 h of TNF-α treatment, but the increases did not reach statistical significance (**Fig. 3**; *[5]*).
7. Taken together, these data indicate that endothelial cells from different vascular beds do exhibit differences in their constitutive expression of VCAM-1. In addition, vascular endothelium isolated from lung and heart and propagated in vitro does retain certain of these properties and can be triggered by TNF-α to upregulate adhesion molecules that mediate leukocyte recruitment.

4. Notes

1. Collagenase preparation. Dissolve collagenase in HBSS by incubation at 37°C for 1 h before sterile filtering with 0.22-μm filter. Collagenase solution should be prepared fresh and excess solution used within the week.
2. The efficacy of FCS to support endothelial cell growth and proliferation varies between commercial vendors and also from lot to lot from the same vendor. Whenever possible, FCS should be batch tested for endothelial cell growth support before use and optimal FCS purchased in bulk.
3. Add 1% goat serum to the wash buffer/diluent if there is noticeable non-specific background staining. Anti-E-selectin and anti-P-selectin antibodies are also available from BD Pharmingen (clone 10E9.6, Pharmingen cat. no. 553749 and clone RB40.34, Pharmingen cat. no. 553742, respectively).
4. Although our previous studies *(5)* used mice from 6 to 7 wk of age, our new studies (Lamont and Luscinskas, unpublished observations) show endothelial cells isolated from pups 7 to 9 d of age routinely display better growth and morphology characteristics and longer resection (beyond passage 4) of phenotypic markers as assessed by FACS of VE cadherin.
5. 0.02% sodium azide may be added to the anti-CD31 (or anti-CD102)-coated beads as preservative but must be removed by wash before use in cell purification outlined here.
6. It is not necessary to add any medium to the tissue while mincing because there will be sufficient fluid carry-over from the wash to keep the tissue moist.
7. Avoid the natural tendency to be too gentle with the resuspension steps in **Subheadings 3.1.4.** and **3.1.5.** The cells/beads clumps must be properly "broken up" to get rid of any unbound cells (i.e., nonendothelial cells) caught within the clumps. However, overvigorous pipetting may dislodge beads from the cells!
8. Overexposure to trypsin will affect the viability of cells. If so desired, the detachment of rounded cells can be sped up with a sharp tap on the side of the flask with an open palm.
9. Murine PECAM-1 (CD31) is trypsin sensitive and anti-CD31 beads should not be used for the second-sort.
10. After the second-sort, plate the cells on a single T75 flask. It is important to maintain high cell density plating to avoid large numbers of senescent cells.

11. Alternatively, the monolayers can be detached by brief trypsin-versene EDTA treatment of no longer than 2 min at 37°C.
12. A confluent monolayer of endothelial cells from a T75 flask contains 1–3 × 10^6 cells. For fluorescence-activated cell sorting analysis, the cells are used at 0.3 to 0.5 × 10^6 cells/100 μL reaction volume.
13. Check five random wells under the phase contrast microscope to confirm monolayers remained intact during the wash steps. We used unfixed monolayers in this protocol because the anti-E-selectin (RME-1) and anti-P-selectin (RMP-1) antibodies are not suitable for use on fixed cells.

Acknowledgments

We thank the late William Atkinson for his invaluable early contributions to this project. We appreciate the advice, assistance, and helpful suggestions of our colleagues Drs. Andy Connolly (Department of Pathology, Stanford University, Palo Alto, CA), Jennifer Allport (Novartis, Cambridge, MA), Michael Gimbrone, and Guillermo Garcia-Cardena (Brigham and Women's Hospital, Boston, MA). We thank Dr. Andrew Issekutz for his kind gifts of reagents and Mr. Matsuni bin Hamzah of Medical Illustration Unit, Department of Pathology, National University of Singapore for his assistance in generating **Fig. 1**. This work was supported by National Institutes of Health grants HL36028, HL56985, HL65090 and HL53993.

References

1. Adams, D. H. and Shaw, S. (1994) Leucocyte-endothelial interactions and regulation of leucocyte migration. *Lancet* **343,** 831–836.
2. Springer, T. A. (1994) Traffic signals for lymphocyte recirculation and leukocyte emigration: the multistep paradigm. *Cell* **75,** 301–314.
3. Gerritsen, M. E. (1987) Functional heterogeneity of vascular endothelial cells. *Biochem. Pharmac.* **36,** 2701–2711.
4. Stevens, T., Rosenberg, R., Aird, W. C., et al. (2001) NHLBI workshop report: endothelial cell phenotypes in heart, lung, and blood diseases. *Am. J. Physiol. Cell Physiol.* **281,** C1422–C1433.
5. Lim Y-C., Garcia-Cardena, G., Allport, J. R., et al. (2003) Heterogeneity of endothelial cells from different organ sites in T-cell subset recruitment. *Am. J. Pathol.* **162,** 1591–1601.
6. Rajotte, D., Arap, W., Hagedorn, M., Koivunen, E., Pasqualini, R., and Ruoslahti, E. (1998) Molecular heterogeneity of the vascular endothelium revealed by in vivo phage display. *J. Clin. Invest.* **102,** 430–437.
7. Goodarzi, K., Goodarzi, M., Tager, A. M., Luster, A. D., and von Andrian, U. (2003) Leukotriene B4 and BLT1 control cytotoxic effector T cell recruitment to inflamed tissues. *Nature Immunol.* **4,** 965–973.
8. Aird, W. C., Edelberg, J. M., Weiler-Guettler, H., Simmons, W. W., Smith, T. W., and Rosenberg, R. D. (1997) Vascular Bed-specific expression of an endothelial

cell gene is programmed by the tissue microenvironment. *J. Cell Biol.* **138,** 1117–1124.
9. Savinov, A. Y., Wong, F. S., Stonebraker, A. C., and Chervonsky, A.V. (2003) Presentation of antigen by endothelial cells and chemoattraction are required for homing of insulin-specific CD8+ T cells. *J. Exp. Med.* **197,** 643–656.
10. Gerritsen, M. E., Shen C-P., Mchugh, M. C., et al. (1995) Activation-dependent isolation and culture of murine pulmonary microvascular endothelium. *Microcirculation* **2,** 151–163.
11. Dong, Q. G., Bernasconi, S., Lostaglio, S., et al. (1997) A general strategy for isolation of endothelial cells from murine tissues—Characterization of two endothelial cell lines from the murine lung and subcutaneous sponge implants. *Arterioscler. Thromb. Vasc. Biol.* **17,** 1599–1604.
12. Marelli-Berg, F. M., Peek, E., Lidington, E. A., Stauss, H. J., and Lechler, R. I. (2000) Isolation of endothelial cells from murine tissue. *J. Immunol. Methods* **244,** 205–215.
13. Allport, J. R., Lim Y-C., Shipley, J. M., Senior, et al. (2002) Neutrophils from MMP-9-or neutrophil elastase-deficient mice show no defect in transendothelial migration under flow in vitro. *J. Leukoc. Biol.* **71,** 821–828.
14. Tager, A. M., Bromley, S. K., Medoff, B. D., et al. (2003) Leukotriene B4 receptor BLT1 mediates early effector T cell recruitment. *Nat. Immunol.* **4,** 982–990.
15. Peng, X., Ueda, H., Zhou, H., et al. (2004) Overexpression of focal adhesion kinase in vascular endothelial cells promotes angiogenesis in transgenic mice. *Cardiovasc. Res.* **64,** 421–430.
16. Luscinskas, F. W., Ding, H., and Lichtman, A. H. (1995) P-selectin and vascular cell adhesion molecule 1 mediate rolling and arrest of CD4+ T lymphocytes on TNF-α-activated vascular endothelium under flow. *J. Exp. Med.* **181,** 1179–1186.

13

Expression Cloning of Signaling Proteins Regulated by Cell Adhesion

Michelle L. Matter and Joe W. Ramos

Summary

Many proteins involved in cell–cell and cell–matrix adhesion are regulated by signal transduction pathways and can activate signal transduction on ligation. Adhesion-related signal transduction is important throughout development, hemostasis, immunity, and in diseases such as cancer. Therefore, the identification of the various signaling pathways that are involved is crucial. Expression cloning is an unbiased way to isolate proteins with specific biological functions. This methodology has been adapted for the identification of proteins involved in cell signaling pathways that are mediated by cell–cell and cell–matrix interactions. We have successfully developed and used a novel expression cloning strategy to isolate the integrin-regulated apoptosis signaling protein BIT-1. This screen was based on previous observations that integrin-mediated adhesion upregulates the anti-apoptotic protein bcl-2. Our strategy described in this chapter uses flow cytometry and a reporter construct in which the bcl-2 promoter is linked to enhanced green fluorescence protein. The advantage of using flow cytometry in expression cloning is that it increases the sensitivity of the screen by enabling us to examine function quantitatively at the level of a single cell millions of times in one experiment. The following protocol provides a detailed method for the isolation of proteins that are regulated by cell adhesion.

Key Words: Expression cloning; integrin; cell–cell adhesion; cell–matrix adhesion; flow cytometry; EGFP; bcl-2; Bit-1.

1. Introduction

Cell surface proteins that mediate cell–cell and cell–matrix adhesion such as cadherins and integrins are regulated by signal transduction pathways and can themselves activate signal transduction in response to adhesion *(1–3)*. One approach that has been used to successfully identify some of the proteins that mediate these signaling events is expression cloning. Expression cloning allows for the isolation of complementary DNAs (cDNAs) that encode proteins with a

From: *Methods in Molecular Biology, vol. 341: Cell–Cell Interactions: Methods and Protocols*
Edited by: S. P. Colgan © Humana Press Inc., Totowa, NJ

specific biological activity. The investigator designs the expression cloning strategy around a robust assay for a function of interest. This can be, for example, a change in adhesion, transcription, protein expression, or cell survival.

A variety of expression cloning strategies have been developed to identify proteins with a particular function. Seed et al. *(4,5)* developed novel expression cloning methods to clone CD antigens using antibodies to the extracellular face of the antigens. Adams et al. *(6)* used an expression cloning strategy that relied on a cell adhesion assay to identify the intracellular protein, muskelin, which is involved in the cellular response to thrombospondin-1 adhesion. In collaboration with others, we have used expression cloning screens to clone proteins involved in integrin-related signaling *(7–9)*. Integrins are cell surface receptors that mediate cell–cell and cell–matrix adhesion *(1)*. Expression cloning screens were used to isolate the integrin regulatory proteins CD98 *(7)* and PEA-15, a protein that also suppresses the Ras mitogen-activated protein kinase pathway *(8)*. Each of these screens relied on measuring integrin activation by flow cytometry with an antibody that binds only active integrin *(9)*.

Recently, in collaboration with Erkki Ruoslahti, Minoru Fukuda, and others, Matter et al. *(10)* cloned the integrin-regulated signaling protein BIT-1 using expression cloning. This method is the one we describe in this chapter. Integrin-mediated cell adhesion is necessary for the survival of many cell types, including neuronal, endothelial, and epithelial cells *(11,12)*. Loss of cell attachment to the extracellular matrix causes apoptosis in these cells. Integrins suppress apoptosis in attached cells by activating signaling pathways that promote survival and inactivating the ones that promote apoptosis. A number of these pathways have been described and are of varying importance in different types of cells *(11)*. It has been proposed that perturbation of the integrin-activated cell survival pathways is a contributing factor to oncogenesis *(11)*. We have previously shown that matrix-bound integrins promote cell survival in part by activating focal adhesion kinase, which activates phosphatidylinositol 3-kinase, leading to the upregulation of bcl-2 transcription and translation *(13)*. To further examine the mechanisms by which integrins promote cell survival, we used the unbiased expression cloning approach described here.

2. Materials

2.1. Mammalian Host Cells

1. A COS or CHO cell line. We used CHO-B2Δ cells (these cells contain a shortened α5 cytoplasmic tail; these cells may be requested from E. Ruoslahti (ruoslahti@burnham.org).
2. Complete Dulbecco's Modified Eagle's Medium (DMEM; Invitrogen, Carlsbad, CA) for maintaining cell conditions. To prepare: 10% fetal bovine serum (HyClone, Ogden, UT), 1% nonessential amino acids, 1% glutamine, and 1%

penicillin–streptomycin. Prepare in DMEM and sterilize through a 0.2-micron filter. Store at 4°C (*see* **Note 1**).
3. L-glutamine (cat. no. G9273, Sigma, St. Louis, MO).
4. Nonessential amino acids (Sigma, cat. no. M7145).
5. Penicillin–streptomycin (Invitrogen).

2.2. cDNAs, cDNA Library, and cDNA Preparation

1. 3 *M* sodium acetate, pH 7.0 (Sigma, cat. no. 52404).
2. Ampicillin (AMP; 25 μg/mL; Sigma).
3. Tetracycline (Tet; 10 μg/mL; Sigma).
4. A cDNA library that has been directionally cloned into a mammalian expression vector (*see* **Note 2**).
5. Falcon no. 2059 tubes.
6. Glycerol (Sigma).
7. MC1061/P3 cells at a competency of 10^9 or more colony-forming units per μg (cfu/μg) DNA (Invitrogen, San Diego, CA; *see* **Note 3**).
8. DNA maxi and mini-prep kits (Qiagen, Valencia, CA).
9. Promoterless enhanced green fluorescence protein (EGFP-1; BD Biosciences Clontech, Palo Alto, CA, cat. no. 6086-1).
10. Phenol:chloroform:isoamyl alcohol (25:24:1; Sigma, cat. no. P3803).
11. Lysogeny Broth (LB) Medium (Invitrogen).
12. SOC Medium (Invitrogen).
13. Nitrocellulose membranes (15 cm, BA-85, Schleicher & Schuell, Keene, NH).
14. Polyoma large T-antigen, if not present in cell line (*see* **Note 2**).
15. 100-mm tissue culture dishes (Falcon).
16. 15-cm plates (Falcon).

2.3. Cell Transfection and Cell Isolation

1. LipofectAMINE (Invitrogen) or other appropriate and efficient transfection reagent.
2. Polystyrene tubes (Falcon, cat. no. 2054).
3. 1X PBS (10X solution): 87.7 g of NaCl, 7.1 g of Na_2HPO_4 (anhydrous), and 6.0 g of NaH_2PO_4 (anhydrous). Dissolve in 1 liter of double-distilled (dd)H_2O, pH 7.2. Dilute 1:10 in ddH_2O for a 1X solution.
4. Hirt's solution: 0.6% sodium dodecyl sulfate, 10 m*M* ETDA in 5 mL of ddH_2O.
5. 5 *M* NaCl (Sigma).
6. Buffer-saturated phenol (Invitrogen).
7. High-performance liquid chromatography-grade H_2O.

2.4. Fluorescence-Activated Cell Sorting

1. Cell dissociation solution (enzyme-free Hank's based Cat S-004-H, Specialty Media Division of Cell and Molecular Technologies Inc., Phillipsburg, NJ).
2. EDTA (10 m*M*; Sigma).
3. 0.2 mg/mL; Propidium iodide (Sigma).
4. 5% fetal calf serum (Invitrogen).

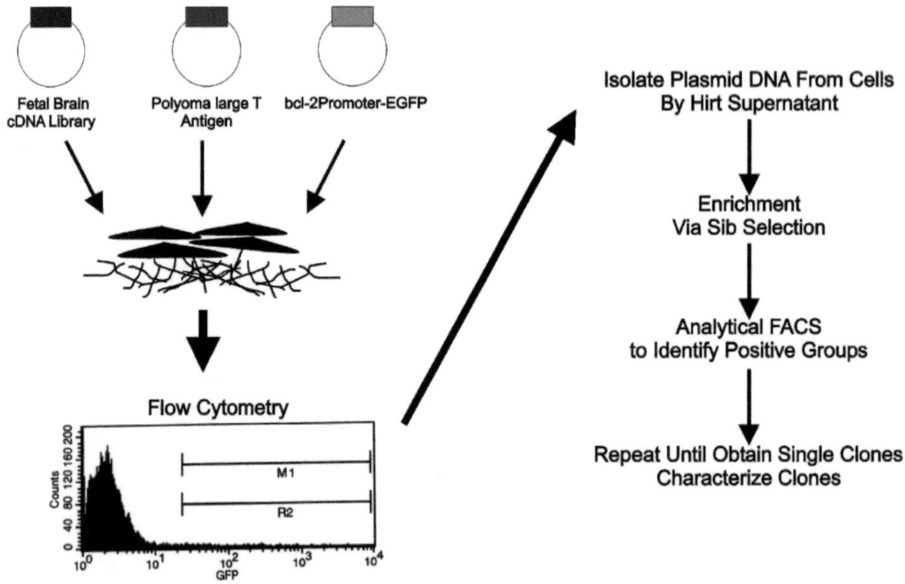

Fig. 1. Expression cloning of transcriptional regulators of bcl-2. CHOΔ cells attached to fibronectin were transfected with a fetal brain cDNA library, the polyoma large T-antigen and the bcl-2 promoter EGFP marker. Cells expressing EGFP were collected by a sterile fluorescence-activated cell sorter and plasmid DNA was isolated by Hirt supernatant. Sibling selection using analytical fluorescence-activated cell sorting analysis to identify positive groups of cells was repeated five times until a single clone that upregulated the bcl-promoter EGFP marker was isolated.

2.5. Equipment

1. Fluorescence-activated cell sorter (FACS; FACSTAR, Becton Dickinson, Franklin Lakes, NJ).
2. Flow cytometer (FACScalibur, Becton Dickinson).
3. 42°C water bath.
4. Electroporator.

3. Methods

The following strategy describes an expression cloning screen to identify proteins involved in signaling mediated by cell–matrix adhesion. This method is briefly described in **Fig. 1**. This method also may be adapted to identify proteins in other signaling pathways including cell–cell adhesion. Expression cloning has numerous advantages for the identification of adhesion proteins and proteins that regulate their signaling. It provides an unbiased approach to clone proteins that are involved in adhesion or that may affect adhesion related signal transduction. Every protein expressed in the cDNA library is tested in the screen and the assay determines the functional relevance of the clones. Moreover, proteins can be isolated that have a functional relationship to adhesion but do not interact directly with specific adhesion proteins. The main disadvantages of expression cloning are that functional alterations that require more than one protein may not be isolated and functions resulting from posttranslational modifications can not be identified.

The expression cloning strategy we describe was successful in isolating the integrin-regulated signaling protein BIT-1 *(10)*. We previously reported that in serum-free conditions cell attachment through the α5β1 integrin promotes upregulation of bcl-2 at both protein and transcriptional levels by activating Shc, which activates focal adhesion kinase, which activates phosphatidylinositol 3-kinase *(13)*. Our screening strategy relied on inactivating this integrin-mediated signaling pathway and isolating clones that increased bcl-2 levels in these cells. Because expression cloning relies on the expression of a library of cDNAs, it is essential that it be done in cells that are highly transfectable and that express exogenous protein well. We used a CHO cell line with an α5 cytoplasmic deletion that inhibits upregulation of bcl-2. These cells were transfected with a fetal brain-derived cDNA library. The cells were also transfected with a plasmid that expressed EGFP (a green fluorescent protein) under the control of the bcl-2 promoter. Thus cells in which bcl-2 transcription was upregulated would fluoresce at 511 nm. This fluorescence was monitored in a flow cytometer, and cells in which bcl-2 transcription was upregulated were isolated (as detected by EGFP expression). This system allows for rapid and quantitative measurement of millions of cells in a single experiment and thus increases the sensitivity of the screen.

The remainder of our expression cloning method involves sibling selection as described herein and was adapted from a method that was used to isolate a novel sulfotransferase using antibody binding and flow cytometry *(14)*. It is also important to note that the cDNAs we isolated were not necessarily specific for integrin-activated signaling. Anything that could generally upregulate bcl-2 might also be cloned. Further experiments are usually required to verify the relevance of a newly identified cDNA. In this way, BIT-1 was identified as a novel regulator of apoptosis by integrins.

3.1. Cell Transfection

It is important to have adequate controls in addition to having a sufficient number of experimental samples for sorting the cells and recovery of plasmid cDNA. Controls should include (1) control plate 1 containing co-transfected transfection marker plasmid (promoter bcl-2 EGFP), empty vector plasmid, and the polyoma large T-antigen; and (2) control plate 2 containing co-transfected EGFP under its own promoter, the empty vector plasmid (pcDNA3.1; *see* **Note 2**), and polyoma large T-antigen (this control allows assessment of overall transfection efficiency). These two controls allow the baseline fluorescence of the cells to be measured (control plate 1) and the overall efficiency of the transfection (control plate 2). In addition, they are helpful when setting the collection gate on the cell sorter.

1. Twenty-four hours before transfection, plate cells to be transfected at 2×10^6 cells per plate in 100-mm tissue culture dishes (at least 20 plates total; *see* **Note 4**).
2. Transfect cells with 20 µL of LipofecAMINE in 100 µL of serum-free DMEM, according to the manufacturer's direction along with the following plasmids or DNAs:
   Control plate 1: 4 µg og bcl-2 promoter EGFP, 6 µg of vector DNA plasmid (pcDNA3.1; *see* **Note 2**), and 4 µg of polyoma large T-antigen.
   Control plate 2: 4 µg of EGFP-C1 (this has its own promoter; Clontech), 6 µg of vector DNA plasmid (pcDNA3.1), and 4 µg of polyoma large T-antigen.
   Sort plates: 4 µg of bcl-2 promoter EGFP, 6 µg of cDNA library (fetal brain cDNA library) and 4 µg of polyoma large T-antigen.
3. Incubate for 24 hours at 37°C, in 5% CO_2 and 95% humidity.
4. Wash all plates with 2 mL of serum-free DMEM media.
5. Add 10 mL of serum-free DMEM and incubate cells for an additional 24 h at 37°C in 5% CO_2 and 95% humidity.

3.2. Fluorescence-Activated Cell Sorting

1. After 24-h incubation time, aspirate media from the dishes.
2. Rinse with 5 mL of 1X PBS. Remove supernatant.
3. Add 10 mL of Cell Dissociation Solution + 10 m*M* EDTA to the sort plates and control plates.
4. Incubate for 5 min at room temperature.
5. Collect cells from sort plates and combine into one 50-mL tube (labeled sort cells).
6. Centrifuge the sort cells and the control cells at 120*g* in a tabletop centrifuge at room temperature to pellet the cells.
7. Aspirate supernatant.
8. Resuspend the sort cells in 1.4 mL of DMEM media by gentle agitation of each tube.
9. Divide the sort cells into six tubes (Falcon, cat. no. 2054) of 225 µL each.
10. Resuspend the control cells separately, each in 200 µL of DMEM.
11. Add propidium iodide (0.1 µg to 0.5 µg per 1×10^5 cells) to the samples. Incubate all tubes on ice in the dark for 5 min (*see* **Note 5**).
12. Wash the cells with 2 mL of cold 1X PBS.
13. Centrifuge the cells for 5 min at 120*g* at room temperature.
14. Resuspend and combine sort cells in a total of 4 mL of 1X PBS.

15. Resuspend each tube of control cells in 500 µL of 1X PBS.
16. Set the parameters for FACS such that highly transfected control cells from control plate 2 fall within the gate, whereas similarly transfected control cells from control plate 1 do not.
17. Collect the EGFP positive sort cells by FACS, into 1 mL of complete DMEM media (*see* **Note 6**).

3.3. cDNA Recovery From Hirt Supernatant

1. Add an equal volume of 2X lysis buffer (1.2% sodium dodecyl sulfate/20 m*M* EDTA, pH 7.4) to the tube containing 1 mL of sorted cells.
2. Mix by inversion and incubate for 15 min at room temperature.
3. Add 500 µL of 5 *M* NaCl, mix and incubate overnight at 4°C.
4. Transfer sorted cells into two sterile microcentrifuge tubes, centrifuge for 20 min at 18,000*g* at 4°C.
5. Carefully transfer and combine supernatant into a 15-mL Falcon polypropylene tube.
6. Phenol:choloroform extract each supernatant twice (*see* **Note 7**).
7. Transfer aqueous phase equally into four microtubes.
8. Transfer content of each tube into a Microcon Unit (Microcon 30; Amicon) that is on a filtrate collection tube.
9. Close cap and centrifuge at 18,000*g* for 10 min (or until there is 10 to 20 µL of retentate remaining).
10. Add 400 µL of high-performance liquid chromatography-quality H_2O to the Microcon unit and spin at 18,000*g* for 10 min. Repeat this step three times to ensure salt removal because salt can negatively affect the subsequent electroporation step.
11. Remove Microcon unit and invert into a new collection tube (*see* **Note 8**).
12. Add 8 µL of sterile DI H_2O directly into the six small holes of the Microcon unit and pulse spin at high speed for 5 s (*see* **Note 9**).
13. Remove unit from each collection tube and discard Microcon unit. Keep plasmid samples on ice.

3.4. Electroporation

1. Prepare LB AMP/Tet plates (AMP at 25 µg.mL; Tet at 10 µg/mL).
2. Take 1 µL of plasmid cDNA and electroporate into MC1061/P3 cells (efficiency is usually from 1×10^7 to 1×10^8 colony-forming units/µg DNA).
3. Titer to see how many colonies per 1 µL. Rescue cells in 500 µL of SOC media and grow for 1 h at 37°C.
4. Plate 10, 100, and 250 µL onto 15-cm LB AMP/Tet plates.
5. Place remaining samples at –20°C until required.

3.5. Sib-Selection of Clones and Detection of Positive Clones

1. Calculate the amount of plasmid DNA required for electroporation into bacterial cells to obtain 500 to 1000 colonies per 15-cm LB AMP/Tet plate. Enough plasmid samples should be electroporated to obtain 8000 colonies, which require using between 10 to 16, 15-cm LB AMP/Tet plates (*see* **Note 10**).

2. Label plates. Incubate overnight (<16 h) at 37°C.
3. Prepare 16 nitrocellulose membranes and carefully layer over the agar. Incubate upside down for 5 h at 37°C.
4. Carefully remove membrane from each plate. Place each membrane (colony side up) into a labeled 15-cm Petri dish.
5. Prepare 170 mL of LB AMP/Tet (50 μg AMP and 10 μg Tet/mL). Transfer 10 mL of LB AMP/Tet to each plate containing a membrane.
6. Using a sterile cell scraper, scrape off the colonies from each membrane. Transfer culture into a labeled sterile 50-mL centrifuge tube. Secure lid with a piece of tape (do not tightly close lid) and incubate overnight at 37°C.
7. Make a permanent stock of each of the cultures by removing 900 μL of culture into a sterile microfuge tube containing 100 μL of dimethyl sulfoxide. Store bacteria at –80°C.
8. Isolate and purify the plasmids from the remaining 9 mL using a Qiagen mini plasmid kit (*see* **Note 11**).
9. Dissolve each plasmid pool into 100 μL of 1X Tris-EDTA buffer, pH 7.5.
10. Determine the concentration of plasmid by absorbance at A260 nm and store plasmid preps at –20°C.

3.6. Sib-Selection of Clones and Detection of Positive Clones by FAC Analysis

At this step, the goal is to identify the pools of cDNAs that induce the signal that is being measured. The example described here is the ability of the protein from one of the cDNAs to activate bcl-2 promoter signaling and thereby turn on EGFP. These isolated cDNAs will substitute for the library in the initial screening process. The cells will have to be collected and subjected to FACS analysis following the same procedure as that described for the library screening method above (*see* **Note 12**).

1. Co-transfect 8 μg of each pool of DNA into the cells along with 6 μg of activator (bcl-2 promoter-EGFP) and 6 μg of polyoma large T-antigen. In addition, transfect the two control plates as described previously.
2. Isolate the cells as described previously.
3. Perform analytical FACS to identify the cDNA pools that produce the desired change in the activity of the pathway (here it is the upregulation of bcl-2 transcription that is visualized by EGFP expression).
4. From positive FACS analysis of EGFP expression prepare DNA by Qiagen miniprep according to standard protocol.
5. Repeat sib-selection on the plasmid pool that showed positive results in the FACS analysis by overlaying a labeled nitrocellulose membrane onto the master plate for that plasmid group. Mark the membrane into eight equal quadrants with a pencil. These marked quadrants should then be traced onto the agar-side (bottom) of the 15-cm LB master plate using a black felt tip pen and labeled 1 to 8 accordingly

(*see* **Note 13**). Incubate the master plates with the marked nitrocellulose membranes for 5 h at 37°C.
6. Carefully peel nitrocellulose membrane off the master plates and cut into the eight marked quadrants.
7. Scrape the colonies growing on each quadrant into 10 mL of LB-AMP/Tet using a sterile cell scraper.
8. The plasmids from each quadrant of each master plate are then isolated and purified using a Qiagen mini plasmid kit.
9. Cell transfection is repeated until another sub-pool of positive clones is isolated by analytical FACS.
10. Repeat this procedure until one positive clone is isolated.
11. The plasmid DNA from this clone is then isolated, purified and sequenced (*see* **Note 14**).

4. Notes

1. During expression cloning experiments using our system, no fetal bovine serum is added to the DMEM.
2. Although there are several options in preparing a cDNA library, we chose to purchase a premade cDNA library from a commercial source. We used a fetal brain cDNA library (Clontech) cloned into pcDNA1, which is no longer available commercially. A replacement vector such as pcDNA3.1 may be used; however, it lacks the polyoma origin of replication and instead contains an SV40 origin. Therefore, the cell line will need to express the SV40 T-antigen to obtain high plasmid amplification. If a murine host cell line is used, then the cDNA encoding the polyoma large T-antigen should be co-transfected. This will promote increased amplification of the desired plasmid (the plasmid should harbor the polyoma origin of replication).
3. The pcDNA1 vector contains the supF suppressor tRNA, which allows the MC106/P3 cells containing pcDNA1 to be resistant to both AMP and Tet. MC1061/P3 cells containing the polyoma large T-antigen (pPSVE1-PyE) are resistant to AMP but not to Tet. This fact allows for the isolation of plasmids derived from pcDNA1 to be obtained and amplified.
4. We transfect 16 sort plates per screen and two control plates.
5. Propidium iodide stains the dead cells and allows them to be removed from the sort gate.
6. We sorted more than 6×10^6 cells and collected at least 70,000 cells. Make a note of the number of cells sorted and collected for comparison between multiple screens.
7. Make sure that you no longer see the interface. If it is still apparent, then do one more extraction.
8. It is important to use new tubes at this step.
9. It is important to place the H_2O directly into the small holes using a P10 pipet.
10. To obtain 8000 colonies, 6 μL of plasmid DNA is electroporated into 6X 20-μL cells. These plates must be dried for 15 min at 42°C to have a distribution of colonies over the entire plate surface.

11. For a higher yield, use 2X the volume of P1, P2, and P3 solutions when using a pcDNA1-derived cDNA library.
12. Changes in the activity of the signaling pathway may be very low during the initial rounds of screening, but the signal should increase as the individual cDNA clone is isolated.
13. Careful concise labeling is imperative for each of the 16 master plates and subsequent overlays with nitrocellulose.
14. Isolation of a single active clone usually requires three to five rounds of sib-selection. Each subsequent round should demonstrate an enrichment of positive cells and an increase in the FACS signal. We performed five rounds of sib-selection to obtain Bit-1.

Acknowledgments

The authors would like to thank Drs. M. Fukuda and E. Ong for their instrumental help in the practical considerations of the strategy. We would also like to thank Dr. E. Ruoslahti for his advice in developing the initial screen. M. L. Matter is supported by a grant from the National Institutes of Health (P20-RR016453). J.W. Ramos is supported in part by a grant from the National Institutes of Health (CA93849).

References

1. Hynes, R. O. (2002) Integrins: bidirectional, allosteric signaling machines. *Cell* **110,** 673–687.
2. Juliano, R. L. (2002) Signal transduction by cell adhesion receptors and the cytoskeleton: functions of integrins, cadherins, selectins, and immunoglobulin-superfamily members. *Annu. Rev. Pharmacol. Toxicol.* **42,** 283–323.
3. Wheelock, M. J. and Johnson, K. R. (2003) Cadherin-mediated cellular signaling. *Curr. Opin. Cell Biol.* **15,** 509–514.
4. Seed, B. (1995) Developments in expression cloning. *Curr. Opin. Biotechnol.* **6,** 567–573.
5. Tanaka, T., Camerini, D., Seed, B., et al. (1992) Cloning and functional expression of the T cell activation antigen CD26. *J. Immunol.* **149,** 481–486.
6. Adams, J. C., Seed, B., and Lawler, J. (1998) Muskelin, a novel intracellular mediator of cell adhesive and cytoskeletal responses to thrombospondin-1. *EMBO J.* **17,** 4964–4974.
7. Fenczik, C. A., Sethi, T., Ramos, J. W., Hughes, P. E., and Ginsberg, M. H. (1997) Complementation of dominant suppression implicates CD98 in integrin activation. *Nature* **390,** 81–85.
8. Ramos, J. W., Kojima, T. K., Hughes, P. E., Fenczik, C. A., and Ginsberg, M. H. (1998) The death effector domain of PEA-15 is involved in its regulation of integrin activation. *J. Biol. Chem.* **273,** 3897–3900.
9. Matter, M. L., Ginsberg, M. H., and Ramos, J. W. (2001) Identification of cell signaling molecules by expression cloning. *Sci. STKE* **9,** L9.

10. Jan, Y., Matter, M., Pai, J. T., et al. (2004) A mitochondrial protein, Bit1, mediates apoptosis regulated by integrins and Groucho/TLE corepressors. *Cell* **116,** 751–762.
11. Frisch, S. M. and Screaton, R. A. (2001) Anoikis mechanisms. *Curr. Opin. Cell Biol.* **13,** 555–562.
12. Frisch, S. M. and Ruoslahti, E. (1997) Integrins and anoikis. *Curr. Opin. Cell Biol.* **9,** 701–706.
13. Matter, M. L. and Ruoslahti, E. (2001) A signaling pathway from the alpha5beta1 and alpha(v)beta3 integrins that elevates bcl-2 transcription. *J. Biol. Chem.* **276,** 27,757–27,763.
14. Ong, E., Yeh, J. C., Ding, Y., Hindsgaul, O., and Fukuda, M. (1998) Expression cloning of a human sulfotransferase that directs the synthesis of the HNK-1 glycan on the neural cell adhesion molecule and glycolipids. *J. Biol. Chem.* **273,** 5190–5195.

14

Assays for the Calcium Sensitivity of Desmosomes

Anita J. Merritt, Anthea Scothern, and Tanusree Bhattacharyya

Summary

Epithelial cells in vivo exist as confluent cell sheets, but this confluence is disrupted if the sheets are wounded, if the cells are undergoing morphogenesis, or if they are taking part in invasion and metastasis. Desmosomes are one of the principal types of adhesive junctions in epithelia and are responsible for maintaining tissue integrity. It is likely that modulation of desmosomal adhesion is required to facilitate cell motility in response to alterations in the tissue architecture. Desmosomal adhesion changes from a calcium-dependent state to a calcium-independent state when cells become confluent. Our laboratory has shown that the α isoform of protein kinase C is involved in signaling the response of desmosomes to calcium concentration and wounding, in cultured epithelial cells and in mouse epidermis (in vivo).

Key Words: Desmosomes; calcium dependence; calcium independence; protein kinase Cα (PKCα); desmoplakin; cell adhesion; MDCK cells.

1. Introduction

Desmosomes are intercellular adhesive junctions that provide tensile strength to epithelial tissues and cardiac muscle, and are particularly abundant in the epidermis. Their major adhesive molecules are the desmocollins and desmogleins, members of the calcium (Ca)-dependent cadherin family *(1)* The Ca switch assay has been used extensively for the study of desmosome assembly *(2–5)*. Cells grown in low Ca medium (≤0.1 m*M*) begin to assemble desmosomes when the Ca concentration is raised to physiological levels (~1.8 m*M*). Desmosomes can exist in two adhesive states; Ca-dependent and Ca-independent, and the transition between the two is regulated by protein kinase C (PKC)α *(6)*. Cells in culture become Ca-independent after they reach confluence, and Ca-independence is the natural state of desmosomes in vivo. When confluent cell monolayers are wounded or treated with activators of

From: *Methods in Molecular Biology, vol. 341: Cell–Cell Interactions: Methods and Protocols*
Edited by: S. P. Colgan © Humana Press Inc., Totowa, NJ

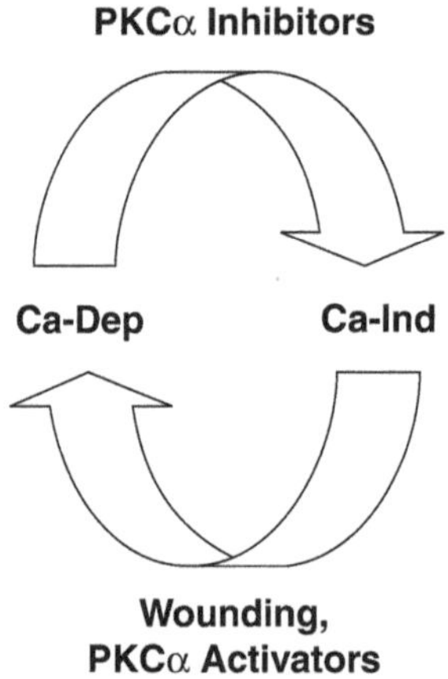

Fig. 1. Diagram to illustrate factors that modulate desmosomes between Ca-dependence (Ca-Dep) and Ca-independence (Ca-Ind).

PKC, cells rapidly revert to Ca-dependence (**Fig. 1**). In this chapter, we illustrate, using the Madin–Darby canine kidney (MDCK) model epithelial cell line, how desmosomal adhesion can be switched between Ca-dependence and Ca-independence. This chapter includes protocols for wounding and treatment with PKC activators/inhibitors. We also will describe how immunofluorescence and electron microscopy can be used to determine the adhesive state of desmosomes and will demonstrate how the assays can be extended to study desmosomal adhesion in mouse epidermis (in vivo).

2. Materials

2.1. Cell Culture

1. Tissue culture hood and incubators at 37°C, 5% CO_2.
2. T75 tissue culture flasks, 24-well plates, 0.2-μm filters, haemocytometer.
3. MDCK II cells (www.ecacc.org.uk).
4. Dulbecco's Modified Eagle's Medium (DMEM) with 10% FBS (Normal Calcium Medium [NCM], Gibco).
5. Ca-free DMEM (Gibco).
6. Phosphate-buffered saline (PBS) or Hank's Balanced Salt Solution (HBSS) without Ca or magnesium (Gibco).
7. Trypsin–EDTA (Gibco).
8. EGTA (Sigma, St. Louis, MO).
9. Fetal bovine serum (FBS; Gibco).
10. Chelex resin (Sigma).
11. Whatman filter paper.
12. Sterile 13-mm circular glass cover slips (Scientific Laboratory Supplies, Nottingham, UK).

13. 100% methanol.
14. Sterile scalpel blades (size 24, Swann-Morton) and cell scrapers (Costar).
15. 50:50 methanol: acetone (store at –20°C).
16. Dimethyl sulfoxide (Sigma).
17. Phorbol-12,13-didecanoate, phorbol-12,13-dibutyrate (TPA; Sigma).
18. Gö6976 (Calbiochem).
19. Chelerythrine chloride (Calbiochem).
20. 100-cm^2 plastic dishes (Sterilin).

2.2. Immunofluorescence

1. Antibody to desmosomal protein such as 11–5F mouse monoclonal Ab to desmoplakin I and II *(7)*.
2. Rabbit anti-PKCα polyclonal antibody (Sigma).
3. Fluorescein isothiocyanate (FITC)-conjugated anti-rabbit immunoglobulin (Ig)G, and rhodamine red-conjugated anti-mouse IgG (Jackson Labs).
4. PBS tablets (Sigma).
5. 0.5% Triton X-100 (Sigma).
6. Wax pen (DAKO).
7. Gelvatol mounting medium: Gradually add 10 g of gelvatol vinyl alcohol (Fisons) to 20 mL of 0.02 *M* Na_2HPO_4, 0.02 *M* KH_2PO_4, and 0.14 *M* NaCl at pH 7.2 with 0.04% w/v sodium azide and 1% v/v DABCO. Stir the solution overnight and adjust pH to 6.7. Remove insoluble material by centrifugation at 12,000*g* for 30 min. Store at 4°C.
8. Microscope slides and 22 × 50-mm glass cover slips (Scientific Laboratory Supplies, Nottingham, UK).
9. Parafilm.
10. Clear nail polish.
11. Poly-L-lysine-coated slides (BDH).
12. Humidified box (*see* **Note 1**).
13. Fluorescence microscope with FITC and rhodamine red filters such as Zeiss Axiophot or Axioplan.

2.3. Wounding of Mouse Epidermis

1. Hair clippers.
2. Scalpel with number 11 blades (Swann-Morton).
3. Marker pen or surgical pen.
4. 70% ethanol.
5. Anaesthesia apparatus.
6. Dissection instruments.
7. Ruler.
8. Aluminium foil.
9. OCT cryoprotectant (RA Lamb Ltd, Eastbourne, UK).
10. Liquid nitrogen.

2.4. Conventional Electron Microscopy

1. Ultramicrotome.
2. Platinum wire loops (0.5–3 mm).
3. Molds for embedding.
4. Razor blade.
5. Tungsten-coated glass knives.
6. 100 mesh hexagonal grids of copper.
7. Cacodylate buffer: 1 *M* sodium cacodylate, 250 m*M* sucrose, and 2 m*M* $CaCl_2$, pH 7.5.
8. Paraformaldehyde (PFA) solution: add 8 g of PFA powder to 100 mL of double distilled (dd)H_2O in a glass flask. Cover with a glass funnel and an inverted glass Petri dish to form a reflux system. Heat with stirring to 60 to 70°C for at least 30 min. Clear the remaining cloudiness with a few drops of 4 *M* sodium hydroxide solution. Allow to cool and then filter through Whatman paper. Dilute to 4% w/v with ddH_2O.
9. PFA in cacodylate buffer: prepare double-strength cacodylate buffer above (i.e., 0.2 *M* sodium cacodylate) and mix 50:50 with fresh 4% aqueous PFA solution. Re-adjust the pH to 7.5.
10. Uranyl acetate stain: prepare a 2% w/v solution of uranyl acetate dissolved in 77% ethanol. Centrifuge 1.5-mL aliquots at 1000*g* for 20 min before use.
11. Lead citrate stain: prepare a 0.3% w/v solution of lead citrate in 0.1 *M* sodium hydroxide. Centrifuge 1.5-mL aliquots at 1000*g* for 20 min before use.
12. Spurr's resin (medium hardness).
13. Ethanol concentrations 70, 80, and 100%.
14. Toluidine blue stain.

2.5. Immunogold Labeling

1. Cryoultramicrotome.
2. Tungsten-coated glass knives.
3. Diamond knife.
4. Eyelash probe (*see* **Note 2**).
5. Platinum loops.
6. 100 mesh nickel hexagonal grids coated with formvar and carbon (Agar Scientific Ltd).
7. Liquid nitrogen storage.
8. Aluminium cryopins (Reichert-Jung Ltd).
9. Light microscope.
10. HEPES buffer: 0.2 *M* HEPES, pH 7.4.
11. PFA in HEPES buffer: prepare double-strength HEPES buffer above (i.e., 0.4 *M* HEPES) and mix 50:50 with fresh 4% aqueous PFA solution. Re-adjust the pH to 7.4.
12. Sucrose/polyvinyl pyrrolidone (PVP *[8]*): prepare a solution of 2 *M* sucrose in 0.2 *M* HEPES buffer. Heat the mixture with stirring to 60 to 70°C until the sucrose has

dissolved completely. Mix the sucrose solution with a further 100 mL of HEPES buffer and then add 4 mL of 1.1 *M* sodium carbonate solution. Add 20 g of PVP-10 (molecular weight 10,000) to the solution. Leave the paste overnight at room temperature covered with aluminum foil. The bubbles in the paste will escape leaving a clear solution.

13. Retrieval buffer: 0.2 *M* HEPES pH 7.4, 2 *M* sucrose, 0.75% gelatin.
14. PBS.
15. 2% gelatin in PBS poured into sterile Petri dishes.
16. 0.02 *M* glycine in PBS.
17. Normal goat serum diluted to 5% in PBS containing 1% bovine serum albumin (BSA).
18. Normal goat serum diluted to 10% in PBS containing 1% BSA.
19. 11–5F antibody *(7)*.
20. 0.1% BSA in PBS.
21. Gold-conjugated goat anti-mouse secondary antibody.
22. 2.5% glutaraldehyde in PBS.
23. 2% uranyl acetate oxalate: mix equal volumes of a 4% (w/v) stock of uranyl acetate and 0.3 *M* oxalic acid. Adjust the pH to 7.5 with 5% ammonium hydroxide. Store in the dark at 4°C and filter through Whatman no. 1 filter paper before use.
24. 2% aqueous polyvinyl alcohol containing 0.2% uranyl acetate.

3. Methods

To examine the effect of Ca concentration on desmosomal adhesion, low Ca medium (LCM) must first be prepared by chelation of the Ca from FBS before adding to Ca-free DMEM (LCM) MDCK cells should then be grown on glass cover slips to the required level of confluence before performing the Ca switch assay. To measure the percentage of Ca-dependent and Ca-independent cells, the cells are first stained with a desmosomal antibody, such as one to desmoplakin, before microscopic examination for internalization of desmoplakin staining (Ca-dependent) or cell membrane staining (Ca-independent) and counting the numbers of each type of cell. Ca-dependence also can be induced in confluent (Ca-independent) monolayers by wounding (scraping the monolayers) or by treatment with PKC activators. Similarly, Ca-independence can be induced by treatment with PKC inhibitors. The Ca switch assay also can be used for untreated or wounded mouse skin tissue, in a similar way to MDCK cells. The effects on desmosomes can be investigated further by electron microscopy to see whether desmosomes have normal ultrastructure with intact midlines (Ca-independent) or have split into half desmosomes (Ca-dependent). Because PKCα is an important regulator of the adhesive state of desmosomes, it often is useful to look for colocalization of PKCα with desmosomes and the ideal way to confirm this is by immunogold labeling of desmosomes for PKCα and desmoplakin.

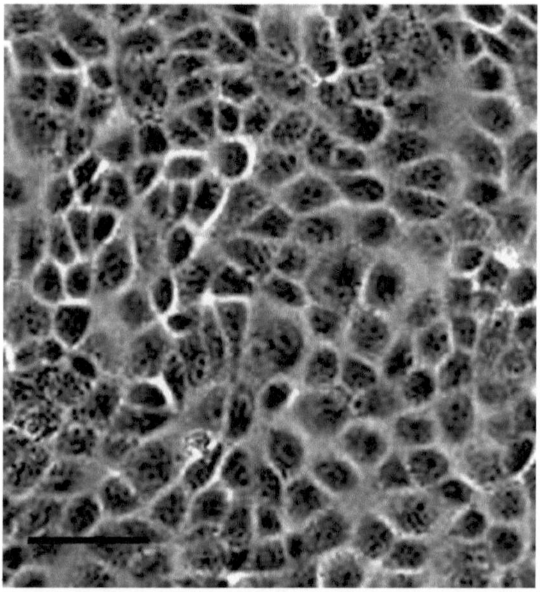

Fig. 2. Phase contrast image of a confluent monolayer of MDCK cells. MDCK cells were seeded at a density of 4.5×10^5 cells/well and maintained for 6 d in NCM. The cobble-stone epithelial morphology is readily visible. Scale bar = 25 μm.

3.1. Ca Switch Assay for MDCK Cells

3.1.1. Preparation of LCM

1. Stir 20 g of Chelex resin with 50 mL of heat-inactivated FBS (56°C for 30 min) for 2 h (or overnight if more convenient) at 4°C.
2. Decant liquid from beads and sterilize by passing through 0.2-μm filter.
3. Add the chelated FBS to Ca-free DMEM to a concentration of 10% w/v.
4. Add EGTA to a concentration of 3 m*M* (pH 7.5; *see* **Note 3**).
5. Store at 4°C until needed for Ca switch assay (*see* **Subheading 3.1.3.**).

3.1.2. Culture of MDCK Cells on Glass Cover Slips

1. Culture MDCK II cells in NCM (DMEM with 10% FBS), until confluent in a T75 flask (~1×10^7 cells/flask). MDCK cells have a characteristic "cobblestone" epithelial morphology when grown as confluent monolayers (*see* **Fig. 2**).
2. Prepare 24-well plates in which one 13-mm diameter circular glass cover slip can occupy each well (*see* **Note 4**). The cover slips can be sterilized by autoclaving or by immersing in 100% ethanol (*see* **Note 5**).
3. Passage the cells with trypsin–EDTA, perform cell counts using a haemocytometer, and seed onto the glass cover slips; 4–5×10^5 cells should be seeded per well to achieve confluence during the next 24 to 48 h.

3.1.3. Ca Switch (NCM to LCM; see **Note 6**)

1. Grow MDCK cells (*see* **Notes 7** and **8**) to the required level of confluence (*see* **Subheading 3.1.2.**). If the recommended 4–5 × 10^5 cells/cover slip are seeded, then to guarantee a majority of Ca-dependent desmosomes, cells should be grown to 95% confluence (~1–2 d): if a majority of Ca-independent desmosomes is required, cells should be maintained at confluent density for 6 d. At approx 4 to 5 d, you will get a mix of Ca-dependent and Ca-independent desmosomes.
2. To perform the Ca switch, remove the NCM from the cells and wash three times with Ca-free PBS or HBSS (1 mL per well per wash).
3. Incubate cells with the preprepared chelated LCM (*see* **Subheading 3.1.1.**) for 90 min at 37°C.
4. Remove LCM, wash briefly in Ca-free PBS or HBSS, and fix with ice-cold 50:50 methanol:acetone for 10 min.
5. Remove fixative, cover plate, and store at –20°C until required for immunofluorescence (*see* **Subheading 3.3.1.**).

3.2. Ca Switch Assay for Mouse Epidermis

1. Sacrifice mouse and remove hair from back skin using clippers.
2. Dissect mouse skin from back, trim fat and dermis tissue, wash in Ca-free PBS, and cut into 1-mm^3 pieces.
3. Submerge tissue in LCM and incubate for up to 6 h in tissue culture incubator (37°C, 5% CO_2) Incubate control tissue in NCM.
4. Remove medium and wash in Ca-free PBS.
5. Place tissue in aluminum foil molds containing OCT cryoprotectant and freeze in liquid nitrogen.
6. Label molds and store in a –80°C freezer.
7. Cut sections (7 μm) using a cryostat onto poly-L-lysine-coated slides.
8. Air-dry slides for 1 h at room temperature, then cover and place in –80°C freezer until required for immunofluorescence (*see* **Subheading 3.3.2.**).

3.3. Immunofluorescence for Desmoplakin

3.3.1. Cell Monolayers (see **Fig. 3**)

1. If cover slips have been stored in –20°C, allow to reach room temperature for 15 min.
2. Wash cells in PBS, 3 × 10 min.
3. Block nonspecific binding in 5% normal goat serum, 2% BSA, and 0.05% Triton-X-100 in PBS for 30 min.
4. Incubate in mouse anti-desmoplakin antibody (11–5F), at a dilution of 1/50 in PBS, for 1 h at room temperature (*see* **Notes 9** and **10**).
5. Wash in PBS, 3 × 10 min.
6. Incubate in secondary anti-mouse IgG antibody, such as Donkey anti-mouse IgG FITC at a dilution of 1/100 in PBS, for 30 min at room temperature (*see* **Note 11**).
7. Wash in PBS, 3 × 10 min, then wash off the PBS by washing with dH_2O.
8. Mount onto microscope slides with gelvatol and allow to solidify.

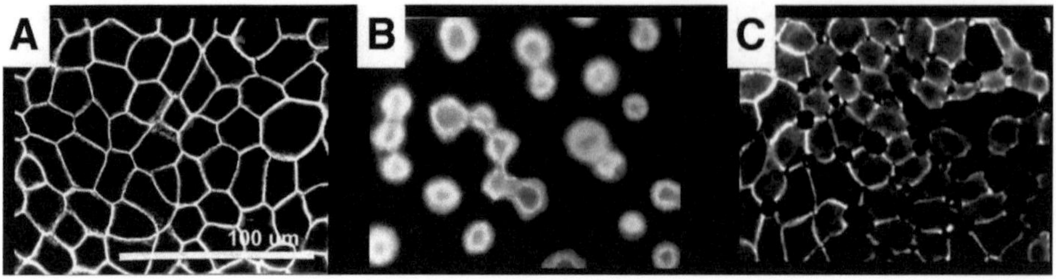

Fig. 3. Immunofluorescence for desmoplakin on MDCK cell monolayers. MDCK cells were seeded at confluent density (4.5×10^5 cells/well) and maintained for 2 d (**A, B**) or 6 d (**C**). Cells were then either left in NCM (**A**), or cells were switched to LCM (**B, C**) Cells were then fixed and stained for desmoplakin using the antibody 11-5F. In (**A**), cells have staining at the cell membranes. In (**B**), the cells have separated and desmoplakin has been internalized, indicative of Ca-dependent desmosomes. In (**C**), the cells have pulled apart, but processes remain connecting the cells and those stain brightly, indicative of Ca-independent desmosomes.

9. To stop drying out, seal the cover slips onto the slides with clear nail polish.
10. When nail polish is dry, wrap slides in aluminium foil and store at –20°C until viewing under fluorescence microscope (*see* **Notes 12** and **13**) and performing counts of Ca-dependent and Ca-independent cells (*see* **Subheading 3.6.**).

3.3.2. Tissue Sections (see *Fig. 4*)

1. Take frozen sections out of the –80°C freezer and fix in ice-cold 50:50 methanol:acetone for 20 min at –20°C.
2. Draw a ring around each section on the slide using a wax pen to prevent antibody running from one section to another.
3. Stain the sections using the same protocol for monolayers with two exceptions: (a) add 20 µL of diluted antibody per section and (b) incubate slides in a humidified box (*see* **Note 1**).
4. Perform counts of Ca-dependent and Ca-independent cells (*see* **Subheading 3.6.**). Similar protocols can be used for PKCα immunofluorescence (*see* **Notes 14** and **15**).

3.4. Wounding

3.4.1. Wounding Cell Monolayers

1. Grow MDCK cell monolayers on glass microscope slides within 100-cm^2 plastic dishes. The cells should be seeded at a density of 1.5×10^5 cells/cm^2. Incubate at 37°C for 6 d.
2. Make a template for wounds by drawing a rectangular grid on paper and place under the slide and view under microscope.

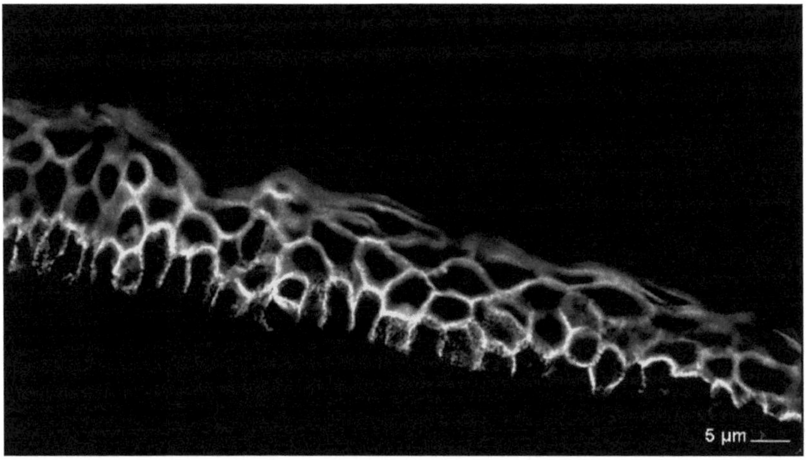

Fig. 4. Immunofluorescence for desmoplakin on mouse epidermis. Immunofluorescence image of 2-wk-old Balb-C mouse skin stained with 11-5F antibody for desmoplakin. Positive staining is evident around the cell membranes of epidermal cells. Scale bar = 5 μm.

3. Create simple wounds by drawing a scalpel blade across the monolayer by hand, tracing the predrawn grid.
4. For multiple wounds, score parallel lines 5 to 20 mm apart, rotate the monolayer 90° and draw a second set of parallel lines.
5. Prepare cell islands by removing cells between alternative sets of parallel score lines using a cell scraper.
6. Place cells in LCM or NCM (for controls).
7. Perform immunofluorescence for desmoplakin (*see* **Subheading 3.3.1.**), and count the percentage of Ca-independent and Ca-dependent cells (*see* **Subheading 3.6.**).

3.4.2. Wounding Mouse Epidermis

1. Anesthetize mice using inhalation anaesthetic such as halothane.
2. Sterilize back of mouse with a swab of 70% ethanol.
3. Shave back with electric clippers.
4. Mark 1-cm regions for wounding using a marker pen (*see* **Fig. 8**; **Note 16**).
5. Using a scalpel with a no. 11 blade, wound the mice on these marks. The wounds should be of full thickness, cutting through epidermis, dermis, and muscle.
6. Let the mouse recover and give analgesia as required.
7. To harvest the wounds, sacrifice the mice at the appropriate time.
8. Dissect at least 1 cm around the wound area and process for Ca switch assay (**Subheading 3.2.**) conventional electron microscopy (**Subheading 3.7.**), or immunogold labeling (**Subheading 3.8.**) as required.

3.5. Treatment of Cell Monolayers With PKC Activators and Inhibitors

3.5.1. Treatment With PKC Activators to Revert Ca-Independent Desmosomes to Ca-Dependent

1. Seed MDCK cells at 4–5 × 10^5 cells/cm^2 on to cover slips and culture for 6 d.
2. Treat the cells with the drug as follows: 5 n*M* TPA for 1 h; 50 nM phorbol-12, 13-didecanoate for 1 h; and 50 n*M* phorbol-12, 13- dibutyrate for 1 h. With different batches of cells, the optimal incubation times and concentrations of the drugs may vary slightly.
3. Switch to drug-free LCM for 90 min.
4. Fix the cells and stain for desmoplakin using protocol described in **Subheading 3.3.1.**
5. Calculate the proportion of cells with Ca-independent desmosomes as described in **Subheading 3.6.**

3.5.2. Treatment With PKC Inhibitors to Revert Ca-Dependent Desmosomes to Ca-Independent

1. Seed MDCK cells at 4–5 × 10^5 cells/cm^2 on to cover slips and culture for 24 h.
2. Treat the cells with the drug as follows: (a) 10 n*M* Gö6976 for 1 h; (b) 8.7 μ*M* chelerythrine chloride for 15 min. With different batches of cells, the optimal incubation times and concentrations of the drugs may vary slightly.
3. Switch to drug-free LCM for 90 min.
4. Fix the cells and stain for desmoplakin using protocol described in **Subheading 3.3.1.**
5. Calculate the proportion of cells with Ca-independent desmosomes as described in **Subheading 3.6.**

3.6. Scoring Cells for Ca-Independence

1. Select at least 10 fields of view at random.
2. Count the number of cells that remain attached by at least one desmoplakin-staining projection (*see* **Fig. 3**), using a hand-held counter. (These are Ca-independent cells.)
3. Count the total number of cells using a hand-held counter.
4. Express the number of Ca-independent cells as a proportion of the total number of cells in the fields.
5. Count several duplicates of cultures/tissue for each experimental point.

3.7. Conventional Transmission Electron Microscopy Examination of Desmosome Ultrastructure (see *Figs. 5 and 6*)

3.7.1. Fixation and Embedding

1. Fix 1-mm^3 pieces of tissue (**Subheading 3.2.** or **Subheading 3.4.2.**) in 2% PFA plus 2% glutaraldehyde in cacodylate buffer pH 7.5 for 2 h at room temperature, changing the fixative after 1 h. Leave in fixative overnight at 4°C.
2. Wash the samples in several changes of cacodylate buffer for 1 h at room temperature.

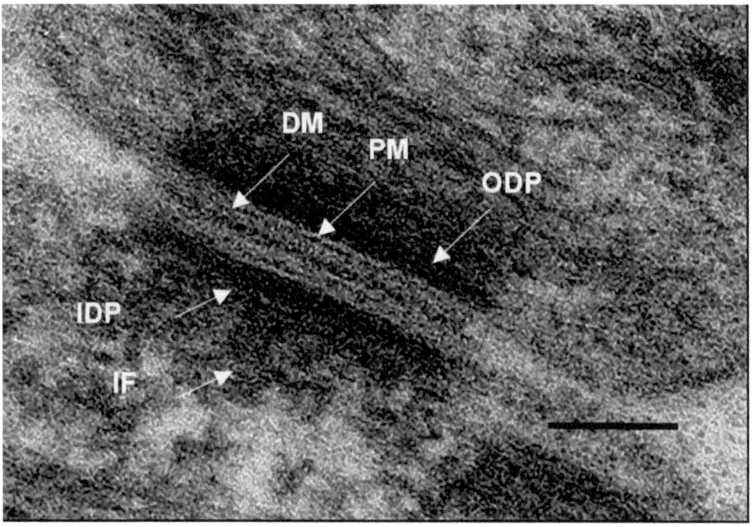

Fig. 5. Electron micrograph showing the ultrastructure of a desmosome. DM, dense midline; PM, plasma membrane; ODP, outer dense plaque; IDP, inner dense plaque; IF, intermediate filaments. Scale bar = 100 nm.

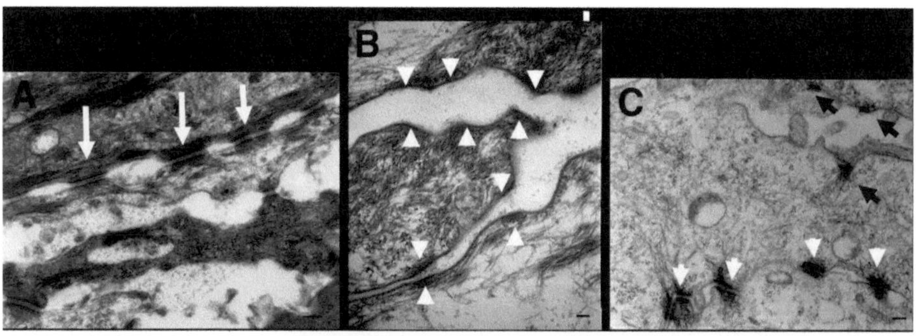

Fig. 6. Electron micrographs showing Ca-independent and Ca-dependent desmosomes after incubation in low Ca medium. **(A)** Ca-independent desmosomes in mouse epidermis (white arrows) Note the enlarged intercellular spaces. **(B)** Ca-independent desmosomes in wounded mouse epidermis. White arrows show the desmosomes split into half-desmosomes. **(C)** Wounded mouse epidermis with Ca-dependent desmosomes (black arrows) and Ca-independent desmosomes (white arrows) further away from the wound edge. Scale bars = 100 nm.

3. Postfix in 2% osmium tetroxide in cacodylate buffer for 1 h at room temperature.
4. Rinse several times in ddH_2O.
5. Dehydrate through an ethanol series: 70% for 15 min, 80% for 15 min, and 100% three times for 20 min each.

6. Infiltrate tissue with Spurr's resin (medium hardness): 50% resin in dry ethanol overnight, 70% resin in dry ethanol for 1 h, 100% resin in dry ethanol for 3 h three times.
7. Embed tissue in molds containing fresh resin and polymerize at 60°C for 72 h.
8. Embedded tissue can be stored at room temperature before sectioning.

3.7.2. Sectioning

1. Trim resin blocks first with a razor blade and then finely trim on ultramicrotome to remove excess resin (*see* **Note 17**).
2. Cut semithin sections (250–500 nm) and collect on poly-L-lysine-coated glass slides for staining with toluidine blue and examination under the light microscope to ensure that epidermis or wound edges are included in the section.
3. Cut ultrathin sections (50–70 nm) using a diamond knife. The sections will float away from the knife edge on the water reservoir.
4. Collect sections from the water on 100 mesh grids and leave to air-dry (*see* **Note 18**).

3.7.3. Staining

1. Wet grids on drops of ddH_2O.
2. Float grids on drops of 2% uranyl acetate solution for 30 min at room temperature.
3. Wash each grid three times in ddH_2O.
4. Incubate on 0.3% lead citrate solution for 5 min.
5. Wash thoroughly in ddH_2O and then allow to air-dry overnight before examination in the transmission electron microscope (TEM).

3.8. Immunogold Electron Microscopy to Localize Desmoplakin or PKCα Within the Desmosome (see Fig. 7 and Notes 19 and 20)

3.8.1. Fixation and Infiltration

1. Fix 1-mm^3 pieces of tissue (**Subheadings 3.2.** or **3.4.2.**) in 2% PFA in HEPES buffer for 45 min at room temperature.
2. Wash four times with HEPES buffer, 10 min per wash.
3. Infiltrate with sucrose/PVP for 4 h at room temperature or overnight at 4°C (*see* **Notes 21** and **22**).
4. Mount the infiltrated tissue on freshly cleaned aluminium cryopins and remove any excess sucrose/PVP.
5. Manually plunge freeze the tissue in liquid nitrogen and then store under liquid nitrogen prior to sectioning.

*3.8.2. Cryoultramicrotomy Sectioning (see **Notes 23** and **24**)*

1. Precool the chamber of a cryoultramicrotome to –80°C and place the 45° glass and diamond knives inside to equilibrate.
2. Transfer the frozen tissue from liquid nitrogen to the cryochamber and allow it to equilibrate for 15 min.

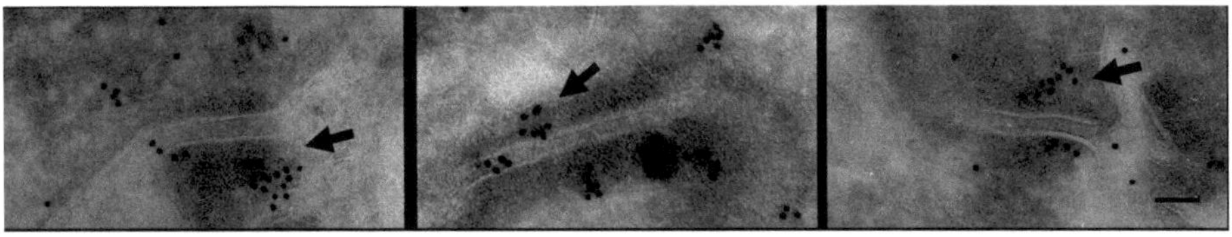

Fig. 7. Electron micrographs of wounded mouse epidermis immunogold labeled for PKCα. Note the localization of PKCα to the desmosomes (black arrows). Scale bar = 100 nm.

3. Using the glass knife, trim the sucrose/PVP block to the smallest possible area.
4. Cut semithin sections for staining with toluidine blue and collect on poly-L-lysine-coated glass slides. Examine under the light microscope to ensure visualization of the epidermis or wound edge.
5. Reduce the temperature of the cryochamber to –120°C and allow it and the knives to equilibrate.
6. Once the temperature of the tissue has reached –110°C, ultrathin sections can be cut using the diamond knife.
7. Adjust the cutting speed to allow sections of the required thickness (50–60 nm) to be obtained and to maintain the temperature of the specimen.
8. The sections will form a ribbon on the knife. Use an eyelash probe to arrange them in groups of three or four.
9. Retrieve the sections using a 1.5- to 3-mm wire loop containing a droplet of retrieval buffer. Work quickly to avoid the buffer freezing. The sections will melt instantaneously and adhere to the buffer. Rapidly withdraw the loop (*see* **Note 25**).
10. Transfer the cryosections to 100 mesh formvar-coated grids.
11. Invert the grids onto a plate of 2% gelatin in PBS and store overnight at 4°C.

3.8.3. Immunogold Labeling

1. Immediately before labeling, liquefy the gelatin block by incubating at 37°C for 20 min (*see* **Notes 26** and **27**).
2. Transfer the grids across three drops of 0.02 *M* glycine in PBS for 1 h at room temperature (*see* **Note 28**).
3. Block nonspecific labeling by incubating grids on droplets of 5% normal goat serum plus 1% BSA for 1 h at room temperature.
4. Prepare the monoclonal antibody 11–5F at a dilution of 1/20 in PBS plus 1% BSA.
5. Transfer the grids through three drops of the diluted primary antibody at 37°C (20 min per drop).
6. Wash the grids five times (10 min per wash) in PBS plus 0.1% BSA.
7. Dilute gold conjugated goat anti-mouse secondary antibody 1:200 in 10% normal goat serum plus 1% BSA in PBS.
8. Incubate grids on the secondary antibody for 45 min at room temperature.
9. Wash five times (10 min per wash) in PBS.

10. Post fix in 2.5% glutaraldehyde in PBS for 5 min at room temperature.
11. Wash in five changes of ddH_2O for 5 min (*see* **Note 29**).
12. Contrast sections on drops of 2% uranyl acetate oxalate and then wash briefly on three drops ddH_2O (*see* **Note 30**).
13. Incubate the grids on 2% aqueous polyvinyl alcohol (PVA)/0.2% uranyl acetate for 15 min (*see* **Note 31**).
14. Use wire loops to retrieve the grids and draw across filter paper to remove excess liquid.
15. Allow the grids to air-dry before examination in the TEM (*see* **Note 31**).

4. Notes

1. A humidified box can be made from a plastic sandwich box containing paper towels wetted with distilled water or PBS.
2. An eyelash probe is made by attaching an eyelash to a wooden stick.
3. The Ca concentration of LCM should be less than 0.05 m*M*, and the Ca concentration of NCM is approx 1.8 m*M*. The Ca concentration can be measured by atomic absorption spectroscopy if required. *See* **ref. *10*** for details.
4. Each time point in an experiment requires a separate 24-well plate because all cells in one plate should be fixed simultaneously. For each time point, some cover slips should be kept in NCM as controls.
5. If the ethanol method is chosen for sterilizing cover slips, the ethanol should be completely removed before adding cells by either flaming cover slips in a Bunsen flame or allowing to air-dry in the tissue culture hood for 30 min.
6. This method is for a downward Ca switch, i.e. from NCM to LCM. Cells can also be switched from LCM to NCM as described in ***(4)***.
7. The Ca switch assay also can be performed with primary keratinocytes ***(3)***, Caco-2 colon cells ***(9)***, HaCaT cells, and A431 cells (our unpublished data).
8. MDCK cells stably transfected with GPF or YFP-labeled desmosomal proteins can also be used to directly visualize desmosomes during Ca switch assays.
9. Other antibodies than the mouse monoclonal antibody to desmoplakin (11–5F) can be used as a marker for desmosomes. Some excellent antibodies are commercially available from Progen (www.progen.de) or Research Diagnostics Inc (www.researchd.com) Requests for our antibody to desmoplakin (11–5F) should be addressed to Professor David Garrod, University of Manchester, UK.
10. To perform the immunofluorescence of the cover slips, cover a large glass plate with Parafilm. Place drops of solutions directly onto the Parafilm and invert the cover slips in the drops. Allow 15 μL for blocking reagent and antibodies and 200 μL for PBS washes. Use forceps to transfer the cover slips from one drop to the next.
11. Nuclei can be counterstained blue by including a 1/1000 dilution of Hoechst 33258 stock (5 mg/mL) with the secondary antibody.
12. Fluorescently stained slides can be kept for up to 2 wk wrapped in foil in –20°C freezer and sometimes longer.

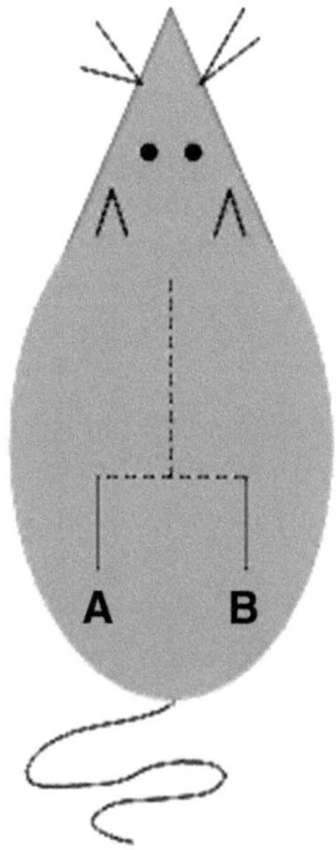

Fig. 8. Diagram to show the regions for wounding mouse epidermis. Wounds were made 3 cm posterior to the animal's neck, measured from the base of the scull, 1 cm long and 2 cm apart. Each mark was 1 cm lateral to the animal's midline. **A** and **B** are 1-cm wound sites.

13. All our microscopy is conducted using a Zeiss Axioplan microscope with a 50W HBO bulb for epi-fluorescence, using ×16, ×40, and ×63 objectives. The filter sets we use are: blue (G365, FT 395, LP420), green (450–490, FT510, 515–565), and red (530–585, FT 600, LP615).
14. To analyze the distribution of PKCα by immunostaining: seed the cells at confluent density and maintain for 6 d to acquire Ca-independent desmosomes then stain exactly as for desmoplakin (DP) but use the anti-PKCα antibody at a dilution of 1/50–1/100, and a goat-anti-rabbit secondary antibody conjugated to cy3 or rhodamine red.
15. To perform double staining for desmplakin and PKCα, add anti-DP antibody for 1hr, wash in PBS, add anti-PKCα for 1 h, wash in PBS, then add secondary antibodies together for 30 min. We recommend using a FITC-labeled secondary antibody for detecting DP and either rhodamine red or cy3 for PKCα. Take care when analyzing the double staining under the fluorescence microscope, not to get bleedthrough from red channel into FITC channel.
16. The regions for wounding on mouse epidermis were made 3 cm posterior to the animal's neck, measured from the base of the scull, 1 cm long, and 2 cm apart. Each mark was 1cm lateral to the animal's midline. *See* **Fig. 8** and ***(11)*** for further details. In the UK, a Home Office licence is required for this procedure.

17. We conduct our sectioning for conventional electron microscopy on an Ultracut ultramicrotome (Reichert-Jung Ltd).
18. An indication of a good section is a silver/gold interference color.
19. Before trying any antibody in immunogold labeling, we strongly recommend trying it on PFA fixed tissue with the immunofluorescence method. If the antibody does not work with this method, it almost certainly will not work in immunogold labeling.
20. Our cryoultramicrotomy protocol is based on that developed by Tokuyasu ***(8,12–15)***.
21. The addition of PVP to the sucrose infiltrating solution confers greater plasticity on the blocks than sucrose alone.
22. Tissue sections for immunogold labeling can be embedded in PVA; *see* **ref. *15*** for a protocol.
23. The cryoultramicrotome we use is a Leica Ultracut S/FCS.
24. Glass knives are coated with tungsten as described by Roberts ***(16)*** and Griffiths et al. ***(17)***.
25. To transfer a grid to a droplet from the loop: hold the loop containing the grid perpendicular to the drop and lower into the liquid. The surface tension will release the grid.
26. The immunogold labeling is conducted on drops of each reagent placed on a sheet of clean parafilm.
27. It is very important to keep the back of the grids dry and not to immerse them in reagent. The grids should float on top of the drops.
28. The glycine wash is to quench any free aldehydes in the fixed tissue.
29. It is important to wash the grids in ddH_2O afrer postfixing to remove all traces of PBS that would precipitate with the uranyl acetate forming large, electron-dense crystals of uranyl phosphate.
30. The uranyl acetate oxalate step is thought to stabilize membranes against the low pH of unbuffered uranyl acetate.
31. The final PVA embedding step is to protect the cryosections from air-drying artefacts and should be sufficient to leave only a thin film over the section. The addition of uranyl acetate to the PVA adds contrast to the sections.
32. We use Philips 400 and Philips Tecnai Biotwin transmission electron microscopes.

Acknowledgments

We acknowledge Drs Sarah Wallis, Mohamed Berika, and Melanie Blisset for protocols obtained from their PhD theses. We also thank Maybo Chiu, Tomomi Kimura, and Svetlana Riazanskaia for providing some of the figures, and Professor David Garrod for his advice on the manuscript. Work in this laboratory is funded by the Medical Research Council and the Welcome Trust.

References

1. Garrod, D. R., Merritt, A. J., and Nie, Z. (2002) Desmosomal cadherins. *Curr. Opin. Cell Biol.* **14,** 537–545.

2. Hennings, H. and Holbrook, K. A. (1983) Calcium regulation of cell-cell contact and differentiation of epidermal cells in culture: an ultrastructural study. *Exp. Cell Res.* **143,** 127–142.
3. Watt, F. M., Mattey, D. L., and Garrod, D. R. (1984) Calcium-induced reorganization of desmosomal components in cultured human keratinocytes. *J. Cell Biol.* **99,** 2211–2215.
4. Mattey, D. L. and Garrod, D. R. (1996a) Calcium-induced desmosome formation in cultured kidney epithelial cells. *J. Cell Sci.* **85,** 95–111.
5. Mattey, D. L. and Garrod, D. R. (1986b) Splitting and internalization of the desmosomes of cultured kidney epithelial cells by reduction in calcium concentrations. *J. Cell Sci.* **85,** 113–124.
6. Wallis, S., Lloyd, S., Wise, I., Ireland, G., Fleming, T. P., and Garrod, D. R. (2000) The α isoform of protein kinase C is involved in signaling the response of desmosomes to wounding in cultured epithelial cells. *Mol. Biol. Cell* **11,** 1077–1092.
7. Parrish, E. P., Steart, P. V., Garrod, D. R., and Weller, R. O. (1987) Antidesmosomal monoclonal antibody in the diagnosis of intracranial tumors. *J. Pathol.* **153,** 265–273.
8. Tokuyasu, K. T. (1989) Use of poly(vinylpyrrolidone) and poly(vinyl alcohol) for cryoultramicrotomy. *Histochem. J.* **21,** 163–171.
9. Collins, J. E., Taylor, I., and Garrod, D.R (1990) A study of desmosomes in colorectal carcinoma. *Br. J. Cancer* **62,** 796–805.
10. Brennan, J. K., Mansky, J., Roberts, G., and Lichtman, M. A. (1975) Improved methods for reducing calcium and magnesium concentrations in tissue culture medium: application to studies of lymphoblast proliferation in vitro. *In Vitro* **11,** 354–360.
11. Tomlinson, A. and Ferguson, M. J. (2003) Wound healing: a model of dermal wound repair, in *Inflammation Protocols* (Winyard, P. and Willoughby, D. A., eds.), Humana, Totowa, NJ, pp. 249–260.
12. Tokuyasu, K. T. (1973) A technique for ultracryotomy of cell suspensions and tissues. *J. Cell Biol.* **57,** 551–565.
13. Tokuyasu, K. T. (1980) Immunochemistry on ultrathin frozen sections. *Histochem. J.* **12**, 381–403.
14. Tokuyasu, K. T. and Singer, S. J. (1976) Improved procedures for immunoferritin labelling of ultrathin frozen sections. *J. Cell Biol.* **71,** 894–906.
15. North, A. J., Chidgey, M. A. J., Clarke, J. P., Bardsley W. G., and Garrod, D. R. (1996) Distinct desmocollin isoforms occur in the same desmosomes and show reciprocally graded distribution in bovine nasal epidermis. *Proc. Natl. Acad. Sci. USA* **93,** 7701–7705.
16. Roberts, I. M. (1975) Handling and staining epoxy resin sections for light microscopy. *J. Microsc.* **103,** 113–119.
17. Griffiths, G., McDowall, A., Back, R., and Dubochet, J. (1984) On the preparation of cryosections for immunocytochemistry. *J. Ultrastruct. Res.* **89,** 65–79.

15

Tight Junctions and Cell–Cell Interactions

Markus Utech, Matthias Brüwer, and Asma Nusrat

Summary

Chronic inflammation in mucosal tissues can influence epithelial barrier function via pro-inflammatory cytokines such as interferon (IFN)-γ and tumor necrosis factor-α. Increased mucosal levels of these cytokines have been observed in mucosal biopsies from patients with a chronic inflammatory condition referred to as inflammatory bowel disease. Paracellular permeability across epithelial cells is regulated by tight junctions (TJs), which are the apical most junctions in epithelial cells. Given that pro-inflammatory cytokines modulate the epithelial barrier and that TJs regulate epithelial permeability, we analyzed the influence of IFN-γ on TJ function/structure. Our results suggest that IFN-γ induced a time-dependent increase in paracellular permeability that was associated with internalization of TJ transmembrane proteins, occludin, junction adhesion molecule A, and claudin-1. In this chapter, we focus on selected methods used to investigate the influence of IFN-γ on epithelial barrier function.

Key Words: Tight junction, interferon-γ; confocal microscopy; immunofluorescence; transepithelial electrical resistance; paracellular flux.

1. Introduction

Epithelial cells lining luminal organs, such as the gastrointestinal tract, interface luminal contents and underlying tissue compartments. Polarized epithelial cells create a highly efficient and selective barrier between different tissue compartments *(1)*. Epithelial paracellular permeability is regulated by the apical intercellular junctional complex referred to as the apical junctional complex. The major constituents of this complex are tight junctions (TJs) and subjacent adherens junctions. TJs represent a multiprotein complex consisting of transmembrane and cytosolic components *(2,3)*. Three major TJ transmembrane proteins include occludin, members of the claudin family, and the immunoglobulin-like superfamily members, such as junctional adhesion molecule (JAM)-A and coxackie-adenovirus-receptor. These transmembrane proteins affiliate with the underlying

From: *Methods in Molecular Biology, vol. 341: Cell–Cell Interactions: Methods and Protocols*
Edited by: S. P. Colgan © Humana Press Inc., Totowa, NJ

perijunctional F-actin ring via cytoplasmic plaque proteins, primarily *zonula occludens* (ZO) proteins ZO-1, ZO-2, and ZO-3 ***(2,4,5)***. Increased permeability of the epithelial barrier is a key contributor to various gastrointestinal diseases such Crohn's disease and ulcerative colitis ***(6,7)***, which are characterized by increased mucosal levels of proinflammatory cytokines such as interferon (IFN)-γ ***(8)***.

Given the complex composition of mucosal tissues and diverse array of cytokines released in inflammatory states, we have used a reductionistic in vitro cell culture approach to examine the influence of a single proinflammatory cytokine on epithelial barrier function. A model intestinal epithelial cell line, T84 and human recombinant IFN-γ was used to analyze the effects of this cytokine on structure and function of the epithelial TJs.

2. Materials

2.1. Collagen Coating of Permeable Polycarbonate Filter Supports

1. Rat tails (store at –80°C; Pet-Freez Biologicals, Rogers, AK).
2. Hemostat (Daigger, Vernon Hills, IL).
3. 1% (v/v) acetic acid solution in water (*see* **Note 1**).
4. Dialysis tubing: Spectra/Por no. 1 (Spectrum Laboratories Inc., Rancho Dominguez, CA).
5. Parafilm "M" (pechiney plastic packing, Chicago, IL).
6. Permeable polycarbonate membrane filters with surface areas of 0.33 cm^2 (5-μm pore size) and 5 cm^2 (3-μm pore size; Costar Transwell, Corning Incorporated, Corning, NY).

2.2. Cell Culture

1. T84 epithelial cells (ATCC, Rockville, MD).
2. Cell medium: 1:1 Dulbecco's Modified Eagle's Medium and Ham's F-12 medium supplemented with 15 m*M* HEPES, pH 7.5, 14 m*M* $NaHCO_3$, 0.4% (w/v) penicillin, 0.08% (w/v) ampicillin, 0.9% (w/v) streptomycin, and 6% (v/v) newborn calf serum.
3. IFN-γ stock solution (2500 U/mL; kind gift of Genentech, San Francisco, CA).

2.3. Transepithelial Electrical Resistance

1. EVOM™ Epithelial Voltohmmeter (World Precision Instruments, Sarasota, FL).
2. STX2 electrode set (World Precision Instruments, Sarasota, FL).
3. 0.15 *M* KCl.

2.4. Fluoresceinated Dextran-3 Flux

1. Fluoresceinated dextran (FD)-3 solution: 1% (w/v) fluorescein-labeled dextran 3 kDa (Molecular Probes, Eugene, OR) in Hank's balanced salt solution (HBSS).
2. 96-well plate (NUNC, Rochester, NY).

2.5. Cell Preparation for Immunoblotting

1. Cell lysate buffer: 100 m*M* KCl, 3 m*M* NaCl, 3.5 m*M* $MgCl_2$, 10 m*M* HEPES, pH 7.4, 1% (v/v) Triton X-100, protease inhibitor cocktail (1:100, Sigma, St. Louis, MO).
2. Cell scraper: 1.3-cm blade, total length 16 cm. (Sarstedt Inc., Newton, NC).
3. Glass homogenizer (douncer; Weathon Inc., Millville, NJ).
4. Protein concentration assay: BCA Protein assay kit (Pierce, Rockford, IL).

2.6. Differential Detergent Extraction of TJ Transmembrane Proteins

1. TX 100 extraction buffer: 100 m*M* KCl, 3 m*M* NaCl, 3.5 m*M* $MgCl_2$, 10 m*M* HEPES, pH 7.4, 1% (v/v) Triton X-100, protease inhibitor cocktail (1:100, Sigma).
2. Running buffer (2X) (pH 8.3): 3% (w/v) sodium dodecyl sulfate (SDS), 0.75 *M* Tris-HCl, pH 8.8, 20% (v/v) glycerol, and 20 m*M* dithiothreitol.

2.7. SDS-Polyacrylamide Gel Electrophoresis

1. Separating buffer: 1.5 *M* Tris-HCl, pH 8.6. Store at 4°C.
2. Stacking buffer: 1 *M* Tris-HCl, pH 6.8. Store at 4°C.
3. 30% (w/v) acrylamide:bisacrylamide solution (37.5:1; National Diagnostics, Atlanta, GA).
4. TEMED (Bio-Rad, Hercules, CA).
5. SDS solution: 10% (w/v) solution in water (store at room temperature).
6. Ammonium persulfate (AP) solution: 10% (w/v) AP solution in water (prepare solution before use).
7. Running buffer: 50 m*M* Tris-HCl, 190 m*M* glycine, 1% (w/v) SDS. Store at room temperature.
8. Prestained protein markers, broad range (6–175 kDa; New England Biolabs, Beverly, MA).

2.8. Western Blotting

1. Transfer buffer: 25 m*M* Tris-HCl (do not adjust pH), 190 m*M* glycine, 20% (v/v) methanol.
2. Trans-Blot Transfer membrane, pure nitrocellulose membrane (0.2 μm; Bio-Rad, Hercules, CA).
3. Tris-buffered saline (TBS): 200 m*M* NaCl, 50 m*M* Tris-HCl, pH 7.4.
4. TBS with Tween (TTBS): Add 1% (v/v) Tween-20.
5. Blocking buffer: 5% (w/v) nonfat dry milk in TTBS.
6. Primary antibodies are diluted in blocking buffer: polyclonal antibodies: anti-occludin (3 μg/mL), claudin-1 (3 μg/mL), and JAM-A (12 μg/mL) pAbs (Zymed Laboratories, San Francisco, CA).
7. Secondary antibodies are diluted in blocking buffer: horseradish peroxidase-conjugated goat anti-rabbit 0.16 μg/mL (Jackson Immunoresearch Laboratories, West Grove, PA).

8. ECL Western blot detection reagent (Amersham Biosciences, Piscataway, NJ).
9. Hyperfilm ECL film (Amersham Biosciences).
10. Trans-blot paper (Bio-Rad).

2.9. Immunofluorescence Labeling of TJ Proteins and Confocal Microscopy

HBSS+: 0.95% (w/v) HBSS (Sigma) and 10 m*M* HEPES, pH 7.4.

1. Fixation solution: absolute ethyl alcohol stored at –20°C.
2. Blocking solution and antibody dilution buffer: 1.5% (w/v) bovine serum albumin (BSA; Sigma) in HBSS+.
3. Primary antibodies: anti-occludin (8 μg/mL), claudin-1 (8 μg/mL), and JAM-A (8 μg/mL) pAbs (Zymed Laboratories).
4. Filter paper: 90-mm diameter (Whatman, Florham Park, NJ).
5. Secondary antibodies: goat anti-rabbit and goat anti-mouse secondary antibodies conjugated to Alexa-488 or Alexa-546 dyes (4 μg/mL; Molecular Probes, Eugene, OR).
6. Mounting medium: Slow Fade anti-fade kit (Molecular Probes, Eugene, OR).
7. Cover glasses: 18 mm^2 (Corning Inc., Corning, NY).

3. Methods

Epithelial cells are grown as monolayers on permeable polycarbonate supports. In this system epithelial cells can polarize and form well-developed intercellular junctions and are amenable to analysis by numerous techniques. Functional analysis of paracellular permeability across epithelial monolayers can be determined by measuring the transepithelial electrical resistance (TEER) to passive ion flow and by measuring the rate of paracellular solute flux of tracers from the apical to the basolateral compartment of epithelial cells. Functional studies can be complemented by morphological analysis of TJ protein localization by immunofluorescence labeling/confocal microscopy and by biochemical determination of protein expression and their subcellular compartmentalization. Because such an approach has been widely used to study epithelial TJs, we have highlighted some of the practical aspects of these techniques that we have used to examine the influence of IFN-γ on epithelial barrier function. More biophysical explanations and a general overview of these methods can be found in other reviews ***(9–11)***.

3.1. Collagen Coating of Permeable Polycarbonate Filter Supports

1. A razor blade is used to make a longitudinal incision through the skin of a rat-tail. The skin is stripped from the underlying tendons, which are extracted using hemostats. Mince and weigh the material extracted. This material can be frozen in 0.75-g aliquots at –80°C.
2. Soak 0.75 g of collagen in 70% ethanol for 10 to 20 min to dehydrate.

3. Solubilize collagen in 100 mL of 1% (v/v) acetic acid at 4°C. Stir the mixture slowly at 4°C overnight. The next day, the solution will be viscous and some undissolved material will remain.
4. Spin the collagen solution at 25,000*g* for 30 min at 4°C.
5. Collect the collagen solution, avoiding sedimented material, and dialyze against 2 L water for 24 h; use dialysis tubing, presoaked, for 1 h. Change water three times during dialysis.
6. Remove the collagen from the bag and dilute the solution with water if necessary (a little thinner than honey).
7. Store at 4°C.
8. Before use, dilute stock collagen 1:100 with 70% ethanol.
9. Put 50 μL of diluted collagen on each 0.33-cm^2 filter, and 500 μL on 5-cm^2 filters.
10. Allow to dry before use.

3.2. Cell Culture

1. The intestinal epithelia cell line, T84 cells is grown in 5% CO_2 at 37°C within humidified incubators. These cells are passaged when approaching confluence with trypsin/EDTA to provide new maintenance cultures in 162-cm^2 flasks and experimental cultures on different collagen coated filters.
2. For cell culture we plate the cells at 50% density (*see* **Note 2**).
3. Confluent monolayer of T84 cells in 162-cm^2 flask will yield 1.7×10^7 cells. Place 2×10^5 cells on 0.33 cm^2 and 2×10^6 cells in 5 cm^2. At such a density, epithelial cells achieve confluence and become polarized after 7 days on 0.33-cm^2 filter and 10 d on 5-cm^2 filters.
4. Before each experiment, measure the TEER of each filter.
5. For IFN-γ incubation in 0.33-cm^2 (5-cm^2) filters, remove 40 μL (80 μL) of media in the lower chamber and add 40 μL (80 μL) of IFN-γ stock solution basolaterally (*see* **Note 3**).

3.3. Transepithelial Electrical Resistance

1. To determine TEER, plate cells on 0.33-cm^2 filters and let them grow for 7 d (*see* **Note 4**).
2. The total resistance of the monolayer is defined as follows:

$$R_{(total)} = R_{(monolayer)} + R_{(blank)}$$

3. Measure the blank resistance through a 0.15 *M* KCl solution and across the membrane support lacking cells.
4. Before measuring TEER incubate the STX2 electrode in 70% ethanol for 10 min and let it air-dry.
5. Measure TEER by dipping simultaneously the electrode in the upper and lower chambers.
6. Calculate the TEER (*see* **Note 5**).

Record TEER of filters with or without cytokine incubation over a period of 48 h are presented in **Fig. 1A**.

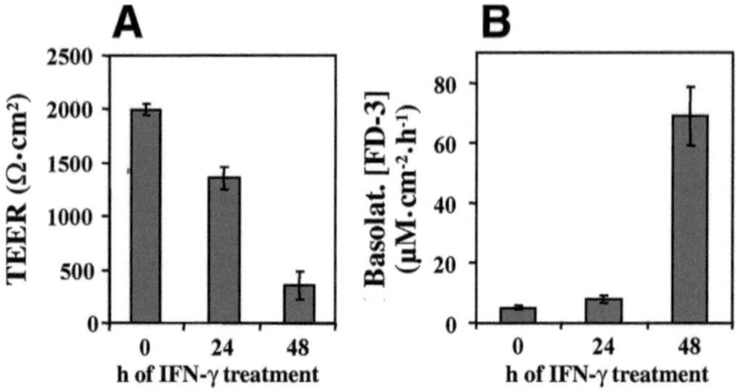

Fig. 1. Influence of IFN-γ on epithelia barrier function. Barrier function was monitored by measuring TEER and paracellular flux of FD-3. **(A)** TEER was measured after basolateral incubation of polarized epithelial monolayers with IFN-γ. **(B)** Paracellular flux of FD-3 in confluent T84 monolayers was measured after basolateral exposure to IFN-γ.

3.4. Paracellular Flux

1. For flux experiments plate cells on 0.33-cm^2 filters and let them grow for 7 d (*see* **Note 6**).
2. Check TEER of all monolayers.
3. Wash cells three times with warm HBSS+.
4. Add 200 μL of HBSS+ in the upper and 1 mL in the lower chamber.
5. Equilibrate cells in HBSS+ for 10 min on an orbital shaker and keep the cells on a heating stage at 37°C.
6. Aspirate HBSS+ from the upper chamber (avoid disturbing the monolayer).
7. Add 200 μL of FD-3 solution to the upper chamber (*see* **Note 7**). Shake the plate on orbital shaker at 37°C.
8. Remove 50 μL from the lower chamber at T = 120 min. Place in 96-well plate (*see* **Note 8).**
9. Create a standard curve with the FD-3 (serial dilution of mg/mL).
10. Read plate in fluorescence plate reader. Excitation: approx 490 nm, emission: approx 520 nm.
11. Plot standard curve to convert arbitrary fluorescence units to concentration. Flux is expressed as concentration/observed area/time (μ*M*/cm^2/h).

Paracellular flux of fluorescein isothiocyanate dextran (3 kDa) measurements across epithelial cells exposed to IFN-γ is shown in **Fig. 1B**.

3.5. Cell Preparation for Immunoblotting

1. T84 cells are grown on 5-cm^2 permeable supports for 10 d and incubated with IFN-γ for 48 h.
2. Wash cells twice with HBSS+.

3. Scrape cells into cell lysis buffer by using cell scraper at 4°C.
4. Break cells by using a douncer. Use for every sample the same number of strokes (e.g., 15) at 4°C.
5. Centrifuge cell lysates (1500*g*, 5 min, 4°C).
6. Determine the protein concentration of each sample and equalize protein concentrations (10 µg/lane).
7. Add the sample buffer and boil the sample for 5 min at 95°C.
8. Analyze equal amount of each fraction by SDS-polyacrylamide gel electrophoresis (PAGE)/Western blotting.

3.6. Differential Detergent Extraction of TJ Proteins

1. T84 cells are grown on 5-cm^2 permeable supports for 10 d and incubated with IFN-γ for 48 h.
2. Wash cells twice with HBSS+.
3. Incubate cells for 30 min with TX-100 extraction buffer. Shake the plate on orbital shaker at 4°C.
4. Carefully remove supernatant.
5. Subject the TX-100-soluble fraction to low-speed centrifugation (1500*g*, 5 min, 4°C) to remove cellular debris.
6. Add the TX-100-soluble fraction to an equal amount of sample buffer.
7. Scrape the TX-100-insoluble fraction into an equal amount of SDS sample buffer.
8. Boil samples for 5 min at 95°C.
9. Analyze equal volumes of each fraction by SDS-PAGE/Western blotting.

3.7. SDS-Polyacrylamide Gel Electrophoresis

1. These instructions assume the use of a Mini-PROTEAN 3 Electrophoresis System (Bio-Rad).
2. For occludin and JAM-A Western blotting prepare a 1.5-mm thick, 8% gel by mixing 2.5 mL of separating buffer with 2.7 mL of acrylamide/*bis* solution, 4.6 mL of water, 100 µL of AP solution, 100 µL of SDS solution, and 6 µL of TEMED. For claudin-1 Western blotting, prepare a 1.5-mm thick, 12% gel by mixing 2.5 mL of separating buffer with 4.0 mL of acrylamide/*bis* solution, 3.3 mL of water, 100 µL of AP solution, 100 µL of SDS solution, and 4 µL of TEMED. Pour the gel, leaving space for a stacking gel, and overlay it with water. The gel should polymerize in about 30 min.
3. Pour off the water.
4. Prepare the stacking gel by mixing 0.75 mL of stacking buffer with 1 mL acrylamide/*bis* solution, 4.1 mL of water, 50 µL of AP solution, 50 µL of SDS solution, and 10 µL of TEMED. Pour the stack and insert the comb. The stacking gel should polymerize within 30 min.
5. Once the stacking gel has set, carefully remove the comb and use a 3-mL syringe fitted with a 22-gage needle to wash the wells with running buffer.
6. Add the running buffer to the upper and lower chambers of the gel unit and load the samples.

7. Complete the assembly of the gel unit and connect to a power supply. The gel can be run at 30 mA.

3.8. Western Blotting

1. Switch the refrigerated/circulating water bath to maintain a temperature around 4°C.
2. Prechill the transfer buffer.
3. The samples that have been separated by SDS-PAGE are transferred electrophoretically to nitrocellulose membranes. These directions assume the use of a Bio-Rad transfer tank system.
4. Assemble a transfer cassette consisting of the gel and nitrocellulose.
5. Ensure that no bubbles are trapped in the gel, nitrocellulose sandwich.
6. The cassette is placed into the transfer tank such that the gel is between the nitrocellulose membrane and the cathode.
7. Assemble the lid on the transfer tank and activate the power supply. Transfers can be accomplished at 1 A for 1 h at 4°C.
8. Once the transfer is complete the cassette is taken out of the tank and disassembled carefully.
9. The nitrocellulose is incubated in 50 mL of blocking buffer for 1 h at room temperature on a rocking platform.
10. The blocking buffer is discarded and the membrane is quickly shrink-wrapped so that a small aperture is left to fill with 1 mL of primary antibody solution.
11. Air bubbles are smoothed out and the plastic foil is finally sealed.
12. Rotate the membrane for 1 h at room temperature.
13. The primary antibody is then removed and the membrane washed three times for 5 min each with 50 mL of TTBS.
14. The secondary antibody is freshly prepared for each experiment and the membrane is placed in 10 mL of secondary antibody solution for 1 h at room temperature on a rocking platform.
15. The secondary antibody is discarded and the membrane washed twice for 20 min with TTBS and finally with TBS for 20 min.
16. ECL reagent is assembled according to manufacture instructions.
17. After rotating the membrane in 2 mL of ECL reagent by hand for 1 min, the blot is removed, slightly desiccated by using Kim-Wipes, and then placed between the leaves of an acetate sheet protector that has been cut to the size of an X-ray film cassette.
18. The blot is then placed in an X-ray film cassette with film for a suitable exposure time.

Examples of proteins expression and differential detergent extraction results are shown in **Fig. 2**.

3.9. Confocal Immunofluorescence for TJ Proteins

1. For immunofluorescence labeling confluent T84 cells are plated on 0.33-cm^2 filters.
2. The cells should be rapidly rinsed twice with ice-cold HBSS+.
3. Precooled ethanol solution is then added to inserts for 20 min and stored at –20°C (200 μL in the upper and 1 mL in the lower chamber).

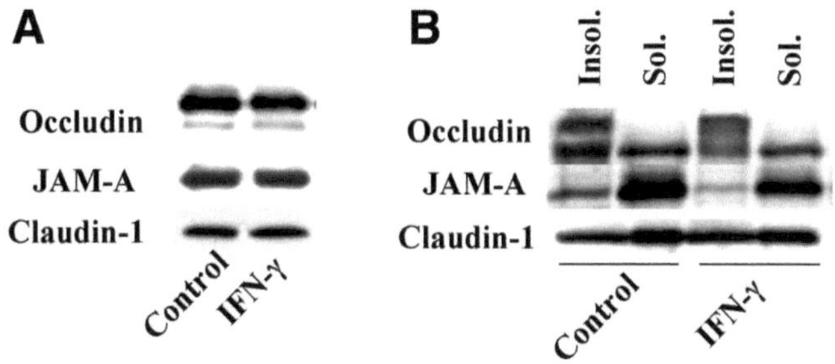

Fig. 2. Influence of IFN-γ on protein expression level and differential detergent solubility of TJ transmembrane proteins. **(A)** Protein expression levels of occludin, JAM-A and claudin-1 in cells incubated with IFN-γ (100 U/mL) for 48 h or control were determined. By using SDS-PAGE and immunoblotting for occludin, JAM-A, and claudin-1, no alteration in their overall expression was observed after treatment with IFN-γ. **(B)** T84 monolayers treated with or without IFN-γ (100 U/mL) for 48 h were incubated at 4°C with extraction buffer containing 1% TX-100. The TX-100-soluble (Sol.) and -insoluble (Insol.) fractions were analyzed by SDS-PAGE and immunoblotted for occludin, JAM-A, and claudin-1. Significant pools of occludin were TX-100 insoluble, whereas JAM-A and claudin-1 had a significant TX-100-soluble pool. In IFN-γ-treated cells, we observed a change in the TX-100 insolubility profiles of JAM-A. The small fraction of JAM-A in the TX-100-insoluble pool was further diminished by cytokine exposure.

4. Ethanol is discarded and the samples are blocked by incubation in blocking buffer for 1 h at room temperature.
5. Disassemble the filters supports and place filters in buffer drops on parafilm in humidity chambers.
6. Transfer filters to 50-μL drops containing primary antibody for 1 h at room temperature.
7. After incubation with the respective first antibody the filters are washed three times and incubated with secondary antibodies.
8. The filters are washed three times for 5 min with HBSS+.
9. Filters are mounted in mounting solution on glass slides. Nail varnish is used to seal the cover glasses and slides. The sample can be viewed immediately after the varnish has dry, or be stored in the dark at –20°C for as long as a month.

Confocal images showing internalized TJ transmembrane proteins after 48 h of IFN-γ treatment are shown in **Fig. 3**.

4. Notes

1. Unless stated otherwise, all solutions should be prepared in water with a resistance of 18.2 *M* Ohm*cm.
2. T84 cells should not be plated too thinly, otherwise they do not grow properly. A way to increase the density of cells is to trypsinize the cells and to place them in a smaller flask.

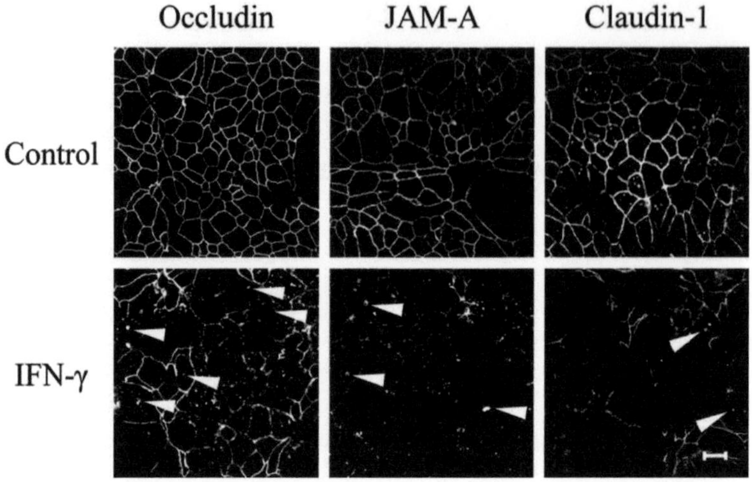

Fig. 3. IFN-γ stimuli induce internalization of TJ transmembrane proteins. Localization of occludin, JAM-A and claudin-1 was determined by immunofluorescence labeling/ confocal microscopy. In control T84 cells occludin, JAM-A and claudin-1 reveal typical distribution of these proteins in TJs. In T84 cells treated with 100 U/mL IFN-γ (for 48 h, occludin, JAM-A, and claudin-1 were internalized from the lateral membrane [arrowheads]). Scale bar = 10 μm.

3. T84 cells express the IFN-γ receptors exclusively basolaterally. Furthermore we observed that T84 cells responsed to IFN-γ if the cells are plated on 5-μm pore size filters.
4. Before each experiment check TEER of each filter to maintain equal starting point for each experiment. The TEER should not be lower than 1000 $\Omega \cdot cm^2$.
5. The TEER is definite as the resistance of a 1-cm^2 area:

$$TEER = R_{(monolayer)} \cdot \pi \cdot d^2 \div 4$$

   where d = diameter. Resistance readings for 24 mm or lager diameter inserts obtained by using the EVOM with the STX2 electrode should not be converted to unit area resistance. The Endohm is recommended for these large inserts.
6. Fluorescent-labeled dextran instead of radiolabeled solute was used for experiments. A consistent picomolar flux of FD in a range of sizes (3–70 kDa) across T84 epithelial monolayers has been demonstrated. FD flux is time- and dose-dependent ***(12)***.
7. Keep 1mL of HBSS+ in the lower chamber without FD-3 solution.
8. Protect plate from light.

References

1. Madara, J. L. and Stafford, J. (1989) Interferon-gamma directly affects barrier function of cultured intestinal epithelial monolayers. *J. Clin. Invest.* **83,** 724–727.

2. Gonzalez-Mariscal, L., Betanzos, A., and Nava, P. (2003) Tight junction proteins. *Prog. Biophys. Mol. Biol.* **81,** 1–44.
3. Tsukita, S., Furuse, M., and Itoh, M. (2001) Multifunctional strands in tight junctions. *Nat. Rev. Mol. Cell Biol.* **2,** 285–293.
4. Fanning, A. S., Jameson, B. J., and Jesaitis, L. A. (1998) The tight junction protein ZO-1 establishes a link between the transmembrane protein occludin and the actin cytoskeleton. *J. Biol. Chem.* **273,** 29,745–29,753.
5. Wittchen, E. S., Haskins, J., and Stevenson, B. R. (1999) Protein interactions at the tight junction. Actin has multiple binding partners, and ZO-1 forms independent complexes with ZO-2 and ZO-3. *J. Biol. Chem.* **274,** 35,179–35,185.
6. Sandle, G. I., Higgs, N., Crowe, P., Marsh, M. N., Venkatesan, S., and Peters, T. J. (1990) Cellular basis for defective electrolyte transport in inflamed human colon. *Gastroenterology* **99,** 97–105.
7. Katz, K. D., Hollander, D., Vadheim, C. M., et al. (1989) Intestinal permeability in patients with Crohn's disease and their healthy relatives. *Gastroenterology* **97,** 927–931.
8. Stallmach, A., Giese, T., and Schmidt, C. (2004) Cytokine/chemokine transcript profiles reflect mucosal inflammation in Crohn's disease. *Int. J. Colorectal Dis.* **19,** 308–315.
9. Madara, J. L. (1998) Regulation of the movement of solutes across tight junctions. *Annu. Rev. Physiol.* **60,** 143–159.
10. Clausen, C. (1989) Impedance analysis in tight epithelia. *Methods Enzymol.* **171,** 628–642.
11. Stoddard, J. S. and Reuss, L. (1988) Voltage- and time dependence of apical membrane conductance during current clamp in Necturus gallbladder epithelium. *J. Membr. Biol.* **103,** 191–204.
12. Sanders, S. E., Madara, J. L., McGuirk, D. K., Gelman, D. S., and Colgan, S. P. (1995) Assessment of inflammatory events in epithelial permeability: a rapid screening method using fluorescein dextrans. *Epithelial Cell Biol.* **4,** 25–34.

16

Simultaneous Immunofluorescence Detection of Cadherin and β-Catenin in Mouse Tissue Sections

Nobuko Obara

Summary

Both cell–cell adhesion and secreted signaling molecules are involved in the regulation of normal tissue development and maintenance. β-catenin seems to link cadherin-mediated cell–cell adhesion and Wnt signaling. Thus, its activity in particular cells or tissues, either during normal development or in tumorigenesis, has become a fascinating topic of investigation. Because the multiple functions of β-catenin intimately relate to its amount and localization within a cell, immunohistochemical detection of this molecule simultaneously with the cell adhesion molecule cadherin should provide helpful information on its potential activities. This detection can be conducted on tissue sections by using appropriate tissue preparation procedures and primary antibodies from different species.

Key Words: Cadherin; β-catenin; cell adhesion; Wnt signaling; subcellular localization; immunofluorescence; confocal laser microscopy.

1. Introduction

Direct cell–cell adhesion mediated by cell-adhesion molecules, as well as indirect cell–cell interaction mediated by secreted signaling molecules, constitutes the fundamentals of normal tissue development and maintenance *(1)*. Classical cadherins are the best-studied family of cell adhesion molecules that mediate Ca^{2+}-dependent cell–cell adhesion *(2)*, and alterations in cadherin function have been implicated in tumorigenesis *(3)*. β-catenin is one of the cytoplasmic components of the molecular assembly required for cell–cell adhesion mediated by cadherin *(4)*, and this molecule also participates in the canonical Wnt signaling pathway *(5,6)*. During embryonic development, the Wnt signaling regulates cell proliferation, cell fate specification, and differentiation. Misregulation of this Wnt signaling pathway caused by defects in Wnt components

From: *Methods in Molecular Biology, vol. 341: Cell–Cell Interactions: Methods and Protocols*
Edited by: S. P. Colgan © Humana Press Inc., Totowa, NJ

leads to abnormality in embryonic development and to a variety of diseases, including cancer in adults ***(6)***. Because β-catenin has been presumed to be a key molecule integrating cadherin-mediated cell–cell adhesion with Wnt signaling ***(5,7,8)***, the activity of this molecule is an attractive subject to be studied in both normal development and tumorigenesis.

The amount and localization of β-catenin in a cell can be regulated by Wnt signaling. In the absence of Wnt signaling, β-catenin free from cadherin is phosphorylated and targeted for ubiquitination through interaction with axin, adenomatous polypopsis coli, and the protein kinases casein kinase-1 (CK1) and glycogen synthase kinase-3 and is then degraded by the proteasome. In this situation, the level of β-catenin in the cytoplasm is kept to a low level. When cells are exposed to Wnt signals, activation of Wnt signaling inhibits β-catenin phosphorylation and, hence, its degradation. As a result, β-catenin accumulates in the cytoplasm and enters into the nucleus, where it interacts with T-cell factor/lymphoid enhancer factor (LEF/TCF) transcription factor to activate Wnt target genes ***(5,6)***. In this way, the amount of β-catenin changes depending on the Wnt signaling and its localization within a cell intimately relates to its function. Using confocal laser microscopy, we have demonstrated that abundant nuclear accumulation of β-catenin and upregulation of P-cadherin simultaneously take place in the enamel knot, the signaling center in the developing tooth germ ***(9)***. The use of primary antibodies from different species allows simultaneous detection of two or three antigens in general. Provided that the appropriate fixatives and antibodies are used, cadherins and β-catenin can be detected at the same time on a fixed and frozen tissue section, which has the advantage of providing tissue morphology of good quality. For this purpose, we have used periodate-lysine 2% paraformaldehyde (PLP) fixative solution ***(10)***. β-catenin colocalized at the cell surface along with cadherins most likely participates in cadherin-mediated cell–cell adhesion, whereas the same molecule accumulated in the cytoplasm or nucleus probably plays another role. Nuclear localization of β-catenin can be distinguished from the cytoplasmic one by use of nuclear counterstaining.

2. Materials

2.1. Preparation of Frozen Tissues

1. PLP fixative solution: To 50 mL of 200 m*M* lysine-HCl, add 100 m*M* dibasic sodium phosphate (Na_2HPO_4) until the pH has been adjusted to 7.4 and then adjust the volume to 100 mL with 100 m*M* sodium phosphate buffer, pH 7.4 (stock solution A). Suspend 8 g of paraformaldehyde (extra pure, Merck) in 100 mL of distilled water and heat in a fume hood on a stirring hot plate at 60°C. Add a few drops of 1 *N* NaOH to the solution and stir until it clears (stock solution B). Both stock solutions should be stored at 4°C and used within 10 d (solution A) or 1 mo (stock solution B). Just before use, mix three volumes of stock solution A and 1 volume of

stock solution B, and add solid sodium *m*-periodate (Na IO_4) to achieve a 10 m*M* concentration (do not re-adjust the pH).
2. NH_4Cl-PB: Prepare a solution of 100 m*M* sodium phosphate buffer (pH 7.4) containing 50 m*M* NH_4Cl. Store at 4°C.
3. 20% sucrose–phosphate-buffered saline (PBS): To prepare PBS containing 20% sucrose, dissolve 20 g of sucrose in 80 mL of PBS (150 m*M* NaCl in 10 m*M* sodium phosphate buffer, pH 7.4), and add more PBS to make a volume of 100 mL. Store at 4°C. To make 10 and 15% sucrose–PBS solutions, mix this solution with PBS (*see* **Note 1**).
4. Embedding medium for tissue freezing: Tissue-Tek O. C. T. Compound (Sakura Finetek).
5. Tissue-embedding molds (*see* **Note 2**).
6. Spray freezers (*see* **Note 3**).

2.2. Indirect Immunofluorescence Staining for Cadherin, β-Catenin, and the Nucleus

1. Glass slides coated with silane (*see* **Note 4**).
2. Glass cover slips.
3. Delimiting pen (Dako Cytomation, Denmark; *see* **Note 5**).
4. Moist chambers for glass slides (*see* **Note 6**).
5. Staining jars.
6. PBS: 150 m*M* NaCl in 10 m*M* sodium phosphate buffer, pH 7.4. Prepare 3 *M* NaCl and 200 m*M* sodium phosphate buffer, pH 7.4 (200 m*M* PB), and store at room temperature. To prepare PBS, mix 50 mL of 3 *M* NaCl, 50 mL of 200 m*M* PB, and 900 mL of distilled water.
7. PBS containing 0.1 % Triton X-100 (PBS-TX): prepare 10 % (v/v) Triton X-100 in distilled water. Mix 50 mL of 5 *M* NaCl, 50 mL of 200 m*M* PB, 10 mL of 10% Triton X-100, and 890 mL of distilled water.
8. Primary antibody dilution buffer: prepare Tris-buffered saline (150 m*M* NaCl in 20 m*M* Tris-HCl, pH 7.4) and supplement it with 1% bovine serum albumin (fraction V) and 0.1 % Triton X-100. Store at –20°C (*see* **Note 7**).
9. Rat monoclonal antibody against mouse E-cadherin (clone no. ECCD-2, Takara Bio, Otsu, Japan): reconstitute the lyophilized material according to the manufacturer's instruction, aliquot and store at –20°C. Avoid repeated freeze–thaw cycles to prevent denaturing the antibody (*see* **Note 8**).
10. Rabbit polyclonal anti-human β-catenin (BioSource International, Camerillo, CA, cat. no. AHO0462). Store at 4°C (*see* **Note 9**).
11. Goat anti-rat immunoglobulin (Ig)G conjugated with Alexa Fluor 488 (Molecular Probes, Eugene, OR): protect from light and store at 4°C (*see* **Note 9**).
12. Goat anti-rabbit IgG conjugated with Alexa Fluor 568, highly cross-adsorbed (Molecular Probes, Eugene, OR): protect from light and store at 4°C (*see* **Note 9**).
13. Igs for negative control staining: normal rat IgG and normal rabbit IgG. Aliquot and store at –20°C (*see* **Note 10**).
14. Normal goat serum. Aliquot and store at –20°C (*see* **Note 10**).

15. DAPI: make 100 μ*M* solution in distilled water (stock solution). Protect from light and store at 4°C. To make a working solution, add 3 μL of stock solution to 1 mL of PBS just before use (working concentration is 300 n*M*).
16. PermaFluor, aqueous mounting medium (Shandone, Pittsburg, PA): store at 2 to 8°C.

3. Methods

In general, fixation before tissue freezing makes it easier to cut frozen sections and preserve tissue morphology of good quality in the sections. In some cases, however, fixation may lead to the loss of antigenicity. The conditions of tissue preparation suitable for immunohistochemistry depend on the nature of the individual antigen determinants. Furthermore, several antibodies are commercially available for a particular molecule and sometimes they react with different antigen determinants. It is thus crucial to choose an appropriate combination of tissue preparation procedures and antibodies to obtain good results. Both of the primary antibodies used here react with the antigens in frozen sections of mouse tissues fixed with the PLP fixative solution without any antigen retrieval procedures.

3.1. Preparation of Frozen Sections

1. To obtain mouse embryos, mother mice are anesthetized with ether and killed by cervical dislocation. After sacrifice of the mother, embryonic tissues are dissected and immediately immersed in the ice-cold PLP fixative solution and fixed for 6 h in the same solution at 4°C. For sufficient penetration of the fixative solution, the size of the dissected tissues should be as small as possible (*see* **Note 11**).
2. To inactivate residual aldehyde groups, wash the fixed specimens three times for 30 min each in NH_4Cl-PB at 4°C and then overnight in the same solution. Immerse the specimens sequentially in 10, 15, and 20% sucrose–PBS solutions, for 2 h for each concentration, at 4°C (*see* **Note 12**).
3. Pour a small amount of Tissue-Tek O. C. T. Compound in the embedding mold and put the specimen in it. If necessary, information on the specimens (name of the tissue, date of preparation, orientation of the tissue, and so on) can be marked on the mold before embedment.
4. Quickly freeze the tissues with a spray freezer. Spray the freezing solution in a Petri dish until the bottom of the dish is covered with the solution, and place the embedding mold containing the specimen and Tissue-Tek O. C. T. Compound in it. The clear Tissue-Tek O. C. T. Compound turns white when it is frozen. Immediately put the frozen tissues into a freezer at –80°C and store them at the same temperature before use (*see* **Note 13**).

3.2. Indirect Immunofluorescence Staining and Observation

1. Cut frozen sections at 8 μm with a cryostat at a temperature from –25 to –28°C (*see* **Note 14**). When the tissues stored at –80°C are used, they should be placed for a while in the cryostat before cutting to allow equilibration of their temperature.

2. Pick up the sections on silane-coated glass slides, and air-dry them for 30 min at room temperature (*see* **Note 15**).
3. Draw a circle around the sections with a delimiting pen and wait until the drawn line becomes dry. To rehydrate the sections, immerse the glass slides in PBS for 10 min.
4. Blocking: to make blocking solution, dilute normal goat serum to 10% with PBS-TX. Remove remaining PBS in the circles on the glass slides, and cover the sections with the blocking solution. Place the glass slides in a moisture chamber and incubate them for 10 min at room temperature.
5. Make a mixture of diluted primary antibodies as follows: Add 0.5 μL of rat anti-mouse E-cadherin monoclonal antibody (*see* **Note 16**) and 1 μL of rabbit anti-human β-catenin polyclonal antibody (*see* **Note 17**) to 100 μL of primary antibody dilution buffer. The final concentration of each antibody is 10 μg/mL. For control staining, make a mixture of purified normal rat IgG and normal rabbit IgG in the same way.
6. Remove the blocking solution from the glass slides and carefully wipe the remaining blocking solution from the line drawn with the delimiting pen (*see* **Note 18**).
7. Cover the sections on the glass slides with the mixture of diluted primary antibodies (or for control staining, with that of normal Igs) and incubate them for either 2 h at room temperature or overnight at 4°C. In either case, the glass slides should be kept in moist chambers to avoid drying of the sections.
8. Make a mixture of secondary antibodies, Alexa Fluor-conjugated goat anti-rat IgG and anti-rabbit IgG, at a dilution of 1/500 in PBS-TX (*see* **Note 19**).
9. Wash the sections twice with PBS-TX and then once with PBS for 5 min each time.
10. Incubate the sections with the mixture of secondary antibodies at room temperature for 2 h. The glass slides should be kept in a moisture chamber and protected from the light.
11. Wash the sections twice with PBS-XT for 5 min each time.
12. Cover the sections with DAPI working solution (300 n*M* in PBS) and incubate for 10 min at room temperature.
13. After a brief wash with PBS, apply mounting medium to the sections and mount cover slips. Because the aqueous mounting medium PermaFluor solidifies and immobilizes the cover slip, sealing with nail varnish is not necessary.
14. The stained sections are examined under a confocal microscope to acquire digital images. Blue, green, and red fluorescence channels give the images of DAPI, E-cadherin, and β-catenin, respectively. Images for each of the three channels can be merged with software when necessary. An example of a triple-stained section illustrating both membrane-associated and nucleus-localized β-catenin is shown in **Fig. 1**.

4. Notes

1. For long-term storage, autoclave the solution at 121°C for 15 min.
2. Instead of commercial embedding molds, cups made from aluminum foil can alternatively be used for the same purpose (**Fig. 2**).
3. Spray freezers used for cutting sections in a cryostat are available from many suppliers dealing in materials for histology and pathology.

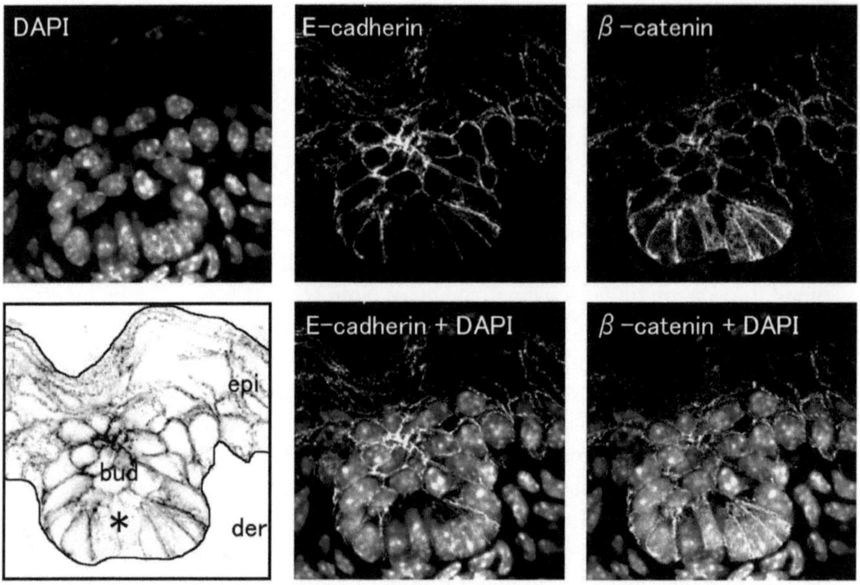

Fig. 1. Simultaneous immunolocalization of E-cadherin and β-catenin in the developing skin of a mouse embryo at embryonic day 17. Images for each of the three elements, i.e., DAPI, E-cadherin, and β-catenin, were acquired separately (upper images) and merged later with Photoshop (Adobe Systems) as shown in lower ones (E-cadherin + DAPI or β-catenin + DAPI). Tissue structure in the images is illustrated in the left lower box. E-cadherin is expressed only in the epidermis (epi) and epithelial bud of the developing hair follicle (bud), whereas β-catenin is expressed in both the dermis (der) and epidermis. At the tip of the epithelial bud (*), fluorescence indicating E-cadherin expression is rather weak; in contrast, abundant accumulation of β-catenin in the cytoplasm and the nucleus is evident.

4. Glass slides coated with silane can be replaced by ones coated with poly-L-lysine. Both are available from Sigma (Silane-Prep Slides or Poly-Prep Slides). Alternatively, glass slides can be coated in the laboratory by using poly-L-lysine solution (also available from Sigma) according to the instructions provided by the manufacturer.
5. The water-repelling circle drawn with the pen provides a barrier to liquids such as antibody solutions applied to the sections, thus helping to obtain more uniform immunocytochemical staining and allowing a reduction in the amount of reagents used. Similar pens are available also under the designation "PAP pen" from Sigma and some other suppliers.
6. Although some moist chambers designed for immunohistochemistry are commercially available, the use of plastic boxes with some additional laboratory materials is sufficient. An example of such a chamber is shown in **Fig. 2**.
7. This solution should contain 0.1% NaN_3 as a preservative when stored at 4°C.

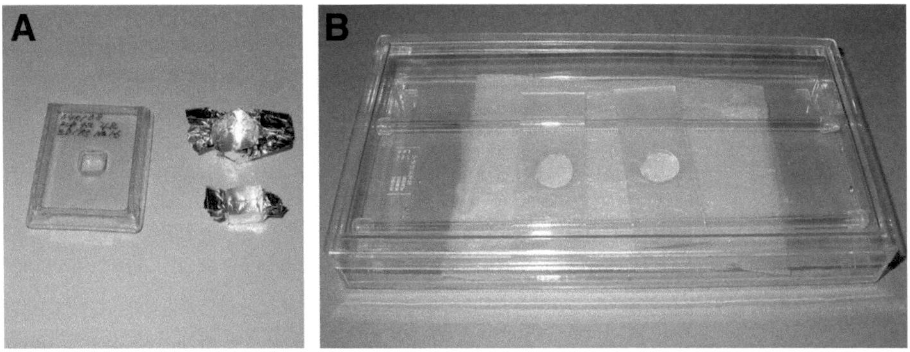

Fig. 2. **(A)** Tissue-embedding molds and **(B)** a moisture chamber. **(A)** A commercial tissue-embedding mold **(left)** and two examples for cups made from aluminum foil **(right)** are depicted. **(B)** An example for a homemade moisture chamber. In a plastic box with a cover, a piece of wet paper wiper is placed, and two glass tubes are aligned in parallel on it. For incubation with antibodies, glass slides bridge the glass tubes, as shown.

8. Reconstituted solution should be supplemented with 0.1% NaN_3 as a preservative when stored at 4°C.
9. For long-term storage, aliquot and store at –20°C.
10. Both purified normal immunoglobulins for negative control staining and normal sera for immunohistochemical blocking applications are available from many suppliers (Santa Cruz, Dako Cytomation, and so on).
11. The length of fixation should be changed depending on the size and the kind of tissues to be examined. Homogeneous and high-quality fixation can be achieved when the tissue is small enough to allow penetration of the fixative solution within a short time. A thickness of less than 2 mm may be appropriate for most kinds of tissues when they are fixed only by immersion in PLP solution. Alternatively, tissues of postnatal animals can be prepared by perfusion with PLP solution followed by excision and post-fixation with the same solution.
12. The time required for this treatment may depend on the size and the kind of tissues.
13. Alternatively, liquid nitrogen can be used for freezing samples. Pour a small amount of liquid nitrogen into a box made of styrene foam to fill the box with the vapor of liquid nitrogen. Put the embedding molds containing the specimens and Tissue-Tek O. C. T. Compound on a small styrene foam plate, place the plate in the vapor-filled box, and keep it there until the Tissue-Tek O. C. T. Compound has turned white.
14. The appropriate temperature differs depending on the kind of tissues. Also, the temperature should be changed when sections are cut at a different thickness. In general, a higher temperature is required when sections are cut thicker.
15. The dried frozen sections can be stored at –20 or –80°C for a few weeks or more. However, the stability of individual antigens may vary greatly from one to another.
16. Original concentration of the rat monoclonal antibody against mouse E-cadherin is 2 mg/mL when it is reconstituted according to the manufacturer's instructions;

and the recommended working concentration for immunohistochemistry indicated in these instructions is 10 μg/mL.

17. Concentration of the undiluted antibody is 1 mg/mL, and the recommended working concentration for immunohistochemistry is 10 to 20 μg/mL according to the manufacturer's instructions.
18. The water-repelling line drawn with the pen is ineffective if it is covered with liquid containing detergent (e.g., Triton X-100).
19. The original concentration of Alexa Fluor-conjugated secondary antibodies is 2 mg/mL, and the recommended working concentration is 10 to 20 μg/mL according to the manufacturer's instructions.

Acknowledgment

The author thanks Dr. Hervé Lesot, Institut de Biologie Médicale, Université Louis Pasteur, for critically reviewing the manuscript.

References

1. Gumbiner, B. M. (1996) Cell adhesion: the molecular basis of tissue architecture and morphogenesis. *Cell* **84,** 345–357.
2. Takeichi, M. (1988) The cadherins: cell–cell adhesion molecules controlling animal morphogenesis. *Development* **102,** 639–655.
3. Wheelock, M. J. and Johnson, K. R. (2003) Cadherins as modulators of cellular phenotype. *Annu. Rev. Cell Dev. Biol.* **19,** 207–235.
4. Ozawa, M., Baribault, H., and Kemler, R. (1989) The cytoplasmic domain of the cell adhesion molecule uvomorulin associates with three independent proteins structurally related in different species. *EMBO J.* **8,** 1711–1717.
5. Nelson, W. J. and Nusse, R. (2004) Convergence of Wnt, beta-catenin, and cadherin pathways. *Science* **303,** 1483–1487.
6. Logan, C. Y. and Nusse, R. (2004) The Wnt signaling pathway in development and disease. *Annu. Rev. Cell Dev. Biol.* **20,** 781–810.
7. Vleminckx, K. and Kemler, R. (1999) Cadherins and tissue formation: integrating adhesion and signaling. *Bioessays* **21,** 211–220.
8. Jamora, C. and Fuchs, E. (2002) Intercellular adhesion, signalling and the cytoskeleton. *Nat. Cell Biol.* **4,** E101–E108.
9. Obara, N. and Lesot, H. (2004) Subcellular localization of beta-catenin and cadherin expression in the cap-stage enamel organ of the mouse molar. *Histochem. Cell Biol.* **121,** 351–358.
10. McLean, I. W. and Nakane, P. (1974) Periodate-lysine-paraformaldehyde fixative: A new fixative for immunoelectron microscopy. *J. Histochem. Cytochem.* **22,** 1077–1083.

17

In Vitro Neutrophil Transepithelial Migration

Winston Y. Lee, Alex C. Chin, Susan Voss, and Charles A. Parkos

Summary

Polymorphonuclear leukocyte (PMN) transmigration into tissues is a highly regulated process and plays a central role in host defense. In inflammatory human diseases such as ulcerative colitis and Crohn's disease, the infiltration of intestinal mucosa by large numbers of PMNs contributes to epithelial pathophysiology. The sequence of events that fine-tune PMN migration across epithelial cells is not well-understood. In this chapter, we describe a method to study PMN transmigration across intestinal epithelial T84 monolayers using a modified Boyden chamber system. This in vitro model system consists of three main components: the epithelium, purified PMN, and a chemoattractant gradient. Intestinal epithelial cells are cultured as inverted monolayers on permeable filter supports to facilitate the study of PMN transmigration in the physiologically relevant basolateral-to-apical direction. PMNs are isolated from human blood using dextran sedimentation followed by Ficoll density gradient centrifugation. PMN transmigration is elicited using *N*-formyl-methionyl-leucyl-phenylalanine gradients and is quantified by assaying for myeloperoxidase activity. The advantages of this model are its reductionist approach and the fact that the system can be easily manipulated. Studies using this model system will shed more light on the mechanisms regulating PMN responses in acute inflammatory diseases.

Key Words: Neutrophil; PMN; intestinal epithelial cells; T84; chemotactic gradients; fMLP; Boyden chamber; myeloperoxidase; transmigration.

1. Introduction

As a first line of defense, polymorphonuclear leukocytes (PMNs) play a critical role in many inflammatory processes. For example, in acute illnesses such as bronchial pneumonia and bacterial enterocolitis and in chronic diseases such as inflammatory bowel disease and asthma, large numbers of PMNs infiltrate through tissues and across mucosal surfaces at sites of inflammation *(1–4)*. Furthermore, in ischemic injuries of the heart, kidney, and intestine, subsequent PMN infiltration can result in reperfusion injury *(5–8)*. In these

From: *Methods in Molecular Biology, vol. 341: Cell–Cell Interactions: Methods and Protocols*
Edited by: S. P. Colgan © Humana Press Inc., Totowa, NJ

examples, the severity of symptoms correlates with the degree of PMN infiltration. In mucosal organs, the presence of PMN transepithelial migration correlates strongly with disease pathophysiology.

For PMNs to exit the circulation and migrate across mucosal surfaces, they must cross vascular endothelium, interstitial matrix (including mesenchymal tissue) and, finally, the epithelium. This process involves a complex series of signaling events between PMN and other cells within the tissue. At sites of inflammation, activated vascular endothelial cells secrete mediators, such as tumor necrosis factor-α and interleukin-8, that are important in PMN recruitment. In addition, the endothelium also expresses surface adhesion molecules that facilitate adhesion of activated PMN ***(9,10)***. Selectin-mediated rolling of PMN along the vascular endothelium within the microcirculation at inflammatory sites precedes firm, integrin-mediated adhesion that is followed by extravasation either transcellularly or paracellularly through the vascular endothelium ***(11,12)***. Afterward, the movement of subendothelial PMN in the interstitial matrix is guided by gradients of signals derived from invading microorganisms such as bacterial peptides, or agents elicited as cellular responses to injury/invasion such as interleukin-8 ***(13–16)***. To migrate across mucosal epithelia, PMNs must first adhere to the basolateral membrane, an event mediated by β_2-integrins ***(17–19)***. Subsequently, PMNs migrate along the paracellular space between epithelial cells where a series of ill-defined signaling events facilitate the passage of the PMN across tight junctions and into the lumen. Recent studies have highlighted the importance of junctional adhesion molecule protein family members ***(20)*** and CD47-SIRPα interactions ***(21,22)*** as mediators of PMN migration across the epithelial paracellular space.

Most of what is known about PMN recruitment, extravasation, and transmigration is derived from studies conducted with modified Boyden chamber systems (**Fig. 1**). This in vitro model of transepithelial or transendothelial migration consists of three components: an epithelial or endothelial monolayer cultured on a permeable support of defined pore size, purified PMNs, and a chemotactic gradient. For epithelial monolayers, cells are grown on such permeable filter supports until confluency is reached, as determined by a number of methods. Freshly isolated PMNs are added to one side of the monolayer and stimulated to transmigrate across the monolayer/filter in the presence of a chemotactic gradient. The utility of this model is its reductionist nature, where the minimal necessary components are assembled to interact under simple conditions. Another utility is the ease with which this system can be manipulated. For instance, antibodies or pharmacological agents can be added to the system to see their effects on transmigration ***(17,20–23)***. Cell monolayers can also be stimulated with cytokines and evaluated for alterations in characteristics of PMN

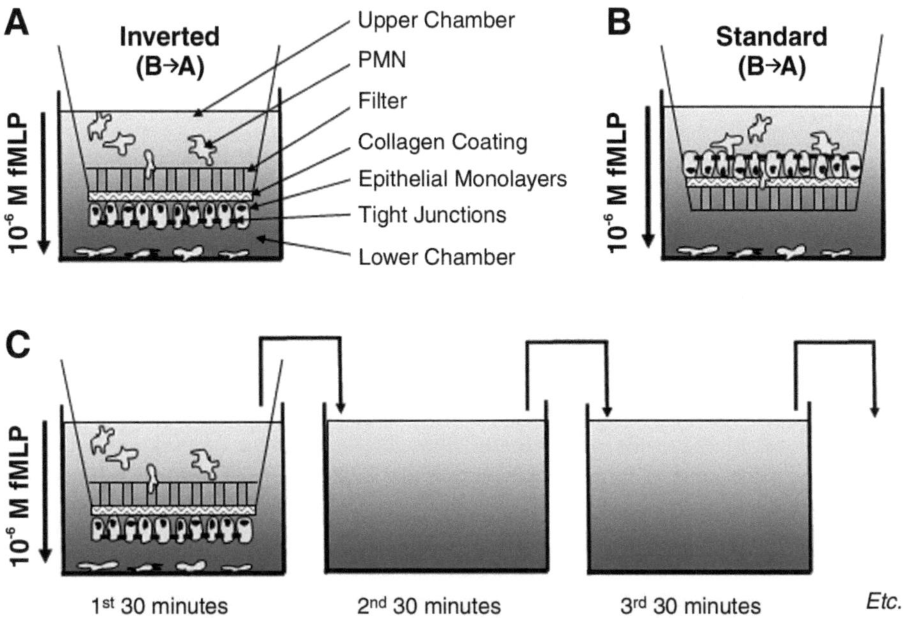

Fig. 1. Illustrations of different setups of the PMN transepithelial migration assay. **(A)** In this inverted configuration, the apical sides of epithelial monolayer are grown facing the bottom of the well. The different shades of gray represent the chemotactic gradients of fMLP, which dictates the directionality of PMN migration. Therefore, in this setup, PMN will migrate from basolateral side to the apical side of the monolayer (B→A), then detach and settle at the bottom of the well. This setup recapitulates PMN migration across the mucosal surfaces during inflammation. **(B)** In this "standard" configuration, the apical side of the monolayer is facing away from the bottom of the well. PMN will migrate from the apical side to the basolateral side of the monolayer (A→B), then detach and settle at the bottom of the well. **(C)** To perform a kinetic assay of PMN transmigration across the epithelial monolayer, the transwell is serially transferred into new wells containing chemotactic gradients of fMLP for each time point needed.

transepithelial migration *(24)*. Although many variations of this system exist, this discussion will focus on the most commonly used transwell setup, in which freshly isolated human PMN transmigrate across T84 intestinal epithelial monolayers in response to *N*-formyl-methionyl-leucyl-phenylalanine (fMLP) gradients.

2. Materials

2.1. Plating T84 Monolayers

1. Transwell inserts with surface areas of 0.33 cm^2 and pore size of 5 μm placed in a 24-well flat bottom cell culture plate (Costar).

2. T84 cell medium is a 1:1 mixture of Dulbecco's Modified Eagle's Medium and F-12 Nutrient Mixture, supplemented with 0.014 *M* $NaHCO_3$, 0.015 *M* HEPES, penicillin, streptomycin, and 6% fetal calf serum (Life Technologies/Gibco).
3. 0.05% Trypsin–EDTA (Cellgro).
4. Transepithelial resistance is measured by using an EVOM™ Epithelial Voltohmeter (World Precision Instruments).
5. Collagen stock can be prepared from rat tail tendon using methods modified from Cereijido and others ***(25)***. Dissect the rat tail tendon by pealing off the shiny white layer underneath skin with a hemostat. The tendon is then minced and divided into 75-mg aliquots. At this point, the tendon can be stored at –80°C. Dehydrate the 75 mg of tendon in 70% ethanol for 15 minutes. Solubilize tendons in 100 mL of 1% (v/v) acetic acid at 4°C. Stir the mixture at 4°C overnight. The next morning, the mixture will be viscous, and some undissolved materials will remain. Spin the collagen at 25,000*g* for 30 min. Dialyze the spun collagen solution against at least 2 L of cold distilled water with two to three changes. The dialyzed collagen will be referred to as the collagen stock and stored at 4°C.

2.2. PMN Isolation From Human Blood

1. Sterile 10-mL and 30-mL plastic syringes and 21- and 20.5-gage needles (Becton, Dickinson, and Company).
2. Cell culture grade distilled and deionized water (ddH_2O) with impedance greater than 18Ω.
3. Anti-coagulation buffer: 4% sodium citrate (Sigma) w/v in ddH_2O.
4. Normal saline: 0.9% sodium chloride (Sigma) w/v in ddH_2O.
5. Sedimentation buffer: 2% dextran 500 (Amersham Bioscience) w/v in normal saline.
6. Chilled Ficoll-Paque™ Plus (Amersham Bioscience).
7. Lysis buffer: 0.8% ammonium chloride supplemented with 1.9 mg of EDTA and 0.05 g of sodium bicarbonate per 50 mL of buffer.
8. Hank's balanced buffered saline (HBSS) with calcium (HBSS+) and without calcium (HBSS–; Sigma).

2.3. Transmigration and Myeloperoxidase Assay

1. 10^{-2} *M* fMLP (Sigma) stock dissolved in dimethyl sulfoxide. Subsequent dilutions of the stock are done in HBSS+.
2. 1 *M* sodium citrate in water, pH 4.5.
3. 10% Triton-X 100 in water.
4. 2,2′-azino-bis(3-ethylbenzthiazoline-6-sulfonic acid) (ABTS) buffer: for every 50 mL, add 0.036 g of ABTS, 5 mL of 1 *M* sodium citrate, 45 mL of water, and 100 µL of 30% w/v H_2O_2.
5. Flat bottom 96-well plates (Falcon).
6. Microplate reader with 405 nm filter (Molecular Devices).

3. Methods

3.1. Culturing T84 Intestinal Epithelial Cell Monolayers on Transwell Filter Supports

1. The night before seeding T84 cells on transwells, coat the sides of transfilters (*see* **Note 1**) onto which cells will be seeded with collagen in a 150 × 25-mm Petri dish in a laminar flow hood (*see* **Note 2**). To do so, dilute collagen stock 100-fold in 60% (v/v) ethanol in water, and transfer 50 μL to each filter. Let the filters dry overnight.
2. As T84 cells reach near confluency in a T162 flask, the cells are first washed with phosphate-buffered saline then detach with trypsin–EDTA. One T162 flask should be treated with 1 mL of trypsin–EDTA and incubated at 37°C for 30 min. To culture monolayers in a normal configuration, in which the apical membranes of epithelial cells face upward, trypsin–EDTA-treated cells are diluted 1 : 20 in media and 162 μL of the cell suspension is seeded on the collagen coated filter support, which is placed in a 24-well tissue culture dish containing 1 mL of media in each well. To culture monolayers in an inverted configuration, in which the apical membranes of the cells will be facing downward, the trypsin–EDTA elicited cells are diluted 1 : 10 in media and 81 μL of the suspension is placed on the underside of collagen coated transfilters that are resting upside down in a 15-cm^2 sterile Petri dish (*see* **Note 2**). The cells are incubated at 37°C overnight to allow for adhesion. Afterwards, the inverted transwells are simply "flipped" and placed into 24-well tissue culture wells filled with 1 mL of fresh T84 medium. The upper chamber is then supplemented with 162 μL of medium.
3. Like many other epithelial cell lines that are maintained in the laboratory, T84 cells will, over time, begin to exhibit aberrant phenotypes, presumably because of genetic drift. To monitor for such "drift," we routinely measure transepithelial resistance (TER) as T84 cells are becoming confluent in transwells (**Fig. 2A**). Typically, TER increases over time after plating of cells and plateaus between days 7 and 10. For T84 cells, during this 3-d window, TER typically ranges from 800 to 2500 $\Omega \cdot cm^2$, which is considered satisfactory (*see* **Note 3**).

3.2. Isolation of PMNs

1. From experience, we have observed that substantial care must be taken to avoid preactivation of PMNs during the process of isolation. Such preactivation is likely to occur secondary to exposure to trace amounts of microbial-derived products in buffers, glassware, and so on. Assays with preactivated PMNs typically yield altered results. It is recommended to make fresh solutions with sterile water and reagents from dedicated chemical sources each time before isolation.
2. Blood collection and plasma separation: fill a syringe with enough 3.8% sodium citrate so that the final concentration after blood collection will be 0.38%. For example, to collect 27 mL of blood, fill the syringe with 3 mL of 3.8% sodium citrate. After drawing blood, invert the syringe several times, and

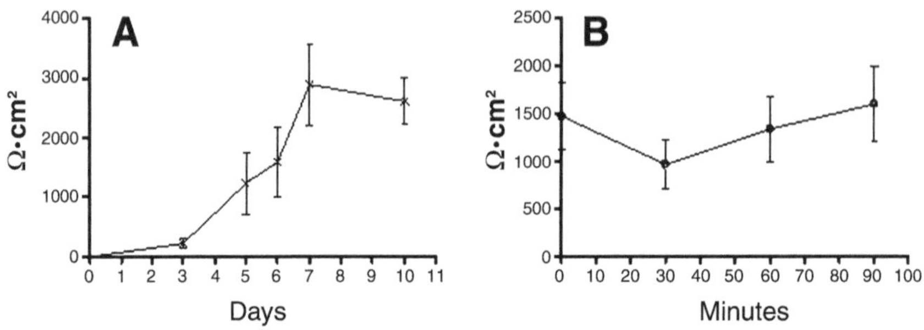

Fig. 2. Measuring TER to monitor the growth of epithelial monolayers. **(A)** TER was measured from the day of seeding epithelial cells onto the permeable support (day 0) to day 10 in the inverted configuration. From day 3 to 7, TER increased rapidly, indicating that epithelial cells were growing rapidly. Then, TER stabilized as the monolayer becomes confluent. It was during the 3-d window that we performed the transmigration assay. **(B)** TER was monitored every 30 min for 90 min after washing epithelial monolayers with HBSS+. Notice how the TER decreased during the first 30 min and then recovered in the next hour. It is recommended to let the monolayers "recover" before performing the transmigration assay.

transfer the blood to a 50-mL conical tube. Centrifuge blood at 162*g* for 10 min without braking at room temperature, which will separate the blood into three layers: a yellow plasma layer, a thin white buffy coat, and a dark layer of red blood cells at the bottom.

3. Dextran sedimentation of red blood cells (RBCs): from the aforementioned centrifuged blood sample, remove the plasma layer with a clean plastic pipet, taking care not to disturb the white buffy coat. It is not a problem if a small amount of plasma remains on top of the buffy coat. Then, add an equal volume of chilled sedimentation solution and gently invert the tube several times until the dark red layer becomes suspended. Place the conical tube upright at 4°C for 45 min. As the RBCs sediment, forming a thick red layer at the bottom, a pink turbid fluid phase will appear on top. Transfer the pink upper phase to a new 50-mL conical tube and centrifuge at 162*g* for 10 min at 4°C with the brake turned on. At the same time, aliquot 3 mL of chilled Ficoll into a 15-mL conical tube. After centrifugation, remove the supernatant with a sterile pipet and gently resuspend the pellet in 10 mL of normal saline.
4. Separation of PMN/RBC from monocytes/lymphocytes with a Ficoll gradient: carefully layer the resuspended cell pellet onto the 3 mL of Ficoll using a syringe with a 21-gage needle. After centrifugation at 162*g* for 30 min at 4°C, four distinct layers will appear. On top there will be a clear fluid phase of approx 8 mL. Immediately below will be a thin cloudy layer consisting of monocytes and lymphocytes. Underneath the thin cloudy layer is a clear Ficoll phase, which overlies a red pellet consisting of RBCs and PMNs.

5. RBC lysis: from the aforementioned Ficoll separation, remove all layers except the red pellet. Gently resuspend the pellet in 50 mL of cold lysis buffer and incubate on ice for no more than 2 min. As RBCs undergo lysis, the suspension will become less turbid. Immediately centrifuge the lysis suspension at 162*g* for 10 min at 4°C followed by a wash with 50 mL of cold HBSS–.
6. Calculating the yield of PMNs: For each 30 mL of blood drawn, resuspend washed PMN in 1 mL of HBSS–. Mix 10 µL of PMN with 990 µL of 2% acetic acid. The acetic acid will facilitate in the counting of PMN nuclei, which look multi-lobular, whereas the nuclei of monocytes and lymphocytes will be round (*see* **Note 4**). Estimate the yield of the cells with a hemacytometer (*see* **Note 5**).
7. Assessing the state of activation of PMNs: PMNs can easily become primed if at any point during the preparation they are exposed to microbial product contaminants. PMNs from contaminated preparations will migrate across collagen coated filters in significant numbers within the first 30 min of exposure to an fMLP gradient, whereas we have observed that unprimed PMN barely migrate. However, over a time course of 2 h, primed PMNs will migrate significantly less than unprimed PMNs. To perform this assay, place a collagen-coated 5-µm pore-sized transwell filter in a 24-well tissue culture plate. Transfer 100 µL of HBSS+ and 10^6 PMNs in 20 µL of HBSS– into the upper chamber; then place 500 µL of 10^{-7} *M* fMLP in HBSS+ in the lower chamber. Incubate for 25 min at 37°C and visually inspect the number of PMNs migrated using a ×50 field of view. We routinely use this assay before major experiments as a way to evaluate whether PMN preparations are primed or not.

3.3. Transmigration of PMNs Across Intestinal Epithelial Cell Monolayers

1. Because epithelial cells are known to secrete growth factors and other cytokines that may modulate transepithelial migration of PMNs, it is important to wash epithelial monolayers with 200 µL of warm HBSS+ in the upper chamber and 1 mL in the lower chamber. Immediately after washing, we have observed that the TER may fluctuate; therefore, allow monolayers to "recover" from washing for 45 min at 37°C. Measure the TER (**Fig. 2B**). At this point, the TER should have equilibrated, usually to a lower resistance (*see* **Note 6**). If kinetic assays are desired, one additional well filled with 1 mL of 10^{-6} *M* fMLP in HBSS+ should be added per time point for each monolayer. For example, if the experiment involves a kinetic assay with a 30-min time-point over 2 h, it will be necessary to fill four additional wells with chemoattractant.
2. After equilibration of T84 monolayers, remove the HBSS+ from the upper and lower chambers and replace with 100 µL and 500 µL of HBSS+, respectively. Then, apply the number of PMN needed in a volume of 20 µL of HBSS–. For example, if applying 10^6 PMNs, add 20 µL of PMN suspension at the density of 5×10^7 cells/mL. If desired, one can add antibodies or pharmacological agents into the top chamber (*see* **Note 7**). It is important not to vary volumes of buffers used in upper and lower chambers between experiments. Differences in volumes used can change chemotactic gradients and result in different hydrostatic pressures, which can alter the pattern of transmigration. To initiate transmigration,

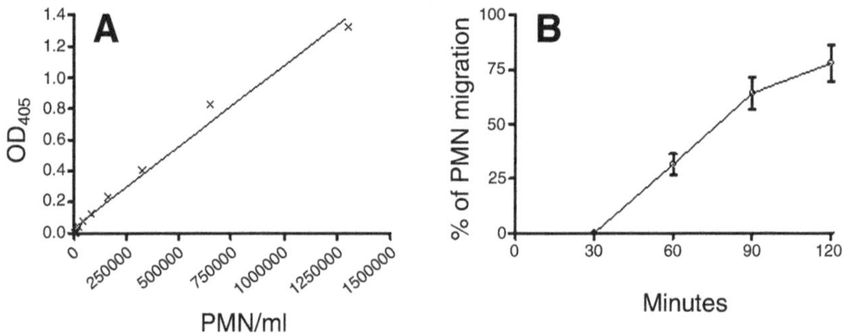

Fig. 3. Using myeloperoxidase assay to quantify PMN in transmigration assay. (**A**) Serial dilutions of PMNs are quantified using myeloperoxidase assays. The OD_{405} correlates with the PMN concentration linearly. This linear correlation ($r^2 = 0.987$) allows us to calculate unknown concentrations of PMNs. (**B**) A typical 2-h course of PMN transepithelial migration quantified using myeloperoxidase assays.

add 5 µL of 10^{-4} *M* fMLP to all bottom chambers so that the final concentration of fMLP in the bottom chambers will be 10^{-6} *M*. Incubate the transwells at 37°C and serially transfer the monolayers at each time-point into one of the chemo-attractant containing wells that have been prepared for the assay. Swirl the trans-well plate lightly before transferring. After the last time point, place the insert in a well with 500 µL of HBSS+ (**Fig. 1**).

3. To quantify migrated PMN, the azurophil granule enzyme myeloperoxidase activity is assayed using a simple colorimetric assay. To prepare the standard curve, serially dilute PMN suspension before transmigration (*see* **Note 8**). The standard curve should range from 1.5×10^6 to 2×10^5 PMNs/mL (*see* **Note 9**). Add 1 *M* sodium citrate buffer and 10% Triton-X100 solution in a 1:10 ratio to each well and dilution of the standard. Puncture the filter supports with 100-µL pipet tips to solubilize migrated PMNs. Place the plates on a rocking or swirling platform for 15 min at room temperature to permeablize the cells. Transfer 100 µL of PMN lysate from each transwell to a well in a flat bottom 96-well plate. With a multi-channel pipet, aliquot 100 µL of ABTS solution to the PMN lysate in the plate. Allow color to develop until the highest OD_{405} reading in the standard is approx 2.5. With an OD of 2.5 for the highest reading, the standard curve should be linear. Standard software such as MS Excel® can be used to perform calculations. The standard curve allows both calculation of cell numbers from OD readings and normalization among samples (**Fig. 3**).

4. Notes

1. The pore size of transwell filters should be no less than 3 µm in diameter. Otherwise, PMNs will not be able to migrate through the filters.
2. T84 intestinal epithelial cells can be seeded either on the top or bottom sides of transwell filters. If T84 cells are seeded on top, the apical membranes will face

upwards resulting in a "standard" configuration. However, with the standard configuration, PMNs will transmigrate across the epithelial monolayers in an apical-to-basolateral direction which is opposite to what occurs in vivo. To recapitulate transmigration in a physiologically relevant direction, T84s can be cultured on the underside of filters in an "inverted" configuration in which the basolateral aspect of the monolayer is facing upwards. Using such inverted monolayers, PMNs will migrate in a basolateral-to-apical direction which is consistent with the direction of PMN transmigration in vivo.

3. When T84 cells or other epithelia are cultured for many passages (typically more than 15 or 20 passages), monolayer properties commonly change as manifested by a faster growth rate and altered barrier function. Epithelial cells may grow and divide through the filter pores and form monolayers on both sides of the transwell filters resulting in artificially high TER (>3000 $\Omega \cdot cm^2$). Conversely, cells can form poor barriers with low TER values. Thus, when culturing monolayers on permeable supports, if TER values start to drift, it may be worth considering bringing up a new batch of cells.
4. A Trypan Blue exclusion test also can be used to determine the viability of the cells in addition to estimating the yield of the cells in the preparation. However, it is more difficult to distinguish monocytes from PMNs with Trypan Blue. Typically, we obtain more than 99% viability of cells isolated by these methods.
5. For every 30 mL of blood, the yield is generally in the range of 5×10^7 PMNs. However, it may be worth considering to maintain a donor log to obtain a reasonable prediction of the expected PMNYield among individual donors.
6. After the initial wash, TER often will increase by 10% and subsequently decrease as the cells are equilibrating.
7. To test the effects of antibodies on transmigration, generally it is sufficient to add antibody at a concentration of 10 to 20 μg/mL in the upper chamber. The diffusion of antibody to the lower chamber can be assumed to be minimal during the time course of the transmigration assay. However, for pharmacological agents, one should consider adding the agent in both upper and lower chambers. It may be necessary to optimize the assay for each antibody or agent.
8. Once PMNs are exposed to calcium in HBSS+, they are often more difficult to handle because of cell clumping and aggregation. Therefore, it is better to serially dilute PMN in cold calcium free HBSS– before use in transmigration assays.
9. After subtracting the blank OD values from the standard wells, the OD readings of the standards should be proportional to the dilution factor used for the samples. For instance, if 10^6 cells give an OD reading of 2.2, then the OD reading of 5×10^5 cells should be 1.1.

References

1. Weiland, J. E., Davis, W. B., Holter, J. F., Mohammed, J. R., Dorinsky, P. M., and Gadel J. E. (1986). Lung neutrophils in the adult respiratory distress syndrome. Clinical and pathophysiologic significance. *Am. Rev. Respir. Dis.* **133,** 218–225.

2. Hawker, P. C., McKay, J. S., and Turnberg, L. A. (1980) Electrolyte transport across colonic mucosa from patients with inflammatory bowel disease. *Gastroenterology* **79,** 508–511.
3. Nusrat, A., Parkos, C. A., Liang, T. W., Carnes, D. K., and Madara, J. L. (1997) Neutrophil migration across model intestinal epithelia: monolayer disruption and subsequent events in epithelial repair. *Gastroenterology* **113,** 1489–1500.
4. Koyama, S., Rennard S. I., Leikauf, G. D., et al. (1991) Endotoxin stimulates bronchial epithelial cells to release chemotactic factors for neutrophils. A potential mechanism for neutrophil recruitment, cytotoxicity, and inhibition of proliferation in bronchial inflammation. *J. Immunol.* **147,** 4293–4301.
5. Winn, R. K., Vedder, N. B., Sharar, S. R., and Harlan, J. M. (1997) Leukocyte-endothelial cell interactions in ischemia-reperfusion injury. *Ann. NY Acad. Sci.* **832,** 311–321.
6. Kakkar, A. K. and Lefer, D. J. (2004) Leukocyte and endothelial adhesion molecule studies in knockout mice. *Curr. Opin. Pharmacol.* **4,** 154–158.
7. Jordan, J. E., Zhao, Z.-Q., and Vinten-Johansen, J. (1999) The role of neutrophils in myocardial ischemia-reperfusion injury. *Cardiovasc. Res.* **43,** 860–878.
8. Frangogiannis, N. G., Smith, C. W., and Entman, M. L. (2002) The inflammatory response in myocardial infarction. *Cardiovasc. Res.* **53,** 31–47.
9. Liu, Y., Shaw, S. K., Ma, S., Yang, L., Luscinskas, F. W., and Parkos, C. A. (2004) Regulation of leukocyte transmigration: cell surface interactions and signaling events. *J. Immunol.* **172,** 7–13.
10. Bianchi, E., Bender, J. R., Blasi, F., and Pardi, R. (1997) Through and beyond the wall: late steps in leukocyte transendothelial migration. *Immunol. Today* **18,** 586–591.
11. Feng, D., Nagy, J. A., Pyne, K., Dvorak, H. F., and Dvorak, A. M. (1998) Neutrophils emigrate from venules by a transendothelial cell pathway in response to FMLP. *J. Exp. Med.* **187,** 903–915.
12. Carman, C. V. and Springer, T. A. (2004) A transmigratory cup in leukocyte diapedesis both through individual vascular endothelial cells and between them. *J. Cell Biol.* **167,** 377–388.
13. McCormick, B., Hofman, P., Kim, J., Carnes, D., Miller, S., and Madara, J. (1995) Surface attachment of *Salmonella typhimurium* to intestinal epithelia imprints the subepithelial matrix with gradients chemotactic for neutrophils. *J. Cell Biol.* **131,** 1599–1608.
14. Loike, J. D., el Khoury, J. Cao, L., et al. (1995) Fibrin regulates neutrophil migration in response to interleukin 8, leukotriene B4, tumor necrosis factor, and formyl-methionyl-leucyl-phenylalanine. *J. Exp. Med.* **181,** 1763–1772.
15. Loike, J. D., Cao, L., Budhu, S., et al. (1999) Differential regulation of beta 1 integrins by chemoattractants regulates neutrophil migration through fibrin. *J. Cell Biol.* **144,** 1047–1056.
16. Loike, J. D., Cao, L., Budhu, S., Hoffman, S., and Silverstein, S. C. (2001) Blockade of $\alpha5\beta1$ integrins reverses the inhibitory effect of tenascin on chemotaxis of human monocytes and polymorphonuclear leukocytes through three-dimensional gels of extracellular matrix proteins. *J. Immunol.* **166,** 7534–7542.

17. Parkos, C. A., Delp, C., Arnaout, M. A., and Madara, J. L. (1991) Neutrophil migration across a cultured intestinal epithelium. Dependence on a CD11b/CD18-mediated event and enhanced efficiency in physiological direction. *J. Clin. Invest.* **88,** 1605–1612.
18. Tosi, M., Hamedani, A., Brosovich, J., and Alpert, S. (1994) ICAM-1-independent, CD18-dependent adhesion between neutrophils and human airway epithelial cells exposed in vitro to ozone. *J. Immunol.* **152,** 1935–1942.
19. Agace, W., Patarrayo, M., Svensson, M., Carlemalm, E., and Svanborg, C. (1995) *Escherichia coli* induces transuroepithelial neutrophil migration by an intercellular adhesion molecule-1-dependent mechanism. *Infect. Immun.* **63,** 4054–4062.
20. Zen, K., Babbin, B. A., Liu, Y., Whelan, J. B., Nusrat, A., and Parkos, C. A. (2004) JAM-C is a component of desmosomes and a ligand for CD11b/CD18-mediated neutrophil transepithelial migration. *Mol. Biol. Cell* **15,** 3926–3937.
21. Liu, Y., Buhring, H. -J., Zen, K., et al. (2002) Signal regulatory protein (SIRPalpha), a cellular ligand for CD47, regulates neutrophil transmigration. *J. Biol. Chem.* **277,** 10,028–10,036.
22. Liu, Y., Merlin, D., Burst, S. L., Pochet, M., Madara, J. L., and Parkos, C. A. (2001) The Role of CD47 in neutrophil transmigration. Increased rate of migration correlates with increased cell surface expression of CD47. *J. Biol. Chem.* **276,** 40,156–40,166.
23. Parkos, C., Colgan, S., Liang, T., et al. (1996) CD47 mediates post-adhesive events required for neutrophil migration across polarized intestinal epithelia. *J. Cell Biol.* **132,** 437–450.
24. Colgan, S., Parkos, C., Delp, C., Aranaout, M., and Madara, J. (1993) Neutrophil migration across cultured intestinal epithelial monolayers is modulated by epithelial exposure to IFN-gamma in a highly polarized fashion. *J. Cell Biol.* **120,** 785–798.
25. Cereijido, M., Robbins, E., Dolan, W., Rotunno, C., and Sabatini, D. (1978) Polarized monolayers formed by epithelial cells on a permeable and translucent support. *J. Cell Biol.* **77,** 853–880.

18

The Blot Rolling Assay

A Method for Identifying Adhesion Molecules Mediating Binding Under Shear Conditions

Robert Sackstein and Robert Fuhlbrigge

Summary

Adhesive interactions of cells with blood vessel walls under flow conditions are critical to a variety of processes, including hemostasis, leukocyte trafficking, tumor metastasis, and atherosclerosis. We have developed a new technique for the observation of binding interactions under shear, which we have termed the "blot rolling assay." In this method, molecules in a complex mixture are resolved by gel electrophoresis and transferred to a membrane. This membrane can be rendered semitransparent and incorporated into a parallel-plate flow chamber apparatus. Cells or particles bearing adhesion proteins of interest are then introduced into the chamber under controlled flow, and their interactions with individual components of the immobilized substrates can be visualized in real time. The substrate molecules can be identified by staining with specific antibodies or by excising the relevant band(s) and performing mass spectrometry or microsequencing of the isolated material. Thus, this method allows for the identification, within a complex mixture and without previous isolation or purification, of both known and novel adhesion molecules capable of binding under shear conditions.

Key Words: Adhesion molecules; shear stress; shear conditions; flow conditions; parallel plate flow chamber; Western blot; selectins; selectin ligands; blot rolling assay; gel electrophoresis.

1. Introduction

The adhesion of cells to blood vessel walls under shear stress is central to many important physiological and pathological processes. Importantly, shear forces critically influence the adhesion of circulating cells and platelets to vessel walls in physiological cell migration and hemostasis, as well as in inflammatory and thrombotic disorders, cancer metastasis, and atherosclerosis. Although various techniques for analyzing cell adhesion have been described, most involve binding

From: *Methods in Molecular Biology, vol. 341: Cell–Cell Interactions: Methods and Protocols*
Edited by: S. P. Colgan © Humana Press Inc., Totowa, NJ

assays under static conditions and thus do not assess dynamic function. This limitation can introduce bias and exaggerate the relative role(s) of specific adhesion molecules. Notably, certain adhesive receptor–ligand interactions occur preferentially if not exclusively under physiological shear stress (e.g., binding of L-selectin to its ligands) or depend on low affinity and rapidly reversible interactions to serve their functions (e.g., rolling of leukocytes via selectins). Under static binding assay conditions, these types of interactions may be overtly neglected or overshadowed by molecules specialized to form more stable adhesions.

Methods for in vitro study of adhesion under shear conditions have been described, including the Stamper-Woodruff ***(1)*** and the parallel plate flow chamber ***(2)*** assays. However, these methods are constrained in that they require the availability of reagents (e.g., antibodies) that can specifically interfere with the activity of relevant receptors and/or ligands or require purified substrate materials that can be immobilized on the chamber wall. Thus, the applicability of these methods to examine the structure or function of previously uncharacterized ligands is highly limited.

To address these issues, we developed a method for direct real-time observation of adhesive interactions between cells or particles in flow and proteins separated by sodium dodecyl sulfate polyacrylamide gel electrophoresis (SDS-PAGE) and immobilized on membranes ***(3)***. This technique, called the "blot rolling assay," has allowed for the identification of several new glycoprotein ligands for selectins ***(4,5)***, and this approach recently has been extended to the study of glycolipid selectin ligands ***(6)***. This method allows the real-time assessment and measurement of interaction parameters (e.g., rolling vs firm attachment, specificity, and reversibility with inhibitors) in both physiological and nonphysiological shear conditions, thus permitting a unique user interface for the observation of adhesive events on membrane-immobilized materials. Ligands under investigation can be immobilized directly or segregated by gel electrophoresis (e.g., SDS-PAGE, isoelectric focusing) or other methods before transfer to the membrane. This method also provides for the real-time manipulation of interaction conditions, including wall shear stress, ion requirements, temperature, the influence of metabolic inhibitors, and the presence of activating agents or inhibitors of cell function. Blot-immobilized substrates can be used repeatedly, allowing the *in situ* manipulation of the substrate under continuous direct visualization or direct comparison of different conditions or different cell populations in shear flow. The capacity to observe sequential experimental and control conditions on a single substrate and to observe physiological behaviors and responses to manipulations in real time, provides particularly powerful advantages of this method over conventional static binding assays. Similarly, the ability to observe interaction with individual components of a complex mixture without requiring purification or previous knowledge

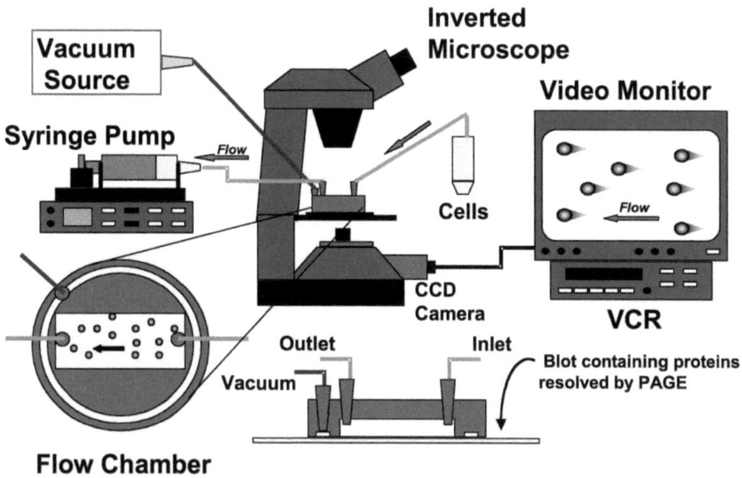

Fig. 1. Diagram of blot rolling assay assembly. The components of the blot rolling assay are shown, including the inverted microscope, flow chamber, CCD camera, syringe pump, vacuum source, video monitor and blot membrane (resting on a Petri dish located on the microscope stage, with the flow chamber positioned on top of it).

of the nature of the components provides advantages over conventional flow based assays. In this chapter, we describe a detailed protocol for performing blot rolling assays of membrane proteins. A video recording of a representative experiment is available at http://dermatology.bwh.harvard.edu/leukocyte.html.

2. Materials

1. The required components for performing blot rolling assays include an inverted microscope and a parallel plate flow chamber apparatus. **Figure 1** shows the setup used in our laboratories. The parallel plate device we use has internal chamber diameters of $2 \times 0.5 \times 0.025$ cm. A commercial product with similar properties is available (Circular flow chamber, GlycoTech Corp., Rockville, MD, product no. 31-001). Flow is adjusted by negative pressure regulated by a high-precision, programmable syringe pump (Harvard Apparatus). Wall shear stress (dyne/cm^2) is calculated according to the formula: wall shear stress = $6\ \mu Q/wh^2$, where w = chamber width (cm), h = chamber height (cm), and Q = volumetric flow rate (mL/s). μ = viscosity (poise, or dyne $\times$ s/cm^2; for water at 25°C, $\mu = 0.009$), which is adjusted for contribution of 10% glycerol (*see* **step 8**), according to the equation $\mu = 1.37 \times$ medium ($\mu = 0.0123$ at 25°C; *see* **Subheading 3.4., step 8**, for calculations based on our chamber dimensions). Interaction events can be captured on videotape by using a standard videomicroscopy camera and video recording assembly (**Fig. 1**), for later off-line analysis.

2. PAGE gel system.
3. Sequi-Blot™ polyvinylidene difluoride (PVDF) membrane (Bio-Rad, cat. no. 162-0184).
4. Fetal bovine serum.
5. 0.1% Tween-20–phosphate-buffered saline (PBS) solution.
6. Prestained molecular weight markers.
7. Western Blue stabilized substrate for alkaline phosphate (Promega, cat. no. S3841).
8. Blot rolling medium (H/H medium): Hank's balanced salt solution (Gibco, cat. no. 14170-112) with 10 m*M* HEPES, pH 7.4. Additions to H/H medium: 2 m*M* $CaCl_2$, 5 m*M* Na_2 EDTA, and 10% glycerol (American BioAnalytical).
9. Monoclonal or polyclonal antibody reagents for identifying any epitope(s) of interest.

3. Methods

3.1. Preparation of Western Blots/Visualization of Protein Bands

1. Material of interest can be prepared for PAGE (with or without SDS) using standard methods. In general, we prepare lysates of cells at 4×10^8 cell equivalents per milliliter of lysate buffer. A number of standard detergents have been used for the preparation of lysates, although we have found 1% n-octylglucoside (Roche Molecular Biochemicals, Indianapolis, IN) to provide low nonspecific background binding. Acrylamide gradient gels (e.g., Criterion 4–20%; Bio-Rad) provide improved resolution in higher molecular weight ranges. The amount of protein required will vary with the material under study, but we have found 25 to 100 μg of protein per lane to be sufficient for most of our studies. Both reducing and nonreducing gel conditions have been utilized with good results. It is helpful to include prestained molecular weight standards in the lanes adjacent to the material of interest to assist in alignment and localization of regions supporting adherence.
2. When gel electrophoresis is complete, the component proteins are transferred to PVDF membrane using standard transfer methods; although other membrane materials may work as well, we have preferred Bio-Rad Sequi-Blot™, which has 0.2-μm pore size and minimal surface irregularities, thus promoting laminar flow over the membrane.
3. After transfer, the blot is placed in a blocking solution to block nonspecific interactions and incubated with gentle agitation (on a rocker platform or rotating platform) for 1 h at room temperature. It is important to place the blot in the dish with the protein side facing up (i.e., the side of the membrane that was facing the gel). The blot is then washed in PBS–0.1% Tween-20 twice for 5 min each with gentle agitation.
4. Immunostaining, if desired, can be performed using standard procedures. Membranes typically are stained with primary antibody for 1 h under gentle agitation, washed twice in PBS–0.1% Tween-20 for 5 min, and then incubated with secondary alkaline phosphatase-conjugated antibody in Tris-buffered saline for 1 h under gentle agitation. The blot is again washed twice in PBS–0.1% Tween-20 for 5 min, rinsed

in PBS followed by Tris-buffered saline (to remove residual PBS and Tween-20), and then developed with alkaline phosphate substrate (i.e., Western Blue™). After developing blots to an appropriate signal-to-background perspective, reactions are stopped with PBS and the blot is washed with PBS twice. Blots to be used for flow experiments may be stored at 4°C. They should be kept in buffer until use. Blots will maintain binding fidelity for several days, though best results are obtained with freshly prepared materials.

5. The membranes are again blocked by incubation in 50% newborn calf serum or fetal bovine serum in binding media for at least 1 h at 4°C and equilibrated in binding media before use in the flow assay.

3.2. Preparation of Selectin-Expressing Cells

To assay blots for selectin ligand activity, we typically use human peripheral blood lymphocytes (PBLs) expressing L-selectin or Chinese hamster ovary (CHO) cells permanently transfected to express P-selectin or E-selectin.

1. Human PBLs can be prepared from whole blood by Ficoll density gradient separation. For analysis of L-selectin function on lymphocytes, monocytes may be removed by plastic adherence or other methods to reduce nonspecific binding (e.g., magnetic bead separation). Murine spleen and lymph node T-cells and rat thoracic duct lymphocytes also have been used in the same fashion. E-selectin and P-selectin transfected cell lines are typically split the day before use to achieve log phase growth and are harvested by replacing the media with H/H 5 m*M* EDTA. The cells can then be released from the flask by manual agitation or scraping. We do not use trypsin to release the cells as this may alter the function of surface adhesion molecules.
2. After counting, the cells are suspended at 10X the concentration desired for use in the flow chamber (e.g., 10–20 × 10^6 cells/mL in H/H without Ca^{2+} or Mg^{2+} and without glycerol) and stored on ice. Clumping of adherent cell populations (i.e., CHO cells) may develop after prolonged storage on ice. This can be remedied by washing again in H/H 5m*M* EDTA before use.
3. Controls to confirm specificity for selectin binding to blots include use of mock transfected CHO cells and/or preincubation of cell aliquots with function-blocking anti-selectin MAb.

3.3. Inverted Microscope Setup

1. Setup will vary with type of microscope. In general, we do not use any filters.
2. **Figure 2** shows the alignment of the parallel plate chamber onto the blot. Begin visualization at low power to align the chamber and identify the area of interest. Many observations are best performed at relatively low magnification (e.g., ×10 objective) to allow observation of larger areas and numbers of cells. Observation under increased magnification can be performed to observe individual cell characteristics. Adjust the shutter arm so that the light narrows specifically over area of interest on the blot through the chamber.

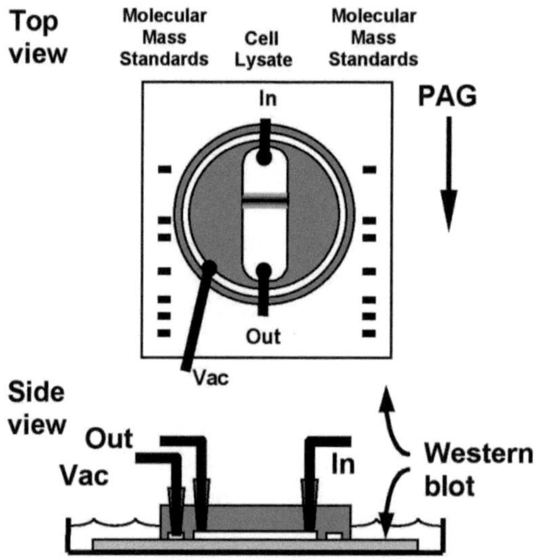

Fig. 2. Blot rolling apparatus. Western blots stained for components of interest are equilibrated in binding media containing 10% glycerol and placed in a Petri dish. A parallel plate flow chamber mounted over the area of interest is held in place by low continuous vacuum pressure (Vac). The inlet (In) is connected to a buffer reservoir and the outlet (Out) to a high precision syringe pump. The apparatus is mounted on the stage of an inverted phase-contrast microscope equipped for video microscopy. Molecular mass(es) of interacting protein(s) resolved in the polyacrylamide gel are estimated by alignment with colored mass standards in adjacent lanes. (From **ref.** ***3*** with permission.)

3.4. Blot Rolling Assay

1. All media should contain 10% glycerol. Equilibration in dilute glycerol alters the opacity of the PVDF membrane sufficiently to allow transmission of light and the direct visualization of cells interacting with the surface of the blot by standard light microscopy. Assessment of some functions (e.g., integrin activation) requires maintaining the chamber and media at 37°C. This can be achieved with a stage warming device and placement of supply media in water baths adjacent to the apparatus.
2. The chamber should be mounted onto the blot so that the flow channel aligns in parallel over the lane of interest. The flow chamber is secured by application of low vacuum pressure in the same fashion as is used for attachment to glass or plastic surfaces. We have not found it necessary to perform any special maneuvers to maintain an adequate seal on Sequi-Blot PVDF membrane.
3. The flow field is rinsed with media introduced through the 20-mL syringe, attached to the three-way stopcock, to remove bubbles and prepare the blot surface for cell input.

4. Immediately before use, aliquots of cells (e.g., PBL or transfected CHO cells) are diluted 10-fold into binding medium (H/H w/ Ca^{2+} with 10% glycerol; final concentration is typically 1–2 × 10^6 cells/mL) at room temperature (or 37°C as indicated). This minimizes exposure to glycerol and divalent cations, as these may affect cell viability and promote the formation of cell aggregates, respectively.
5. The cell input line is flushed with assay medium and placed into the tube containing cells. Flow is regulated by function of a downstream syringe pump. The use of a programmable pump allows for automation of the assay technique and more reproducible results. We typically initiate flow at a high rate of flow (2 mL/min or approx 7.5 dyne/cm^2) to bring cells through the sample tubing and into the chamber. Upon arrival of cells in the field of view, the flow rate is reduced (typically to 0.5–1 dyne/cm^2) to allow cells to interact with the blot surface. The flow rate can be adjusted up or down to increase or decrease, respectively, the stringency of binding interactions. One can program step-wise increases in flow (shear stress) without interruption. Maintaining continuous flow while cells are in the chamber minimizes nonspecific binding to the blot surface. This is critical with adherent cell populations, such as CHO cells or monocytes, although lymphocytes typically do not bind when flow is stopped for brief periods (<30 s).
6. Observation of cell interactions with substrate molecule(s) are made in real time and recorded via the camera/video camera record for later analysis. Typically, two types of analyses are performed. First, a "scanning" analysis is performed by moving the microscope stage to view the entire length of the lane in which proteins have been resolved while maintaining a constant shear rate. Typically, tethering and rolling interactions are observed at the leading edge of a "band" on the blot, with rolling in the middle and discontinuation of rolling interactions observed at the trailing edge. Once bands or areas of the blot of interest are identified, we typically perform additional studies comparing control cell preparations (mock-transfected or cells incubated with function blocking antibodies) and varying shear conditions while viewing a fixed area of the blot. In this analysis, tethering is observed at low physiological wall shear stresses (0.5–1.5 dyne/cm^2) for short periods (one to several minutes). Bound cells are then subjected to timed stepwise increases in wall shear stress to assess shear resistance as an estimate of relative strength of binding. The dependence of interactions on the presence of calcium (typical of selectin-mediated binding) can be confirmed by perfusion of the chamber with H/H with 10% glycerol and 5 m*M* EDTA and observing release of bound cells, or by repeating the analysis using H/H with 10% glycerol and 5 m*M* EDTA for both dilution of the cells and perfusion through the chamber.
7. Antibody inhibition experiments using antibodies to the membrane-bound ligand can be performed *in situ* by first observing interactions on an area of defined interest, then perfusing the chamber with a solution of antibody (e.g., 1–10 μg/mL for 30 min) and repeating the observation on the same site. If the antibody blocks binding interactions, the blot is unusable for further studies of that ligand. However, if the applied antibody does not block adhesive interactions, the blot may be reused, i.e., successive rounds of antibodies or reagents may be screened in this fashion.

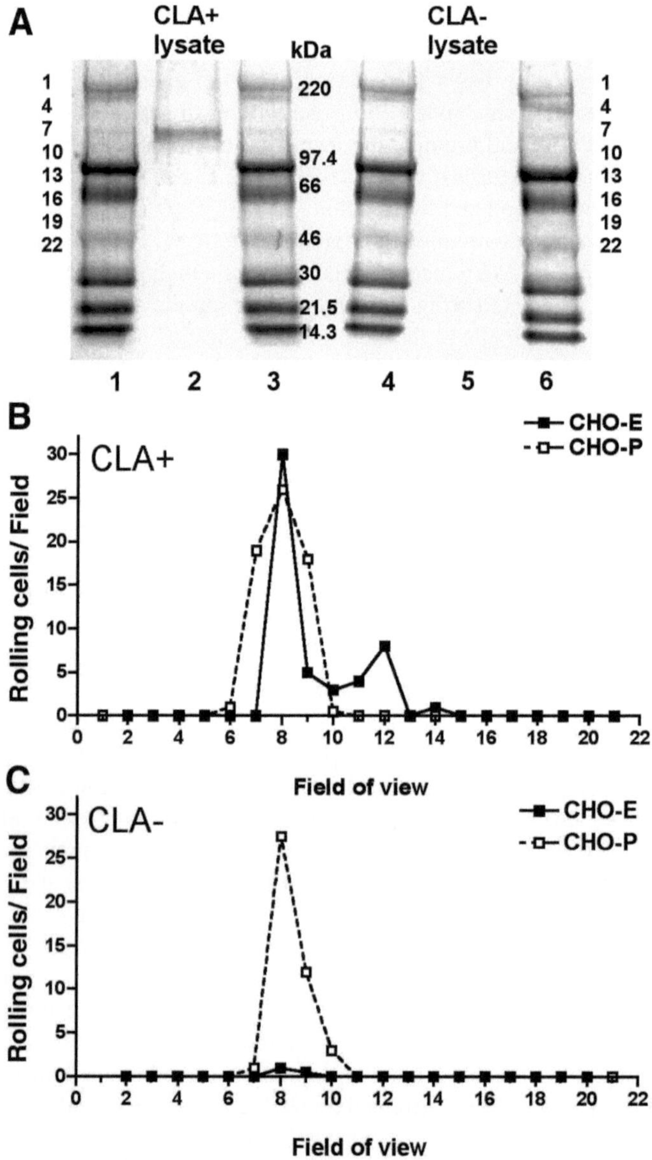

Fig. 3. Cutaneous lymphocyte antigen (CLA) immobilized on Western blots supports both E-selectin- and P-selectin-mediated rolling. **(A)** Western blots of T-cell lysates stained for CLA expression. Fifty micrograms of cell lysate protein from CLA-positive **(left panel)** or CLA-negative **(right panel)** T-cells were subjected to SDS-PAGE, electroblotted onto PVDF membranes, and immunostained with HECA-452 MAb. The single major reactive glycoprotein identified at approx 140 kDa in CLA-positive cells (lane 2) corresponds to the monomer form of CLA/PSGL-1.

8. For selectin-mediated binding, cells must tether and roll in shear flow to be considered specific. Tethering in our studies is defined as reduction of forward motion below the hydrodynamic velocity lasting a minimum of two video frames (0.07 s), and rolling is defined as more than 5 cell diameters of lateral translation below the hydrodynamic velocity. The majority of tethered cells in such studies are observed to roll smoothly across the entire field of view. Nonspecific interactions (i.e., cellular collisions with the substrate that did not lead to tethering and/or rolling) are defined as interactions lasting less than 0.07 s and are not included in the analysis. Nonspecifically bound cells (not rolling and/or not released by perfusion with EDTA) also are discounted from analysis. In general, there is a time-dependent increase in nonspecific (nonrolling) attachments in the absence of flow, though very few cells formed firm (nonrolling) attachments to the blot in continuous shear flow. Tethering rate is calculated as the number of cells that tether per field per time (usually 30–60 s) at a defined shear stress and adjusted to per minute values. As stated above, wall shear stress (τ) values are calculated according to the formula τ (dynes/cm^2) = 6 μQ/wh^2, where μ is the viscosity of the solution in the chamber (poise), Q is the volumetric flow rate (mL/s), w is the channel width (e.g., 0.5 cm for chamber we use), and h is the channel height (e.g., 0.025 cm for chamber we use *[7]*). A value of 0.009 poise is used for the viscosity (μ) of water at 25°C *(8)* and a value of 0.0123 poise is used for the viscosity of 10% glycerol (v/v) in water at 25°C ***(9)***. By these values, wall shear stress in 10% glycerol at 25°C is approx 1.37-fold greater than in water at the same temperature.
9. Regions (bands) of interest can be aligned against the molecular weight markers run in adjacent lanes to estimate molecular weight. Additionally, one may subsequently probe the blot with antibody to specific proteins of interest to determine whether band(s) supporting binding represent known proteins. One can excise regions of blot supporting binding and send sample for mass spectrometry and/or protein micro sequencing. This approach was used to identify HCELL, a glycoform

---

Fig. 3. *(continued)* Similar blots stained with anti-PSGL-1 mAb show approx equal quantities of 140-kDa PSGL-1 protein in each sample (data not shown). **(B,C)** Rolling cells per visual field (×10 objective) at 1.75 dyne/cm^2 wall shear stress. The number of cells observed to bind and roll was observed in overlapping fields of view extending from approx 250 kDa to 40 kDa (identified alongside blot images in **A**). Both CHO-E and CHO-P cells are observed to tether and roll across the area of the Western blot corresponding to CLA/PSGl-1 from CLA-positive T-cells (**B**), whereas only CHO-P cells are noted to form significant rolling adhesions on PSGL-1 from CLA-negative T-cells (**C**). Both CHO-E and CHO-P cells were noted to tether primarily over the area corresponding to CLA/PSGL-1. Mock-transfected CHO cells did not form rolling tethers on any areas of the blots observed (data not shown). Results presented are the means of two independent experiments on a single substrate and are representative of observations on numerous blots of CLA-positive and CLA-negative T-cell. (From **ref. *3*** with permission.)

of CD44 that functions as a high affinity L-selectin and E-selectin ligand *(4–6)*. We have also used this method to examine human T-lymphocyte glycoprotein(s) that support E-selectin and P-selectin binding under shear flow *(3)*. **Figure 3** is a representative histogram of a blot rolling assay of T-cell lysates probed with stably transfected CHO cells expressing P-selectin (CHO-P) and E-selectin (CHO-E). A videotape of these studies is available for viewing on the Harvard Skin Disease Research Center web page, under the Leukocyte Migration Core subheading (http://dermatology.bwh.harvard.edu/leukocyte.html).

Acknowledgments

This work was supported by National Institutes of Health grants HL60528 (RS), CA84156 (RS), and AR02115 (RCF). The authors thank Drs. Sandra L. King and Monica M. Burdick for critical review of the manuscript.

References

1. Stamper, H. B. and Woodruff, J. J. (1976) Lymphocyte homing into lymph nodes: in vitro demonstration of the selective affinity of recirculating lymphocytes for high-endothelial venules. *J. Exp. Med.* **144,** 828–833.
2. Lawrence, M. B. and Springer, T. A. (1991) Leukocytes roll on P-selectin at physiologic flow rates: distinction from and prerequisite for adhesion through integrins. *Cell* **65,** 859–873.
3. Fuhlbrigge, R. C., King S. L., Dimitroff, C .J., and Sackstein, R. (2002) Direct, real-time observation of E- and P-selectin mediated rolling on cutaneous lymphocyte-associated antigen immobilized on Western blots. *J. Immunol.* **168,** 5645–5651.
4. Dimitroff, C. J., Lee, J. Y., Fuhlbrigge, R. C., and Sackstein, R. (2000) A distinct glycoform of CD44 is an L-selectin ligand on human hematopoietic cells. *Proc. Natl. Acad. Sci. USA* **97,** 13,841–13,846.
5. Dimitroff, C. J., Lee, J. Y., Rafii, S., Fuhlbrigge, R. C., and Sackstein, R. (2001) CD44 is a major E-selectin ligand on human hematopoietic progenitor cells. *J. Cell Biol.* **153,** 1277–1286.
6. Hanley, W. D., Burdick, M. M., Konstantopoulos, K., and Sackstein, R. (2005) CD44 on LS174T colon carcinoma cells possesses E-selectin ligand activity. *Cancer Res.* **65,** 5812–5817.
7. Glasby, M. A. (1995) Principles of haemodynamics, in *Applied Physiology for Surgery and Critical Care* (M. A. Glasby and C. L.-H. Huang, eds.), Oxford, Boston, MA, p. 131.
8. Sengers, J. V. and Watson, J. T. R. (1986) Improved international formulations for the viscosity and thermal conductivity of water substance. *J. Phys. Chem. Ref. Data* **15,** 1291–1314.
9. Borchers, H. (ed.) Material values and mechanical behaviour of non-metals (1995) *Landolt-Bornstein Numerical Data and Functional Relationships in Physics, Chemistry, Astronomy, Geophysics, and Technology*, Vol. 4 (Part 1).

19

Cell–Cell Interaction in the Transcellular Biosynthesis of Novel ω-3-Derived Lipid Mediators

Nan Chiang and Charles N. Serhan

Summary

Omega-3 polyunsaturated fatty acids (PUFAs) such as eicosapentaenoic acid (EPA) and docosahexaenoic acid (DHA) display beneficial actions in human diseases. The molecular basis for these actions remains of interest. We recently identified novel mediators generated from ω-3 PUFA during cell–cell interactions that displayed potent anti-inflammatory and proresolving actions. Compounds derived from EPA are designated resolvins of the E series (RvE1), and those biosynthesized from DHA are denoted resolvins of the D series (RvD) and docosatriene, such as protectin D1 (PD1), which belongs to the family of protectins. In addition, treatment using aspirin initiates a related epimeric series by triggering endogenous formation of the 17*R*-RvD series, denoted as aspirin-triggered (AT)-RvDs. These compounds possess potent anti-inflammatory actions in vivo that essentially are equivalent to their counterpart generated without aspirin, namely the 17*S*-RvDs. In this chapter, we provide an overview and detail protocols of the biosynthesis and bioactions of these newly uncovered pathways and products that include three distinct series: 18*R*-resolvins of the E series derived from EPA (i.e., RvE1); 17*R*-resolvins of the D series from DHA (AT-RvD1 through RvD4); and 17*S*-resolvins of the D series from DHA (RvD1 through RvD4).

Key Words: Resolvin; aspirin; neuroprotectin; resolution; ω-3; lipid mediator; inflammation.

1. Introduction

Inflammation has emerged as playing a central role in many prevalent diseases, including Alzheimer's disease, cardiovascular disease *(1)*, and cancer *(2,3)* in addition to those chronic disorders that are well known to be associated with inflammation, such as arthritis and periodontal disease *(4,5)*. In parallel, the beneficial impact of essential ω-3 fatty acids was uncovered as early as 1929 *(6)* and was consistently observed in many human diseases for the next several decades *(7–10)*. Recently, we identified novel oxygenated products generated from the precursors eicosapentaenoic acid (EPA) and docosahexaenoic acid (DHA) that

From: *Methods in Molecular Biology, vol. 341: Cell–Cell Interactions: Methods and Protocols*
Edited by: S. P. Colgan © Humana Press Inc., Totowa, NJ

possess potent biological activities within resolving inflammatory exudates *(11–13)*. Hence, the terms Resolvin (Rv; resolution phase interaction products) and docosatrienes were introduced because the new compounds displayed both potent anti-inflammatory and immunoregulatory properties.

These compounds include Rvs of the E series (RvE1) derived from EPA that carry potent biological actions and Rvs of the D series (RvD) from DHA, as well as bioactive members from these biosynthesis pathways carrying conjugated triene structures denoted as docosatrienes (DTs). DTs recently were termed protectin D1 (PD1) because they are both anti-inflammatory ***(12,13)*** and neuroprotective ***(14,15)***. Unlike other products identified earlier from ω-3 fatty acids that are similar in structure to eicosanoids but less potent or devoid of bioactions, these Rvs, DTs, and neuroprotectins evoke potent biological actions in vitro and in vivo ***(11–15)***. In this chapter, we focus on the novel compounds derived from cell–cell interactions (i.e., RvE1, RvDs, and aspirin-triggered [AT]-RvDs) and provide detailed protocols for their biosynthesis as well as bioassays.

1.1. Novel Lipid Mediators Identified During Resolution of Acute Inflammation: Arachidonic Acid vs EPA and DHA

We now appreciate that intimate cell–cell interactions within vessel walls, that is, adherent platelets that are studied with polymorphonuclear leukocytes (PMNs), converge on the endothelium and can be visualized by intravital microscopy ***(16,17)*** and promote transcellular lipid mediator biosynthesis ***(18–20)***. During platelet–leukocyte interactions, arachidonic acid is converted to lipoxins (LX), which are generated to "stop" PMN and stimulate monocytes ***(21)*** and macrophages to promote resolution ***(22)***. These forms of cell–cell interactions in the vasculature are impinged on by aspirin ***(23,24)***. Aspirin inhibits thromboxane production by platelets and prostacyclin biosynthesis in vascular endothelial cells ***(25,26)***. During PMN–endothelial and/or PMN–epithelial interactions, aspirin triggers the biosynthesis of 15-epi-lipoxins (aspirin-triggered lipoxin [ATL]; **Fig. 1**; *[23]*). Both LX and ATL and their respective stable analogs are potent regulators of transendothelial–transepithelial migration of PMN across these cells and endothelial cell proliferation in vitro and in vivo ***(27–29)***.

LX and ATL act in the nanomolar range via specific receptors we identified ***(30,31)*** and confirmed by others ***(32,33)***, namely LXA_4 receptor. Transgenic mice overexpressing the LXA_4 receptor with myeloid-specific promoter display reduced PMN infiltration in peritonitis and heightened sensitivity to LXA_4 and ATL ***(34)***. Transgenic rabbits overexpressing 15-lipoxygenase (LOX) type I generate enhanced levels of LXA_4, have an enhanced anti-inflammatory status, and are protected from the inflammatory bone loss of periodontal disease ***(35)***. Hence, the results of these studies have heightened our awareness that PMN, in addition to their host defense position and the possibility that they can spill

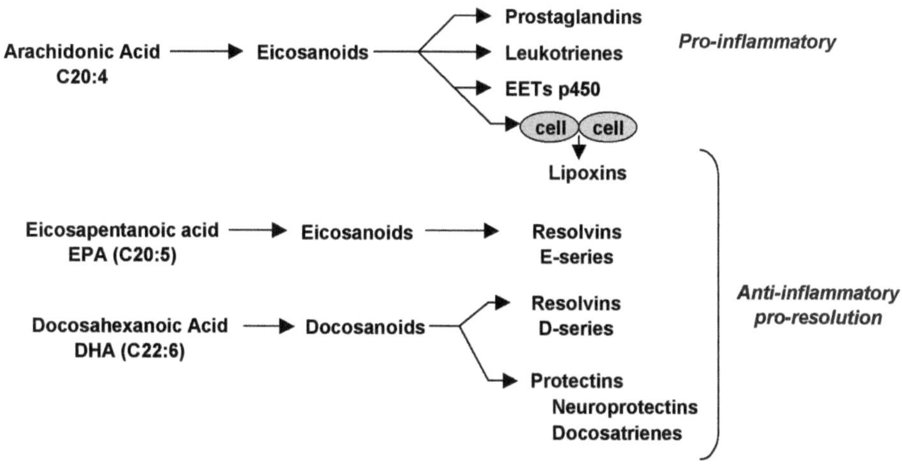

Fig. 1. Essential polyunsaturated fatty acids are precursors to bioactive lipid mediators.

pro-inflammatory agents *(36)* and mediators such as the classic eicosanoids, prostanoids, and leukotrienes *(37,38)*, which amplify inflammatory responses, also can produce novel protective lipid mediators that actively counter-regulate the inflammation (**Fig. 1**).

It often is questioned whether essential fatty acids, such as the ω-3 EPA or DHA, are converted to potent lipid mediators, as is the case with arachidonic acid. In short, both DHA and EPA are important precursors. In view of the compelling results from the GISSI study, which showed improvements in more than 11,000 cardiovascular patients *(39,40)*, namely, reduction in sudden death by approx 45% by taking almost 1 g of ω-3 per day, we recently addressed a potential role of ω-3 polyunsaturated fatty acids (PUFAs). The inspection of their methods indicated that all the patients also took daily aspirin, which was unaccounted for in the authors' analysis. Despite very large doses (milligrams to grams daily), most literature regarding ω-3 PUFAs suggests beneficial actions in many human diseases, including periodontal *(41)*, anti-inflammatory, and antitumor actions *(7,42)*.

The three major lipoxygenase activities (5-LOX, 12-LOX, 15-LOX) can each convert DHA to various monohydroxy-containing products; however, at the time their in vivo functions were either not apparent or they did not display bioactivity *(43,44)*. Also, DHA can be nonenzymatically oxygenated to isoprostane-like compounds termed neuroprostanes that reflect oxidative stress in the brain *(45)* or autooxidized to products that are monohydroxy racemates *(46)* of the compounds that are now known to be enzymatically produced during biosynthesis of the Rvs and DT *(12–15)*. Overall, it is noteworthy that the molecular basis and mechanisms underlying ω-3 PUFA's immunoprotective actions remained to be established and appreciated, and their direct connection to human disease and treatment are still important biomedical challenges ***(47)***.

1.2. Aspirin Unveils New Directions: AT Lipid Mediators

Aspirin is an active ingredient in more than 60 over-the-counter remedies, making it a difficult substance to control for in some human studies. What is the molecular basis for the protective action of ω-3, and is there overlap(s) in its actions? To address this in an experimental setting, we used murine dorsal skin pouches ***(11,12)***. This model of inflammation is known to spontaneously resolve in rats ***(48)***. We adapted this for mice to include both genetics and to set up lipidomics, using liquid chromatography-ultraviolet-tandem mass spectrometry (LC-UV-MS-MS)-based analyses geared to evaluating whether potential novel lipid mediators are indeed generated during the resolution phase of inflammation ***(11,12)***. In this pouch (experimental contained inflammation) after 4 h, PMN numbers begin to decrease within exudates ***(11,12)***. Exudates were taken at timed intervals, focusing on the period of "spontaneous resolution," and lipid mediator profiles were determined using tandem LC-UV-MS-MS. We constructed lipid mediator libraries with physical properties (i.e., MS and MS/MS spectra, elution times, ultraviolet spectra, and so on) for matching and to assess whether known and/or potential novel lipid mediators were present within the exudates, and are presently expanding these libraries and the software for their matching ***(49)***. If novel lipid mediators were encountered, their structures were elucidated by conducting retrograde analysis for both biogenic enzymatic synthesis and total organic synthesis. This approach permitted assessment of structure–activity relationships as well as the scale-up required to confirm the bioactions of novel compounds identified ***(11,12)***.

1.3. Anti-Inflammatory and Proresolving Properties

The main bioactive resolvins and docosatrienes as representative members are shown in **Figs. 2–4**. The bioactions of Rv and DT are listed in **Table 1**. Direct comparisons of Resolvin E vs the D series (17*R* and 17*S* epimer series) for their ability to regulate PMN in vivo were conducted ***(12,13,35)***. Both the D and E classes of Rvs are potent regulators of PMN infiltration. 17*R*-RvD series, triggered by aspirin, and 17*S*-RvD series give essentially similar results, indicating that the *S* to *R* switch does not diminish their bioactions. When injected intravenously at 100 ng/mouse, they both gave approx 50% inhibition, and the RvE1 gave approx 75 to 80% inhibition. In comparison, indomethacin at 100 ng/mouse (or ~3 μg/kg) gave roughly 25% inhibition (*see* **refs.** ***12*** and ***13***). In addition, with microglial cells that liberate cytokines in the brain, the Rv D series block tumor necrosis factor (TNF)-α-induced interleukin (IL)-1β transcripts and are potent regulators of PMN infiltration in brain, skin, and peritonitis in vivo ***(13,14)***.

(text continued on p. 237)

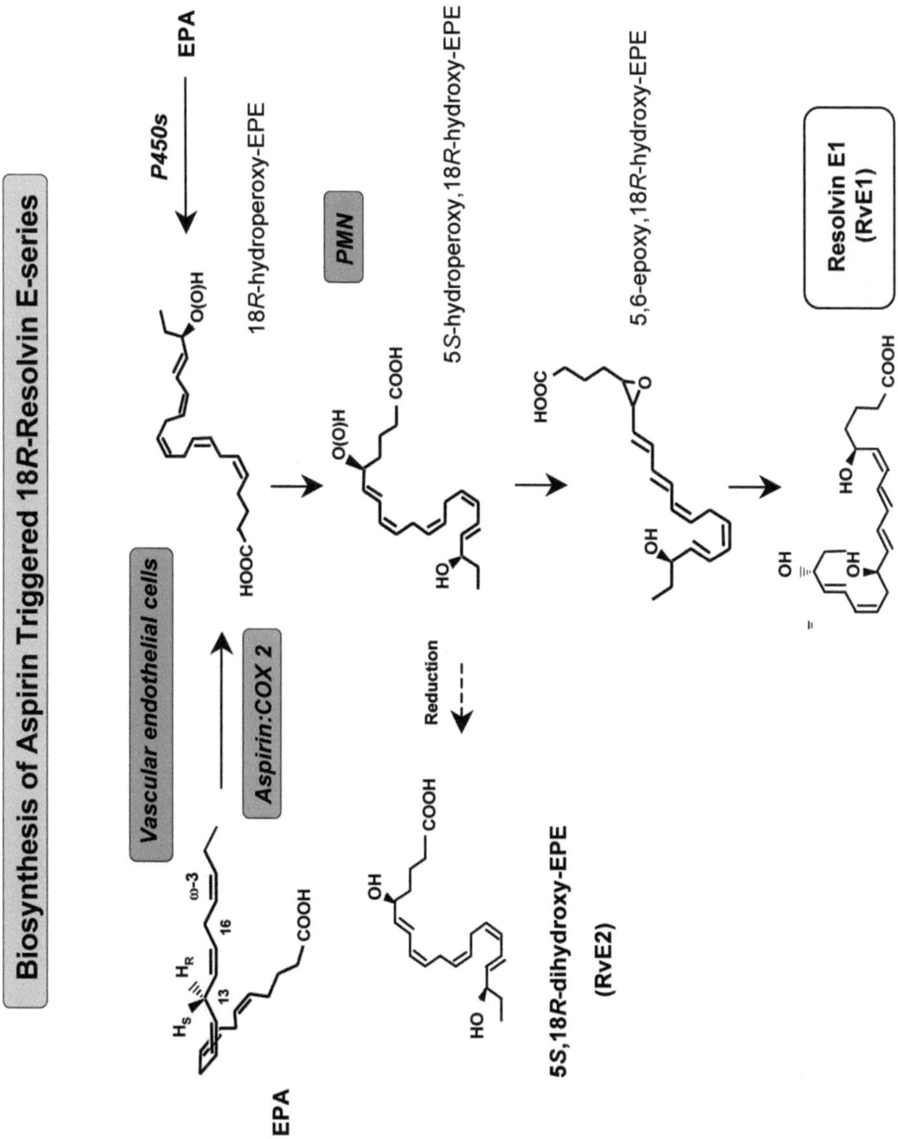
Biosynthesis of Aspirin Triggered 18R-Resolvin E-series
Vascular endothelial cells
EPA
Aspirin:COX 2
EPA
P450s
18R-hydroperoxy-EPE
PMN
5S-hydroperoxy,18R-hydroxy-EPE
Reduction
5S,18R-dihydroxy-EPE
(RvE2)
5,6-epoxy,18R-hydroxy-EPE
Resolvin E1
(RvE1)

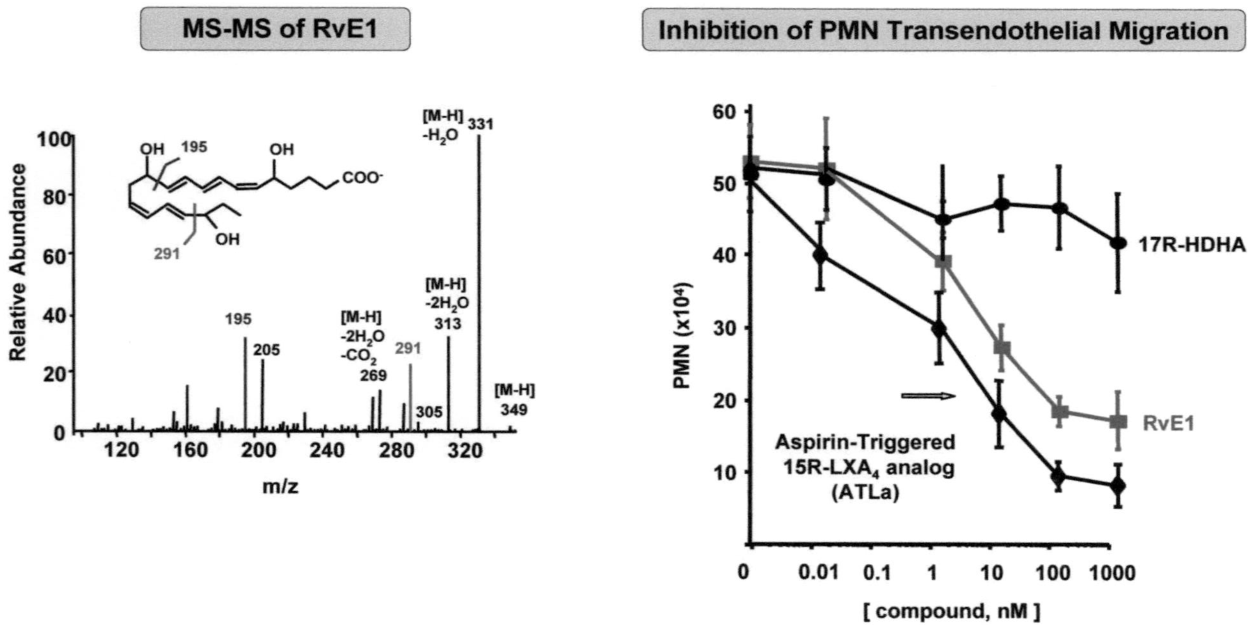

Fig. 2. EPA conversion to RvE1. Biosynthesis of AT 18*R*-RvE-series. Human endothelial cells expressing COX-2 treated with aspirin transform EPA by abstracting hydrogen at C16 to give R insertion of molecular oxygen to yield 18*R*-hydroperoxy-EPE, which is reduced to 18*R*-HEPE. They are further converted via sequential actions of leukocyte 5-LO and lead to formation of RvE1 ***(12,50)***. MS-MS analysis of RvE1:murine inflammatory exudate. Selected ion monitoring at *m/z* 349.5 [M-H]. Structure depicted in the inset giving diagnostic ions at *m/z* 305, 233, 195, and 291. Inhibition of PMN transendothelial migration. PMN were incubated with isolated compounds [10^{-11} to 10^{-6} *M*]: 17*R*-HDHA, ●; RvE1, ■; or ATL analog used for reference, ◆; transmigration was initiated using an optimal amount of LTB_4 [10 n*M*] (100%) and HUVEC coincubations (90 min, 37°C). Results are the mean ± SEM; $n = 3$.

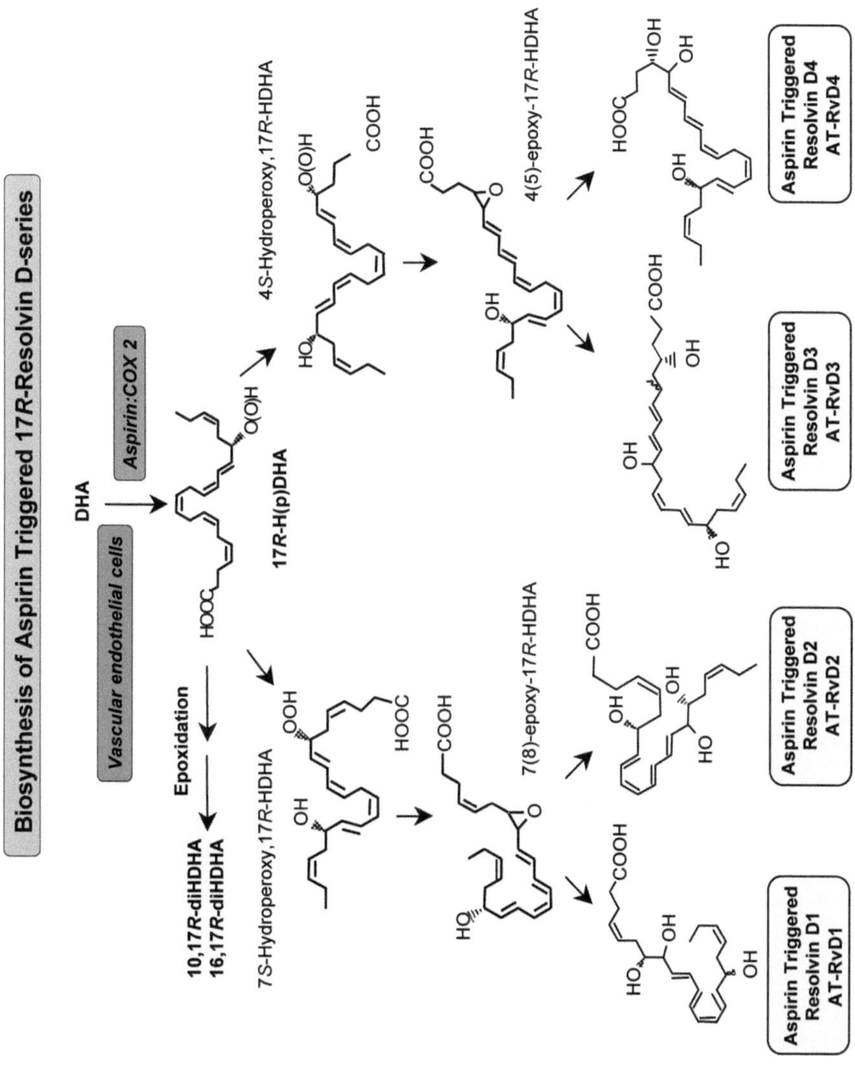
Biosynthesis of Aspirin Triggered 17R-Resolvin D-series
DHA
Vascular endothelial cells
Aspirin:COX 2
Epoxidation
10,17R-diHDHA
16,17R-diHDHA
17R-H(p)DHA
7S-Hydroperoxy,17R-HDHA
4S-Hydroperoxy,17R-HDHA
7(8)-epoxy-17R-HDHA
4(5)-epoxy-17R-HDHA
Aspirin Triggered Resolvin D1 AT-RvD1
Aspirin Triggered Resolvin D2 AT-RvD2
Aspirin Triggered Resolvin D3 AT-RvD3
Aspirin Triggered Resolvin D4 AT-RvD4

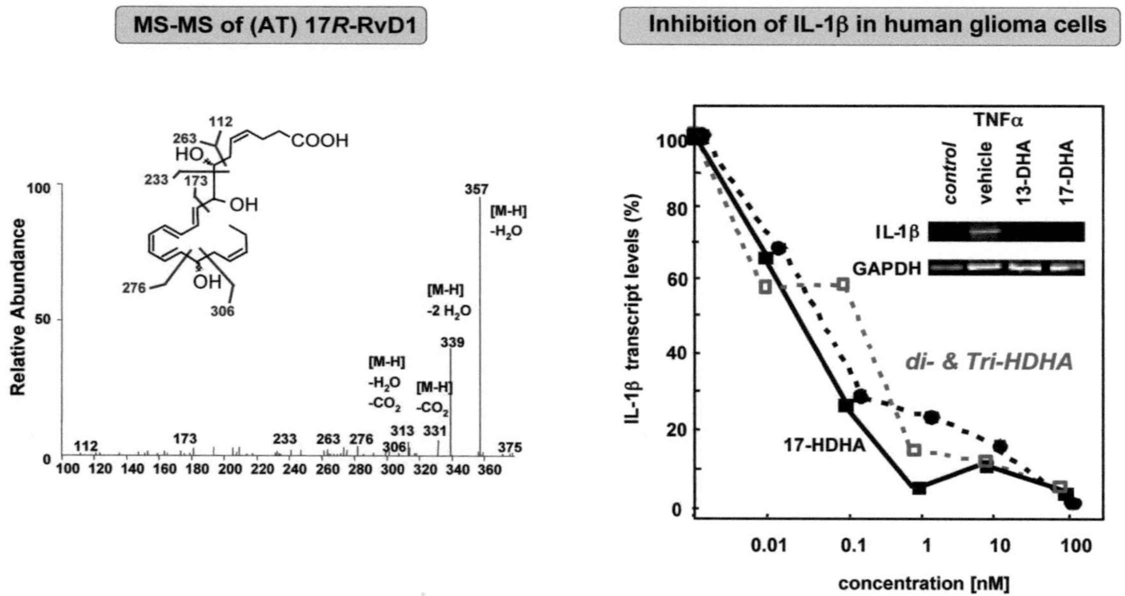

Fig. 3. DHA conversion to AT 17*R*-RvD-series. Biosynthesis of AT 17*R*-RvD-series. The 17*R* series Rvs are produced from DHA in the presence of aspirin. Human endothelial cells expressing COX-2 treated with aspirin transform DHA to 17*R*-HDHA. Human PMNs convert 17*R*-HDHA to two compounds via 5-lipoxygenation (5LO ***[11,12]***) that are each rapidly transformed into two epoxide intermediates: a 7(8)-epoxide **(left)** and the other a 4(5)-epoxide. These two novel epoxide intermediates open to bioactive products denoted 17*R* series AT-RvD1 through AT-RvD4. Note: stereochemistry depicted in likely configuration based on results with recombinant enzymes (*see* **refs.** ***12*** and ***13***). MS-MS of (AT) 17*R*-RvD1. Selected ion monitoring at *m/z* 375 [M-H]. Structure depicted in the inset giving diagnostic ions at *m/z* 357 [M-H-H_2O], 339 [M-H-2H_2O], 331 [M-H-CO_2], 313 [M-H-CO_2-H_2O], 306, 303, 276, 273, 255 [273-H_2O], 210, 195, and 180. Inhibition of IL-1β in human glioma cells. DBTRG-05MG cells (1×10^6/mL) were stimulated with 50 ng/mL of human recombinant TNF-α for 16 hours to induce expression of IL-1β transcripts. Concentration dependence with COX-2 products: 17-HDHA (■), 13-HDHA (●), and di-/tri-HDHA (□). Results are representative of reverse transcription polymerase chain reaction gels of MG cells exposed to 100 n*M* of 13-HDHA or 17-HDHA and graphed after normalization of the IL-1β transcripts using GAPDH.

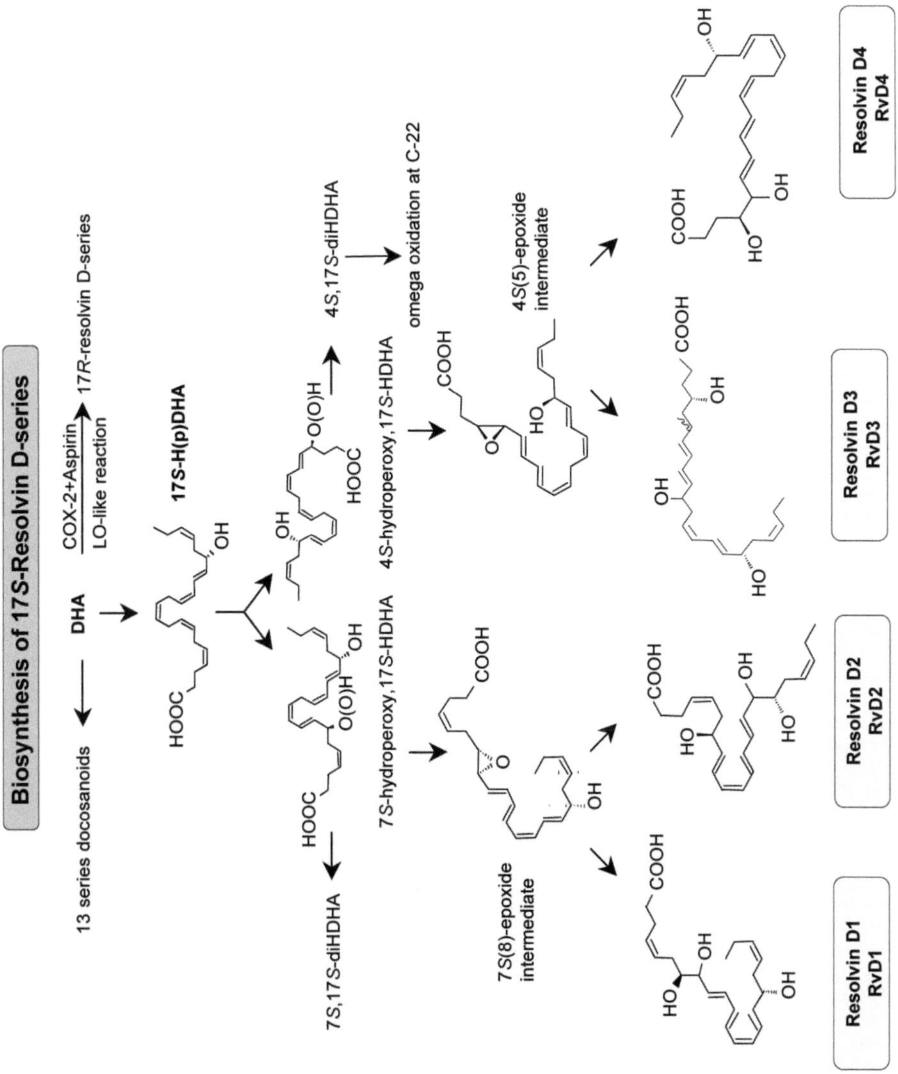

Biosynthesis of 17S-Resolvin D-series
13 series docosanoids
DHA
COX-2+Aspirin
LO-like reaction
17R-resolvin D-series
17S-H(p)DHA
7S,17S-diHDHA
7S-hydroperoxy,17S-HDHA
4S-hydroperoxy,17S-HDHA
4S,17S-diHDHA
omega oxidation at C-22
7S(8)-epoxide intermediate
4S(5)-epoxide intermediate
Resolvin D1
RvD1
Resolvin D2
RvD2
Resolvin D3
RvD3
Resolvin D4
RvD4

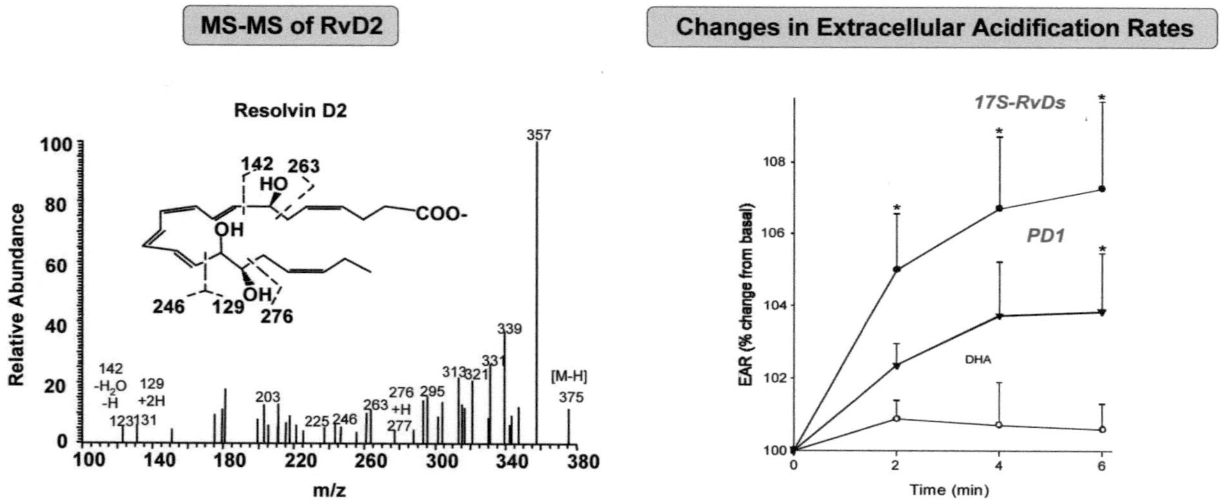

Fig. 4. DHA conversion to 17*S*-RvD series. Biosynthesis of 17*S*-RvD-series. These transformations may involve novel enzymes in addition to those that serve in the biosynthesis of eicosanoids (*see* **Subheading 3.3.** for further detail). The complete stereochemistries for the 17*S*-RvDs remain to be established and are depicted in their tentative configurations based on biogenic total synthesis, lipidomic analyses and alcohol trapping profiles. MS-MS of RvD2. Selected ion monitoring at *m/z* 375 [M-H]. Structure depicted in the inset giving diagnostic ions at *m/z* 123 (142-H_2O-H), 131(129+2H), 203(246-CO_2-H), 225 (263–2H_2O-2H), 246, 263, 277 (276+H), 295 (M-H-2H_2O-CO_2), 313 (M-H-H_2O-CO_2), 321(M-H-3H_2O), 331 (M-H-CO_2), 339 (M-H-2H_2O), 357 (M-H-H_2O-CO_2), and 375 (M-H) for 7*S*, 16, 17*S*-triHDHA (17*S*-RvD2). Changes in extracellular acidification rates. Glial cells: 17*S*-RvDs evoke ligand operated extracellular acidification. Changes in extracellular acidification rates were analyzed using Cytosensor® microphysiometry. Cells were superfused (100 µL/60 s) with DHA or 17*S*-RvDs ($n = 3$). Values are expressed as extracellular acidification rate (µV/s) normalized to baseline (100%), and ligand additions were at 2 min.

Table 1
Biosynthesis and Biological Activity of Resolvins

Compound	Biosynthesis	Biological activity
RvE1	Murine exudates ***(11)***	Inhibit PMN transmigration
	Human blood ***(50)***	Inhibit PMN infiltration (air pouch and peritonitis)
	Coincubations of human endothelial cells and PMN ***(11)***	Inhibit IL-12 production in dendritic cells
AT-RvDs	Murine inflammatory exudates ***(12)***	Inhibit PMN infiltration (air pouch and peritonitis)
	Murine brain stroke ***(14)***	Inhibit TNF-α-induced IL-1β in human glioma cells
		Inhibit PMN transmigration
RvD2-4	Human whole blood ***(13)***	Inhibit PMN infiltration (air pouch and peritonitis)
	Murine exudates	Inhibit TNF-α-induced IL-1β in human glioma cells
		Changes in extracellular acidification rate

PMN, polymorphonuclear leukocyte; IL, interleukin; TNF, tumor necrosis factor; AT, aspirin-triggered 17*R*-containing series.

2. Materials

2.1. Cell Culture

Human glioma cells (DBTRG-05MG cells) and *Sf*9 (*Spodoptera frugiperda*) insect cells are from American Type Culture Collection (ATCC; Rockville, MD). Cell culture media and reagents are from Cambrex Biosciences (Walkersville, MD).

2.2. Biosynthesis

1. Aspirin and A23187 were from Sigma (St. Louis, MO). EPA was from Cayman Chemical Co. (Ann Arbor, MI) and other synthetic standards, hydroxy fatty acids, and intermediates used for identification were purchased from Cascade Biochem Ltd. (Reading, UK).
2. 17*R/S*-hydroxy-DHA (HDHA) from Penn Bio-Organics (Bellefonte, PA).
3. IL-1β is from PeproTech (Rocky Hill, NJ).
4. Extract-clean C18 cartridges were purchased from Alltech Associates (Deerfield, IL).

2.3. In Vitro Bioassays

1. Recombinant murine and human TNF-α from R&D Systems (Minneapolis, MN).

2. Trizol reagent is from GIBCO BRL Life Technologies (Grand Island, NY).
3. Cytosensor® microphysiometer is from Molecular Devices (Sunnyvale, CA).

2.4. Murine Models of Acute Inflammation

1. Male mice of FVB strain (6 to 8 wk) were from Charles River (Wilmington, MA).
2. Zymosan A was from Sigma (St. Louis, MO).

3. Methods

3.1. 18R-Resolvin E Series

Resolving exudates in mice contain 18*R*-hydroxy-eicosapentaenoic acid (HEPE) as well as several related bioactive compounds ***(11)***. These novel compounds are produced from EPA by at least one biosynthetic pathway operative in human cells. This pathway is shown in **Fig. 2**; blood vessel-derived vascular endothelial cells treated with aspirin convert EPA to 18*R*-hydro (peroxy)-EPE that is reduced to 18*R*-HEPE. 18*R*-HEPE is then released from endothelium and rapidly converted by activated human PMN in the proximity to a 5(6) epoxide-containing intermediate that is further transformed to the bioactive 5,12,18*R*-tri-HEPE, termed RvE1.

Organic synthesis was achieved and its complete stereochemical assignment was recently established as 5*S*,12*R*,18*R*-trihydroxy-6*Z*,8*E*,10*E*,14*Z*,16*E*-eicosapentaenoic acid ***(50)***. RvE1 was identified in the resolution phase in mice and appeared as a transcellular biosynthetic product during cell–cell interaction with isolated human cells. More importantly, it proved to be a potent regulator of PMN transendothelial migration and infiltration in vivo. Recent results indicate that ChemR23 is a specific receptor for RvE1 ***(50)***.

3.1.1. Biosynthesis In Vitro and In Vivo

3.1.1.1. Human Endothelial Cell Incubation With PMN

1. Human endothelial cells are incubated with IL-1β (1 ng/mL, 24 h; *see* **Note 1**) to induce COX-2 (*see* **Note 2**).
2. Cells are then washed in Hank's balanced salt solution and treated with ASA (500 μ*M*; *see* **Note 3**) for 20 min at 37°C, followed by EPA (20 μ*M*).
3. Incubations are terminated with two volumes of ice-cold methanol and then extracted (*see* **Note 4**) for LC/MS/MS analysis (*see* **Note 5**) and further incubation.
4. Selected ion monitoring (i.e., at *m/z* 317 and 259) revealed that human umbilical vein endothelial cells (HUVECs) treated with ASA converted EPA to 18*R*-HEPE. Also, human microvascular endothelial cells (HMVECs) treated with ASA and EPA generated 18-HEPE (10.6 ng/10^6 cells).
5. Human PMNs (30 × 10^6 cells/mL; *see* **Note 6**) are then incubated (30 min, 37°C) with serum-treated zymosan A (100 ng/mL) and ASA-acetylated-COX-2-derived products from EPA (from **step 3**).

6. Incubations are terminated with two volumes of ice-cold methanol and then extracted for LC/MS/MS analysis.
7. Selected ion monitoring at [M-H] = *m/z* 349.5 revealed that PMN converted 18*R*-HEPE to RvE1 (*see* **Fig. 2**).

3.1.1.2. Formation of RvE1 in Inflammatory Exudates

1. Murine dorsal air pouches are raised on day 0 (*see* **Note 7**).
2. On day 6, inflammation is initiated by intrapouch injection of recombinant mouse TNF-α (100 ng/pouch).
3. ASA (500 μg) is injected at 3.5 h followed by injection of 300 μg of EPA (*see* **Note 8**) per pouch at 4 h.
4. At 6 h, pouches are lavaged twice with 3 mL of phosphate-buffered saline (PBS) and the exudate cells enumerated.
5. Exudate cells are then activated by calcium ionophore A_{23187} (4 μ*M*) at 37°C for 20 min.
6. The supernatants are collected and extracted for LC-MS-MS analysis.

3.1.2. Biological Activity In Vitro

3.1.2.1. PMN Transendothelial Migration

1. HUVECs were used at passages 1 and 2 and isolated by collagenase digestion (0.1% collagenase) and propagated on gelatin-coated (1%) tissue culture plates in RPMI 1640 cell culture medium supplemented with 15% bovine calf serum, 15% NU-serum, 50 μg/mL endothelial mitogen, 8 units/mL heparin, 50 units/mL penicillin, and 50 μg/mL streptomycin.
2. Human PMN are exposed to RvE1 (10^{-11} to 10^{-6} *M* at 37°C for 15 min), washed twice with Dulbecco's phosphate-buffered saline (DPBS), and coincubated with confluent endothelial monolayers grown on 96-well plates (4×10^5 cells in 0.2 mL of DPBS supplemented with 1% BCS) in the presence of agonist LTB4 (10^{-7} *M*, 30 min).
3. After incubation, the medium is aspirated and monolayers washed gently with 1 mL of DPBS to remove adherent cells.
4. The contents of each well are solubilized with 0.1% sodium dodecyl sulfate/0.025 *M* NaOH.
5. Fluorescence is assessed in a Cytofluor 2300 fluorescence plate reader, and the number of adherent PMN is calculated from the fluorescence of the original PMN suspension.

3.1.3. Bioactions In Vivo

3.1.3.1. Murine Dorsal Air Pouch: Inhibition of PMN Infiltration

1. Murine dorsal air pouches are raised on day 0 with male FVB mice.
2. On day 6, mice are anesthetized with isoflurane. Compounds (RvE1 or RvE2, 100 ng; *see* **Note 8** for handling Rvs) or vehicle (0.1% ethanol; *see* **Note 9**) were delivered as a bolus injection either into the tail vein in 100 μL of sterile saline or locally into pouches in 200 μL of sterile saline.

3. TNF-α (100 ng/pouch) is then injected locally into pouches to stimulate PMN infiltration.
4. Inflammatory pouch exudates are collected at 4 h and cells enumerated.

3.1.3.2. Murine Peritonitis (*see* **Note 10**)

1. Male FVB mice (6–8 wk) are anesthetized with isoflurane.
2. RvE1 (300 ng) or vehicle (0.1% ethanol) alone is administered intraperitoneally in 200 μL of sterile saline.
3. Zymosan A (1 mg in 1 mL of normal saline) is then administrated intraperitoneally 5 min later.
4. Mice are euthanized with an overdose of isoflurane 2 h later and peritoneal exudates collected by lavaging with 5 mL of sterile saline.
5. Exudate cells are collected and enumerated.

3.2. 17R-Series RvDs

In resolving exudates from mice given aspirin and DHA, we found novel 17*R*-HDHA and several related bioactive compounds **(Fig. 3)**. HMECs, also aspirin-treated in hypoxia, generate 17*R*-HDHA. DHA is converted by human recombinant COX-2, which was surprising, because earlier literature indicated that DHA is not a substrate of cyclooxygenase ***(51,52)***. However, these investigations were before knowledge of the COX-2 isoform and used organs rich in COX-1. Human recombinant COX-2 converts DHA to 13-HDHA. With aspirin, this switches to 17*R*-oxygenation to give epimeric AT forms, also in brain ***(12,13)***, of Rvs (AT-RvD1 through RvD4; *see* **Fig. 3**).

3.2.1. Biosynthesis

3.2.1.1. Incubation of DHA and ASA With Endothelial Cells

1. HUVECs or HMVECs are cultured for transendothelial migration.
2. HUVEC or HMVEC monolayers (1, 2, or 3 passages) are seeded (~2×10^5 cells/cm^2) on polycarbonate permeable supports precoated with 0.1% gelatin.
3. For hypoxia experiments, plates of HUVEC are treated with TNF-α and IL-1β (both 1 ng/mL), placed in a hypoxic chamber (3 h, 37°C) and returned to normoxia (21 h, 37°C).
4. ASA (500 μ*M*, 30 min) was added followed by DHA (~5 μ*M*) and calcium ionophore A23187 (2 μ*M*; 60 min, 37°C).
5. Incubations are terminated and extracted for LC-MS-MS analyses. Selected ion monitoring at [M-H] = *m/z* 343 revealed that endothelial cells treated with ASA converted DHA to 17*R*-HDHA (*see* **Fig. 3**).

3.2.1.2. Incubation of DHA and ASA With Recombinant COX-2

1. Human recombinant COX-2 is overexpressed in *Sf*9 insect cells.
2. The microsomal fractions (~8 μL) are suspended in 100 m*M* Tris-HCl, pH 8.0.

3. ASA is incubated (~2 m*M*, 37°C, 30 min) with COX-2 before addition of DHA (10 μ*M*).
4. Incubations are terminated and extracted with deuterium-labeled internal standards for LC-MS-MS analysis (*see* **Note 11**).
5. Selected ion monitoring at [M-H] = *m/z* 343 revealed that recombinant COX-2 treated with ASA converts DHA to 17*R*-HDHA (*see* **Fig. 3**).

3.2.1.3. Incubation of 17-HDHA With Human PMNs

1. Human PMNs are freshly isolated from venous blood of healthy volunteers (who declined taking medication for approx 2 wk before donation; BWH protocol no. 88-02642) by Ficoll gradient and enumerated.
2. Cells are divided into 50 × 10^6 cells/1 mL DPBS with Ca^{2+} and Mg^{2+} ($DPBS^{+/+}$), and incubations (40 min, 37°C) performed with either 17*R/S*-HDHA or 17*R*-HDHA with zymosan A (100 ng/mL).
3. Incubations are terminated and extracted with deuterium-labeled internal standards (15-HETE and C20:4) for LC-MS-MS analysis.
4. Selected ion monitoring at [M-H] = *m/z* 359 and 375 revealed that human PMN converts 17*R*-HDHA to di- and tri-HDHA (AT-RvDs), respectively (*see* **Fig. 3**).

3.2.1.4. Incubation With Murine Whole Brain

1. Mouse whole brains are rapidly isolated from sacrificed mice immediately before incubations.
2. Mouse brains are washed with cold $DPBS^{+/+}$ and incubated with ASA (45 min, 37°C, 500 μ*M*).
3. Incubations are terminated with 5 mL of cold methanol; brain is homogenized and extracted for LC-MS-MS analyses.

3.2.1.5. Formation of 17*R*-Containing Aspirin-Triggered RvDs in Inflammatory Exudates

1. Inflammatory exudates are initiated with intrapouch injection of recombinant mouse TNF-α into dorsal air pouches of 6- to 8-wk-old male FVB mice.
2. ASA (500 μg) is injected locally at 3.5 h followed by injection of 300 μg DHA/pouch at 4 h after the injection of TNF-α.
3. At 6 h (within the resolution phase), pouches were lavaged (using 3 mL of saline) and exudate cells were enumerated.
4. Cell-free supernatants are collected and extracted for LC-MS-MS analyses.

3.2.2. Biological Activity In Vitro

3.2.2.1. Regulation of IL-1β in Human Glioma Cells

1. Human glioma cells (DBTRG-05MG cells) are cultured as recommended by ATCC.
2. 1 × 10^6 cells per well in six-well plates are stimulated for 16 h with 50 ng/mL of human recombinant TNF-α in the presence of test compounds (i.e., 17*R*-HDHA and di-/tri-HDHA, 10^{-12} to 10^{-7} *M*) or vehicle (0.04% ethanol).

3. Cells are washed in DPBS$^{+/+}$ and harvested in 1 mL of Trizol reagent.
4. RNA is purified and reverse transcription polymerase chain reaction is performed. Primers used in amplifications are: 5′-GGAAGATGCTGGTTCCCTGC-3′ and 5′-CAACACGCAGGACAGGTACA-3′ for IL-1β; 5′-TCCACCACCGTGTTGC TGTAG-3′ and 5′-GACCACAGTCCATGACA TCACT-3′ for glyceraldehyde phosphate dehydrogenase (GAPDH).
5. PCR products obtained with these primers are confirmed by sequencing. Analyses are performed for both genes (i.e., GAPDH and IL-1β) in the linear range of the reaction. Results were analyzed using the NIH Image program (http://rsb.info.nih.gov/nih-image).

3.2.2.2. PMN Transmigration

Protocols in this subheading are essentially identical to **Subheading 3.1.2.1.**

3.2.3. Bioactions In Vivo

3.2.3.1. Murine Dorsal Air Pouch and Peritonitis

Protocols in this subheading are essentially identical to **Subheadings 3.1.3.1.** and **3.1.3.2.**

3.3. 17S-Series RvDs

DHA (C22: 6) is highly enriched in brain, synapses, and retina and is a major ω-3 fatty acid. Deficiencies in this essential fatty acid are reportedly associated with neuronal function, cancer, and inflammation. Using new lipidomic analyses using tandem LC-PDA-MS-MS, a novel series of endogenous mediators was identified in blood, leukocytes, brain, and glial cells as 17*S* hydroxy-containing docosanoids, which were denoted protectins (PD) that also contain docosatrienes (the main bioactive member of the series was 10,17*S*-docosatriene; the complete stereochemistry was recently established [*see* **ref. *53***]) and 17*S* series resolvins. These novel mediators were biosynthesized via epoxide-containing intermediates and proved potent (pico- to nanomolar range) regulators of both leukocytes, reducing infiltration in vivo, and glial cells, blocking their cytokine production. These results indicate that DHA is precursor to potent protective mediators generated via enzymatic oxygenations to novel docosatrienes and 17*S* series resolvins that each regulates events of interest in inflammation and resolution.

3.3.1. Biosynthesis

3.3.1.1. Incubation of DHA With Human Whole Blood

1. Human whole (venous) blood is collected from healthy volunteers (who declined taking medication for approx 2 wk before donation; BWH protocol no. 88-02642).

2. After collection, whole blood was immediately incubated with DHA (10 μg) in a covered water bath for 40 min (37°C).
3. Incubations are terminated and extracted for lipidomic and lipid mediator profile analyses by LC-MS-MS.
4. Selected ion monitoring at [M-H] = *m/z* 343, 359 and 375 revealed that human whole blood converts DHA to mono-, di-and tri-HDHA, respectively (*see* **Fig. 4**).

3.3.1.2. Incubation of DHA With Human PMN

1. Human PMN are freshly isolated from the whole blood by Ficoll gradient. PMN (30–50 × 10^6 cells/mL) are exposed to zymosan A (100 μg/mL) followed by addition of either 17*S*-HDHA (3 μg/mL) or DHA (3 μg/mL).
2. Cell suspensions are incubated for 40 min at 37°C in a covered water bath.
3. Incubations are terminated and lipidomic analyses carried out by LC-MS-MS.
4. Selected ion monitoring at *m/z* 359 and 375 revealed that human whole blood converts 17*S*-HDHA to di- and tri-HDHA (*see* **Fig. 4**).

3.3.2. Biological Activity In Vitro

3.3.2.1. Cytosensor Analyses

1. Changes in extracellular acidification rate are evaluated using a Cytosensor® microphysiometer and computer workstation.
2. Glial cells are equilibrated for 30 min in Dulbecco's Modified Eagle's Medium.
3. Extracellular acidification rates are determined by 30-s potentiometric rate measurements (μV/s; pump off cycle) after an 80-s pump cycle with a 10-s delay (120 s total cycle time).
4. Cells are perfused (100 μL/60 s) with agonist (DHA or 17*S* RvD series compounds, 100 n*M*) or vehicle (0.1% vol/vol ethanol) alone before the first rate measurements.
5. Perfusion duration depends on the ligand being evaluated and ranged from 5 to 15 min. Acidification rates (μV/s) are normalized to basal rates (100%) three cycles before agonist addition.

3.3.2.2. TNF-α-Induced IL-1β

Protocols in the subheading are identical to **Subheading 3.2.2.1.**

3.3.3. Bioactions In Vivo

3.3.3.1. Acute Inflammation: Murine Dorsal Air Pouch and Peritonitis

Protocols in the subheading are identical to **Subheadings 3.1.3.1.** and **3.1.3.2.** (*see* **Fig. 5** for the bioaction of 17*S*- and 17*R*-series RvDs).

3.4. Summary

Rvs are comprised of five separate chemical series of lipid-derived mediators, each with unique structures and apparent complementary anti-inflammatory properties and actions. Rvs are also generated in their respective epimeric forms

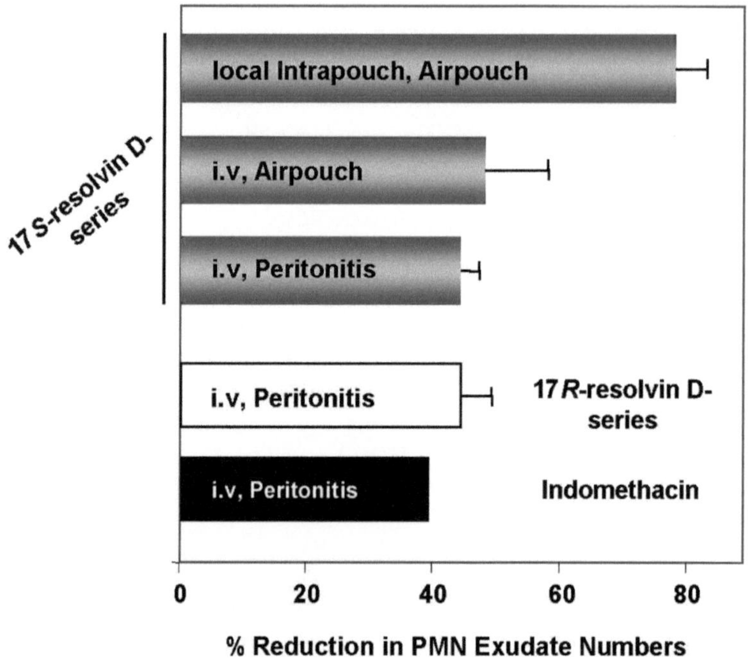

Fig. 5. Anti-inflammatory properties of 17*R*- and 17*S*-containing RvD series. Murine peritonitis and dorsal skin pouch after systemic and topical application. Peritonitis: RvDs (100 ng) or indomethacin (100 ng) was injected intravenously into mouse tail followed by zymosan A into the peritoneum. For direct comparison, mice were treated with 17*R*-RvDs via intravenous injection followed by zymosan A. Mice were sacrificed and peritoneal lavages were collected (2 h) and cells enumerated. Air pouch: RvDs (100 ng) were injected intrapouch or intravenously followed by intrapouch injection of TNF-α. At 4 h, air pouch lavages were collected and enumerated. Values represent mean ± SEM from three to four different mice; all gave $p < 0.05$ when infiltrated PMN was compared with vehicle control.

when aspirin is given in mammalian systems ***(11,12)***. The Rvs each dampen inflammation and PMN-mediated injury from within, key culprits in many human diseases. The results of these initial studies underscore a role in resolution as well as catabasis and give a possible therapeutic potential for this new arena of immunomodulation and host protection. Hence it is likely that the Rv and their AT-related forms may play roles in other tissues and organs involved in physiological and pathological processes as host-protective mediators. In view of the important roles of DHA and EPA in human biology and medicine uncovered to date ***(7–11)***, the physiological relevance of the Rvs is likely to extend well beyond our current appreciation ***(11–15)***.

4. Notes

1. Handling cytokines: It usually is advisable to aliquot cytokines because they are in general dissolved in water and kept at –20°C. Keep and dissolve cytokine stocks according to the manufacturer's instructions. Repeated freezing and thawing can lead to degradation of these protein mediators.
2. COX-2 expression: expression of COX-2 should be verified in the current experimental conditions. COX-2-specific antibodies as well as COX-2 nucleotide primers are available from most vendors that supply eicosanoids. Thus, verification of cytokine COX-2 induction by either Western blot or reverse transcription polymerase chain reaction is recommended.
3. Handling aspirin: aspirin hydrolyses rapidly in an aqueous environment. Therefore, prepare aspirin stocks in ethanol immediately before each experiment and store solid aspirin in a desiccator to prevent decomposition. **Note** that aspirin is the only known non-steroidal anti-imflammatory drug that acetylates COX-2.
4. Extraction protocol optimized for LX and RV: Terminate the incubation by adding two volumes of ice-cold methanol and place the mixture at –20°C for at least 20 min for protein precipitation. Centrifuge at 800*g* for 20 min at 4°C. Collect the supernatant in a 50-mL tube and dilute the sample with at least five volumes of pure water. Rapidly adjust the pH of this aqueous suspension to pH 3.5 with HCL (1 *N*). Load the acidified samples into solid-phase extraction cartridges (C_{18} Sep-Pak), wash the cartridge with 10 mL of water, and elute compounds with 8 mL of hexane, followed by 8 mL of methyl formate, and a final elution with 8 mL of methanol. LX, RV, HETEs, and most leukotrienes elute in the methyl formate fraction, whereas peptide leukotrienes (LTC_4, LTD_4, LTE_4, etc.) elute in the methanol fraction. Take the methyl formate fraction to dryness under a gentle stream of nitrogen and then resuspend compounds in a small volume of methanol (i.e., 100 µL).
5. LC-MS-MS analyses: LC/MS/MS analysis is conducted on an LCQ (Finnigan MAT, San Jose, CA) quadrupole ion trap mass spectrometer system equipped with an electrospray atmospheric pressure ionization probe. Compounds were dissolved in methanol and injected into the high-performance liquid chromatography (HPLC) component, which consisted of a Spectra SYSTEM P4000 quaternary gradient pump (Thermo Separation Products, San Jose, CA), LUNA C18-2 (150 × 2 mm, 5 µm) or C18-2 (100 × 2 mm, 5 µm) columns (Phenomenex, Torrance, CA), and a rapid spectra scanning SpectraSYSTEM UV2000 UV/VIS absorbance detector (Thermo Separation Products). The column is eluted at 0.2 mL/min isocratically for 20 min with methanol/water/acetic acid (65/35/0.01, v/v/v) followed by a 20 min linear gradient to 99.99/0.01 methanol/acetic acid (v/v). The spray voltage is set to 8 kV and the heated capillary to 250°C, and a maximum ion injection time of 350 ms. Over a 2.4 s scan cycle, full-scan MS are acquired by scanning between *m/z* 95-450 in the negative ion mode, followed by the acquisition of product ion MS/MS of the most intense molecular anions (e.g., [M-H] = *m/z* 349.5 for RvE1, [M-H] = *m/z* 343, 359 and 375 for mono-, di- and tri-HDHA, respectively). Also,

a Chiralcel OB-H column (J. T. Baker) was used to determine R and S alcohol configurations of monohydroxy-PUFA using isocratic (hexane:isopropanol; 96:4 vol:vol).

6. Human PMNs: PMNs should be used immediately after isolation (i.e., ~3 h after venipuncture, accounting for isolation) or placed in an ice bath (4°C). This will slow the decrease of enzymatic activity that starts once the cells are removed from blood. Note that careful records should be kept on the blood donors. Unlike cultured cells, peripheral blood cells are exposed to a diverse environment that can include cytokines and mono-HETEs that "prime" leukocytes. Such diversity results in noticeable donor variations in the data. For example, in endogenously "primed" PMN such as from asthmatic patients, 5S,15S-diHETE and LX, which normally are associated with transcellular biosynthesis, are generated endogenously from a single cell type (PMN).
7. TNF-α-initiated murine air pouches. Male FVB mice (6 to 8 wk) fed with laboratory rodent diet 5001 (Lab Diet, Purina Mills; containing less than 0.25% arachidonic acid, 1.49% EPA and 1.86% DHA) were anesthetized with isoflurane. Dorsal air pouches were raised by injecting 3 mL of sterile air subcutaneously on day 0 and 3. On day 6, inflammation in the air pouch was induced by local injection of recombinant murine TNF-α (20 ng) dissolved in 100 μL of sterile PBS. At indicated times, air pouches were lavaged twice with 3 mL of PBS. Lavage cells were collected by centrifugation and enumerated by light microscopy. Cell-free lavage fluids were used for determination of lipid mediators by lipidomic analysis.
8. Handling lipid mediators: RVs are sensitive to light, oxygen, and heat and should be stored at 20 to –80°C. This is especially critical for native EPA and DHA, which upon exposure to oxidative conditions will degrade to a variety of compounds. Thus, prepare a concentrated stock and a "working or daily stock" that you will use as reference materials for the experiments. As a general rule, each time a stock is used, pass a gentle stream of nitrogen into the vial before closing. This will slow the oxidative degradation. Determine the concentration of RV stocks by measuring the ultraviolet absorbance and using the specific extinction coefficient. This is especially important if stocks are used on a regular basis. If possible, HPLC analysis of your stocks is an invaluable tool to determine the purity and, for example, percent isomerization.
9. Vehicle controls: the total amount of vehicle (ethanol) added to your cell incubation or coincubations should not exceed 0.1%. Higher concentrations of alcohol can affect cell integrity (i.e., enzyme activity and membrane integrity). It is therefore desirable to prepare stocks of compounds at 1000 to 5000 times the final concentration so that only small aliquots are added directly to cell suspensions.
10. Zymosan A-initiated murine peritonitis: Male FVB mice (6–8 wk old) were anesthetized with isoflurane. Compounds were delivered as a bolus injection either into the tail vein in 100 μL of sterile saline or intraperitoneally in 200 μL of sterile saline and followed by intraperitoneal injection of zymosan A (1 mg/mL). Vehicle controls were performed in parallel. At indicated time points, mice were

euthanized with an overdose of isoflurane, peritoneal lavages collected and centrifuged. Aliquots of the cells were stained with Trypan blue and numerated by light microscopy.

11. Internal standards: it is essential to terminate incubations with solutions containing an internal standard since this will serve as an invaluable tool for correction of product profiles (e.g., HPLC retention times) and for calculating recoveries of LX and Rv after extraction and sample workup. The internal standard should be a stable measurable compound that is not generated during the cell incubations and does not interfere with the product analysis. PGB_2 usually meets those criteria for analysis of LX and Rv. Depending on the sensitivity of the analytical systems to be used (i.e., LC-MS, GC-MS, or PDA-UV-HPLC), 50 pg to 200 ng are added to the incubations.

Acknowledgments

We thank Mary Halm Small for assistance in preparing the manuscript. This work was supported in part by National Institutes of Health grants GM38675, P01-DE13499, and P50-DE016191.

References

1. Helgadottir, A., Manolescu, A., Thorleifsson, G., et al. (2004) The gene encoding 5-lipoxygenase activating protein confers risk of myocardial infarction and stroke. *Nat. Genet.* **36,** 233–239.
2. Erlinger, T. P., Platz, E. A., Rifai, N., and Helzlsouer, K. J. (2004) C-reactive protein and the risk of incident colorectal cancer. *JAMA* **291,** 585–590.
3. Pasche, B. and Serhan, C. N. (2004) Is C-reactive protein an inflammation opsonin that signals colon cancer risk? *JAMA* **291,** 623–624.
4. Gallin, J. I., Snyderman, R., Fearon, D. T., Haynes, B. F., and Nathan, C. (eds.) (1999) *Inflammation: Basic Principles and Clinical Correlates.* Lippincott Williams & Wilkins, Philadelphia.
5. Van Dyke, T. E. and Serhan, C. N. (2003) Resolution of inflammation: a new paradigm for the pathogenesis of periodontal diseases. *J. Dent. Res.* **82,** 82–90.
6. Burr, G. O. and Burr, M. M. (1929) A new deficiency disease produced by the rigid exclusion of fat from the diet. *J. Biol. Chem.* **82,** 345–367.
7. Lands, W. E. M. (ed.) (1987) *Proceedings of the AOCS Short Course on Polyunsaturated Fatty Acids and Eicosanoids.* American Oil Chemists' Society, Champaign, IL.
8. Bazan, N. G. (1990) Supply of n-3 polyunsaturated fatty acids and their significance in the central nervous system, in *Nutrition and the Brain* (Wurtman, R. J. and Wurtman, J. J., eds.), Vol. 8, Raven Press, New York, pp. 1–22.
9. Simopoulos, A. P., Leaf, A., and Salem, N., Jr. (1999) Workshop on the essentiality of and recommended dietary intakes for omega-6 and omega-3 fatty acids. *J. Am. Coll. Nutr.* **18,** 487–489.
10. Salem, N., Jr., Litman, B., Kim, H.-Y., and Gawrisch, K. (2001) Mechanisms of action of docosahexaenoic acid in the nervous system. *Lipids* **36,** 945–959.

11. Serhan, C. N., Clish, C. B., Brannon, J., Colgan, S. P., Chiang, N., and Gronert, K. (2000) Novel functional sets of lipid-derived mediators with antiinflammatory actions generated from omega-3 fatty acids via cyclooxygenase 2-nonsteroidal antiinflammatory drugs and transcellular processing. *J. Exp. Med.* **192,** 1197–1204.
12. Serhan, C. N., Hong, S., Gronert, K., Colgan, S. P., Devchand, P. R., Mirick, G., and Moussignac, R.-L. (2002) Resolvins: a family of bioactive products of omega-3 fatty acid transformation circuits initiated by aspirin treatment that counter pro-inflammation signals. *J. Exp. Med.* **196,** 1025–1037.
13. Hong, S., Gronert, K., Devchand, P., Moussignac, R.-L., and Serhan, C. N. (2003) Novel docosatrienes and 17S-resolvins generated from docosahexaenoic acid in murine brain, human blood and glial cells: autacoids in anti-inflammation. *J. Biol. Chem.* **278,** 14,677–14,687.
14. Marcheselli, V. L., Hong, S., Lukiw, W. J., et al. (2003) Novel docosanoids inhibit brain ischemia-reperfusion-mediated leukocyte infiltration and pro-inflammatory gene expression. *J. Biol. Chem.* **278,** 43,807–43,817.
15. Mukherjee, P. K., Marcheselli, V. L., Serhan, C. N., and Bazan, N. G. (2004) Neuroprotectin D1: a docosahexaenoic acid-derived docosatriene protects human retinal pigment epithelial cells from oxidative stress. *Proc. Natl. Acad. Sci. USA* **101,** 8491–8496.
16. Lehr, H.-A., Olofsson, A. M., Carew, T. E., et al. (1994) P-selectin mediates the interaction of circulating leukocytes with platelets and microvascular endothelium in response to oxidized lipoprotein in vivo. *Lab. Invest.* **71,** 380–386.
17. Mora, J. R., Bono, M. R., Manjunath, N., et al. (2003) Selective imprinting of gut-homing T cells by Peyer's patch dendritic cells. *Nature* **424,** 88–93.
18. Serhan, C. N., Hamberg, M., and Samuelsson, B. (1984) Lipoxins: novel series of biologically active compounds formed from arachidonic acid in human leukocytes. *Proc. Natl. Acad. Sci. USA* **81,** 5335–5339.
19. Serhan, C. N. and Sheppard, K. A. (1990) Lipoxin formation during human neutrophil-platelet interactions. Evidence for the transformation of leukotriene A_4 by platelet 12-lipoxygenase in vitro. *J. Clin. Invest.* **85,** 772–780.
20. Marcus, A. J. (1999) Platelets: their role in hemostasis, thrombosis, and inflammation, in *Inflammation: Basic Principles and Clinical Correlates* (Gallin, J. I. and Snyderman, R., eds.), Lippincott Williams & Wilkins, Philadelphia, pp. 77–95.
21. Maddox, J. F. and Serhan, C. N. (1996) Lipoxin A_4 and B_4 are potent stimuli for human monocyte migration and adhesion: selective inactivation by dehydrogenation and reduction. *J. Exp. Med.* **183,** 137–146.
22. Godson, C., Mitchell, S., Harvey, K., Petasis, N. A., Hogg, N., and Brady, H. R. (2000) Cutting edge: lipoxins rapidly stimulate nonphlogistic phagocytosis of apoptotic neutrophils by monocyte-derived macrophages. *J. Immunol.* **164,** 1663–1667.
23. Clària, J. and Serhan, C. N. (1995) Aspirin triggers previously undescribed bioactive eicosanoids by human endothelial cell-leukocyte interactions. *Proc. Natl. Acad. Sci. USA* **92,** 9475–9479.
24. Perretti, M., Chiang, N., La, M., Fierro, I. M., Marullo, S., Getting, S. J., Solito, E., and Serhan, C. N. (2002) Endogenous lipid- and peptide-derived anti-inflammatory

pathways generated with glucocorticoid and aspirin treatment activate the lipoxin A(4) receptor. *Nat. Med.* **8,** 1296–1302.
25. Vane, J. R. (2002) Back to an aspirin a day? *Science* **296,** 474–475.
26. Cheng, Y., Austin, S. C., Rocca, B., et al. (2002) Role of prostacyclin in the cardiovascular response to thromboxane A_2. *Science* **296,** 539–541.
27. Fierro, I. M., Colgan, S. P., Bernasconi, G., Petasis, N. A., Clish, C. B., Arita, M., et al. (2003) Lipoxin A_4 and aspirin-triggered 15-epi-lipoxin A_4 inhibit human neutrophil migration: comparisons between synthetic 15 epimers in chemotaxis and transmigration with microvessel endothelial cells and epithelial cells. *J. Immunol.* **170,** 2688–2694.
28. Fierro, I. M., Kutok, J. L., and Serhan, C. N. (2002) Novel lipid mediator regulators of endothelial cell proliferation and migration: Aspirin-triggered-15*R*-lipoxin A_4 and lipoxin A_4. *J. Pharmacol. Exp. Ther.* **300,** 385–392.
29. Kieran, N. E., Doran, P. P., Connolly, S. B., et al. (2003) Modification of the transcriptomic response to renal ischemia/reperfusion injury by lipoxin analog. *Kidney Int.* **64,** 480–492.
30. Fiore, S., Ryeom, S. W., Weller, P. F., and Serhan, C. N. (1992) Lipoxin recognition sites. Specific binding of labeled lipoxin A_4 with human neutrophils. *J. Biol. Chem.* **267,** 16,168–16,176.
31. Fiore, S., Maddox, J. F., Perez, H. D., and Serhan, C. N. (1994) Identification of a human cDNA encoding a functional high affinity lipoxin A_4 receptor. *J. Exp. Med.* **180,** 253–260.
32. Bae, Y.-S., Park, J. C., He, R., et al. (2003) Differential signaling of formyl peptide receptor-like 1 by Trp-Lys-Tyr-Met-Val-Met-$CONH_2$ or lipoxin A_4 in human neutrophils. *Mol. Pharmacol.* **63,** 721–730.
33. Gewirtz, A. T., Collier-Hyams, L. S., Young, A. N., et al. (2002) Lipoxin A_4 analogs attenuate induction of intestinal epithelial proinflammatory gene expression and reduce the severity of dextran sodium sulfate-induced colitis. *J. Immunol.* **168,** 5260–5267.
34. Devchand, P. R., Arita, M., Hong, S., et al (2003) Human ALX receptor regulates neutrophil recruitment in transgenic mice: Roles in inflammation and host-defense. *FASEB J.* **17,** 652–659.
35. Serhan, C. N., Jain, A., Marleau, S., et al. (2003) Reduced inflammation and tissue damage in transgenic rabbits overexpressing 15-lipoxygenase and endogenous anti-inflammatory lipid mediators. *J. Immunol.* **171,** 6856–6865.
36. Weissmann, G., Smolen, J. E., and Korchak, H. M. (1980) Release of inflammatory mediators from stimulated neutrophils. *N. Engl. J. Med.* **303,** 27–34.
37. Samuelsson, B. (1982) From studies of biochemical mechanisms to novel biological mediators: prostaglandin endoperoxides, thromboxanes and leukotrienes, in *Les Prix Nobel: Nobel Prizes, Presentations, Biographies and Lectures*, Almqvist & Wiksell, Stockholm, pp. 153–174.
38. Vane, J. R. (1982) Adventures and excursions in bioassay: the stepping stones to prostacyclin, in *Les Prix Nobel: Nobel Prizes, Presentations, Biographies and Lectures*, Almqvist & Wiksell, Stockholm, pp. 181–206.

39. GISSI-Prevenzione Investigators (1999) Dietary supplementation with n-3 polyunsaturated fatty acids and vitamin E after myocardial infarction: results of the GISSI-Prevenzione trial. Gruppo Italiano per lo Studio della Sopravvivenza nell'Infarto miocardico. *Lancet* **354,** 447–455.
40. Marchioli, R., Barzi, F., Bomba, E., et al. (2002) Early protection against sudden death by n-3 polyunsaturated fatty acids after myocardial infarction: time-course analysis of the results of the Gruppo Italiano per lo Studio della Sopravvivenza nell'Infarto Miocardico (GISSI)-Prevenzione. *Circulation* **105,** 1897–1903.
41. Rosenstein, E. D., Kushner, L. J., Kramer, N., and Kazandjian, G. (2003) Pilot study of dietary fatty acid supplementation in the treatment of adult periodontitis. *Prostaglandins Leukot. Essent. Fatty Acids* **68,** 213–218.
42. Bazan, N. G. (1992) Supply, uptake, and utilization of docosahexaenoic acid during photoreceptor cell differentiation. *Nestle Nutrition Workshop Series* **28,** 121–133.
43. Lee, T. H., Mencia-Huerta, J.-M., Shih, C., Corey, E. J., Lewis, R. A., and Austen, K. F. (1984) Effects of exogenous arachidonic, eicosapentaenoic, and docosahexaenoic acids on the generation of 5-lipoxygenase pathway products by ionophore-activated human neutrophils. *J. Clin. Invest.* **74,** 1922–1933.
44. Sawazaki, S., Salem, N., Jr., and Kim, H.-Y. (1994) Lipoxygenation of docosahexaenoic acid by the rat pineal body. *J. Neurochem.* **62,** 2437–2447.
45. Reich, E. E., Zackert, W. E., Brame, C. J., et al. (2000) Formation of novel D-ring and E-ring isoprostane-like compounds (D_4/E_4-neuroprostanes) in vivo from docosahexaenoic acid. *Biochemistry* **39,** 2376–2383.
46. VanRollins, M., Baker, R. C., Sprecher, H. W., and Murphy, R. C. (1984) Oxidation of docosahexaenoic acid by rat liver microsomes. *J. Biol. Chem.* **259,** 5776–5783.
47. Lands, W. E. M. (2003) Diets could prevent many diseases. *Lipids* **38,** 317–321.
48. Winyard, P. G. and Willoughby, D. A. (eds.) (2003) *Inflammation Protocols*. Humana, Totowa, NJ.
49. Lu, Y., Hong, S., Tjonahen, E., and Serhan, C. N. (2003) Lipid mediator lipidomics: databases and search algorithms of electric spray ionization/tandem mass and ultraviolet spectra for structural elucidation, in *5th Winter Eicosanoid Conference*, Baltimore.
50. Arita, M., Bianchini, F., Aliberti, J., et al. (2005) Stereochemical assignment, anti-inflammatory properties, and receptor for the omega-3 lipid mediator resolvin E1. *J. Exp. Med.*, **201,** 713–722.
51. Corey, E. J., Shih, C., and Cashman, J. R. (1983) Docosahexaenoic acid is a strong inhibitor of prostaglandin but not leukotriene biosynthesis. *Proc. Natl. Acad. Sci. USA* **80,** 3581–3584.
52. Serhan, C. N. and Oliw, E. (2001) Unorthodox routes to prostanoid formation: new twists in cyclooxygenase-initiated pathways. *J. Clin. Invest.* **107,** 1481–1489.
53. Serhan, C. N., Gotlinger, K., Hong, S., et al. (2006) Anti-inflammatory actions of neuroprotectin D1/protectin D1 and its natural stereoisomers: assignments of dihydroxy-containing docosatrienes. *J. Immunol.* **176,** 1848–1859.

20

Isolation and Analysis of Lipid Rafts in Cell–Cell Interactions

Aimee Landry and Ramnik Xavier

Summary

Lipid rafts are dynamic structures made up of proteins and lipids that float freely within the liquid-disordered bilayer of cellular membranes and have the ability to cluster to form larger, more-ordered platforms. These clustered structures have been identified in all cell types and have been shown to play critical roles in signal transduction, cellular transport, and cell–cell communication. Lipid rafts also have been implicated in facilitating bacterial/viral entry into host cells and in human disease, highlighting the significance of understanding the role lipid rafts play in physiological and pathological signaling outcomes. In this chapter, we provide protocols to isolate lipid rafts from polarized and nonpolarized cells and outline novel technologies to analyze signal transduction cascades in vivo.

Key Words: Lipid rafts; detergent-insoluble membranes; cholesterol; signal transduction; biochemical fractionation; microscopy; proteomics; FRET; FRAP; flow cytometry; lymphocytes; polarized epithelial cells.

1. Introduction

Lipid rafts are membrane microdomains enriched in cholesterol and sphingolipids, and are associated with specific microanatomic features of cells from diverse lineages. They float freely in the resting state, carrying a few passenger proteins, but aggregate to form functional platforms mediating diverse signaling cascades when activated ***(1–4)***. Lipid rafts have been described in all cell types to date. Cholesterol content and sphingolipids play a central role in the genesis of these of these microdomains. Analysis of the consequences of directed mutations in mice have established a role for lipid rafts in signal transduction ***(5,6)***. Caveolae, characteristic flask-shaped invaginations of endodermal cells are one class of lipid raft structure; another is the lumenal surface of polarized

From: *Methods in Molecular Biology, vol. 341: Cell–Cell Interactions: Methods and Protocols*
Edited by: S. P. Colgan © Humana Press Inc., Totowa, NJ

epithelium, which is highly enriched in raft-like structures. In other cells, such as lymphocytes and neurons, rafts are isolated structures 50 nm in diameter and do not appear to be associated with any specific ultrastructural features. In lymphocytes, rafts are estimated to represent 40% of total membrane as determined by fluorescence anisotrophy measurements. For the purposes of this overview we will refer to lipid rafts as detergent-insoluble membranes (DIMs). Both phospatidylinositol-linked proteins and palmitoylated transmembrane proteins are concentrated in rafts, as well as nearly all lipidated intracellular proteins, including myristoylated, palmitoylated, and prenylated molecules.

A protein's affinity to lipid rafts can be modified by enzymatic reactions, which alter the biochemical properties of the given protein, allowing the cell to dynamically regulate molecular mechanics which govern cellular function. Glycosylphospatidylinositol (GPI)-linked proteins often are found located within lipid rafts and can be cleaved by phospholipase. Rab GTPases, key regulators of membrane targeting and fusion, require the covalent attachment of geranyl-geranyl isoprenes near their C terminus in a multi-step process, which can subsequently increase lipid affinity *(7)*. Src family kinases, which are responsible for plasma membrane targeting, contain a unique domain in the N-terminus that is modified by dual acylation and palmitoylation *(8)*.

The ability of lipid rafts to isolate protein receptors from the majority of the plasma membrane enables the compartmentalization of signaling components, concentrating certain elements within lipid rafts while excluding others ***(9,10)***. HaRas, which is palmitoylated and farnesylated, is localized to lipid rich microdomains. In contrast, KiRas, which is farnesylated but not palmitoylated, associates with the inner leaflet of the plasma membrane and is excluded from lipid rafts. Thus, a posttranslational addition of palmitate effectively segregates these two GTPases in the plasma membrane ***(11)***. Recent studies have suggested that an equivalent domain organization is present in the cytoplasmic leaflet as well. In polarized epithelial cells, lipid rafts are distributed asymmetrically at the apical surface and play a role in protein localization, trafficking, internalization, and proteosomal degradation ***(12,13)***.

Proteins are thought to target lipid-rich domains resulting from the fact that the number of lipid sites exceeds the number of binding sites; this system allows lipid rafts to recruit many different proteins to a membrane without saturating protein-binding sites. Lipids also can be produced and degraded rapidly by enzymes in a matter of seconds, allowing the characteristics of a membrane to be rapidly remodeled based on the physiological need of the cell at a given time point. In addition, lipid-binding interactions often occur at a low affinity, enabling rapid protein equilibration and movement inside the cell.

Microdomains also play key roles in the life cycle of microbes by facilitating access to cellular receptors, influencing trafficking to appropriate subcellular sites, and play a critical role in the assembly of enveloped viruses during replication ***(14)***. Lipid rafts also have an important function in the organization of signal transduction during lymphocyte activation ***(15,16)***. It appears that differential partitioning of signaling proteins within the lipid rafts is an important mechanism by which integration of signal transduction pathways is achieved. In T- and B-cells, lipid rafts appear to play an important role in development, signaling and differentiation.

Understanding raft-mediated membrane trafficking is important to cell communication. Lipid rafts have been shown to be involved in fundamental processes of all eukaryotic cells, and 30% of all cellular proteins are membrane proteins. Lipid rafts have drawn the attention of a variety of investigators with broad range of research interests. Neurobiologists have identified lipid rafts as the basis of synaptic transmission ***(17)***. Lipid rafts have also been shown to be essential for antigen presentation and processing and have been implicated in several immunological disorders such as systemic lupus erythematosus ***(18)***. An expanding list of monogenic diseases can traced to altered lipid raft function ***(19,20)***. In addition, lipid rafts play a central role in the pathogenesis of numerous common and complex diseases, such as diabetes, atherosclerosis, cancer, and neurodegenerational disorders such as Alzheimer's disease ***(21)***. Lipid rafts are thought to serve as specific targets for entry of immunomodulators (e.g., toxins or lipopolysaccharides) of bacterial origin into host cells or viruses (e.g., human immunodeficiency and influenza viruses).

Persuasive evidence for the existence of lipid rafts comes from analysis of model membranes, liposomes, and biophysical methods such as fluorescence energy transfer and single particle tracking experiments ***(22)***. Considerable progress has been made in studying and understanding the role lipid rafts play in receptor-mediated signal transduction. Here, we briefly review the role of rafts in a model system familiar to our laboratory and provide protocols to isolate and analyze rafts in nonpolarized and polarized cell signaling.

2. Materials

2.1. Cell Stimulation

1. Purified antibodies of interest.

2.2. Cholesterol Depletion and Enrichment

1. Methyl-β-cyclodextrin (MβCD; Sigma, St. Louis, MO, cat. no. C-4555). Stock solutions made in dH_2O and stored at –20°C.
2. Cholesterol complexed to MβCD (Sigma, cat. no. C-4951). Stock solutions made in dH_2O and stored at –20°C.

2.2.1. Sucrose Gradient

1. Refrigerated ultracentrifuge capable of reaching speeds of 200,000*g* and appropriate rotor.
2. Glass Dounce homogenizer (Bellco Glass Inc.; cat. no. 1984-10007).
3. SW-40 polyalomer tubes (Beckman, cat. no. 331372).
4. Mes-buffered saline (MBS): 25 m*M* Mes (pH 6.5), 150 m*M* NaCl. Store at 4°C.
5. MBS lysis buffer: 22 m*M* MES (pH 6.5), 150 m*M* NaCl, 0.5% Triton X-100, 1 m*M* Na_3VO_4, 2 m*M* EDTA, 1 m*M* phenylmethyl sulfonyl fluoride, and 1 μg/μL aproptintin. Store at 4°C.
6. Sucrose solutions: 80% (w/v) sucrose (80 g of ultra pure sucrose into a final volume of 100 mL of MBS). Heat approx 50 mL of MBS while stirring and gradually add sucrose until dissolved. Do not boil. Increase volume to 100 mL with additional MBS. 35% (w/v) and 5% (w/v) sucrose solutions can be made by diluting 85% stock. Store at 4°C.
7. Apparatus for sodium dodecyl sulfate polyacrylamide gel electrophoresis and Western transfer (Bio-Rad, cat. no.165-3318).
8. Polyvinylidene difluoride membrane (Millipore, cat. no. IPVH20200).

2.2.2. Linear Gradient

1. Buffer A: 0.25 *M* sucrose, 1 m*M* EDTA, 20 m*M* Tris-HCl, pH 7.8.
2. Buffer A+: Buffer A+ 0.2 m*M* AEBSF, 1 μg/mL aprotinin, 10 μ*M* bestatin, 3 μ*M* E-64, 10 μg/mL leupeptin, 2 μ*M* pepstatin, and 50 μg/mL calpain inhibitor I.
3. Percoll (Sigma, cat. no. P-1644).

2.2.3. Sodium Carbonate Gradient

1. SCB: 50 m*M* sodium carbonate, pH 11.0.
2. SCB+: SCB + 0.2 m*M* AEBSF, 1 μg/mL aprotinin, 10 μ*M* bestatin, 3 μ*M* E-64, 10 μg/mL leupeptin, 2 μ*M* pepstatin, and 50 μg/mL calpain inhibitor I.
3. Branson Sonifier 250, or similar.

2.2.4. OptiPrep Gradient

1. Base buffer: 20 m*M* Tris-HCl, pH 7.8, 250 m*M* sucrose, 1 m*M* $CaCl_2$, 1 m*M* $MgCl_2$.
2. Base buffer+: Base buffer + 0.2 m*M* AEBSF, 1 μg/mL aprotinin, 10 μ*M* bestatin, 3 μ*M* E-64, 10 μg/mL leupeptin, 2 μ*M* pepstatin, and 50 μg/mL calpain inhibitor I.
3. OptiPrep (Granier BioOne, cat. no. No. 1030061; www.gbo.com).

2.3. Protein Localization and Aggregation of Lipid Rafts

1. Glass slides and cover slips.
2. Clear nail polish.
3. PAP Pen (Sigma, cat. no. Z377821).
4. 0.01% poly-L-lysine (Sigma, cat. no. P-4707).

5. 4% (w/v) paraformaldehyde (PFA): add 4 g of PFA powder to approx 80 mL of dH_2O. Stir and heat solution until solution reaches 60 to 70°C (but do not allow to boil). Add 10 *N* NaOH drop-wise until solution clears. Remove from heat and allow to cool. Add 10 mL of 10X phosphate-buffered saline (PBS) and adjust final volume to 100 mL. Adjust pH to 7.4, filter, and store at 4°C for up to 1 mo. Note: PFA powder is toxic if inhaled. Work in a fume hood or wear a mask while measuring out the powder to avoid inhalation.
6. Tween-20 (Sigma, cat. no. P-9416).
7. Bovine serum albumin (BSA; Sigma, cat. no. A-3059).
8. Primary antibodies of interest.
9. Fluorescent secondary antibodies.
10. VectaMount (Vector Laboratories, cat. no. H-5000, www.vectorlabs.com).

2.4. Molecular Probes and Fluorescent Microscopy

1. Vybrant Lipid Raft Labeling Kit (Molecular Probes; www.probes.com).
2. Bodipy sphingomyelin analog (C5 DMB-SM; Molecular Probes; www.probes.com).

2.5. Immunoisolation

1. Goat-anti-mouse IgG Dynabeads, M-450 4×10^8 beads/mL (Dynal, Oslo, Norway; www.dynalbiotech.com).
2. MPC-E magnet for 1.5-mL microcentrifuge tubes (Dynal).
3. Dynal MPC-1 magnetic partial concentrators for 15- or 50-mL centrifuge tubes (Dynal).
4. Rotating wheel.
5. 100 m*M* sodium vanadate: dissolve 9.5 g of sodium orthovanadate in approx 400 mL of dH_2O. Adjust pH to 10.2 with HCl. Heat in microwave until solution starts to boil. Adjust pH to 10.2 with 1 *M* NaOH. Heat in microwave until solution starts to boil. Check pH is between 9.9 and 10.1. Repeat with new reagents if pH is not correct.
6. 10 m*M* pervanadate: 10 μL of 100 m*M* sodium vanadate stock solution, 10 μL of 30% H_2O_2, and 80 μL of dH_2O. Incubate solution at room temperature (RT) for 10 min before preparing HB+ buffer.
7. 1000X CLAP: 5 mg of chymostatin, 5 mg of leupeptin, 5 mg of antipain, and 5 mg of pepstatin. Dissolve in 500 μL of dimethyl sulfoxide. Store in 50-μL aliquots at –20°C.
8. HB buffer: 250 m*M* sucrose, 10 m*M* Na–HEPES, pH 7.2, 1 m*M* $MgCl_2$, 10 m*M* NaF, and 1 m*M* sodium vanadate. Adjust to a final volume of 500 mL with dH_2O.
9. HB+ buffer: add 1:1500 of 10 m*M* pervandate and 1:1000 of 1000X CLAP to HB.
10. Laemmeli sample buffer (BioRad; cat. no. 161-0737).

2.6. Electron Microscopy

1. 5-mm copper wire loops.
2. Mesh copper grids (Agar Scientific, UK).

Table 1
Volumes for Preparing Reducing Solution for Various-Sized Gold Particles

Gold size (nm)	2–3	4	4.5	5.5	6	7.5
1% tannic acid (mL)	2.5	1.25	0.75	0.5	0.25	0.13
25 m*M* potassium carbonate (mL)	2.5	1.25	0.75	0.5	0.25	0.13

3. Filter paper (Whatman).
4. Fish skin gelatin, 45% aqueous solution (Sigma, cat. no. G-7765).
5. Glutaraldehyde, EM grade, 25% stock (Agar Scientific, UK, cat. no. R1020).
6. Methyl cellulose (Sigma, cat. no. M-6385).
7. Parafilm.
8. Paraformaldehyde, EM grade, 16% stock (Agar Scientific, cat. no. R1026).
9. Adobe Photoshop (San Jose, CA).
10. Image J 1.26 (http://rsb.info.nih.gov/ij/) or NIH Image 1.82 (http://rsb.info.nih.gov/nih-image).
11. Affinity-purified antibodies of interest.
12. Commercially available K-function analysis program.
13. Transmission electron microscope (JEOL 1010 or similar).
14. 1% trisodium citrate: dissolve 0.5 g of sodium citrate in 50 mL of dH_2O. Store at RT.
15. 1% tannic acid: dissolve 0.5 g of tannic acid in 50 mL of dH_2O. Store at RT.
16. 25 m*M* potassium carbonate: dissolve 0.173 g of potassium carbonate in 50 mL of dH_2O. Store at RT.
17. Reducing solution: 2 mL of 1% trisodium citrate, X mL of 1% tannic acid, and X mL of 25 mM potassium carbonate (*see* **Table 1** for quantities). Add dH_2O for a final volume of 10 mL. Make fresh for each use.
18. 1% gold chloride: dissolve 1 g of gold chloride in 100 mL of dH_2O. Store at 4°C. Stable for up to 2 yr.
19. Gold solution: add 0.5 mL of 1% gold chloride to 39.5 mL of dH_2O. Make fresh for each use.
20. 10% BSA: dissolve 2 g of BSA in 20 mL of PBS. (For use in electron microscopy [EM], centrifuge at 100,000*g* for 1 h at 4°C to remove any remaining particles. Store at 4°C for 5 mo.)
21. 0.1% BSA: add 0.5 mL of 10% BSA to 50 mL of PBS. Make fresh for each use.
22. 2% methyl cellulose: dissolve 2 g of methyl cellulose in 98 mL of dH_2O heated to 90°C. Cool on ice while continuing to stir. Store in a sealed container for 72 h at 4°C.
23. Centrifuge at 100,000*g* for 1 h. Decant supernatant and store at 4°C.
24. 10X potassium acetate (KOAc) buffer: add 11.92 g of HEPES, 22.58 g of potassium acetate, and 1.02 g of magnesium chloride to 180 mL of dH_2O. Add six pellets of KOH and continue to adjust pH using additional KOH (1 *M*) to 7.4. Add additional water for a final volume of 200 mL. Aliquot and store at –20°C.

Table 2
Donor/Acceptor Antibodies for FRET

	Donor Imaging Path			Acceptor Imaging Path		
FRET Pair	Excitation Filter	Dichroic	Emission Filter	Excitation Filter	Dichroic	Emission Filter
FITC-TRITC	480/40	505 LP	535/40	545/30	565 LP	610/60
FITC-Cy3	480/40	505 LP	535/40	540/25	565 LP	605/55
Cy3-Cy5	540/25	565 LP	605/55	640/20	660 LP	680/30
BFP-GFP	390/20	420 LP	460/50	475/40	460 LP	535/50
GFP-Cy3	475/40	460 LP	505/40	540/25	565 LP	605/55
CFP-YFP	436/20	455 LP	480/40	500/20	515 LP	535/30
CFP-DsRed	436/20	455 LP	480/40	546/12	560 LP	620/20

All values are expressed as center wavelengths (in nm) with the spread on either side of the peak value. Donor- and acceptor-only images are acquired separately with donor and acceptor filter sets, respectively. FRET images are acquired with donor excitation filter, donor dichroic and acceptor emission filters. These filter combinations were adapted from the catalog of filter sets published by Chroma Technology Corporation. FRET, fluorescence resonance energy transfer; TRITC, tetramethylirhodamine isothiocyanate; BFP, blue fluorescent protein; GFP, green fluorescent protein; CFP, cyan fluorescent protein; YFP, yellow fluorescent protein.

25. 1X KOAc buffer: thaw and add 2 mL of 10X KOAc buffer (above) to 18 mL of PBS. Make fresh for each use.
26. Fixative: add 2 mL paraformaldehyde (16% stock), 32 µL of glutaraldehyde (25% stock), and 0.8 mL of 10X KOAc buffer to 5.17 mL of dH_2O in a fume hood. Store at 4°C for up to 2 wk.
27. 25 m*M* glycine: dissolve 93.8 mg of glycine in 50 mL of PBS. Aliquot (1 mL each) and store at –20°C.
28. Fish skin gelatin: add 3 mL of fish skin gelatin (45% aqueous solution) to 11.5 mL of PBS. Store at 4°C for up to 6 mo.
29. Blocking solution: add 200 µL of 10% fish skin gelatin and 200 µL of 10% BSA to 10 mL of PBS. Make fresh for each use.
30. 3% uranyl acetate: dissolve 1.5 g of uranyl acetate in 50 mL of dH_2O. Aliquot and store at 4°C for 1 yr.

2.7. Fluorescence Resonance Energy Transfer

1. Lipid-based delivery system, such as TransFection (Bio-Rad, cat. no. 170-3250).
2. Donor/acceptor antibodies for FRET (**Table 2**).
3. Normal serum from secondary antibody host diluted in PBS (as recommended by manufacturer).
4. $NaBH_4$ (1 mg/mL).
5. 100% methanol (–20°C) or 0.1% Triton X-100 (4°C) for fixation (if necessary).

6. Confocal laser scanning imaging system with appropriate lasers for excitation of donor–acceptor fluorescence resonance energy transfer (FRET) pairs.

2.8. Fluorescence Recovery After Photobleaching

1. Polybead polystyrene beads (6.0 μm in diameter; Polyscience, cat. no. 07312).
2. Purified antibodies of interest.
3. Blocking Solution: 10 mg/mL in PBS. Filter-sterilize and store at 4°C for several months.
4. Confocal microscope (Carl Zeiss, LSM 510, or similar) and with a 25 mW Argon laser.

2.9. Proteomics

1. Leucine-deficient media with serum.
2. Leucine-deficient media without serum.
3. L-Leucine (Sigma, cat. no. L-8912).
4. L-Leucine-3,3,3-D3 (Sigma, cat. no. 48,682-5).
5. Protein concentration determination kit.
6. 6 *M* urea–2 *M* thiourea (in 10 m*M* HEPES, pH 8.0).
7. 2.5 *M* sodium acetate, pH 5.0.
8. 1,4-dithio-DL-threoitol (DTT; Fluka, cat. no. 43817).
9. Iodoacetamide (Sigma, cat. no. I-1149).
10. LysC enzyme (Sigma, cat. no. A-3422).
11. 50 m*M* ammonium bicarbonate.
12. Trypsin (Promega, cat. no. V5113).
13. Empore disks (3M, Minneapolis, MN, cat. no. 2215).
14. Reverse-phase analytical columns (various manufacturers).
15. 5.6 to 64% acetonitrile (Fluka cat. no. 00682).
16. Mass spectroscopy (MS) (QSTAR-Pulsar quadrupole time-of-flight hybrid; PE-Sciex, Thornhill, Ontario, Canada), or similar.

2.10. Analysis Using the Fluorescence-Activated Cell Sorter

1. Fluorescent antibodies specific against membrane antibodies.
2. Fluorescence-activated cell sorting (FACS) machine (FACSCalibur or similar), corresponding software program (i.e., CellQuest).

3. Methods

3.1. Cell Preparation

3.1.1. Cell Stimulation

Lipid rafts are most commonly isolated from cells based on their differential solubility in nonionic detergents. Raft size is increased by clustering, leading to a new mixture of molecules. This clustering can be triggered at the extracellular side by ligands, antibodies, or lectins ***(23)***; within the membrane by oligomerization ***(24)***; or by cytosolic agents (cytoskeletal elements, adapters, scaffolds ***[25,26]***). Raft clustering occurs at the plasma membrane as well as intracellularly, such as in the endosomal lumen. Ligand binding or oligomerization can alter the partitioning of proteins in and out of rafts. Increased raft affinity of a given protein and its activation within rafts (e.g., phosphorylation by Src-family kinases can initiate a cascade of signaling events, leading to further increase of raft size by clustering ***(27)***. In the example presented below, Jurkat cells are stimulated with anti-CD3 and CD28 antibodies to activate the T-cell receptor (TCR) complex. This method can easily be adapted to other cells.

1. Count cells and determine the total number needed for the experiment (*see* **Note 1**).
2. Collect required cells by centrifugation (800*g* for 5 min at 4°C) and wash once in 1 mL ice-cold PBS. Resuspend in 200 µL of PBS per sample.
3. Transfer cells to labeled microcentrifuge tubes and add 1 µg of both CD3 and CD28 antibodies and incubate in a 37°C water bath for various time points (*see* **Note 2**). If the samples will be in the water bath for more than 5 min, agitate intermittently.
4. Stop the reaction by adding 1 mL of ice-cold PBS to each tube and wash twice (800*g* for 5 min at 4°C).

3.1.2. Cholesterol Depletion and Enrichment

Because of the important role cholesterol plays in the stabilization of lipid rafts, reversible depletion or enrichment of plasma membrane cholesterol are two useful techniques that can be used to examine the biological roles of functional and non-functional lipid rafts. If a protein is found to be associated with DIM, treatment with cyclodextrin, saponin, and sphingomyelinase may be applied to test whether the association depends on cholesterol and sphingomyelin. MβCD extracts cholesterol from membranes and disrupts domains whose integrity depends on cholesterol. Saponin complexes cholesterol and is thought to sequester it away from other interactions.

3.1.2.1. Cholesterol Depletion

1. Collect cells by centrifugation and wash once in MBS. Incubate cells with 10 m*M* MβCD at 37°C for 30 min.
2. Cells can now be used in functional studies, or washed, repelleted, lysed, and subjected to sucrose-gradient fractionation.

3.1.2.2. Cholesterol Enrichment

When precomplexed with cholesterol, cyclodextrins can deliver cholesterol back into the plasma membrane and restore cellular functions previously interrupted by cholesterol depletion.

1. Wash cholesterol-depleted cells. Resuspend cells in MBS containing 3 m*M* MβCD: cholesterol and incubate at 37°C for 30 min.
2. Wash cells once and use in subsequent studies (*see* **Notes 3–5**).

3.2. Gradient Isolation of DIM Fractions

3.2.1. Isolation of DIM Fractions

This protocol takes advantage of the differential solubilization of lipid rafts and the plasma membrane to isolate populations of DIMs. In our experience, Triton X-100 is the most effective detergent in membrane solubilization and recovery of proteins enriched in lipid microdomains. Composition of isolated DIMs depends on the temperature and on the nature and concentration of detergent used. An elegant study recently demonstrated that DIMs can be isolated at 37°C in the presence of Brij 98 ***(28)***. Commonly used detergents are Triton X-100, Brij 58, Brij 96, LubrolWX, CHAPS, and Brij 98. Composition of lipid rafts depends on the detergent used. Recent studies have suggested that all rafts isolated do not reflect the same aspect of membrane organization ***(28,29)***. Keep in mind that proper controls for incomplete solubilization of membrane fractions are important for each experiment. A common problem with this method in to analyze lipid rafts organization in cells is that lipid reorganization occurs during solubilization so that detergent-resistant membrane structures may differ from membranes before detergent extraction. Despite this caveat, however, the isolation of DIMs will remain a valuable tool for the analysis of biological membranes.

3.2.1.1. Sucrose Gradient

1. Precool the ultracentrifuge, rotor, and all solutions to 4°C (*see* **Note 6**).
2. Resuspend the pellet in 2 mL of MBS buffer. Incubate the cells on ice for 30 min. Disrupt cell membranes by homogenizing for ten strokes in a glass Dounce homogenizer at 4°C.
3. Transfer 2 mL of the total cell homogenate to the bottom of a SW-40 centrifuge tube and add 2 mL of 80% sucrose; mix well.
4. Slowly layer 6 mL of the 35% sucrose solution on top of the sample layer; to prevent the disruption of the sucrose gradient, hold the pipet at a 45° angle on the side of the centrifuge tube.
5. Layer 4 mL of the 5% sucrose solution on top of previous two layers. Make sure that the tube is filled to within 2 mm of the top.
6. Balance the tubes to within 0.01 g and place them in the rotor. Centrifuge samples for 16 h at 200,000*g* at 4°C.
7. Using a pipet, carefully collect 12 fractions of 1 mL each, starting from the top of the gradient. Alternatively, extract the visible band at the 35%: 5% sucrose interface with a syringe. Proteins localized to lipid rafts are usually recovered in the low-density sucrose gradient fractions of two, three, and four (*see* **Notes 7–9**).

8. Prepare aliquots for sodium dodecyl sulfate polyacrylamide gel analysis and separate the proteins by electrophoresis. Transfer to a polyvinylidene difluoride membrane. Identify proteins of interest by immunoblotting with antibodies according to manufacture's directions (*see* **Note 10**).

3.2.1.2. Detergent-Free Isolation of Lipid Rafts

Detergents have routinely been used to isolate lipid rafts over the past several years, although questions have recently been raised surrounding this technique. It is thought by some that the extraction of lipid rafts with detergent may generate artificial clusters of lipid rafts that do not exist in vivo. The following is a list of three protocols that avoid the use of detergents, but care should be taken in deciding which protocol will best serve your research needs.

3.2.1.3. Linear Gradient

Originally developed to isolate caveolae from cells or tissues, this protocol uses sonication to disrupt the cell membrane and was used to identify novel proteins associated with lipid rafts based on their unique buoyant density ***(28,30)***. The main advantage of this protocol is the purified lipid rafts obtained without the use of harsh detergents. This purity, however, also results in low overall yields.

1. Wash four confluent 10-cm plates of cultured adherent cells (~7–8 mg of total protein) with PBS. Remove PBS and scrape cells into Buffer A+. Transfer into a 15-mL centrifuge tube.
2. Centrifuge cells (1000*g* for 5 min at 4°C) and resuspended in an additional 1 mL of Buffer A+.
3. Lyse cells with 10 strokes of a Dounce homogenizer using a tight-fitting pestle.
4. Follow homogenization by shearing, passing lysates through a 23-gage needle 10 times.
5. Centrifuge cells (1000*g* for 5 min at 4°C). Collect the supernatant and transfer to a separate tube.
6. Repeat lysing of the remaining pellet with 1 mL of additional Buffer A+. Lyse cells with 10 strokes of a Dounce homogenizer using a tight-fitting pestle.
7. Follow homogenization by shearing by passing lysates through a 23-gage needle 10 times.
8. Centrifuge cells (1000*g* for 10 min at 4°C). Combine both cleared lysates.
9. Carefully layer the combined lysates on top of 8 mL of 30% Percoll and centrifuge for 30 min at 84,000*g*.
10. A visible band should be seen one-third of the way down the gradient. Collect band with a needle and syringe.
11. Adjust total volume of the plasma membrane fraction to 2 mL with additional Buffer A+.

12. Sonicate fraction with six pulses at 50% duty cycle using a Branson Sonifier 250. Allow the sample to settle for 1 min. Repeat sonification two more times for a total of 18 pulses.
13. Mix sonicate with 2 mL of Buffer A+ containing 50% OptiPrep.
14. Carefully layer an 8 mL OptiPrep gradient (10–20%) on top of the 25% sonicate and centrifuge for 90 min at 52,000*g*.
15. Isolate fractions of equal volume as above (*see* **Subheading 3.2.1.1.**) and determine the distribution of key lipid raft proteins via Western blotting.

3.2.1.4. Sodium Carbonate Gradient

The following is another detergent-free method of isolating lipid rafts that preserves the interactions between resident prenylated caveolin-associated proteins while generating a purified population of caveolin-rich membrane domains; using this protocol, Song et al. ***(28,31)*** were first to identify novel interactors of caveolin, including Ras. The main advantage of this protocol is its simplicity; disadvantages include a long spin time and potentially significant contamination of the isolated raft fraction with other membranes.

1. Wash two 10-cm plates of cultured adherent cells with PBS. Scrape cells and transfer into a 15-mL centrifuge tube.
2. Centrifuge cells (800*g* for 5 min at 4°C) and resuspended in 2 mL of SCB+.
3. Lyse cells with 20 strokes of a Dounce homogenizer using a tight fitting pestle.
4. Follow homogenization by shearing by passing lysates through a 23-gage needle 10 times.
5. Sonicate lysate three times 15 s using a Branson Sonicator 250.
6. Combine lysate with 2 mL of MES buffer containing 90% sucrose. Transfer mixture to the bottom of a 12-mL centrifuge tube.
7. Carefully layer 4 mL of MES buffer containing 35% sucrose on top of previous layer.
8. Add an additional 4 mL of MES buffer containing 5% sucrose and centrifuge for 16 to 20 h at 175,000*g*.
9. Take twelve 1-mL fractions and determine the distribution of key lipid raft proteins via Western blotting.

3.2.1.5. OptiPrep Gradient

Here, we outline a recent isolation method for the preparation of detergent-free lipid rafts where the lysis of cells by shearing in an isotonic buffer containing calcium and magnesium yields a highly purified raft fraction and optimal protein yield ***(28)***.

1. Wash four 10-cm plates of cultured adherent cells (~7–8 mg of total protein) with PBS. Remove PBS and scrape into 3 mL of Base Buffer+. Transfer into a 15-mL centrifuge tube.
2. Centrifuge cells (1000*g* for 5 min at 4°C) and resuspended in 1 mL of Base Buffer+.

3. Lyse cells by passing through a 23-gage needle 10 times, being careful to avoid air bubbles.
4. Centrifuge cells (1000*g* for 10 min at 4°C). Transfer the cleared lysate to a separate tube.
5. Lyse the remaining pellet for a second time with 1 mL of additional Base Buffer+.
6. Lyse cells by passing through a 23-gage needle 10 times, being careful to avoid air bubbles.
7. Centrifuge cells (1000*g* for 10 min at 4°C). Combine both cleared lysates.
8. Add 2 mL of Base Buffer + 50% OptiPrep to the combined post-nuclear (now 25% OptiPrep) supernatants and place in the bottom of a 12-mL centrifuge tube.
9. Carefully layer an 8 mL of OptiPrep gradient (0–20%) on top of the combined lysates.
10. Centrifuge for 90 min at 52,000*g*.
11. After centrifugation, a diffuse band will appear one-third of the way down the gradient and a distinct band will be apparent at the 20:25% interface.
12. Isolate 0.67-mL fractions as mentioned previously (*see* **Subheading 3.2.1.1.**) and determine the distribution of key lipid raft proteins via Western blotting.

3.3. Useful Protocols for DIM Analysis

3.3.1. Protein Localization and Aggregation of Lipid Rafts

Lipid-enriched microdomains are principally composed of sphingolipids and cholesterol. The operational definition of rafts is solubility in cold detergent and the functional definition of lipid raft-dependent signaling has been traditionally based on cholesterol depletion studies. However, cholesterol depletion, in addition to affecting lipid rafts structure, can also alter cellular processes such as actin reorganization, receptor mediated signaling and endocytosis. Introducing fluorescent tags to proteins is a valuable way of assessing the subcellular distribution of proteins in response to receptor activation ***(32,33)***. This permits one to assess both the spatial and temporal aspects of protein expression. Two different approaches have been used to identify lipid rafts in cell membranes. One is to observe the random or directed motion of membrane lipids and lipid-anchored proteins in the plane of a bilayer. This approach is useful to identify the size of DIMs. Further microscopic probes can track molecule mobility on the nanometer scale and following lipid raft association induced by ligand receptor interaction. The second approach is to measure the proximity of molecules in rafts by FRET. In addition, fluorescent lipid analogs have been used to define rafts in other lipid microdomains in cells in culture.

The use of fusion proteins, such as green fluorescent protein (GFP), allows an investigator to examine protein distribution in live cells while avoiding the potential artifacts of cell fixation and permeabilization. However, the concerns include the fact that GFP-tagged proteins are often expressed in levels higher than the endogenous protein, and could have an effect on protein localization.

Furthermore, the GFP tag is large, approx 27 kDa in size, and could interfere with tracking recruitment to microdomains. This is further compounded by GFP-dependent dimerization. Despite these limitations, fluorescent tags are a very useful approach to monitor lipid domain based signaling events. It is essential to ensure that the fluorescent-tagged protein co-localizes to the same compartment as the endogenous protein. The availability of different fluorescent proteins (yellow fluorescent protein and cyan fluorescent protein) allows two proteins to be tagged with distinct spectral variants. In these studies, expressed proteins are visualized before and after activation by digital imaging using confocal and low light charge-coupled device (CCD) camera-based microscopes. In addition to live imaging, cells can be visualized after fixation and staining with Cy3-labeled cholera toxin B-subunit.

Lipid rafts in the plasma membrane outer leaflet are enriched in sphingolipids and cholesterol, but the exact lipid composition of the inner membrane raft is unknown. To determine the segregation and kinetics of recruitment of proteins to the outer and inner layer lipid enriched domains, Zacharias et al. generated fusion of GFP spectral variants linked to consensus sequence of different lipids ***(34)***. In an elegant series of experiments, the authors demonstrated that lipid acyl (myristoylated and palmitoylated) proteins were localized to the outer membrane, whereas proteins containing the geranyl-geranyl tail were excluded from the outer lipid enriched caveolae. In follow-up experiments the authors were able to recover the acylated proteins from the DIM fraction. Zacharias found that premylated proteins were insensitive to cholesterol depletion, demonstrating that this approach is a valuable tool to monitor the differential recruitment of proteins to lipid enriched microdomains ***(35)***.

The following is a protocol for a conjugation assay, which enables one to identify physiological events that take place at the point of contact between two cells, such as the site of the immune synapse between T-cells and antigen-presenting cells (APCs). It is also helpful to track the migration of various proteins within the cells with the aid of GFP-tagged proteins or proteins stained with fluorescence antibodies. Here, we outline an experiment using Jurkat T-cells and a cultured Raji B-cell line loaded with staphylococcal enterotoxin E (SEE; Toxin Technology, Sarasota, FL) toxin to investigate the activity of proteins before, during, and after stimulation with TCR. It is important to keep in mind that the protocol can be easily adapted to other cells lines to best suit your own specific research needs.

3.3.1.1. Slide Preparation

1. Encircle a dime-sized area on a glass microscope slide with a PAP Pen to define the experimental area. Add approx 100 μL of 0.01% poly-L-lysine to center of the

circle and incubate in a moist chamber for 30 min at RT. Wash the slides four times with deionized (d)H_2O before use.

3.3.1.2. Cell Preparation

1. Count cells and determine the total number needed for the experiment (*see* **Note 11**).
2. Resuspend the APC cells (~1 × 10^5 cells/50 μL; *see* **Note 12**).
3. Resuspend the T-cells (~1 × 10^5 cells/50 μL; *see* **Note 13**).
4. Mix the T-cells and APCs ± SEE in a 1:1 ratio and centrifuge briefly to ensure proper conjugate formation.
5. Incubate cell lines together for various time points at 37°C (*see* **Note 14**).
6. Resuspend the cell mixture well and spot 100 μL of on lysine-coated slides. Incubate at RT for 10 min to allow complexes to adhere to the lysine. Carefully aspirate media.
7. Check for conjugation efficiency under bright light microscopy.
8. Fix the conjugates by adding approx 150 μL of 4% paraformaldehyde–0.1% Tween-20 in PBS to the slide.
9. Incubate the slides at RT for an additional 10 min.
10. Wash the slides twice in a large volume of PBS at RT for 10 min.
11. Block the cells by adding approx 150 μL of 1% BSA in PBS and incubate for 30 min at RT.
12. Dilute the primary antibody as recommended by manufacturer in a total volume of 100 μL of PBS. Add antibody solution to slides and incubate for 1 h at RT or overnight at 4°C.
13. Wash cells twice in a large amount of PBS at RT for 10 min.
14. Dilute the fluorescent-conjugated secondary antibody 1:5000 in 100 μL of PBS and incubate for 1 h at RT (*see* **Note 15**).
15. Wash cells twice in a large amount of PBS at RT for 10 min.
16. Aspirate the majority of the PBS and add a small amount of VectaMount to the cells. Add the cover slip and seal edges with nail polish to prevent the samples from drying out.
17. Wait at least 30 min before viewing lipid raft aggregates and protein localization under the microscope (*see* **Note 16**).
18. Use confocal laser scanning microscopy to visualize lipid rafts in T-cell membranes before and after cross linking with anti-TCR.

3.3.2. Molecular Probes and Fluorescent Microscopy

Commercially available kits allow for convenient labeling and crosslinking of lipid rafts, such as Molecular Probe's Vybrant Lipid Raft Labeling Kit, which uses a fluorescent cholera toxin subunit B conjugate (in one of three colors to accommodate multiple staining parameters) to selectively bind to the GM1 proteins embedded with lipid rafts. An antibody specific for cholera toxin subunit B then crosslinks the labeled lipid rafts into distinct patches that are then easily visualized by fluorescence microscopy. These kits allow an investigator to identify physiologically significant membrane proteins that associate with lipid rafts. Detailed instructions are included with each kit and can vary from one manufacture to another; it is highly recommended that you familiarize yourself with the directions for the particular kit you will be using before starting your experiments. Another fluorescent microscopic technique uses changes in the emission spectrum of a bodipy sphingomyelin analog (C5 DMB-SM) to identify where lipid probes concentrate within the plasma membrane. When the probe becomes more concentrated, its emissions shifts from green to red; therefore, the existence of lipid rafts would have been expected to lead to small regions of red fluorescence superimposed on the green emission. Unfortunately, some studies have proven the ineffectiveness of this particular probe, where with low concentrations of C5 DMB-SM, the plasma membrane appears entirely green, indicating the absence of highly concentrated sphingolipid domains ***(36)***. On the other hand, it is entirely possible that the presence of the bodipy moiety on the end of a shortened acyl chain of this sphingolipids analogue creates a bulky probe that does not pack well into the ordered environment suggested for lipid rafts ***(37)***. In an attempt to circumvent such technical limitations, Kuerschner et al. recently has devised a method of labeling lipids for fluorescent microscopy that closely mimics the natural conformation ***(38)***. This new reporter lipid is a product of a modified Wittig olefination reaction and consists of five conjugated double bonds in the long chain of the fatty acid and has excitation and emission wavelengths of 350 nm and 470 nm, respectively. This molecule is also able to incorporate into the structure of membrane lipids such as sphingomyelin and have been shown to have no drastic effects on the phase behavior, making it ideal for in vivo studies ***(39)***.

3.3.3. Immunoisolation of Proteins Associated With Lipid Rafts

An elegant approach to isolate raft-associated proteins was first put forward by Harder et al. ***(9)***. Here, DIM fractions from unstimulated and stimulated Jurkat cells were coupled with magnetic beads coated in specific antibodies. The proteins are isolated with a magnet and the raft-associated proteins within the resulting lysates were identified via immunoblotting. This protocol is presented as a technique for isolating lipid rafts in lymphocytes and it can be easily modified for other cell–cell interactions. The interested reader is also referred to Harder et al. for more detailed information ***(9)***.

3.3.3.1. Coating the Dynabeads

1. Determine the number of cells needed. Aliquot half that number of M-450 goat-anti-mouse Dynabeads into a sterile microcentrifuge tube (*see* **Note 17**).
2. Collect beads on an MPC magnet for 1 min. Remove the supernatant and wash the beads twice by resuspending in 1 mL of serum-containing media.

3. Resuspend in 1 mL serum-containing media + 50 m*M* HEPES buffer for up to 4×10^8 beads.
4. Add anti-CD3 and anti-transferrin receptor antibodies at 2 µg per 10^7 beads and incubate for 1 h at RT on a rotating wheel.
5. Collect the beads on the magnet and remove supernatant.
6. Wash the beads twice by resuspending in 1 mL of serum-containing media.
7. Resuspend the beads at a concentration of 4×10^8 beads per milliliter in serum-containing media + 50 m*M* HEPES buffer.

3.3.3.2. Isolation of TCR Complexes

1. Collect the number of cells needed for the experiment by centrifugation (~1–2 × 10^7 cells per data point).
2. Resuspend the cells in 1 mL of serum-containing media + 50 m*M* HEPES buffer per data point and pellet in a microcentrifuge tube. Loosen the cell pellet by gently flicking the tube.
3. Add the antibody-coated Dynalbeads at a ratio of 1:2 (e.g., 0.5×10^7 beads:1×10^7 cells), mix well, and rotate on a lab wheel for 5 to 10 min.
4. Transfer samples to a 37°C water bath for the desired time points.
5. Add 1 mL of ice-cold HB buffer to stop reaction, and pellet cells in a refrigerated microcentrifuge tube for 2 min at 800*g*.

3.3.3.3. Disruption of Cells

1. Harder et al. use a nitrogen cavitation procedure to disrupt the cell sample; for more information, the interested reader is referred to Harder and Kuhn. ***(9)***.

3.3.3.4. Sample Preparation

1. Collect the beads with the magnet for 5 min, on ice.
2. Aspirate supernatant and wash samples twice by resuspending in 1 mL of ice-cold HB buffer.
3. Resuspend beads in 1 mL of HB + buffer and transfer to microcentrifuge tubes.
4. Retrieve beads magnetically and remove supernatant.
5. Add an appropriate amount of 2X Laemmeli sample buffer to the beads and heat samples for 5 min at 95°C.
6. Separate proteins by electrophoresis and transfer onto a polyvinylidene difluoride membrane. Identify proteins of interest by immunoblotting.

3.3.4. Electron Microscopy

In contrast with light microscopy, EM can only provide a snap shot of the cell at one particular time. The resolution of EM, however, is much greater than light microscopy, and particulate markers can provide the means to quantitatively study lipid raft organization at the molecular level ***(40)***. Recently, Prior et al. ***(41)*** developed a novel methodology to examine lipid raft markers on the inner leaflet of the plasma membrane using a combination of EM studies and statistical analysis to screen proteins associated with, and colocalizing within, lipid rafts. This approach has potential to further examine plasma membrane organization and the spatial dynamics of regulated signaling and membrane trafficking events associated with the cell surface. However, it is important to keep in mind that there are limitations with this method; labeling methods for electron microscopy are often harsh and can damage delicate cellular ultrastructures and disrupt lipid organization.

3.3.4.1. Preparing Antibody-Conjugated Collodial Gold

Immunogold electron microscopy allows one to directly visualize membrane-anchored proteins at high resolution. It has been used to determine the size and distribution of lipid rafts, to monitor colocalization, and to observe Ras signaling domains ***(41)***. Briefly, plasma membranes are fixed, labeled with gold-conjugated antibodies, and viewed on a transmission electron microscope. Once the x,y coordinates are extracted from the images, Ripley's K function is used to determine whether the observed gold pattern is random or clustered into lipid rafts. A modification of the K-function also can be used to measure the co-localization of two proteins, each labeled with a different-sized gold particle (i.e., 2 nm and 5 nm).

1. Heat 40 mL of the reducing solution and 10 mL of Gold Solution to 60°C and combine, stirring vigorously. Boil for 5 min (the solution will turn a deep red once colloids form).
2. Cool the solution on ice and adjust the pH to 8.5 with 5 *N* NaOH using pH paper.
3. Identify the optimal antibody concentration by adding 0 to 5 µg of the affinity-purified antibody of interest (diluted in a total volume of 20 µL) to 250 µL of the gold colloid solution made in the previous two steps. Incubate at RT for 5 min.
4. Add 100 µL of 10% NaCl. Titrate for the smallest amount of antibody that stabilizes the gold (where no color change occurs).
5. Incubate this gold-antibody solution for 10 min at RT.
6. Add 10% BSA to a final concentration of 0.1%.
7. To concentrate gold and remove unbound antibody, centrifuge at 120,000*g* for 2–3 nm gold or 100,000*g* for 5 nm gold for 1 h at 4°C.
8. Collect the loose portion of the pellet and store at 4°C.

3.3.4.2. Preparing Plasma Membrane Sheets

1. Incubate pioloform-coated grids with 1 mg/mL poly-L-lysine for 10 min. Wash twice in 1 mL of distilled water for 1 min. Air-dry and store in a dust-free container.
2. Culture cells on 10-mm glass cover slips as below (fluorescence recovery after photobleaching (FRAP) assay protocol); wash cells with PBS to remove cellular debris.
3. Place groups of poly-L-lysine-coated grids onto filter paper with the coated side facing up.

4. Remove cover slip from culture dish and remove excess solution by touching the edge of the cover slip to the filter paper. Place the cover slip face down on the grids.
5. Place filter paper on top of the cover slip and press down using a silicone bung for 1 to 2 s.
6. Remove the filter paper and turn the cover slip over so the grids are on top.
7. Carefully add 1X KOAc buffer around the grids until they float up onto the top of the drop.
8. Wash the grids by transferring them to new drops of 1X KOAc buffer on Parafilm.
9. Fix for 15 min with fixative. Wash twice by transferring the grids to fresh drops of PBS for 1 min each.
10. Quench free aldehyde groups with 25 m*M* glycine for 10 min.
11. Block for 10 min with blocking solution to prevent the antibodies from binding to nonspecific sites.
12. Centrifuge the gold-conjugated antibody at maximum speed in a microcentrifuge for 2 min and dilute in additional blocking solution. Float the grids on top of drops of the diluted antibody for 30 min.
13. Wash the labeled grids with blocking solution five times for 4 min each.
14. Wash the grids with dH_2O five times for 2 min.
15. Incubate the grids in a mixture of nine parts 2% methyl cellulose and one part 3% uranyl acetate for 10 min on ice.
16. Pick up the grids in 5-mm wire loops, drain the excess solution on filter paper, and leave to dry in the loops for 10 min.

3.3.4.3. Imaging and Analysis

1. Using a transmission electron micrograh, scan plasma membrane sheets at low magnification (×4–6000), adjusting field brightness to maximize the contrast between the plasma membrane and background; take micrographs at higher magnification (×20,000).
2. Scan negatives at 1 in. width, 1500 dpi resolution, and 8-bit depth using the imaging software program, and save files as TIFF formats.
3. Using Adobe® Photoshop® or a similar software program, select an area of the negative approx 800 × 800 pixels and paste into a new file.
4. Remove the background from the image using the curves and brightness/contrast tools. Save as a new TIFF file.

3.3.4.4. Calibration and Analysis

1. Identify from an area containing 200 to 400 gold particles; open the file using Image J software or other similar program.
2. Set the image's threshold between 0 and 170.
3. Modify "set measurements" to record center of mass to eight decimal places.
4. Modify "analyze particles" to count the number of gold particles at different sizes from a minimum of 2 pixels to a maximum of 75 pixels for 5 nm of gold.
5. Record the number of particles after each step and repeat the threshold adjustments.

6. Modify "analyze particles" to locate the center of mass of all gold particles within the required pixel range to eight decimal places. Save as a text file.
7. Using Microsoft Excel®, convert the coordinates from pixels (the units generated in the previous step) into nanometers (the conversion factor is determined by the scale bar included with each micrograph or from images of calibrating spheres).
8. At this point, the x,y data points can be entered into any commercially available K-function analysis program to assess protein colocalization; for a more in-depth analysis protocol, the interested reader is referred to Prior ***(41)***.

3.4. Biophysical Analysis

Because lipid rafts are less than 50 nm in size, it is impossible to visualize individual rafts on standard fluorescent microscopes. However, various methods exist that can detect lipid raft aggregates and associated proteins, and these will be explained in detail below. Such experiments provide important information regarding the colocalization of lipid rafts with raft proteins (localized and excluded) and should be used to confirm biochemical data based on detergent insolubility. However, it is essential to keep in mind that not all microscopic evidence clearly supports the theory of lipid rafts. Recently, Glebov and Nichols used FRET analysis to measure the proximity of two different GPI-anchored fluorescent proteins in T-cells. The signal they detected was density-dependent, cholesterol-independent, and similar to data from non-raft proteins, suggesting that resting T-cells do not contain lipid rafts ***(42)***. These findings strengthen work previously performed by Kenworthy and Edidin, who found diffuse distribution, and hence little evidence of clustering, of overexpressed 5′ nucleotidase, a GPI-anchored protein, in polarized MDCK cells; this suggests that this protein is randomly dispersed across the apical plasma membrane and is not found in organized clusters as would be expected with lipid rafts ***(43)***. It is possible, however, that these cells do in fact contain lipid rafts but these rafts are smaller than originally thought (<10 nm in diameter), and contain very few GPI-anchored proteins, making it doubtful that any one raft would contain both the FRET donor and acceptor proteins ***(44)***. With these points in mind, it is essential that all biophysical experiments performed on lipid rafts contain proper positive and negative controls to best analyze ensuing data.

3.4.1. Fluorescence Resonance Energy Transfer

FRET detects the energy transfer between two molecules only a few nanometers apart and is a powerful technique that can be used to identify intermolecular interactions within living cells. Energy transfer from an excited donor fluorophore molecule to an acceptor fluorophore results in quenching of the donor channel signal and an increase in the acceptor signal; because FRET only occurs between molecules in very close proximity; it is an excellent technique to study clustering and co-association of fluorescently-tagged proteins ***(45–47)*** and, although promising, it must be stressed that the absence of interactions is not always indicative of the lack of interaction. Steric constraints between the donor and acceptor could result in a false-positive result ***(44)***. Summarized herein is a simplified outline to FRET assays; for a more in-depth exploration of the theory behind this technique, please refer to the excellent reviews by Gordon ***(45)*** and Herman ***(47)***.

3.4.1.1. Sample Preparation

1. Plate 293 human embryonic kidney cells onto glass cover slips in suitable growing media. Incubate cells at 37°C with 5% CO_2 until the cells are approx 50% confluent.
2. Transfect 1 to 2 μg of plasmid DNA into the cells using a lipid-based delivery system, such as TransFection (Bio-Rad). Overexpressed fluorescent signals can be visualized between 6 and 24 h after transfection.
3. If staining for endogenous or nonfluorescent ove-expressed proteins, wash cells in PBS and choose appropriate donor–acceptor antibody concentrations for labeling (i.e., fluorescein isothiocyanate, Cy3, or Cy5). Typical donor–acceptor molar ratios range from 1:1 to 1:5. Normal serum from the secondary antibody host diluted in PBS should be used as a blocking agent.
4. Prepare the primary antibody in blocking reagent and incubate according to manufacture's instructions (*see* **Note 18**).

3.4.1.2. Cell Fixation and Mounting

1. Wash cells in a large volume of PBS to remove all traces of serum which could interfere with antibody binding.
2. Fix cells in 4% PFA for 30 min at RT.
3. Wash cells twice with additional PBS after fixation.
4. To decrease autofluorescence, incubate the cells with 1 mg/mL $NaBH_4$ for 5 min.
5. Wash cells in PBS. Aspirate most of the excess fluid from the cover slip, and add a drop of Vectashield directly onto the cells.
6. Cover and seal the edges with nail polish to prevent the samples from drying out. Store at 4°C if not viewed on a confocal laser scanning microscope immediately.

3.4.1.3. Imaging Parameters and Data Acquisition

FRET is detected by exciting labeled specimens with light at a corresponding wavelength to the absorption spectrum of the donor and detecting light emitted at the wavelength corresponding to the absorption spectrum of the acceptor ***(45)***. Energy transfer between the two proteins of interest is visible by both the quenching the donor fluorescence signal in the presence of the acceptor and in the sensitized emission spectrum of the acceptor.

1. Three individual samples of each specimen must be prepared to collect FRET measurements. Cells containing only the signal from the donor (d); cells containing

only the signal from the acceptor (a); and cells containing signals from both donor and acceptor (f). Each sample is imaged with a donor filter set (D); an acceptor filter set (A), and a FRET filter set (F). This approach will identify the FRET signal that results from energy transfer that has been transferred from the donor to acceptor molecules while reducing signal bleeding and crosstalk ***(47)***.

2. Exposure time, light intensity, and camera settings for each of the filter sets, or imaging pathways, are determined by scanning a representative area of d with D, a with A, and f with F.
3. Once the imaging parameters are determined for each filter set, they must not be altered during data acquisition. Computer macros can be used to automate acquisition once the parameters are set.
4. Collect at least 10 fields for every specimen with each of the three filter sets.

3.4.1.4. Quantification and Analysis

1. Identify regions on the cover slip that are not covered in cells (background) and calculate the average fluorescence signal from this area.
2. Identify signals emitted from the cells (foreground) and calculate average fluorescence signal from these areas.
3. Normalize results by subtracting background signals from foreground signals for each FRET measurement.

3.4.2. FRAP Analysis

FRAP is an important technique used to measure the diffusion mobility of membrane components labeled with fluorophores, fluorescent antibodies, or GFP fusion proteins. In FRAP assays, a short pulse of intense laser light irreversibly destroys (photobleaches) fluorescence in a defined area of the cell. Recovery of fluorescence into the photobleached area occurs as a result of diffusional exchange between bleached and unbleached molecules. This technique is useful in the study of the dynamics of signaling molecules in Jurkat cells after lipid raft activation as described by Tanimura et al. ***(46)***.

3.4.2.1. Bead Preparation

1. Wash 1 × 10^7 Polybead polystyrene beads (6.0 μm in diameter) with 1 mL of PBS and resuspend in a final volume of 1 mL of PBS. Add 10 μg of the purified antibody to the beads. Poly-L-lysine-coated beads can be used as a negative control in these experiments.
2. Incubate bead:antibody mixture at RT for 90 min on a rotating wheel.
3. Centrifuge in a microcentrifuge at maximum speed for 30 s.
4. Remove supernatant and discard. Wash the beads three additional times with PBS.
5. Add 1 mL of blocking solution to the beads and incubate at RT for 30 min on a rotating wheel.
6. Centrifuge in a microcentrifuge at maximum speed for 30 s.
7. Remove supernatant and discard. Wash the beads three additional times with PBS.
8. Resuspend the coated beads in 1 mL of PBS and store at 4°C for up to 3 wk.

3.4.2.2. Conjugate Formation

1. The cells used in these experiments will need to express a GFP fusion protein. Both stable cell lines and transiently transfected cells can be used.
2. Determine the number of cells needed for the experiment. Harvest by centrifugation and wash once in 10 mLof serum-free media.
3. Resuspend the cells in media at a concentration of 1×10^7 cells per milliliter and transfer 10 µL of the cell suspension (1×10^5 cells) to a microcentrifuge tube.
4. Add 10 µL of antibody-coated beads in PBS (1×10^5 beads for a 1:1 cell:bead ratio) and mix.
5. Incubate the cell:bead mixture at 37°C for 20 min.

3.4.2.3. Sample Preparation for FRAP

1. During the cell:bead conjugate incubation (above), encircle a dime-sized area on a glass microscope slide with a PAP Pen.
2. Add 3 µL of the cell:bead mixture into the center of the circle and place a cover slip on top. Seal edges with nail polish to prevent the media from evaporating.

3.4.2.4. FRAP Assay

FRAP assays are conducted at RT using a confocal microscope with a ×63 objective lens and the 488-nm line of a 25-mW Argon laser. Before data collection can begin, it is necessary to determine how often the bleaching process should be performed with full laser power to reduce the fluorescence in the region of interest to background levels. It is also important to perform preliminary experiments to determine how long recovery of fluorescence takes to reach a plateau after bleaching. A whole cell scan at low laser power (75% power, 10% transmission) before the bleach and photobleach the cell with full laser power (100% power, 100% transmission) 40 times (2 s) is a recommended starting point. Recovery is followed after 2 s with low-laser power at 2-s intervals for 30 s, at 10-s intervals for 120 s, and then 60-s intervals for 6 min.

1. Place the sample on the stage and the scan slide to detect a cell of interest at imaging laser intensity.
2. Define the region of interest for the photobleach with an area of approx 1 µm × 1 µm for bleaching. Program the computer's software to automatically change settings quickly for the photobleaching routine. It should record the image of the cell before photobleaching, execute the bleach, and record an image after the photobleaching and throughout the recovery.
3. Begin the FRAP cycle by executing the program and monitor recovery on the screen.
4. Check the cell during and after completion of the FRAP cycle to ensure that the region of interest has not changed.
5. Save the image as an image database file and calculate fluorescent intensities in the region of interest for the bleached region (I_b) and a nonbleached region (I_r) for each time point.

6. Calculate the relative intensity (rI) for each time point: $rI_{(t)} = I_b(t) - I_r(t)$
7. Calculate the percentage recovery as follows: % recovery = $[rI_{(t)} - rI_{(post)}/rI_{(pre)} - rI_{(post)}]$, where $rI_{(pre)}$ = initial (prebleach) relative intensity; $rI_{(post)}$ = post-bleach relative intensity; $rI_{(t)}$ = relative intensity at each time point.
8. Draw a FRAP curve using a calculation software, such as Microsoft Excel (% recovery vs time in min (*see* **Note 19**).

3.4.3. Proteomics

Proteomics, the large-scale study of the protein complement of a given cell, has proven to be an invaluable technique in the identification of novel proteins associated with lipid rafts. von Haller et al. were the one of the first to use MS to identify proteins that co-purified with DIM fractions of T-cells and found that the proteins identified in the screen essentially fell into one of two categories: cytoskeletal proteins and proteins involved in signal transduction ***(48)***. Since then, proteome analysis has offered evidence that several additional classes of proteins associate with DIM fractions isolated from stimulated T-cells (cell adhesion, heat shock, and GTP-binding proteins ***[49,50]***). Foster et al. ***(51)*** have taken this approach one step further by classifying proteins as "raft" proteins or "raft-associated" proteins based on biochemical properties such as their susceptibility to cholesterol depletion and detergent solubilization. This screen uses stable isotope encoding to add a functional dimension that significantly improves specificity of the data, providing strong evidence that their groupings are real and are not statistical artifacts ***(51)***. Stable isotope labeling by amino acids in cell culture (SILAC) entails the addition of labeled, essential amino acids to amino acid-deficient cell culture media; these labeled amino acids are incorporated into all newly synthesized proteins, or "encoded into the proteome" ***(52)***. The advantages of this technique include the fact that no chemical labeling or affinity purification is required, and the method is convenient and inexpensive to perform on a large scale. This method also allows investigators to identify large numbers of proteins from a single experiment in an unbiased manner.

3.4.3.1. SILAC

1. Culture cells in leucine-deficient media. Supplement one population with normal isotopic abundance L-Leucine (Leu, 105 mg/L) and the other with 99% isotropic abundance L-Leucine-**5**,5,5-D3 (LeuD3; 107.5 mg/L).
2. Passage cells three times for a minimum of seven population doublings.

3.4.3.2. Cholesterol Depletion and Isolation of Detergent-Resistant Fractions

1. Serum-starve 10-cm plates of each cell population (Leu, LeuD3) for 18 h before cholesterol depletion.

2. Disrupt membrane cholesterol in one population (Leu) of cells with 10 m*M* MβCD for 30 min at 37°C (*see* **Subheading 3.1.2.1.**). Treat the second population of cells (LeuD3) with a carrier control only (*see* **Note 20**).
3. Lyse Leu, LeuD3 cells for 1 h at 4°C in MBS lysis buffer.
4. Clear lysates by centrifuging at 10,000*g* for 10 min at 4°C.
5. Determine protein concentration of each sample using a commercially available kit.
6. Equal amounts of protein from the Leu (treated) and LeuD3 (untreated) cell populations were combined and mixed with an equal volume of 90% sucrose in MBS for a final concentration of 45%.
7. Transfer the solution to the bottom of an ultracentrifuge tube. Carefully add additional layers of 35 and 5% sucrose on top.
8. Balance the tubes to within 0.01 g and place them in the rotor. Centrifuge samples for 16 h at 200,000*g* at 4°C.
9. Extract the low-density band (at ~18% sucrose) with a needle and syringe. Dilute the band with four volumes of additional MBS. Centrifuge samples for an additional 2 h at 200,000*g* at 4°C to pellet the detergent-resistant material.

3.4.3.3. High-Performance Liquid Chromatography and Tandem MS

1. Solubilize the pellet from the second centrifugation step in a small volume of 6 *M* urea/2 *M* thiourea (in 10 m*M* HEPES, pH 8.0). Precipitate proteins by diluting in five volumes of absolute ethanol. Adjust to 50 m*M* sodium acetate with a 2.5 *M* stock solution, pH 5.0.
2. Allow the solution to stand at RT for 2 h. Centrifuge at 12,000*g* for 10 min at RT.
3. Resolubilize the pellet in additional urea/thiourea. Reduce the solution by adding 1 μg of DTT per 50 μg of estimated total protein. Incubate for 30 min at RT.
4. Alkylate the solution by adding 5 μg of iodoacetamide per 50 μg of estimated total protein. Incubate for 30 min at RT.
5. Digest the solution by adding 1 μg of LysC per 50 μg of estimated total protein. Incubate for 3 h at RT.
6. Dilute the sample in four volumes of 50 m*M* NH_4HCO_3. Digest with 1 μg of trypsin per 50 μL of estimated total protein. Incubate for 12 h at RT.
7. Prepare the stop and go extraction (STAGE) tips by punching out small disks of C18 Empore filters using a 22-gage flat syringe and plugging the disks into P200 pipet tips.
8. Equilibrate column by forcing approx 100 μL of methanol through the Empore disk with a syringe fitted to the end of the pipet tip.
9. Remove any remaining organic solvent in the column by forcing a small volume of 0.5% acetic acid through twice.
10. De-salt peptide mixture with the STAGE tips and bomb-load onto reverse-phase analytical columns for liquid chromatography (LC). Elute peptides from the columns with three-step linear 2.5-h gradients running from 5.6 to 64% acetonitrile and sprayed directly into the orifice of a MS.
11. Identify peptides by LC/MS/MS by information-dependent acquisition of fragmented spectra for multiply charged peptides.

12. Search acquired proteins against a known database, such as the Human International Protein Index database (http://www.ensembl.org/IPI/), using MASCOT (Matrix Science, London, UK, http://www.matrixscience.com).

3.4.4. Flow Cytometry

Gombos et al. ***(53)*** recently outlined a novel flow cytometry based assay that characterizes surface proteins and raft association by utilizing two defining characteristics of lipid rafts, cholesterol dependency, and detergent resistance. Here, the authors stained specific surface protein and compared treated and untreated groups (MβCD, nonionic detergent, or both); raft-associated proteins will be significantly affected by cholesterol depletion and/or detergent solubilization show dramatic differences in fluorescence in treated groups versus untreated groups. Raft-independent proteins will not be affected by cholesterol depletion and/or detergent solubilization and will therefore show very little difference between treated and untreated groups. This method is easy to perform and allows for rapid screening of multiple surface proteins.

1. Count and pellet cells by centrifugation (*see* **Note 21**).
2. Divide the total number of cells into two populations. Treat one half of the cells with MβCD to disrupt cholesterol in the plasma membrane (*see* **Subheading 3.1.2.1.**). The remaining (untreated) cells will serve as a control population for cholesterol dependence.
3. Divide both MβCD-treated and untreated cell populations into half again. Treat one half of the cells with either 0.5% Triton X-100 or NP-40 for 5 min on ice. Wash with PBS. The remaining cells will serve as a control population for detergent sensitivity.
4. Label all cells with fluorescent antibodies specific against surface markers (*see* **Notes 22** and **23**).
5. Collect the following data for each sample on a flow cytometer:
   Forward scatter (FSC) and Side scatter (SSC).
   Fluorescence intensity (FL1, FL2, or FL3 histograms, depending on the color of the antibody).
6. Analyze FACS data by using the following histograms:
   Autofluorescence of untreated, unlabeled cells (background) (FL_{Bg}).
   Fluorescence of labeled, untreated cells, proportional to total protein expression (FL_{Max}).
   Autofluorescence of detergent-treated, unlabeled cells ($FL_{Bg\text{-}Det}$).
   Fluorescence of cells treated after detergent for 5 min (FL_{Det}).
   Calculate the extent of detergent resistance (FCDR) as follows: $FCDR = (FL_{Det} - FL_{Bg\text{-}Det}) / (FL_{Max} - FL_{Bg})$.

The closer the FCDR value is to 1, the less soluble a given protein is in the detergent. Likewise, a low FCDR value is indicative of a highly-soluble protein.

A significant decrease in the FCDR value after cholesterol depletion (lipid raft disruption) indicates that the observed resistance is a result mainly due to localization of the protein in lipid rafts.

When a high FCDR value does not change significantly after MßCD treatment, it can be assumed that the protein does not bind directly to lipid rafts (e.g. cytoskeletal anchorage only).

4. Notes

1. Both cultured cell lines and primary cells can be used in stimulation experiments. As with all experiments, experimental design is critical, and the inclusion of a negative control (unstimulated cells) is essential. Start with a minimum of 1×10^6 cells per sample.
2. Try 0, 1, 2, 5, 10, and 15 min; if the number of cells is limited, select the critical time points.
3. The use of positive and negative controls is recommended strongly. For example, Src family kinases that are palmitoylated do not associate with DIMs after the depletion of cholesterol.
4. Cyclodextrins also can remove other lipid components from membranes; therefore, it is useful to check that any effects can be reversed by using cholesterol-loaded cyclodextrin carrier to deliver cholesterol back to cells.
5. Not all lipid rafts appear to be equally sensitive to the depletion of cholesterol. Recent studies have suggested that the depletion of cholesterol reduces lateral mobility of membrane proteins because it disrupts the highly regulated interactions of PIP_2 with molecules controlling actin cytoskeleton ***(54)***.
6. It is essential that while extracting proteins associated with DIM fractions that all reagents and buffers be kept at 4°C and the experiments themselves be carried out on ice or in a cold room. It has been shown that the DIM fractions themselves can be lysed by detergents at higher temperatures.
7. Milder detergents such as the Brij family of detergents produce larger raft fractions, whereas 1% Triton X-100 produces less buoyant fractions containing less than 50% of the typical raft molecules and only a few nonpalmitoylated transmembrane proteins.
8. If two proteins that reside in the same membrane float to different densities during DIM preparation it can be concluded that they had different lipid environments before detergent extraction.
9. Regardless of the conditions for detergent treatment, a completely soluble membrane protein localized to the same membrane as the protein of interest should be included in the analysis.
10. For a comprehensive list of proteins associated with DIM fractions from T-cells, please refer to Razzaq et al. ***(55)***; such proteins can be used as markers for cell biological studies.
11. Both cultured cell lines and primary cells can be used in conjugation assays. In the example given here, Jurkat T-cells are conjugated to SEE-loaded B-cells to investigate

the effects of TCR stimulation. This protocol, however, can easily be adapted to other cell systems. An important consideration for these particular experiments is the use of a negative control (Raji B-cells without SEE).

12. To prepare the APC cell population, use 1×10^6 cells per sample. Significant cell loss will occur through subsequent washing steps. Aliquot the cells needed for the negative control and set aside. Resuspend the remaining cells in 500 μL of media + 5 μg/mL (SEE) and incubate at 37°C for 30 min. Resuspend the two cell populations in 50 μL per sample and keep cells on ice.
13. Count the cells needed for the T-cell population (~5×10^5 cells/50 μL) and pellet. Wash once in additional media and resuspend for a final cell count of 1×10^5 cells/50 μL. Keep on ice.
14. The time points used in conjugation assays depend on whether the protein of interest migrates early after activation or not. When in doubt, begin with various time points, such as 0, 1, 2, 5, 10, and 15 min.
15. Fluorescence antibodies are sensitive to light. The signal fades as a result (photobleaching). Be sure to keep slides covered from this point onward
16. For a detailed discussion on conjugate quantification, *see* Burack et al. ***(24)***.
17. Use 1×10^6 cells per sample. Before working with Dynalbeads, read the literature provided with the reagents and become familiarized with the technical aspects of the beads.
18. If cytoplasmic labeling is required, permeablize the plasma membrane to allow antibodies into the cell; 100% methanol at –20°C or 0.1% Triton X-100 at 4°C are both effective techniques.
19. For an excellent review that explains how to interpret FRAP data, *see* Sprague ***(56)***.
20. Total cholesterol content can be measured from a 10-μg aliquot using Amplex red kit (Molecular Probes), according to the manufacturer's instructions.
21. Approximately 1×10^6 cells per sample will be needed.
22. It is important to include proper positive and negative controls in the FACS analysis. In addition to any other proteins of interest, stain cells for GM1 or CD48 as positive controls (both markers are raft residents) and CD2 or CD71 as a negative control (both markers are excluded from lipid rafts).
23. For an excellent protocol on FACS sample preparation and analysis, *see* Vacchio and Shores ***(57)***.

References

1. Horejsi, V. (2005) Lipid rafts and their roles in T-cell activation. *Microbes Infect.* **7,** 310–316.
2. Mayor, S. and Rao, M. (2004) Rafts: scale-dependent, active lipid organization at the cell surface. *Traffic* **5,** 231–240.
3. Mukherjee, S. and Maxfield, F. R. (2004) Membrane domains. *Annu. Rev. Cell Dev. Biol.* **20,** 839–66.
4. Rajendran, L. and Simons, K. (2005) Lipid rafts and membrane dynamics. *J. Cell Sci.* **118,** 1099–1102.

5. Demetriou, M., Granovsky, M., Quaggin, S., and Dennis, J. W. (2001) Negative regulation of T-cell activation and autoimmunity by Mgat5 N-glycosylation. *Nature* **409,** 733–739.
6. Sillence, D. J. (2001) Apoptosis and signalling in acid sphingomyelinase deficient cells. *BMC Cell Biol.* **2,** 24.
7. Watzke, A., Brunsveld, L., Durek, T., et al. (2005) Chemical biology of protein lipidation: semi-synthesis and structure elucidation of prenylated RabGTPases. *Org. Biomol. Chem.* **7,** 1157–1164.
8. van't Hof., W. and Resh, M.D. (1999) Dual Fatty Acylation of p59Fyn is required for association with the T-cell receptor z chain through phosphotyrosine-Src homology domain-2 interactions. *J. Cell Biol.* **145,** 277–389.
9. Harder, T. and Khun, M. (2001) Immunoisolation of TCR signaling complexes from Jurkat T leukemic cells. *SciSTKE* **71,** pI1–pI18.
10. Thomas, S., Preda-Pais, A., Casares, S., and Brumeanu, T. D. (2004) Analysis of lipid rafts in T-cells. *Mol. Immunol.* **41,** 399–409.
11. Parton, R. G. and Hancock, J. F. (2004) Lipid rafts and plasma membrane micro-organization: insights from Ras. *Trends Cell Biol.* **14,** 141–147.
12. Paladino, S., Sarnataro, D., Pillich, R., Tivodar, S., Nitsch, L., and Zurzolo, C. (2004) Protein oligomerization modulates raft partitioning and apical sorting of GPI-anchored proteins. *J. Cell Biol.* **167,** 699–709.
13. Harder, T. and Kuhn, M. (2000) Selective accumulation of raft-associated membrane protein LAT in T-cell receptor signaling assemblies. *J. Cell Biol.* **151,** 199–207.
14. Simons, K. and Vaz, W. L. (2004) Model systems, lipid rafts, and cell membranes. *Annu. Rev. Biophys. Biomol. Struct.* **33,** 269–295.
15. Lafont, F., Abrami, L., and van der Goot, F. G. (2004) Bacterial subversion of lipid rafts. *Curr. Opin. Microbiol.* **7,** 4–10.
16. Harder, T. (2004) Lipid raft domains and protein networks in T-cell receptor signal transduction. *Curr. Opin. Immunol.* **16,** 353–359.
17. Golub, T., Wacha, S., and Caroni, P. (2004) Spatial and temporal control of signaling through lipid rafts. *Curr. Opin. Neurobiol.* **14,** 542–550.
18. Krishnan, S., Nambiar, M. P., Warke, V. G., et al. (2004) Alterations in lipid raft composition and dynamics contribute to abnormal T-cell responses in systemic lupus erythematosus. *J. Immunol.* **172,** 7821–7831.
19. Aridor, M. and Hanna, L. A. (2000) Traffic jam: a compendium of human diseases that affect intracellular transport processes. *Traffic* **1,** 836–851.
20. Pagano, R. E., Puri, V., Dominguez, M., and Marks, D. L. (2000) Membrane traffic in sphingolipid storage diseases. *Traffic* **1,** 807–815.
21. Simons, K. and Ehehalt, R. (2002) Cholesterol, lipid rafts, and disease. *J. Clin. Invest.* **110,** 597–603.
22. Henderson, R. M., Edwardson, J. M., Geisse, N. A., and Saslowsky, D. E. (2004) Lipid rafts: feeling is believing. *News Physiol. Sci.* **19,** 39–43.
23. Horejsi, V. (2003) The roles of membrane microdomains (rafts) in T-cell activation. *Immunol. Rev.* **191,** 148–164.

24. Burack, W. R., Lee, K. H., Holdorf, A. D., Dustin, M. L., and Shaw, A. S. (2002) Cutting edge: quantitative imaging of raft accumulation in the immunological synapse. *J. Immunol.* **169,** 2837–2841.
25. Maine, G. N. and Mule, J. J. (2002) Making room for T-cells. *J. Clin. Invest.* **110,** 157–159.
26. He, H. T., Lellouch, A., and Didier, M. (2005) Lipid rafts and the initiation of T-cell receptor signaling. *Semin. Immunol.* **17,** 23–33.
27. Drevot, P., Langlet, C., Guo, X. J., et al. (2002) TCR signal initiation machinery is pre-assembled and activated in a subset of membrane rafts. *EMBO J.* **21,** 1899–1908.
28. Macdonald, J. L. and Pike, L. J. (2005) A simplified method for the preparation of detergent-free lipid rafts. *J. Lipid Res.* **46,** 1061–1067.
29. Pike, L. J. (2004) Lipid rafts: heterogeneity on the high seas. *Biochem. J.* **378,** 281–292.
30. Smart, E. J., Ying, Y. S., Mineo, C., and Anderson, R. G. W. (1995) A detergent-free method for purifying caveolae membrane from tissue culture cells. *Proc. Natl. Acad. Sci. USA* **92,** 10,104–10,108.
31. Song, K. S., Li, S., Okamoto, T., Quilliam, L. A., Sargiacomo, M., and Lisanti, M. P. (1996) Co-purification and direct interaction of Ras with Caveolin, an integral membrane protein of caveolae microdomains. *J. Biol. Chem.* **271,** 9690–9697.
32. Du, G., Altshuller, Y. M., Vitale, N., et al. (2003) Regulation of phospholipase D1 subcellular cycling through coordination of multiple membrane association motifs. *J. Cell Biol.* **162,** 305–315.
33. Stauffer, T. P. and Meyer, T. (1997) Compartmentalized IgE receptor-mediated signal transduction in living cells. *J. Cell Biol.* **139,** 1447–1454.
34. Zacharias, D. A., Violin, J. D., Newton, A. C., and Tsien, R. Y. (2002) Partitioning of lipid-modified monomeric GFPs into membrane microdomains of live cells. *Science* **296,** 913–916.
35. van Meer, G. (2002) The different hues of lipid rafts. *Science* **296,** 855–856.
36. Chen, C. S., Martin, O. C., and Pagano, R. E. (1997) Changes in the spectral properties of a plasma membrane lipid analog during the first seconds of endocytosis in living cells. *Biophys. J.* **72,** 37–50.
37. Jacobson, K. and Dietrich, C. (1999) Looking at lipid rafts? *Trends Cell Biol.* **9,** 87–91.
38. Kuerschner, L., Ejsing, C. S., Ekroos, K., Shevchenko, A., Anderson, K. I., and Thiele, C. (2005) Polyene-lipids: a new tool to image lipids. *Nat. Methods* **2,** 39–45.
39. van Meer, G. and Liskamp, R. M. J. (2005) Brilliant lipids. *Nat. Methods* **2,** 14–15.
40. Prior, I. A., Parton, R. G., and Hancock, J. F. (2003) Observing cell surface signaling domains using electron microscopy. *SciSTKE* **177,** pI9–pI23.
41. Prior, I. A., Muncke, C., Parton, R. G., and Hancock, J. F. (2003) Direct visualization of Ras proteins in spatially distinct cell surface domains. *J. Cell Biol.* **160,** 165–170.
42. Glebov, O. O. and Nichols, B. J. (2004) Lipid raft proteins have a random distribution during localized activation of the T-cell receptor. *Nat. Cell Biol.* **6,** 238–243.
43. Kenworthy, A. K. and Edidin, M. (1998) Distribution of a glycosylphosphatidylinositol-anchored protein at the apical surface of MDCK cells examined at a

resolution of <100 A using imaging fluorescence resonance energy transfer. *J. Cell Biol.* **142,** 69–84.
44. Pierce, S. K. (2004) To cluster or not to cluster: FRETting over rafts. *Nat. Cell Biol.* **6,** 180–181.
45. Gordon, G. W., Berry, G., Liang, X. H., Levine, B., and Herman, B. (1998) Quantitative fluorescence resonance energy transfer measurements using fluorescence microscopy. *Biophysical J.* **74,** 2702–2713.
46. Tanimura, N., Nagafuku, M., Liddicoat, D. R., Hamaoka, T., and Kosugi, A. (2003) Analysis of the mobility of signaling molecules in lymphocytes using fluorescence photobleaching techniques. *SciSTKE* **185,** pl10–pl18.
47. Herman, B., Krishnan, R. V., and Centonze, V. E. (2004) Microscopic analysis of fluorescence resonance energy transfer (FRET), in *Protein–Protein Interactions: Methods and Applications* (Fu, H., ed.), Vol. 261, Humana Press, Totowa, NJ, pp. 351–370.
48. von Haller, P. D., Donohoe, S., Goodlett, D. R., Aebersold, R., and Watts, J. D. (2001) Mass spectrometric characterization of proteins extracted from Jurkat T-cell detergent-resistant membrane domains. *Proteomics* **1,** 1010–1021.
49. Pompach, P., Man, P., Novak, P., Havlicek, V., Fiserova, A., and Bezouska, K. (2004) Mass spectrometry is a powerful tool for identification of proteins associated with lipid rafts of Jurkat T-cell line. *Biochem. Soc. Trans.* **32,** 777–779.
50. Tu, X., Huang, A., Bae, D., et al. (2004) Proteome analysis of lipid rafts in Jurkat cells characterizes a raft subset that is involved in NF-kappaB activation. *J. Proteome Res.* **3,** 445–454.
51. Foster, L. J., de Hoog, C. L., and Mann, M. (2003) Unbiased quantitative proteomics of lipid rafts reveals high specificity for signaling factors. *Proc. Natl. Acad. Sci. USA* **100,** 5813–5818.
52. Ong, S. E., Blagoev, B., Kratchmarova, I., Kristensen, D. B., Steen, H., Pandey, A., et al. (2002) Stable isotopic labeling by amino acids in cell culture, silac, as a simple and accurate approach to expression proteomics. *Mol. Cell Proteom.* **1,** 373–386.
53. Gombos, I., Basco, Z., Detre, C., Nagy, H., Goda, K., Andrasfalvy, M., et al. (2004) Cholesterol sensitivity of detergent resistance: a rapid flow cytometric test for detecting constitutive or induced raft association of membrane proteins. *Cytometry* **61A,** 117–126.
54. Kwik, J., Boyle, S., Fooksman, D., Margolis, L., Sheetz, M. P., and Edidin, M. (2003) Membrane cholesterol, lateral mobility, and the phosphatidylinositol 4, 5-bisphosphate-dependent organization of cell actin. *Proc. Natl. Acad. Sci. USA* **100,** 13,964–13,969.
55. Razzaq, T. M., Ozegbe, P., Jury, E. C., Sembi, P., Blackwell, N. M., and Kabourids, P. S. (2004) Regulation of T-cell receptor signalling by membrane microdomains. *Immunology* **113,** 413–426.
56. Sprague, B. L. and McNally, J. G. (2004) FRAP analysis of binding: proper and fitting. *Trends Cell Biol.* **15,** 84–91.
57. Vacchio, M. S. and Shores, E. W. (2000) Flow cytometric analysis of murine T lymphocytes, in *T Cell Protocols* (Kearse, K., ed.), Vol. 134, Humana Press, Totowa, NJ, pp. 153–176.

Index

MIX
Papier aus verantwortungsvollen Quellen
Paper from responsible sources
FSC® C105338

If you have any concerns about our products,
you can contact us on
ProductSafety@springernature.com

In case Publisher is established outside the EU,
the EU authorized representative is:
Springer Nature Customer Service Center GmbH
Europaplatz 3, 69115 Heidelberg, Germany

Printed by Libri Plureos GmbH
in Hamburg, Germany

BURLEIGH DODDS SCIENCE: INSTANT INSIGHTS

NUMBER 107

Regulatory frameworks for new agricultural products and technologies

burleigh dodds
SCIENCE PUBLISHING

Published by Burleigh Dodds Science Publishing Limited
82 High Street, Sawston, Cambridge CB22 3HJ, UK
www.bdspublishing.com

Burleigh Dodds Science Publishing, 1518 Walnut Street, Suite 900, Philadelphia, PA 19102-3406, USA

First published 2024 by Burleigh Dodds Science Publishing Limited

British Library Cataloguing in Publication Data
A catalogue record for this book is available from the British Library

ISBN 978-1-83545-006-2 (Print)
ISBN 978-1-83545-007-9 (ePub)

DOI: 10.19103/9781835450079

Typeset by Deanta Global Publishing Services, Dublin, Ireland

Contents

Series list

Title	Series number
Sweetpotato	01
Fusarium in cereals	02
Vertical farming in horticulture	03
Nutraceuticals in fruit and vegetables	04
Climate change, insect pests and invasive species	05
Metabolic disorders in dairy cattle	06
Mastitis in dairy cattle	07
Heat stress in dairy cattle	08
African swine fever	09
Pesticide residues in agriculture	10
Fruit losses and waste	11
Improving crop nutrient use efficiency	12
Antibiotics in poultry production	13
Bone health in poultry	14
Feather-pecking in poultry	15
Environmental impact of livestock production	16
Sensor technologies in livestock monitoring	17
Improving piglet welfare	18
Crop biofortification	19
Crop rotations	20
Cover crops	21
Plant growth-promoting rhizobacteria	22
Arbuscular mycorrhizal fungi	23
Nematode pests in agriculture	24
Drought-resistant crops	25
Advances in detecting and forecasting crop pests and diseases	26
Mycotoxin detection and control	27
Mite pests in agriculture	28
Supporting cereal production in sub-Saharan Africa	29
Lameness in dairy cattle	30
Infertility and other reproductive disorders in dairy cattle	31
Alternatives to antibiotics in pig production	32
Integrated crop-livestock systems	33
Genetic modification of crops	34

Chapter 1

Creating a regulatory framework for cultured meat products: Singapore

Johnny Yeung, Yong Quan Tan, Siew Herng Chan, Kern Rei Chng, Calvin Yeo, Jer Lin Poh, Teng Yong Low and Joanne Sheot Harn Chan, National Centre for Food Science, Singapore

1 Introduction

The Singapore Food Agency (SFA) is Singapore's national competent authority for food safety. In Singapore, all foods, including novel foods, come under the purview of SFA. The SFA was formed on 1 April 2019 under Singapore's Ministry of Sustainability and the Environment, with the vision to ensure and secure a supply of safe food. SFA consolidates all food-related resources and capabilities in the Singapore government for the holistic management of the food industry 'from farm to fork'.

2 Key food safety issues in Singapore

As a small island city-state, Singapore currently imports over 90% of its food supply to meet the demands of the population (Singapore Food Agency, 2019). As food safety challenges continue to evolve with emerging risks and an increasingly complex industry landscape, it is important for SFA to continually develop and enforce standards, guidelines and policies to keep pace with latest scientific knowledge to ensure food safety.

SFA is committed to ensuring and securing a supply of safe food for the nation through the adoption of a risk-based approach to food safety guided by science and consistent with international standards. With food safety ultimately

http://dx.doi.org/10.19103/AS.2023.0130.03

being a joint responsibility among stakeholders in the food supply chain, SFA fosters an enabling environment so that all stakeholders can each play their part. SFA's food safety system is therefore underpinned by three key considerations, as described next.

2.1 Consideration 1: farm-to-fork systems approach

SFA regulates the safety of food from import to local primary production and manufacturing and finally to food retail. This is achieved by identifying key stakeholders and control points along this farm-to-fork food supply chain. The farm-to-fork model of food safety assurance is best exemplified in SFA's import control regime which has been established with three levels of control as shown in Fig. 1. These levels are:

- Pre-import control;
- Import control; and
- Post-import control.

As with any other food, the importation of novel foods such as cultured meat will be subject to the same regime even after it has received SFA's pre-market approval.

First, before food is imported into Singapore, pre-import control measures are established to ensure safety of the food sources through a risk-differentiated approach. Measures range from the accreditation of food sources for high-risk items (such as meat and eggs), to application of licensing and import permits for food consignments or submission of export certificates for lower risk products. These measures serve to enhance food safety assurance by

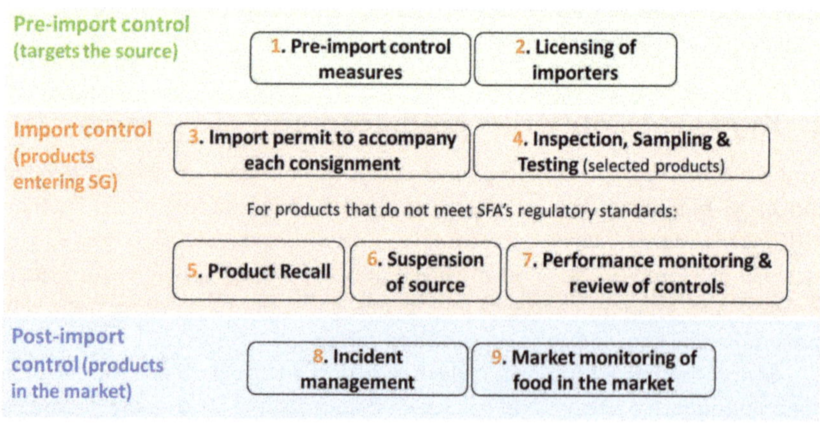

Figure 1 Summary of SFA's pre-import, import and post-import control regimes.

addressing food safety risks prior to import and protect the local food supply by preventing diseases that may affect production in local farms. These measures also help locally manufactured food products to gain overseas market access by ensuring that they are free from animal diseases of trade importance (e.g. Avian influenza, foot and mouth disease, African swine fever) (Ryan, 2016). For novel foods, pre-market safety assessments are also required to establish the safety of such food products. This is because they do not have a history of safe use and any potential safety concerns need to be identified and addressed.

Second, at the point of import, commercial shipments of food may only be imported by licensed or registered importers and must be accompanied by import permits. All importers, whether licensed or registered, need to comply with the relevant legislation. Import permits allow SFA to ensure that the relevant pre-import requirements and import conditions are met, facilitate downstream targeting of specific products for inspection as well as facilitate sampling and information collection for food traceability. Food products are also subject to SFA's inspection, sampling and testing programme whereby higher risk items are targeted and tested for commonly associated food safety hazards (e.g. *Salmonella* spp. in eggs).

Third, even beyond the point of import, food products may be recalled, and suppliers may be suspended if food does not meet regulatory standards, or when SFA is alerted by international/regional platforms (e.g. WHO INFOSAN, EU Rapid Alert System, G2G notifications) and/or horizon scanning. This is complemented by SFA's market monitoring programme which is non-regulatory in nature. Under this programme, food products are sampled and tested from the retail end to enable SFA to better understand the risk profile of imported products, identify high-risk products for more targeted checks and to refine the regulatory sampling and testing programme when necessary.

2.2 Consideration 2: science-based risk assessment and management

SFA uses science and data obtained from fieldwork (e.g. food safety incidents, non-compliances to requirements, food testing results) to assess the level of food safety risks posed at various control points along the farm-to-fork supply chain. During the assessment, SFA aligns to the risk analysis framework that has been established by the Codex Alimentarius Commission (CAC), an international food safety standards-setting body under the governance of the Food and Agriculture Organization (FAO) and United Nations (UN). The framework comprises risk assessment, risk management and risk communication (Fig. 2) (FAO, 2007).

Based on the assessed risk, SFA determines and adopts the appropriate food safety risk management measures in its regulatory regime to reduce and

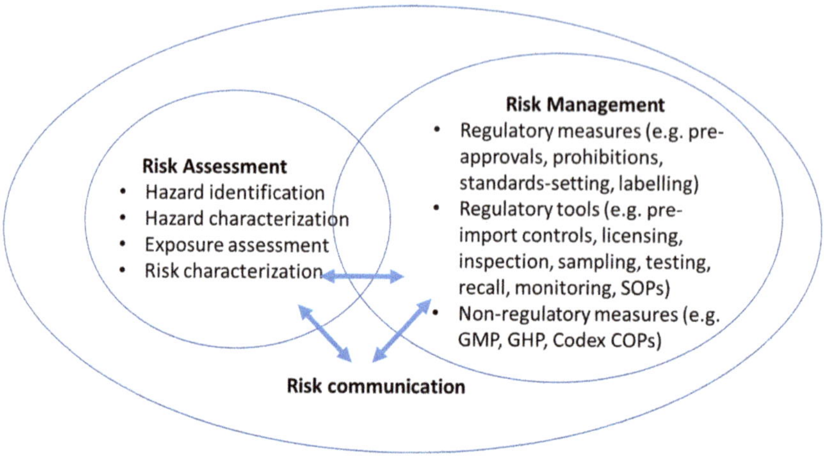

Figure 2 Codex risk analysis framework.

mitigate the foodborne risks to consumers. These risk management measures include the following:

i. *Standard setting*: Food safety standards (i.e. regulatory limits and conditions that the industry must comply with) are set to an appropriate level of protection for Singapore, based on risk assessments conducted and with reference to international standards, guidelines and recommendations such as those issued by the CAC. In the absence of international standards, reference is taken from major developed countries and jurisdictions (e.g. US, Australia, New Zealand, European Union, Japan) as well as Association of Southeast Asian Nations (ASEAN) Standards. SFA also works with stakeholders to develop industry standards that the food industry can choose to adopt to help them in food safety assurance.
ii. *Imposition and enforcement of regulatory requirements*: Regulatory levers (e.g. licensing of food operators) are established and the food safety requirements are enforced through inspection and testing.
iii. *Establishing mechanisms for food safety incident management*: Anticipatory and responsive mechanisms have been put in place. These include issuing food safety advice prior to periods of possible higher food safety risk, food recall plans by industry and suspension of affected food suppliers.

SFA recognizes that a science-based risk assessment and management approach to food safety will only be successful with robust food safety science capabilities in the ecosystem. Hence, SFA has multiple building blocks in place to support the development of such food safety science capabilities. First, SFA keeps a close watch on food safety developments through our horizon scanning programme, and reviews our standards based on developments at regional international forums and developed countries, in which SFA participates in discussions. Second, SFA houses the National Centre for Food Science (NCFS), which is recognized locally, regionally and internationally as a centre for food science excellence. NCFS has been awarded several accolades, such as the OIE Collaborating Centre for Food Safety, WHO Collaborating Centre for Food Contamination Monitoring, as well as having four ASEAN Food Reference Laboratories. Third, under SFA's Laboratory Recognition Program (LRP), a network of private laboratories has been set up, which enables NCFS' food testing capabilities to be scaled to the private industry. Furthermore, SFA has launched the Future Ready Food Safety Hub (FRESH; see later), which is a public-private partnership between SFA, Nanyang Technological University (NTU) and Agency for Science, Technology and Research (A*STAR) tasked with pursuing research and development (R&D) in food safety science.

Inevitably there are broader expertise areas that are not located within SFA. To address this limitation, SFA engages and collaborates with overseas authorities (e.g. signing of Memorandum of Understandings) with the Food Standards Australia and New Zealand (FSANZ) and French Agency for Food, Environmental and Occupational Health and Safety (ANSES), and also convenes scientific advisory committees on specific topics to tap on the domain expertise in the wider community to help fine tune scientific aspects to support SFA's regulatory decisions.

2.3 Consideration 3: joint responsibility for food safety

Food safety is a joint responsibility where the government, industry and consumers have a role to play in ensuring safe food for all. Even as SFA puts in place a regulatory system and enabling environment to ensure food safety, food business operators must play their part to maintain food safety assurance and be responsible for the safety of the food it offers to the public. To this end, SFA has embarked on the following industry-centred initiatives:

i. *Safety Assurance for Food Establishments (SAFE)*: SFA established the SAFE framework for all food establishments to incentivize them to consistently maintain high food safety standards. Food establishments that have demonstrated a good track record of food safety assurance and have put in place capabilities and systems to ensure better food safety standards will be eligible for longer licence durations and higher award tiers.

ii. *Engagements with industry*: SFA holds regular dialogues with traders, food manufacturers and retailers to raise awareness of food safety measures, encourage the implementation of food safety management systems (e.g. hazard analysis and critical control point (HACCP)), and assist them to comply with regulatory requirements.
iii. *Building food testing capability in the private sector*: SFA facilitated the development of a network of third-party laboratories that the food industry could tap on for food testing needs. SFA also introduced the LRP from April 2020 to enhance the recognition of these laboratories among the industry and foreign competent authorities.
iv. *Building capabilities in the food industry*: It is necessary for the food manufacturing and food & beverage industry workforce to be trained to support the adoption of systems and processes and ensure consistent food safety performance. Hence, SFA has rolled out an enhanced training and competency framework to upskill workers in this sector.

At the same time, consumers should be empowered to carry out good food hygiene practices and make informed food safety decisions. Through its public outreach efforts, SFA raises consumer awareness that food safety is a joint responsibility and educates consumers on foodborne hazards and steps on keeping food safe. This is done through Learning Journeys, school/workplace talks, community events as well as mainstream and social media. SFA develops communication materials such as consumer advisories on topics ahead of anticipated public concern. These are proactively pushed out on various platforms such as SFA's website ('Risk at a glance' articles), and SFA's social media platforms (e.g. Facebook and Instagram).

In the longer term, SFA's vision is to shift towards a state of food safety culture, i.e. where the food industry is progressive, self-directed and intrinsically motivated to ensure food safety along the supply chain; where consumers are informed and take proactive food safety measures; and where government enables better food safety, by creating an enabling tech- and data-based environment for industry and consumers to take on greater ownership.

3 The Singapore Food Agency's regulatory framework for cultured meat

Under SFA's novel food regulatory framework, food that has not been consumed as part of an ongoing diet by any significant human population for a period of at least 20 years is considered novel. Therefore, SFA considers cultured meat as a novel food. SFA's novel food regulatory framework has based its principal considerations in its food safety system, as described in Sections 2.1–2.3, which is applied to all foods. For cultured meat and other novel foods, SFA supplements the existing food safety system by considering and focusing on new inputs and processes that can introduce additional food safety risks. Established food safety principles and concepts pertaining to conventional foods are also adapted for novel foods. In addition, SFA utilizes its existing regulatory levers to ensure the safety of novel foods.

SFA considers cultured meat as cells from a whole animal that have been cultured in a bioreactor and supplemented with a growth medium for cell proliferation and differentiation. These cultured cells differentiate into muscles or fat cells and propagate into protein-rich biomass or complex tissue structures, which are ultimately converted into meat products. Under SFA's regulatory framework for novel foods, companies are required to conduct and submit pre-market safety assessments for SFA's review and obtain SFA's approval before they are allowed to produce/manufacture, import, distribute or sell their cultured meat product in Singapore.

There is currently no internationally accepted term to describe products obtained from animal cell culture intended for consumption. Several terms have been proposed to describe cultured meats, including clean, *in vitro*, synthetic, cellular, lab grown, cultivated and cultured. For example, the Good Food Institute (GFI) and FSANZ have adopted the terms 'cultivated meat' and 'cell-based meat', respectively. In Singapore, SFA has adopted the working term 'cultured meat' to describe meat developed from animal cell cultures in bioreactors supplemented by culture media. Other government agencies such as Enterprise Singapore and Singapore Economic Development Board were also consulted on the use of 'cultured meat' terminology to synchronize communications between government bodies as well as the public. Regardless of the terminology, it is mandatory for companies to label their products with qualifying terms such as 'cultured', 'cultivated' or 'cell-based', to accurately convey the true nature of their food product.

SFA's regulatory framework for novel foods is primarily focused on food safety to ensure that novel foods are safe for consumption. SFA understands that new food sources and production systems prompt diverse discussions among numerous stakeholders. Apart from food safety risks, these diverse discussions could include topics such as nutritional, sustainability, socioeconomic, animal welfare and cultural implications of new food sources and production systems. SFA's regulatory framework for novel food is focused on food safety risks and is not intended to capture considerations not related to food safety. SFA is of the view that food safety considerations of novel foods should not be intermingled with other considerations that are not related to food safety. This is to ensure scientific robustness in the safety assessment of novel foods.

SFA's safety assessment framework for cultured meat captures the food safety considerations during the cultured meat manufacturing process and it covers three levels. First, the safety of individual inputs in the production process (e.g. cell lines, culture media and reagents) are assessed to ensure that potential safety risks have been addressed. Second, the production process and quality controls are guided by detailed risk assessments and risk management plans to ensure that there is no introduction of additional food safety risks. As cultured meat typically involves cell culture processes, the framework for cultured meat takes guidance from good practices in biopharmaceutical manufacturing, cell culture, microbiology and molecular biology that ensure stability and traceability of the cell line throughout the manufacturing process. The production process must be properly controlled and adhere to good food safety and hygiene practices. Last, the final product must comply with the Singapore's Food Regulations. This includes compliance to the regulatory limits for additives and contaminants.

As each cultured meat manufacturing process is unique and certain techniques and methodologies are new to food manufacturing, SFA's safety assessment framework utilizes food safety principles and guidelines that can be generally applied to all cultured meat manufacturing processes that are currently conceivable. SFA posits that a safety assessment framework that is based on broad scientific and food safety principles, and that which is not unnecessarily prescriptive, facilitates current and future innovation in food technology while still ensuring food safety. Details of food safety considerations in SFA's safety assessment framework for cultured meat are depicted in Fig. 3 and further elaborated next.

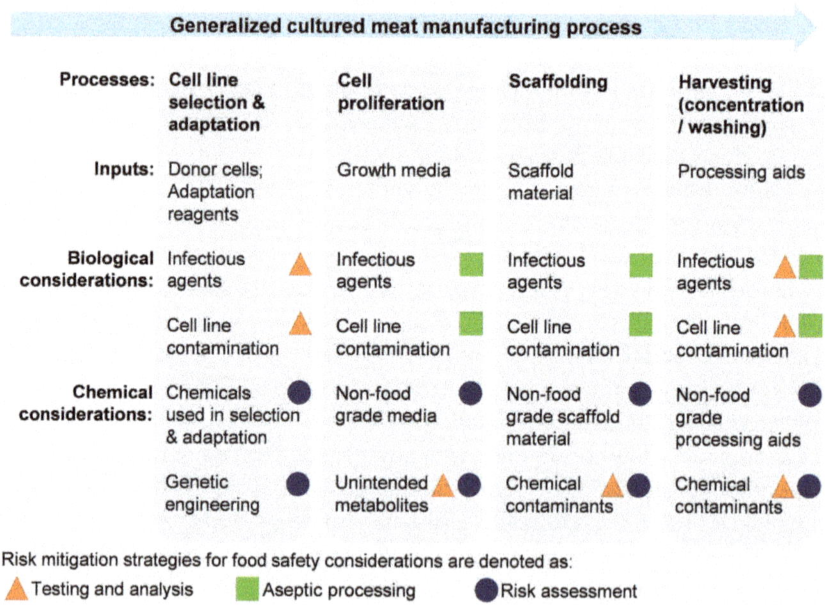

Figure 3 Food safety assessment framework for cultured meat.

i. *Infectious agents*: Foodborne pathogens, which include viruses, bacteria, fungi and other parasites, can be introduced at any stage of the manufacturing process. In the presence of nutrient-rich cell culture growth media, these pathogens can easily proliferate to levels that could infect and cause disease for the consumer. Arguably, the most important control point for removing foodborne pathogens in the manufacturing process is at the cell isolation step. Here, attention is focused on the presence of pathogens, particularly those known to be present in the particular animal type from which the cells are derived. If cells are isolated from whole animals, whether live or slaughtered, it is typically necessary to use antibiotics and/or other drugs that inhibit the growth of pathogens naturally present in the whole animal. Absence of these pathogens must be confirmed through standard immunological, microbiological or molecular biology methods. After ensuring the sterility of the donor cells, the highest standards of cleanliness must be maintained during cell proliferation and processing, and routine analysis of foodborne pathogens is necessary to provide assurance on the absence of infectious agents. In recent years, metagenomics tools, such as 16S rRNA sequencing and shotgun sequencing, have been increasingly recognized for its usefulness in rapidly identifying a wide range of pathogens, including those that are less well characterized (Kovac et al., 2017). However, metagenomics tools currently suffer from lower sensitivity and species/sub-species resolution, and should only be used to complement standard detection methods at this point (Snipen et al., 2021). If bovine products such as foetal bovine serum are introduced into the process, companies should ensure that these products are sourced from regions with no indigenous cases of bovine spongiform encephalopathy (BSE). This is because the presence of prions in the bovine products, which are the infectious agent causing BSE, can lead to the formation and accumulation of prions in cells (Hedlin et al., 2012). Consumption of meat infected by prions has been linked to development of Creutzfeldt-Jakob disease in humans, which is incurable and invariably fatal (Roma and Prayson, 2005).

ii. *Cell line contamination by other cell lines and adventitious agents*: SFA's safety assessment for a cultured meat product is specific to the cell line from which the product is derived from. As such, measures that minimize the possibility of cell line contamination are important considerations in the safety assessment framework. During the isolation of the target primary cell line from a donor animal, scientifically validated cell sorting methodologies such as flow cytometry, magnetic-activated cell sorting and microfluidic cell sorting should be used (Wyatt Shields et al., 2015). Primary cell lines that are directly sourced from a cell line supplier should ensure that the supplier is reputable, and the cell line has been authenticated. It is widely recognized that continual subculturing of cells exerts selective pressure on these cells and this can cause significant genetic drifts, resulting in cells that are not genotypically and/or phenotypically representative of the initial cell line (Hughes et al., 2007). Good cell banking protocols that are widely practised in biopharmaceutical manufacturing, such as establishment of a Master Cell Bank and Working Cell Bank, as well as adhering to good cell subculture protocols, can mitigate the effects of genetic drifts (Zhu et al., 2017). Routine monitoring of cell line stability via genomics (e.g. whole-genome sequencing, single-nucleotide polymorphism microarray) and transcriptomics (e.g. RNA-seq) approaches, as well as phenotypic approaches (e.g. cell doubling time, oxygen consumption rate), contributes to a comprehensive depiction of cell line stability that supports food safety assessment. Where significant differences in gene expression are observed between the initial cell line and the cells in the cultured meat product, companies should investigate if the differences would result in food safety risks. Examples of such risks include increased expression of allergenic proteins or production of unintended metabolites.

iii. *Genetic engineering*: Some companies utilize genetic engineering to endow cell lines with useful traits, predominantly immortalization of cell lines beyond their natural replication limit (termed Hayflick limit) (Shay and Wright, 2000; Wikandari et al., 2021). Genetic engineering may entail the integration of transgenes into the genome of the cell line, or to make precise alterations or deletions of existing genes using more recent genome editing tools (Soice and Johnston, 2021). The process of genetic engineering can introduce food safety risks that must be assessed. If the cell line has been engineered to produce proteins that are novel to the cell line (i.e. foreign proteins), such proteins must be assessed for safety to ensure that they do not present toxicity, allergenicity or digestibility issues when consumed. All intentional genetic alterations, be it gene insertion, alteration or deletion, must also be proven to be stable through consecutive subcultures. In addition, companies must verify that all nucleotide sequences used during the genetic engineering process, but that are not intended to remain in the cells (e.g. plasmid backbone, nuclease gene) are indeed absent from the cell line used in production. Verification must be performed using established molecular biology methodologies, such as Southern blotting and next-generation sequencing. Currently, prevailing methods of cell line immortalization via genetic engineering involve the expression of foreign oncogenes or inactivating the expression of endogenous tumour suppressor genes (Soice and Johnston, 2021), which may increase the oncoprotein content of the cells. Further research is warranted to determine if such genetic engineering methods would introduce further food safety risks to the cultured meat product.

iv. *Chemical contaminants*: The cultured meat manufacturing process can introduce chemical contaminants into the final food product. Some of these contaminants may present food safety risks. For example, chitin, which is used in some cell culture scaffolds, may trigger allergic reactions in some individuals. (Lee, 2009; Campuzano and Pelling, 2019) Certain antibiotics can cause allergic reactions, liver damage and other adverse health effects in sensitive individuals, while the overuse of antibiotics in food production can contribute to antimicrobial resistance, a public health concern (Tang et al., 2017; Bacanlı and Başaran, 2019). Safety assessments of chemical contaminants in the cultured meat product are necessary to ensure food safety. Testing and analysis of chemical contaminants should be done using protocols and guidelines that are established in the scientific literature or supported by OECD, AOAC and other reputable trade or scientific organizations. Acceptable maximum levels of chemical contaminants in all foods, including cultured meat, are stipulated in Singapore's Food Regulations. Food containing chemical contaminants that exceed the stipulated levels will not be allowed for sale in Singapore.

v. *Non-food grade chemicals*: The cultured meat manufacturing process requires the use of a wide range of chemicals that, while routinely used in cell culture, are novel to food processing. Thus, it is not feasible for only food-grade chemicals and reagents to be used in the process. Some of the non-food-grade chemicals used may be found at low levels in the final cultured meat product and may present health risks. Notwithstanding the absence of food-grade certification, chemicals sourced from reputable suppliers come with specification documents that support food safety risk assessment. Some chemicals that are not traditionally used in food processing or manufacturing have associated toxicological data, such as a no-observed-adverse-effect level (NOAEL) and benchmark dose level (BMDL), that are available in the scientific literature. Both NOAEL and BMDL serve as departure points to estimate health-based guidance values such as acceptable daily intake (ADI) of the chemical in the cultured meat product (Hardy et al., 2017). However, many other chemicals do not have associated toxicological data. In the absence of established toxicological data, companies may adopt the threshold of toxicological concern as a pragmatic approach towards food safety assessment of these chemicals (More et al., 2019), but only if the chemical is suitable for assessment by such an approach (one of which is that it must be demonstrated to be non-genotoxic). A number of decision-tree approaches to estimate toxicological effects of small molecules, such as Toxtree, have been developed to support both food and environmental risk assessments (Kroes et al., 2004; Patlewicz et al., 2010; Williams et al., 2017).

SFA's safety assessment framework for cultured meat also takes into consideration the hazard control plan (HCP) for the manufacturing process, regardless of whether the cultured meat product is imported or manufactured locally. Companies should provide information pertaining to food safety management systems that are put in place, and to provide relevant certification or documentation pertaining to HACCP, good manufacturing practices and good cell culture practices. Process flow diagrams may also be supplied to better identify critical control points that contribute to food safety of the product. Companies that intend to manufacture cultured meat products in Singapore are required to first obtain a licence from SFA for a food processing establishment.

On top of existing licensing requirements for food processing establishments that manufacture traditionally consumed food, cultured meat companies may also need to put in place additional control measures to address specific risks associated with cultured meat production. Examples of these control measures include the following:

- a designated storage facility for sterile instruments and tools;
- provision of a designated media preparation room;
- cold storage facility for cell banks and the cultured meat product; and
- provision of a designated cell culture room that is kept at a positive pressure relative to other rooms to minimize entry of infectious agents outside of the room.

4 Strengthening food safety capabilities in cultured meat products through partnerships

SFA's emphasis on regulatory agility (Soumyanarayanan et al., 2021) is key to enabling regulatory compliance to ensure food safety while facilitating new food innovations including cultured meats. Our regulatory framework on novel food aims to bring the industry and the regulator in close coordination based on the firm belief that food safety is a joint responsibility. To build trust between the regulator and the industry, SFA encourages and facilitates early engagement opportunities with the industry through the organization of novel food virtual clinics and pre-submission consultations. Such engagements are beneficial to both parties: the industry can gain access to the regulator's guidance and incorporate safety assurance measures as they develop their food innovations; meanwhile the regulator can obtain early insights from the industry and make timely risk assessments to formulate regulatory considerations in preparation for upcoming food innovations.

Apart from the industry, the academic sector is another key stakeholder that SFA partners to build up the wide-ranging and diverse domain expertise necessary for the effective assessment and regulation of food innovations. This is particularly essential since regulators are unlikely to have the breadth and depth of technical expertise and knowledge in-house to comprehensively evaluate food innovations that are constantly and rapidly evolving. Therefore, SFA has implemented various strategies to tap on the academic sector to partner in shaping research and development directions through targeted grant calls and funding to align academic research to the needs of regulatory science, establish public-private partnership platforms to facilitate synergy and integrate of capabilities within the food science ecosystem, and tap on multidisciplinary experts to contribute scientific advice on deep technical issues for regulatory assessments and considerations. Details on initiatives that SFA has participated in pursuit of these strategies are outlined as follows:

i. *Future REady Food Safety Hub (FRESH)*: The Singapore Food Story (SFS) was initiated to support Singapore's national agenda of strengthening food security. The R&D programme under the SFS focuses serves as a platform to drive the creation of new biotech-based foods and ingredients that are underpinned by a robust future-ready food safety system. One means of securing Singapore's food supply and economic value proposition involves building local food safety capabilities to support growing food innovations in Singapore, including in alternative proteins.

To this end, a public-private partnership platform known as FRESH was set up to drive food safety research to support food innovations. FRESH, with Agency for Science, Technology and Research (A*STAR) as its implementing agency, aims to foster synergistic collaborations between industry, academia and regulators, and aggregate relevant existing scientific expertise in Singapore's food science ecosystem to address emerging challenges in food safety, particularly in the development of alternative proteins and new processing technologies. In addition, FRESH plays an important role in facilitating novel food product development, by providing scientific technical support on safety risk assessment to the local industry. At the same time, FRESH provides a medium for regulatory scientists to collaborate and be exposed to upstream food safety issues in industry research and development. Finally, FRESH also provides relevant scientific and risk assessment training to build up a pipeline of food safety talent pool in Singapore over the longer term.

ii. *Novel Food Safety Expert Working Group*: SFA recognizes the deep and complex science driving new food innovations and is cognizant that is important to work with experts in various specialized scientific domains to ensure that food safety assessments are scientifically robust and carried out in a timely fashion. Therefore, SFA taps onto the expertise outside of the agency by forming a novel food safety expert working group to provide scientific advice to SFA for safety risk assessment reviews of novel foods. The expertise residing in the workgroup is diverse and has high relevance to cutting-edge food technologies and food safety sciences such as food toxicology, nutrition, bioinformatics, epidemiology and public health.

In addition to engaging the academic sector, SFA actively engages other regulatory counterparts to foster consensus-building in novel food regulations. Inaugurated in 2019 as the Regulator's Forum for Novel Foods, the Roundtable for Novel Food Regulations, as it is now known as, is organized under the umbrella of the Singapore International Agri-Food Week. This Roundtable provides an opportunity for the industry, research community, as well as local and overseas regulators to pursue discussions on the rapidly developing novel food sector, including the challenges and possible solutions in the safety assessment of novel food. The third Roundtable, organized in November 2022 (most recent iteration at time of writing), brought together 160 participants from the industry, academia, and governments across 15 jurisdictions to share and discuss about the rapidly developing novel food sector. This included members at the forefront of innovative food, including international organizations such as the CAC, the FAO of the UN and the World Health Organization. The Roundtable also included a panel discussion in which developments on the safety assessment, regulatory approvals and risk communications around novel foods were discussed.

5 Building consumer confidence in the safety of cultured meat products

While food safety risk assessment and management programmes are pivotal to ensuring that food is safe for consumption, it is also key for regulators to conduct risk communication effectively, to facilitate consumers' understanding of food safety risks and to build confidence in the food regulatory system.

SFA anticipated that its regulatory approval, the first in the world, of a cultured chicken product was likely to generate high levels of interest among consumers and industry experts, despite its limited commercial availability at launch. SFA therefore adopted a responsive risk communication strategy to assure the public of the rigorous nature of our safety assessment approach using a range of tools and channels targeting different audiences. These tools and channels included as follows:

i. *Articles for consumers*: Short articles and infographics are regularly published on SFA's social media platforms, which are written in accessible language. Articles covering various alternative proteins including cultured meat were initiated in the lead-up to SFA's announcement of our regulatory approval of the cultured chicken product. The articles explained how SFA assesses and monitors the safety of these foods.
ii. *Articles for industry players*: SFA also published a detailed explanation of our regulatory framework and safety assessment approach for novel foods on the SFA website to provide more information to industry experts and interested members of the public.
iii. *Sharing to stakeholders in workshop/meeting*: In preparation for publicity on Singapore's regulatory approval for cultured chicken, SFA conducted a local media briefing to share on its novel food regulatory framework with local media outlets and safety assessment approach, while A*STAR shared on the technology used in the creation of alternative proteins, Singapore's research and development efforts in future foods, as well as its potential contribution to sustainability and food supply. Meanwhile, other government agencies such as Singapore's Economic Development Board (EDB) also regularly roll out messaging on topics such as the potential of alternative proteins to contribute towards Singapore's economy.

Singapore's regulatory approval for cultured chicken garnered high traction on social media platforms and was widely publicized in international and local news. Online sentiments primarily focused on the 'artificial' nature of cultured meat, and some members of the public expressed concern over potential side effects of consuming cultured meat. Others reacted positively and viewed cultured chicken as a more humane alternative to conventionally farmed meat, and applauded SFA's regulatory agility and rigorous food safety assessment processes. This observation corroborates with published studies. For example, a survey conducted by researchers from NTU showed that a third of respondents found genetically modified food unappealing and that these negative feelings extended to their feelings for 'nano-enabled food' (Ho et al., 2020). On the other hand, a more recent study by the Singapore Management University surveyed Singaporeans and Americans and found that Singaporean participants had greater acceptance of cultured meat as compared to their American counterparts (Chong et al., 2022).

SFA continues to build its risk communication capabilities to raise understanding, acceptance and trust among consumers towards new food production technologies that could play an important role in meeting Singapore's food supply needs in the years ahead. For example, SFA has co-funded research projects on consumer science and risk communication on novel foods through the Singapore Food Story Research and Development Programme. Research outcomes will be translated into risk communication messages that are easily accessible to consumers. Nonetheless, even as SFA continues its risk communications efforts, the cultured meat industry will need to play its part to address public perception and consumer acceptance of cultured meat.

6 Conclusion

As a single agency overseeing all food-related matters in Singapore, SFA has leveraged on its unique position to ensure food safety through an integrated food safety system. To ensure the safety of food arising from innovation in Singapore, SFA has developed a regulatory framework for novel foods based on science and risk. Such policy thinking enabled SFA to ensure adequate space for the industry and research community to innovate while protecting consumers' health.

Nevertheless, it is important for SFA to remain agile in regulating novel foods, especially when the novel food ecosystem is developing rapidly. To this end, SFA actively engages with companies through Novel Food Virtual Clinics and pre-submission consultations. SFA also actively pursues advancements in food safety science through collaborative research with the academic sector and establishing a local ecosystem for novel foods (e.g. SFA's Novel Food Expert Panel and FRESH), as well as maintaining close alliances with SFA's regulatory counterparts outside of Singapore through bilateral and multilateral engagements (e.g. annual Roundtable for Novel Food Regulations).

At the same time, it is important to build up consumer confidence in new food innovations and address any concerns that they might have through an effective communications approach. Leveraging on both mainstream and social media platforms, a communications strategy has been crafted and implemented related to cultured meat, and media and consumer sentiments on this topic are also being monitored. This is an ongoing endeavour with active research efforts being conducted and the research outcomes will help to fine tune the risk communication strategy surrounding cultured meat.

7 References

Bacanlı, M. and Başaran, N. (2019). Importance of antibiotic residues in animal food, *Food and Chemical Toxicology* 125, 462. doi: 10.1016/J.FCT.2019.01.033.

Campuzano, S. and Pelling, A. E. (2019). Scaffolds for 3D cell culture and cellular agriculture applications derived from non-animal sources, *Frontiers in Sustainable Food Systems* 3, 38. doi: 10.3389/FSUFS.2019.00038/BIBTEX.

Chong, M., Leung, A. K. Y. and Lua, V. (2022). A cross-country investigation of social image motivation and acceptance of lab-grown meat in Singapore and the United States, *Appetite* 173, 105990.

EFSA Scientific Committee, Hardy, A., Benford, D., Halldorsson, T., Jeger, M. J., Knutsen, K. H., More, S., Mortensen, A., Naegeli, H., Noteborn, H. and Ockleford, C. (2017). Update: Use of the benchmark dose approach in risk assessment, *EFSA Journal* 1, 15, e04658. doi: 10.2903/J.EFSA.2017.4658.

EFSA Scientific Committee, More, S. J., Bampidis, V., Benford, D., Bragard, C., Halldorsson, T. I., Hernández-Jerez, A. F., Hougaard Bennekou, S., Koutsoumanis, K. P., Machera, K. and Naegeli, H. (2019). Guidance on the use of the threshold of Toxicological Concern approach in food safety assessment, *EFSA Journal* 6(17), e05708. doi: 10.2903/J.EFSA.2019.5708.

Food safety risk analysis. Available at: https://www.fao.org/3/a0822e/a0822e00.htm.

Hedlin, P., Taschuk, R., Potter, A., Griebel, P. and Napper, S. (2012). Detection and control of prion diseases in food animals, *ISRN Veterinary Science*. Hindawi Publishing Limited 2012, 254739. doi: 10.5402/2012/254739.

Ho, S. S., Goh, T. J., Chuah, A. S. F., Leung, Y. W., Bekalu, M. A. and Viswanath, K. (2020). Past debates, fresh impact on nano-enabled food: A multigroup comparison of presumed media influence model based on spillover effects of attitude toward genetically modified food, *Journal of Communication* 70(4), 598.

Hughes, P., Marshall, D., Reid, Y., Parkes, H. and Gelber, C. (2007). The costs of using unauthenticated, over-passaged cell lines: How much more data do we need?, *BioTechniques*. Eaton Publishing Company 43(5), 575. doi: 10.2144/000112598.

Kovac, J., Bakker, H. D., Carroll, L. M. and Wiedmann, M. (2017). Precision food safety: A systems approach to food safety facilitated by genomics tools, *TrAC Trends in Analytical Chemistry*. Elsevier 96, 52. doi: 10.1016/J.TRAC.2017.06.001.

Kroes, R., Renwick, A. G., Cheeseman, M., Kleiner, J., Mangelsdorf, I., Piersma, A., Schilter, B., Schlatter, J., van Schothorst, F., Vos, J. G., Würtzen, G. and European branch of the International Life Sciences Institute. (2004). Structure-based thresholds of toxicological concern (TTC): Guidance for application to substances present at low levels in the diet, *Food and Chemical Toxicology* 42(1), 65. doi: 10.1016/J.FCT.2003.08.006.

Lee, C. G. (2009). Chitin, chitinases and chitinase-like proteins in allergic inflammation and tissue remodeling, *Yonsei Medical Journal*. Yonsei University College of Medicine 50(1), 22. doi: 10.3349/YMJ.2009.50.1.22.

Patlewicz, G., Jeliazkova, N., Safford, R. J., Worth, A. P. and Aleksiev, B. (2008). An evaluation of the implementation of the Cramer classification scheme in the Toxtree software, *SAR and QSAR in Environmental Research*. Taylor & Francis Group 19(5–6), 495. doi: 10.1080/10629360802083871.

Roma, A. A. and Prayson, R. A. (2005). Bovine spongiform encephalopathy and variant Creutzfeldt-Jakob disease: How safe is eating beef?, *Cleveland Clinic Journal of Medicine* 72(3), 185. doi: 10.3949/CCJM.72.3.185.

Ryan, J. (2016). *Biosecurity and Bioterrorism: Containing and Preventing Biological Threats* (2nd edn.). Oxford: Elsevier.

Shay, J. W. and Wright, W. E. (2000). Hayflick, his limit, and cellular ageing, *Nature Reviews: Molecular Cell Biology*. Nature Publishing Group 1(1), 72. doi: 10.1038/35036093.

Singapore Food Agency. (2019). *SFA|Our Singapore Food Story – The 3 Food Baskets, Singapore Food Agency Website*. Available at: https://www.sfa.gov.sg/food-farming

/sgfoodstory/our-singapore-food-story-the-3-food-baskets (Accessed: 18 May 2022).

Snipen, L., Angell, I. L., Rognes, T. and Rudi, K. (2021). Reduced metagenome sequencing for strain-resolution taxonomic profiles, *Microbiome*. BioMed Central Ltd 9(1), 1–19. doi: 10.1186/S40168-021-01019-8/TABLES/1.

Soice, E. and Johnston, J. (2021). Immortalizing cells for human consumption, *International Journal of Molecular Sciences*. Multidisciplinary Digital Publishing Institute 22(21), 11660. doi: 10.3390/IJMS222111660.

Soumyanarayanan, U., Choong, M., Leong, J., Lumpkin, M. M., Rasi, G., Skerritt, J. H., Vogel, S. and Lim, J. C. W. (2021). The COVID-19 crisis as an opportunity to strengthen global regulatory coordination for sustained enhanced access to diagnostics and therapeutics, *Clinical and Translational Science* 14(3), 777. doi: 10.1111/CTS.12954.

Tang, K. L., Caffrey, N. P., Nóbrega, D. B., Cork, S. C., Ronksley, P. E., Barkema, H. W., Polachek, A. J., Ganshorn, H., Sharma, N., Kellner, J. D. and Ghali, W. A. (2017). Restricting the use of antibiotics in food-producing animals and its associations with antibiotic resistance in food-producing animals and human beings: A systematic review and meta-analysis, *Lancet. Planetary Health* 1(8), e316. doi: 10.1016/S2542-5196(17)30141-9.

Wikandari, R., Manikharda, Baldermann, S., Ningrum, A. and Taherzadeh, M. J. (2021). Application of cell culture technology and genetic engineering for production of future foods and crop improvement to strengthen food security, *Bioengineered*. Taylor & Francis 12(2), 11305. doi: 10.1080/21655979.2021.2003665.

Williams, A. J., Grulke, C. M., Edwards, J., McEachran, A. D., Mansouri, K., Baker, N. C., Patlewicz, G., Shah, I., Wambaugh, J. F., Judson, R. S. and Richard, A. M. (2017). The CompTox chemistry dashboard: A community data resource for environmental chemistry, *Journal of Cheminformatics*. BioMed Central 9(1), 61. doi: 10.1186/S13321-017-0247-6.

Wyatt Shields, C., Reyes, C. D. and López, G. P. (2015). Microfluidic cell sorting: A review of the advances in the separation of cells from debulking to rare cell isolation, *Lab on a Chip* 15(5), 1230. doi: 10.1039/C4LC01246A.

Zhu, M. M., Mollet, M., Hubert, R. S., Kyung, Y. S. and Zhang, G. G. (2017). Industrial production of therapeutic proteins: Cell lines, cell culture, and purification, In *Handbook of Industrial Chemistry and Biotechnology*. Nature Publishing Group, p. 1639. doi: 10.1007/978-3-319-52287-6_29.

Chapter 2

The regulation of genome-edited crops

Gregory Jaffe, Center for Science in the Public Interest, USA

1 Introduction

Scientists are applying genome editing to crops with the hope and expectation that some of those new crops will provide benefits when released as commercial products. National governments need to determine whether, and if so, which of those products require regulation because they are made using genome-editing technologies. It is the job of regulators to ensure that only products that are safe for humans, animals, and the environment enter the market. Should they apply existing national laws that regulate genetically modified organisms (GMOs)? Should they regulate genome-edited products in the same fashion as they regulate conventionally bred crops? Or should they develop new regulatory procedures and requirements that are uniquely tailored to genome-edited products? How to regulate genome-edited crops is complicated by the fact that there are many different genome-editing techniques that produce crops with a full spectrum of different types of genomic changes -- including but not limited to deletions, additions, and substitutions. Products of genome editing have different genotypic changes and phenotypic traits. This chapter will provide background on genome editing and the debate over oversight of genome-edited crops. Then, it will describe how the Cartagena Biosafety Protocol might apply to genome-edited products as well as the regulatory oversight adopted by different countries. Finally, it will analyze the different regulatory systems of countries and describe issues that arise from those regulatory oversight policies and procedures.

http://dx.doi.org/10.19103/AS.2020.0082.16

2 Background on genome editing and its regulation

Before discussing the regulation of genome-edited crops, it is important to provide some information about genome editing as well as some information about the international discussions that have occurred in the past few years on the issues surrounding the regulation of genome-edited products. This section will start by describing what is meant by genome editing and the range of changes that can be made to an organism's genome. Next, there will be a discussion about regulation, including how regulation of genome-edited organisms fits into broader regulations of crops, food, and products from agricultural biotechnology. Finally, the section will touch on the complexities of the discussion among government officials and different interested stakeholders on how to appropriately regulate genome-edited products.

2.1 What is genome editing?

Genome-editing (also called 'gene editing') techniques allow scientists to make precise changes to an organism's genetic material at a pre-programmed location. Those techniques can be used to turn genes on or off, modify the regulation and function of a gene, and introduce genes that confer new traits. A neutral definition that encompasses the many different types of genome editing is the definition from a recent US National Academy of Sciences report: 'Specific modification of the DNA of an organism to create mutations or introduce new alleles or new genes' (NASEM, 2017).

To make desired changes in an organism's genome, scientists use genome editing to recognize a specific location in the organism's genome and cut the DNA at that location. Currently, the most widely used genome-editing technique is called clustered regularly interspaced short palindromic repeats (CRISPRs), although transcription activator-like effector nucleases (TALENs) and zinc finger nucleases (ZFNs) are also used at a lower frequency. New editing technologies will most likely be discovered and employed in the coming years.

Once a cell's DNA is cut, it is normally repaired perfectly by its own DNA repair mechanisms. At times, however, the repair mechanisms can make mistakes that create desirable edits. For example, the cell can repair the cut in a somewhat random fashion, resulting in an addition or deletion as small as a single base-pair or as large as hundreds of base-pairs. A random repair can also mistakenly result in small base-pair substitutions. Alternatively, the cell can repair the break in its DNA using a repair template that is designed by the scientist and introduced into the cell along with the genome-editing components. The template can be designed to result in a predetermined deletion of a section of the cell's existing DNA. It can also be designed to result in a substitution of one or a few targeted base-pairs to correct or modify a mutant gene. The template can also include one or more genes from the same or a different species to be inserted into the organism's genome at the location of the cut or to swap one allele of a gene for a different allele. For example, an additional copy of an existing gene could be added to increase production of the product that the gene codes for. Finally, some genome-editing techniques only introduce into the cell the specific guide RNA molecule (without its corresponding DNA) to locate and cut the genome in a particular spot. (Metje-Sprink et al., 2019). The different outcomes from genome editing a crop are summarized in Fig. 1.

The term 'genome editing' in this chapter does not have the same meaning as the commonly used terms 'genetic engineering', 'genetically modified', or 'genetically modified organism' (typically referred to as 'GMOs'). While both the scientific and legal definitions of 'genetic engineering' or 'GMO' vary throughout the world, in this chapter, 'genetic engineering' of plants will refer to a process whereby scientists extract a gene with a useful trait from an organism and insert it into the genome of a different organism most often using either an *Agrobacterium* or the 'gene gun'. There is no control over where this insertion happens in the genome. Genome editing can also insert a new gene into an organism's genome, but it does so more precisely, because breaks are made in the DNA at specific targeted locations and a template with the new additional gene is added to the cell so it is copied into the DNA when repairing the break. Therefore, some genome-edited products can mimic genomic changes that occur through traditional genetic engineering, such as the addition of a single gene from a different species (often referred to as a 'foreign' gene or a transgene). In other instances, the changes caused by genome editing can mimic

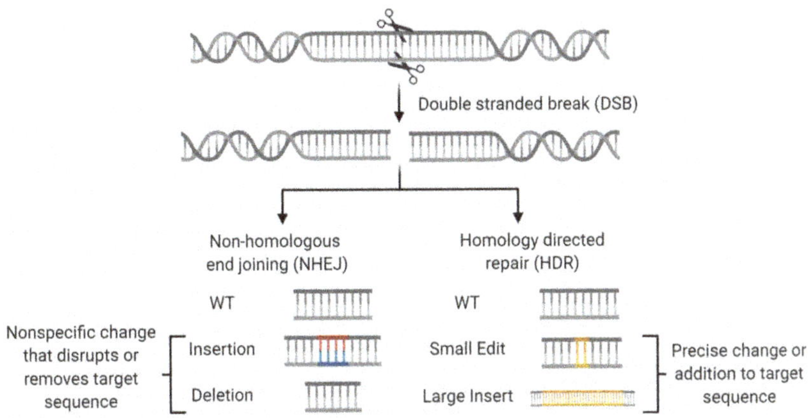

Figure 1 Diagram of most common genome editing and the different genomic outcomes. After the double-stranded break (DSB), non-homologous end joining can result in a return to the wild type (WT) or a random deletion or insertion. Homology directed repair involves a template which introduces specific edits (usually small or large additions). Image courtesy of Morgan Carter and Biorender.

those produced through conventional breeding techniques, such as chemical mutagenesis or irradiation (which result in random changes to an organism's DNA as the result of inaccurate DNA repair). Understanding how genome-edited organisms compare to organisms produced with other breeding techniques has been an important part of the discussion on how those products are regulated.

In addition to making a single change in an organism's genome, genome-editing techniques can produce targeted genomic changes - deletions, substitutions, or additions - at one or more locations in an organism at the same time (Putcha, 2016). A scientist also can stack multiple genome-edited changes into a single organism over several generations (which can also be done with traditional genetic engineering). Therefore, the degree of change in a genome-edited organism's genome can vary tremendously from a single gene edit to multiple edits. The type and degree of change is important both in determining potential risks from the organism as well as how it might be regulated. As stated in a report on the bioeconomy by the World Economic Forum (2018):

> [D]epending on the type of graduated change in an organism's genome, gene editing could produce a product identical to one created with 'traditional' mutagenesis or a product identical to a transgenic GMO...No longer are there clear distinctions for 'transgenic methods,' conventional breeding, biotechnology, and other methods and techniques to develop improved crops and animals. (WEF, 2018)

Thus, genome editing does not lead to monothetic products, but instead can achieve many different outcomes. It is just a tool that can be utilized to make a wide variety of genomic changes. This acknowledgement of varied scientific outcomes is critical for any discussion of regulation because policymakers and regulators need to decide whether to treat all products of genome editing identically or differentiate between them based on the wide-ranging changes and outcomes possible.

Finally, for genome editing, while modification can be achieved at a precise location, modifications, referred to as 'off-targets' can inadvertently occur at other places in the organism's DNA, especially at places which have a similar base-pair sequence as the targeted location. Those off-target changes often have no phenotypic impact on the organism, but investigation usually is required to confirm there is no inadvertent adverse impact. Predicting and analyzing the impact of off-targets is an issue that has been raised by scientists carrying out genome editing as well as stakeholders arguing for and against the need for genome-editing specific regulation.

2.2 What is regulation and why focus on genome-editing specific regulation?

A government's legal authority to regulate comes from laws passed by the Legislative branch of government and enacted by the Executive branch.

Regulatory obligations are either set forth in those laws or in regulations, which are formal orders or rules usually from a government agency in the Executive branch meant to oversee either a product or the conduct of an individual. There are many reasons governments may regulate a product or activity, but this analysis will focus on regulations established to ensure an independent check of a product's or activity's safety to humans, animals, and/or the environment.[1]

Government regulation can take many forms. Many people associate safety regulation with a process in which the government conducts a risk assessment and decides on whether to allow a product to enter commerce based on scientific evidence. However, there are many other forms of government oversight, some that are less stringent and others that are more stringent. Regulation can involve providing information so that the government and the public are informed about products and potential risks. The U.S. Surgeon General's warnings on tobacco and alcohol products are examples of this type of regulation. Other regulations involve the government establishing general conditions for safe conduct (such as research standards) and then requiring individuals and institutions to demonstrate compliance with those standards. Pharmaceutical companies are subject to these types of regulations, which ensure drugs have been adequately tested for safety and efficacy before they can be prescribed. At the other extreme, regulation can prohibit or greatly restrict certain products or conduct. For example, the U.S. Environmental Protection Agency bans certain chemicals, such as polychlorinated biphenyls (PCBs), if they demonstrate excessive risks to human or environmental health.

Food and agricultural products derived from genome-edited crops, like food and agricultural products produced using other techniques, will be subjected to various general laws and regulations. Most countries have laws and regulations to ensure the safety of food that enters commerce. These types of regulations make no distinction between foods produced with genome editing, conventional breeding, or any other method of producing a crop or animal. Similarly, seeds produced using genome editing would be regulated as all new seed varieties are regulated, using a country's varietal protection laws, plant protection laws, or other laws involving the safety, purity, and transport of seed.

This chapter does not discuss laws that will generally apply equally to genome-edited products and other similar products in a product category, such as food, plants, animals, seeds, and so on. Instead, it will focus on regulations that are tailored specifically to genome-edited products, meaning

1 Different countries use different terminology for legislation enacted by the legislative branch, and rules or orders set forth by the executive branch. For example, some countries refer to 'Act' to describe a law while others call similar legal instruments 'Directives.' Legal documents that qualify as regulations might be called regulations in some countries and 'resolutions' in others. In this document, legal documents will be referred to as they are identified by each country. When not discussing a specific document from a country, the terms for legal documents in the United States will be used in the text.

the regulatory requirements only apply because the product was genome edited. The most obvious example of laws and/or regulations that apply to a subcategory of products produced in a specific way is biosafety laws that regulate GMOs. As discussed below, these laws in some countries, but not others, encompass some genome-edited products. Another example of specific regulation of a sub-category of products is the regulation of pesticides. Pesticides are chemicals that could be regulated under a general chemical law. However, many countries often have special regulations for pesticidal chemicals because of their unique properties and purpose, which is to kill a pest.

The issue discussed in this chapter with respect to genome-edited products is not whether they should be subjected to any regulation, but whether they should be subjected to a country's current GMO regulations or have some other genome-editing special regulatory process due to the fact that the product was made with genome editing. A reason for genome-edited products to get special regulation would be that those products and/or the process by which they are made pose potential risks that might not be adequately covered by generally applicable laws. It is outside the scope of this chapter to include a discussion of the potential risks from genome editing and genome-edited products and whether those potential risks are different or similar to potential risks from other breeding methods and the products made from them. However, the next section does briefly describe the different perspectives on how genome-edited products should be regulated.

2.3 International debate about regulating genome-edited agricultural products

In the global discussion over whether and how to regulate genome-edited agricultural products, the question dominating the discussion has been whether and in which cases genome-edited products should be considered 'GMOs' and fall within a country's national biosafety regulatory systems for GMOs (WEF, 2018; Eckerstorfer et al., 2019). If they are not 'GMOs', then the question is whether they should be regulated as conventionally bred products or whether a country should establish new regulations specifically for genome-edited products.

Many scientists and seed developers, and some governments, have argued that if a molecular genome-editing technique is used to alter a crop's existing genome but does not introduce DNA from a different species ('foreign' DNA), the resulting product is no different from a crop with its genome altered using other traditional breeding methods. Chemical mutagenesis and ionizing radiation, which cause random changes (additions and deletions) in an organism's genome, are currently exempt from product-specific regulation in most countries around the world. Some argue that molecular genome-editing techniques that create similar products (but with greater precision) should also be exempt. This position was articulated by the American Seed Trade Association, which wrote: 'Plant varieties developed through the latest breeding methods should not be differentially regulated if they are similar to or indistinguishable from varieties that could have been produced through earlier breeding methods' (ASTA, 2020). For advocates who argue that many genome-edited products are not GMOs, the criteria they advocate using to show they are not GMOs include: (1) whether there is 'foreign' DNA in the final product; (2) whether the genome edit could have been 'found in nature'; or (3) whether the edit could have been produced through 'conventional' or 'traditional' breeding methods, including chemical mutagenesis and irradiation breeding (WEF, 2018).

On the other side of the debate, some scientists, consumer and environmental groups, and other stakeholders assert that genome-editing techniques are a category of 'genetic engineering' that could result in harmful products and should, therefore, meet the substantive legal obligations required for GMOs. As stated by the Non-GMO Project: 'Regardless of whether foreign DNA is used, though, any process where nucleic acid is engineered in the laboratory is genetic engineering, and the resulting products are GMOs' (NGMO, 2020). In fact, some of these stakeholders even argue that genome editing could be more dangerous than genetic engineering and should have more stringent regulatory oversight. A Friends of the Earth report claimed that genome editing had 'the potential for significant unintended consequences' and that CRISPR might cause 'genetic havoc' (FOE, 2018).

There is also a middle ground in the debate. That position acknowledges that genome-edited products may need some sort of specific regulation but that the laws and regulations that have been put in place around the world for GMOs are not appropriate. As stated by the World Economic Forum's Bio-Innovation Report:

> Appropriate science-based and proportionate oversight could provide predictable and transparent rules that enable developers to bring products to market and an independent safety assessment that consumers can trust. Alternatively, overly stringent and costly regulation could prevent many beneficial products from entering the marketplace, while no regulation could cause consumers to question their safety. (WEF, 2018)

In this middle ground, regulation could be established that is tailored to the potential risks, including the level of novelty of the techniques and products, and any other issues that arise from the various genome-editing techniques and the resulting products.

3 Whether the Cartagena Biosafety Protocol applies to genome-edited crops

While most countries in the world have not yet determined what type of regulatory oversight, if any, is or will be required for genome-edited products, many countries are discussing the issue and several countries already have established regulatory policies. This section analyzes whether the Cartagena Biosafety Protocol (CBP) - an international agreement that aims to ensure the safe use and trade of living modified organisms - applies to genome-edited organisms. Many countries around the world have biosafety regulatory systems that implement the requirements of the CBP and directly adopt many of its substantive provisions (including key definitions). If the scope of that international agreement includes some or all genome-edited products, that will determine how many countries regulate such edited products.

The Cartagena Biosafety Protocol, which is a sub-agreement to the United Nations' 1992 Convention on Biological Diversity (CBD), is an international treaty enacted to ensure 'an adequate level of protection in the field of the safe transfer, handling, and use of living modified organisms (LMO)'[2] (CBP, 2000). While the obligation to establish national frameworks for biosafety stems from Article 8g of the CBD, for many countries, the CBP has been the starting point for their GMO oversight (CBD, 1992). The CBP establishes a regulatory framework for carrying out scientifically sound risk assessments to identify and evaluate potential adverse effects of LMOs on biodiversity in the context of transboundary movement of LMOs. It was adopted in 2000 and came into force in 2003 when 50 countries had ratified it. As of the end of 2019, 172 countries and the European Union are parties to the CBP, but the United States is not (CBP, 2020).

Many countries have adopted laws and regulations that implement the CBP's provisions and those national laws and regulations utilize many of the terms and definitions from the CBP. With the development of genome editing, there have been discussions both at international meetings involving the CBP and within countries about to what extent the CBP covers organisms resulting from the different types of genome-editing techniques. Analyzing whether the CBP includes organisms developed through genome editing can help determine whether countries whose biosafety laws mirror provisions from the CBP likely will include organisms developed through genome editing. Examples of countries whose biosafety regulatory systems are based largely on the CBP include Argentina and Kenya.

2 The Cartagena Protocol uses the term 'living modified organism'. Almost all countries implementing the Cartagena Protocol use the term 'genetically modified organism' as well as most other international discussions about this topic. In virtually all instances, there is no scientific or substantive distinction between an LMO and a GMO. Therefore, except when specifically referring to language in the Cartagena Protocol, this article uses GMOs to be consistent with international best practice.

To determine whether the CBP's provisions incorporate genome-edited organisms, one must analyze whether those organisms fall within the CBP's definition of an LMO, which is defined as 'any living organism that possesses a novel combination of genetic material obtained through the use of modern biotechnology'. The two critical phrases in this definition are 'novel combination of genetic material and 'obtained through the use of modern biotechnology'. 'Novel combination of genetic material' is not defined in the CBP or the CBD and so that phrase is let up to individual countries to interpret its scope.

The CBP defines 'modern biotechnology' as the application of:

a In vitro nucleic acid techniques, including recombinant deoxyribonucleic acid (DNA) and direct injection of nucleic acid into cells or organelles, or
b Fusion of cells beyond the taxonomic family,

that overcome natural physiological reproductive or recombination barriers and that are not techniques used in traditional breeding and selection (CBP Article 3, 2000).

Analyzing the definition of 'modern biotechnology', it is difficult to argue that genome editing is not included. Most genome-editing techniques use '*in vitro* nucleic acid techniques' or 'recombinant DNA' so they satisfy part (a) of that definition. Also, it is hard to argue that they are 'traditional breeding and selection techniques'. In addition, they overcome natural recombination barriers because they make specific targeted genomic changes in an organism's genome and those changes can be done at a time in the cell's life cycle when natural recombination does not occur. Therefore, whether a genome-edited organism is subjected to regulation under the CBP as an LMO will hinge on the interpretation of the phrase 'a novel combination of genetic material' in the definition of 'living modified organism'.

To determine if a genome-edited crop is 'novel', one could argue that unless the genome-edited change is identical to an existing allele that has been scientifically documented in an individual organism, it is a novel combination of genetic material. This would make most genome-edited organisms qualify as LMOs, even ones caused by random DNA repair unless the random change results in an allele whose existence already has been documented in that species. Alternatively, one could argue that if the change could arise as a mutation naturally because the genetic alteration is a limited small change to the organism's existing genome without incorporating foreign DNA, it is not a novel combination of genetic material (even if the altered allele has not yet been documented). That interpretation would exempt from the definition of LMOs most genome-edited organisms with insertions or deletions produced during natural DNA repair as well as alleles that already exist within the species. Only recombination that crosses species barriers or involves foreign DNA would qualify as an LMO. However, this argument becomes more difficult to make when multiple genes are edited in an organism, because as more edited genes are placed (or 'stacked') into one organism, the likelihood of that organism's genomic changes occurring naturally decreases. Of course, the issue arises of whether you look at each genomic change individually or the cumulative change in the organism's genome to determine novelty.

Therefore, the CBP definitions do not provide a clear answer about the extent to which they capture gene-edited organisms because 'novel combination of genetic material' could have varied interpretations, some more and some less inclusive. Novelty could also vary with time because genomes are constantly changing. Until there is a decision by the collective membership of all the countries who are parties to the CBP about whether, and to what extent, to include genome-edited organisms, countries will have to choose the interpretation they believe is correct and most appropriate for their situation.

For all genome-edited organisms which do qualify as LMOs, the CBP requires scientific risk assessments and advanced informed consent by a country before the LMO is released into the environment. It also requires that the identity of those organisms and the decision made to approve them be made publicly available. The CBP sets up a Biosafety Clearinghouse mechanism which contains that information and most regulators also make that information available to the public through public notifications, public dockets, and/or websites (CBP Article 20, 2000). Therefore, for countries that signed onto the CBP, if a genome-edited product satisfies the definition of an LMO, the approval of that product will be transparent and available to the public. However, if a genome-edited product is not an LMO and does not fall within the scope of the CBP, the individual country's regulatory system will determine the degree of transparency as discussed in the country case studies below.

4 Country case studies demonstrating regulation of genome-edited crops

This section describes the regulatory policies for genome-edited products of several 'leader' countries.[3] The countries were chosen to represent the range of regulatory systems currently adopted for genome-edited products around the world. Those country systems are likely to become models that other countries around the world look to and potentially replicate when establishing their own regulatory policy.

3 While the focus of this chapter is on the regulation of genome edited crops, the CBP and most country regulatory systems apply equally to crops and animals. Where a country has regulations or other documents that address genome edited animal oversight, they are included briefly in the discussion in this section. If no reference is made to animals, it can be assumed that the regulatory system being discussed treats crops and animals in an identical manner.

Before reading about the different national regulatory systems set forth below, it is important to understand that each country cannot start from scratch and simply ask the question 'What regulation, if any, is needed for genome-edited agricultural products?' Countries already have enacted laws and regulations that may apply to genome-edited products. A country with a biosafety law that defines GMOs broadly may have no choice but to regulate genome-edited products as GMOs (at least until that law is amended or repealed). Another country may have a narrow definition of GMOs in their biosafety law and could determine whether to extend the law to cover genome-edited products or to regulate genome-edited products de novo. A third country's biosafety law may not include genome editing, but a different law, such as a law regulating new technologies, may apply to genome editing. Therefore, when reading about the different national regulatory systems set forth below, keep in mind that how a country decides to regulate genome-edited products must be viewed within the context of that country's existing laws and regulations. At the end of the section, Table 1 summarizes key aspects of the regulatory systems for genome-edited crops in the different countries.

4.1 Argentina - mandatory consultation to determine which products are genetically modified organisms (GMOs)

Argentina has a functional biosafety regulatory system that regulates GMOs and Argentina has been growing and exporting GMOs for many years. That biosafety regulatory system involves multiple laws and regulations implemented by several agencies (Burachik, 2012), including Resolution 763/2011 issued by the Ministry of Agriculture. Argentina's regulatory system defines a GMO using the CBP definitions of 'living modified organism' and 'modern biotechnology'. To obtain a permit to commercialize a GMO, the government conducts and weighs the results of three assessments: (1) an assessment on potential impacts on the agroecosystem; (2) an assessment of food safety; and (3) an evaluation of potential impacts on Argentina's international trade (CRS, 2015).

Argentina was one of the first countries to set forth a regulation establishing how it would regulate genome-edited plants. After several years of internal government discussion and external consultation with stakeholders, Argentina decided that every genome-editing technique utilizes in vitro recombinant DNA and fulfills that portion of the 'modern biotechnology' definition, but not every genome-edited product necessarily contains a 'novel combination of genetic material'. They then established 2015 Resolution no. 173/15, a regulation applicable to Innovative Biotechnology/New Breeding Techniques ('NBT') for plants. (AMALF, 2015), which sets forth a regulatory procedure for genome-edited crops different from that for GMOs. In addition to relying on the definition of a GMO to justify their new regulatory process, they also relied on the definition of an 'event', which is 'the insertion into the plant genome in a stable and joint form, of one or more genes or DNA sequences that are part of a defined genetic construction' (AMALF, 2015). In 2017, Argentina updated its animal biotechnology regulatory system with Resolution no. 79-E/17, which established the same regulatory oversight procedures for animals produced using NBTs (AMALF, 2017).

Resolution no. 173/15 sets up a 60-day consultation process with the National Advisory Commission on Agricultural Biotechnology (CONABIA) to determine if the NBT product has a new combination of genetic material. The developer submits a detailed application with data on (1) breeding methodology; (2) the introduced trait or characteristic; and (3) evidence of genetic change in the final product (Article 2). The regulator then looks at this data and determines whether there is a stable insertion of one or more genes or DNA sequences that were part of a template inserted into the genome with the genome-editing components. The developer must also show that no component molecules used to genome edit the organism remain in the DNA of the final plant product (such as a temporary transgene) (Article 3). The regulation assumes that if the plant is a GMO for any stage of its development, it remains a GMO until there is scientific evidence to show the introduced DNA is gone (Article 3) (AMALF, 2015). Finally, the regulation does not define NBTs, which allows for flexibility to encompass future techniques and arguably prevents the regulation from becoming outdated as the science develops (Whelan and Lema, 2015). Examples of current NBTs that are not genome edited, but are included in the consultation process are grafting, reverse breeding, and agro-infiltration (Whelan and Lema, 2015). The consultation process set up by Resolution no. 173/15 is summarized in Fig. 2.

To submit a consultation package, an applicant must be registered under a National Registry of Operators with Genetically Modified Plant Organisms (Resolution no. 46/04, Article 6). An applicant can submit an inquiry about a hypothetical product for a preliminary ruling, but must then submit a full consultation with scientific data for a final ruling before a product may enter commerce (Article 7) (AMALF, 2015). Argentina's regulatory body established a consultative form in a 2019 Administrative Order to standardize the information to be submitted. That form requires the expected molecular biology information to show that no foreign DNA has been added but it also requires phenotypic information, such as the organism's new phenotype, examples of products on the market with the same phenotype, analysis of any other potential impacts beyond the anticipated phenotype, and any expected changes to manage the resulting organism (AMALF, 2019a). This information is helpful to the regulator because they have the responsibility, even if the genome-edited crop is not a GMO, to recommend, based on scientific grounds, that the Ministry of

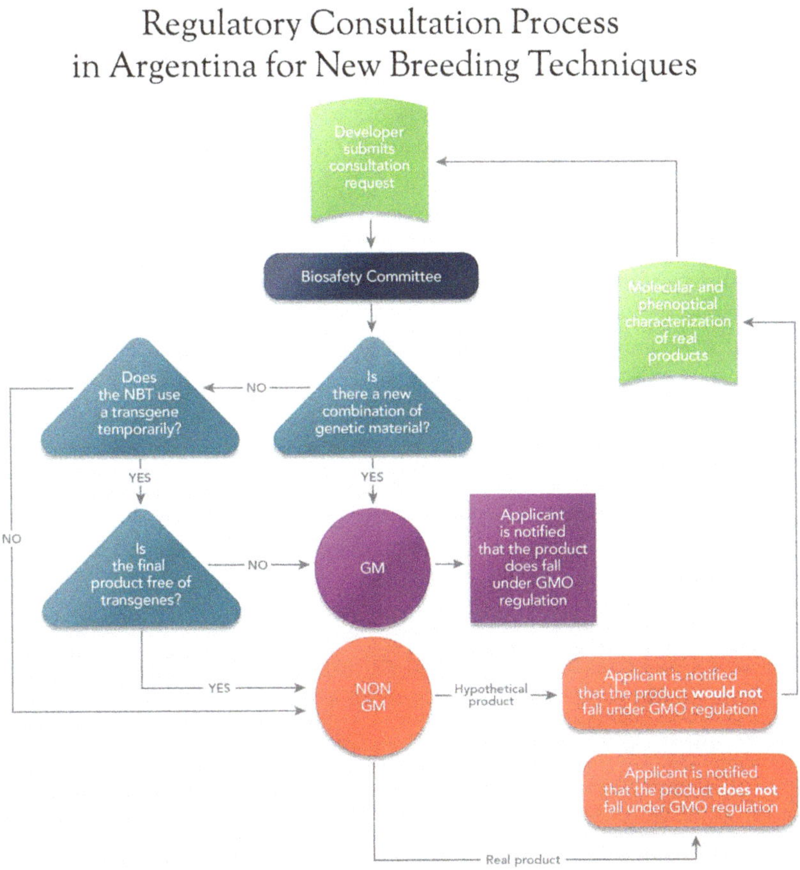

Figure 2 Regulatory consultation process in Argentina for new breeding techniques. Source: Adapted from Whelan, A. I., and Lema, M. A. (2019). Regulation of Genome Editing in Plant Biotechnology: Argentina. In 'Regulation of Genome Editing in Plant Biotechnology' (pp.19-62), Springer, Cham.

Agriculture follow up on the crop where its traits and/or novelty make it worth monitoring. (AMALF, 2019b).

Applying the resolution's language to the different types of changes possible to a crop's genome from genome editing, it is likely that: (1) deletions of any size in an organism's genome will not be considered GMOs; (2) small additions may or may not be considered GMOs depending on whether they were with the addition of an introduced DNA template; and (3) introductions of one or more genes will be considered GMOs, unless the introduced gene can be shown to be a '"perfect allelic replacement" completely indistinguishable from the outcome of a sexual crossing' (Lema, 2019). The genome-edited crops that are determined to be a GMO require government approval after completing the three assessments discussed above.

It is impossible to review how this system has operated in practice because the Argentina regulatory system does not provide detailed information about products reviewed under Resolution no. 173/15. The public does not have access to the product submissions nor the decisions by the regulator. Therefore, it is impossible for the public to know which products are or are not determined to be GMOs and what scientific evidence informed those determinations. It is unclear how the regulator has treated products with multiple edits, or how many base-pairs can be added before the DNA sequence is considered 'novel'. Similarly, one cannot determine what analysis was made about off-targets and any impact they might have. Finally, the lack of transparency makes it impossible to know which products are entering the market and could cause trade problems if those products end up in markets where they have not yet satisfied regulatory requirements. Therefore, while Argentina has a functional and operational case-by-case process for determining oversight of genome editing and has reviewed a number of products, the lack of transparency makes it impossible to judge how well the system is working and how decisions are being made.

4.2 Australia – some products are exempt and others are genetically modified organisms (GMOs)

Australia regulates GMOs under the Gene Technology Act of 2000 and has approved GMO varieties of cotton and canola that are grown in the country. That statute regulates organisms modified by 'gene technology', which is defined as 'any technique for the modification of genes or other genetic material' (GTA, 2000). A plain reading of that definition would include not only genetic engineering, but also genome editing and even some traditional breeding techniques such as mutagenesis. However, in the Gene Technology Regulations of 2001, certain techniques, such as radiation-induced and chemical-induced mutagenesis, are exempt (GTT, 2001).

In 2019, the Gene Technology Regulations were amended to address organisms modified with genome editing (GTA, 2019). Those regulations now exempt two types of organisms: (1) a 'mutant organism in which the mutational event did not involve the introduction of foreign nucleic acid'; and (2) 'an organism modified by repair of single-stranded or double-stranded breaks of genomic DNA induced by a site-directed nuclease, if a nucleic acid template was not added to guide homology-directed repair' (GTA, 2019). This means that when site-directed nuclease techniques are used to modify an organism's genome without any nucleic acid template, the organism is not a GMO. However, for this to apply, the site-directed nuclease technique must involve a protein and not the introduction of a transgene or its expressed product. An organism produced using a transgene is regulated as a GMO if the transgene is integrated into the final product or if it is transiently present. When the transgene and its product are degraded, then the organism is no longer a GMO. So, if the Cas9 and guide RNA come from DNA that is either integrated in the genome or is translated in the cell, it is a GMO. However, if the organism's DNA break is caused by the introduction into the cell of a Cas9 protein and a guide RNA without the DNA that codes for the guide RNA, it is not a GMO. Similarly, using TALENs/ZFNs would be exempt if the proteins are delivered directly into the cell without DNA. With genome editing involving a nucleic acid template, the organism is a GMO and whether the template contains a native or foreign segment of DNA is immaterial.

The 2019 amendments to the regulations also state that an organism is a GMO if the 'organism is modified using oligonucleotide-directed mutagenesis' or if the 'organism is modified with site-directed nuclease techniques involving templates to guide the repair action' (Australia, 2019). As the Australian regulatory office stated:

> In each case, the method used to modify the organism is central to determining whether or not the organism is a GMO. The number of resulting nucleotide changes, whether insertions or deletions, or whether the resulting nucleotide sequence may be found in sexually compatible species is not a deciding factor. (OGTR, 2019)

Therefore, Australia amended its regulations to address genome editing by determining that GMO status will hinge on whether a template has been introduced to change the organism's DNA. It does not depend on whether the introduced DNA is native or foreign. It also specifies which genome editing techniques produce organisms that are exempt from oversight and which ones are treated as GMOs. Those genome-edited crops treated as GMOs under the regulations require submission of an application and the granting of a license by the Office of the Gene Regulator before conducting any activities with the product.

4.3 The European Union – regulates genome-edited products the same as GMOs

The European Union has an elaborate biosafety regulatory system that regulates GMOs using several different Directives and Regulations, including 'EU Directive 2001/18 on the Deliberate Release into the Environment of Genetically Modified Organisms' (EU, 2001) and 'EU Regulation 1829/2003 on Genetically Modified Food and Feed' (EU, 2003). The EU defines GMO as 'an organism, . . . in which the genetic material has been altered in a way that does not occur naturally by mating and/or natural recombination' (EU, 2001). There is mandatory pre-market approval for all GMOs based on detailed science-based risk assessments. In addition, there is an exclusion from the substantive provisions of Direction 2001/18 for organisms produced using mutagenesis if they do not involve the use of recombinant nucleic acid molecules (EU, 2001).

European regulators' first discussion on how to apply these directives to genome-edited products started in 2007 in the New Techniques Working Group (NTWG), established by the European Commission. Parallel to the discussions in the NTWG, the European Food Safety Agency (EFSA) issued a technical report in 2012 about the scientific safety assessment needed for plants developed with the addition of a gene using ZFNs. Their conclusion was that the environmental and food safety of those products was appropriately addressed with the risk assessment requirements set forth for GMOs (EFSA, 2012).

Before European Union regulators could determine administratively whether they would apply the same oversight for genome-edited products that they conduct for GMOs, nine French NGOs brought a court case in France in 2016 that asked, among other things: (1) whether EU Directive 2001/18 applies to organisms developed through mutagenesis; (2) what the scope of the mutagenesis exemption in the Directive is; and (3) how these provisions apply to genome-editing techniques.

The case was elevated to the European Court of Justice (ECJ), who issued a decision in July 2018 (ECJ, 2018). The question before the ECJ was whether the exemption from being required to comply with the substantive requirements of the Directive for certain mutagenesis techniques applies to mutagenesis from modern breeding techniques, such as genome editing. The ECJ confirmed that organisms developed though mutagenesis are GMOs but exempt from any substantive oversight. It also decided that all genome-edited products were GMOs that did not qualify for the mutagenesis exemption (ECJ, 2018). Therefore, genome-edited organisms are GMOs and are required to comply with all the obligations required for GMOs, including the requirements to obtain pre-market approval based on a risk assessment.

To reach this decision, the ECJ found the following. First, the Directive includes as GMOs organisms made by mutagenesis methods and so all organisms developed with mutagenic techniques, including those made from chemical-induced or radiation-induced mutagenesis, are GMOs. Then, the ECJ said that the Directive exempts a subset of mutagenesis techniques 'which have been conventionally used in a number of applications and have a long safety record' from the Directive's substantive requirements (i.e. they are still GMOs but don't need any oversight). That exemption applies to conventional mutagenesis breeding methods, such as chemical mutagenesis, because the Parliament determined they had a long safety record. However, the ECJ found that some mutagenesis techniques, including genome-editing techniques, are new, do not have a long safety record, and thus do not qualify for the exemption. Therefore, they are GMOs under the Directive and must meet all the substantive requirements (ECJ, 2018). This means that the 'mutagenesis exemption' is frozen in time and new techniques cannot qualify for it (at least arguably not until they have a long history of safe use).

As the highest Court in the EU, the ECJ decision is the law and the situation can only be changed if the EU Parliament either amends the Directive or passes a new Directive specifically addressing genome-edited products. Numerous stakeholders in Europe who believe the ECJ decision was wrongly decided are mobilizing to change the regulatory system, but it is unclear whether they will be successful (EUSAGE, 2019). In November 2019, the EU Council of Ministers asked the European Commission to conduct a study on the status under Union Law of organisms developed through genomic techniques (EU, 2020).

The European Commission began consultations in 2020 on how it should regulate organisms developed with genome editing. Any legislative change would take several years to adopt and implement. EU regulators could more quickly make administrative changes that would reduce the regulatory requirements for categories of genome-edited products as a quicker interim measure (Wasmer, 2019). However, until changes are made legislatively or administratively, all genome-edited products will require pre-market approval and be regulated with the same procedures and requirements as GMOs. Those procedures require the submission of an application with detailed information about the new variety, the completion of a detailed risk assessment, an opportunity for public review and comment on the submission, and then a decision by EU officials.

4.4 Japan – mandatory notifications to determine whether a product is a GMO

Japan is a party to the CBP and has three agencies – Ministry of Environment (MOE), Ministry of Agriculture, Forestry and Fisheries (MAFF), and Ministry of Health, Labor and Welfare (MHLW) -- that regulate GMOs. In 2004, it established its Cartagena Act, which defines an LMO as 'living organisms having nucleic acids obtained by utilizing a technique for processing nucleic acids outside the cell for the purpose of transferring or replicating the nucleic acids by transferring them into a cell, virus, or viroid' (Tsuda et al., 2019). Japan does not grow any GMOs, but it has approved them to be imported and to be used in food and feed.

A plain reading of Japan's definition of an LMO suggests that most, if not all, genome-edited plants are GMOs. However, in 2019 and 2020, the three agencies each established policies for genome-edited products. Their policies each require a case-by-case review of genome-edited products to see if they qualify as a GMO, but each agency policy contains different details on what information and analysis the agency will conduct.

The MOE, which implements the CBP in Japan, set forth a policy in 2019 entitled, 'About the Handling of Organisms Produced by the Use of Genome

Editing Technology that Do Not Meet the Definition of "Genetically Modified Organism" in the Cartagena Act' (Tsuda et al., 2019). According to the policy, 'any organism that has inserted extracellularly nucleic acid (including RNA) is an LMO' unless there is 'complete removal of the inserted nucleic acid (including RNA), or its replicated product' (MOE, 2019). Japan interprets this to mean that if the genome is cut with an artificial nuclease without any foreign nucleic acid, and if the repair is random without a template being introduced into the cell, the resulting organism is not a GMO. However, if the product is developed using an artificial nuclease with foreign DNA, even if the DNA repair occurs naturally, it is a GMO. Similarly, if any inserted nucleic acid is incorporated into the final product, it is a GMO regardless of whether the artificial nuclease cutting the nucleic acid has foreign nucleic acid. There is no minimum size requirement for the extracellularly processed nucleic acid (Tsuda et al., 2019).

For genome-edited products not considered as GMOs, the MOE still has some regulatory requirements. If the genome-edited product will be used in a contained facility, it must follow the same containment measures as GMOs. If the genome-edited product is to be released outside the laboratory, then the developer must submit the following information and wait for approval to proceed from MOE:

a the fact that the organism does not possess remnants of extracellularly processed nucleic acid or its replicated product,
b taxonomical species of the modified organism,
c method of genome editing used for the modification,
d modified gene and its function,
e change in traits added by the modification,
f whether there is a change of traits other than those described in (e),
g usage of the organism, and
h discussion on possible influences on biological diversity when the organism is used (MOE, 2019).

If the organism could have any risk to biodiversity, 'the competent national authority will require additional information from the user; then, necessary measures can be taken'. MOE will post information about each organism after it has been reviewed by MOE on the Japan Biosafety Clearinghouse (it will make public items b, e, g, and h in the above information list) (MOE, 2019). The scientific data to be submitted to MOE to establish that there is no introduced nucleic acids and no impacts from off-targets are yet to be determined by MOE. The MOE's regulatory pathway for determining the regulatory status of a genome-edited crop is summarized in Fig. 3.

The Ministry of Agriculture, Forestry and Fisheries (MAFF), which has primary jurisdiction over agriculture, forests, and fish products, released its procedures for the handling of genome-edited organisms in October 2019

(Notification No. 2743) in compliance with the Japan's Cartagena Act (FAS, 2019a). That policy closely tracks the MOE policy and requires notification with the same information that is submitted to the MOE, which is used by the MAFF to decide whether there is no extracellularly processed nucleic acid or a replicated product in the genome-edited organism. The policy also reiterates that containment measures from Article 12 of the Japan's Cartagena Act are required, even for gene-edited organisms that do not qualify as GMOs. Finally, the policy states that if new breeding technologies are developed in the future to modify organisms, the MAFF should be contacted to determine what information is needed (FAS, 2019a).

In 2019, Japan's Ministry of Health, Labour, and Welfare (MHLW) released its final policy titled, 'Food Sanitation Guidelines on the Handling of Genome-edited Food and Food Additives.' Under that policy, 'genome-editing technology' is defined as 'technology that modifies a specific site of a base sequence on a chromosome in order to add a specific function' (FAS, 2019b, 2020). In addition, if a foreign gene or fragments of such genes are contained in the end product, it shall fall under the recombinant DNA techniques and is treated identical to a GMO. If a genome-edited food has no added foreign DNA but has 'deletion of base(s), substitution, and insertion of a few bases from failure to repair at the cleavage site', it is not subject to GMO oversight but instead to a pre-market notification process. The pre-market notification, which is also required for imported products, includes:

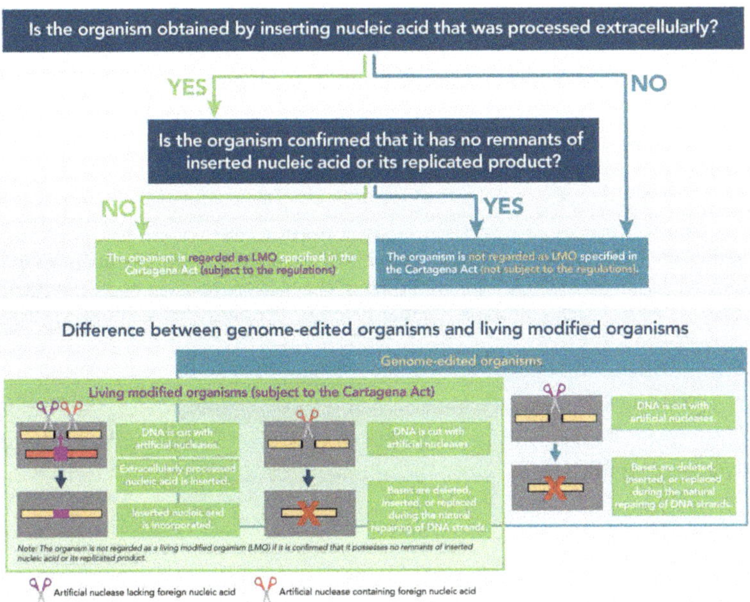

Figure 3 Process for determining regulatory oversight of genome-edited crops in Japan.

1 crop and variety name, summary of the developed food,
2 method of the genome-editing technology applied and details of the modification,
3 information regarding confirmation of absence of introduced foreign gene or residual fragments,
4 information regarding confirmation that the identified changes in the DNA do not produce allergens or enhance known inherent toxic substances, and do not result in adverse effects on human health,
5 in case modifications have been made to change a metabolic pathway to enhance or reduce specific components, information regarding changes in the main components in the targeted metabolic pathway, and
6 the month/year of launch (FAS, 2019b, 2020).

The MHLW then makes its decision to approve the notification and makes some of the information submitted in the notification available to the public.[4] While the notification process is clearly more streamlined and does not require a detailed risk assessment, it does require similar information about the organism - such as allergenicity and toxicity - that is comparable to what is required of a GMO (FAS, 2019b, 2020).

4.5 New Zealand – GMOs and genome-edited products are regulated the same

In New Zealand, the Hazardous Substances and New Organism Act of 1996 regulates the introduction of GMOs into the environment. A GMO is defined as 'any organism in which any of the genes or other genetic material (a) has been modified by in vitro techniques' (New Zealand, 1996). Regulations under the Act identify which organisms are not considered genetically modified. Section 3(1)(b) identifies as not genetically modified:

> Organisms that are regenerated from organs, tissues, or cell cultures, including those produced through selection and propagation of somaclonal variants, embryo rescue and cell fusion (including protoplast fusion or chemical or radiation treatments that cause changes in chromosome number or cause chromosome rearrangements). (New Zealand, 1996)

In 2013, the Environmental Protection Agency (EPA) determined that products resulting from modifications by ZFNs and TALENS were GMOs because they were modified by in vitro techniques but qualified as exempt under

4 Japan's MAFF established a policy entitled 'Feed Safety Guidelines on the Handling of Genome Edited Feed and Feed Additives" (FAS, 2020). In addition, Japan's Consumer Affairs Agency (CAA) established a policy that genome-edited foods that do not contain foreign DNA are not subject to the Food Labeling Standards (FAS, 2019c). Both these policies apply the same triggers and have similar provisions to the other policies discussed in the body of this article.

Section 3(1)(b) (NZ EPA, 2013). That decision was appealed to the High Court of New Zealand by the Sustainability Council of New Zealand Trust.

In 2014, the High Court determined that modifications by ZFNs and TALENS did not qualify for the exemption from being treated as GMOs. The Court found that the only thing in common with all the techniques identified in 3(1)(b) is that they were in common use at the time of the regulation and that it made sense to exempt them because they were 'well understood and established' and there was less likelihood of 'scientific and technical uncertainty about the effects of such techniques' (NZHC, 2014). For ZFNs and TALENs, the Court found that these technologies are 'relatively new' and their environmental effects were not certain. It therefore found that the exclusion only applied to the identified techniques and even if it applied to 'similar techniques', these were not sufficiently similar. Adding new techniques that are exempt is a decision for the legislation, and not for an administrative agency (NZHC, 2014). Therefore, in New Zealand, like in the EU, all genome-edited products will be treated as GMOs for the purposes of environmental safety oversight because they use *in vitro* methods to alter DNA and do not qualify for any exemption.[5] As of July 2020, there is research on GMOs in New Zealand but GMOs manufactured in New Zealand are not commercially available, so treating genome-edited crops as GMOs significantly reduces the likelihood of their commercialization. (NZ, 2020)

4.6 Nigeria - amended biosafety law to include genome editing

The National Biosafety Management Agency Act of 2015 establishes the National Biosafety Management Agency (NBMA) to carry out biosafety regulation in Nigeria. The Act requires NBMA to approve all activities involving GMOs, which are defined with the same definition found in the CBP ('novel combination of genetic material obtained through modern biotechnology') (NBMA, 2015). The NBMA is one of the few biosafety regulatory agencies in Africa to have approved GMOs for use by farmers (e.g. Bt cotton and Bt cowpea).

Nigeria enacted an Amendment to the NBMA Act in 2019 giving the NBMA the authority to approve all activities involving gene editing (as well as gene drives and synthetic biology). That Amendment defines 'gene editing' as 'a type of genetic engineering in which DNA is inserted, deleted, modified or replaced in the genome of a living organism' (NBMA, 2019). A plain reading of that definition suggests a broad legal scope that covers deletions, replacements, and insertions, independent of whether the introduced DNA or gene is from a sexually compatible species or not (i.e. foreign DNA is irrelevant). So, this

5 Australia and New Zealand jointly regulate food safety through Food Standards Australia New Zealand (FSANZ). They issued a report in December 2019 calling for a revision in their definitions related to gene technology to address genome editing. The report acknowledges "divergent views" on the acceptability and risk of genome edited foods and how to regulate them but does not set forth what will eventually be proposed or adopted. (FSANZ, 2019).

Amendment makes it clear that genome-edited products are different from GMOs but that they require approval under the biosafety law. It is anticipated that the NBMA will establish regulations or guidelines that set forth how they will implement this new legal authority and whether they will set up similar or different oversight procedures for different types of genome-edited techniques and/or products.

4.7 United States – generally treats GMOs and genome-edited products the same but exempts certain genome-edited crops

The United States' biosafety regulatory system was described in a policy titled 'Coordinated Framework on Biotechnology' in 1986 (OSTP, 1986). That policy states that products of biotechnology are to be regulated using existing laws and regulations by three primary agencies – the U.S. Department of Agriculture, (USDA), U.S. Environmental Protection Agency (EPA), and the Food and Drug Administration (FDA). Thus, agencies regulate genetically engineered and gene-edited products using laws and regulations that cover a broad range of products.

Plants produced using genetic engineering and the products made from those plants can be regulated by one, two, or all three agencies. USDA regulates genetically engineered plants utilizing the Plant Protection Act and requires permits for interstate movement, importation, and the growing of engineered plants outdoors, both for confined field trials and unconfined releases, unless the engineered plant has been granted non-regulated status (USDA, 2010). However, the USDA only regulates products it considers to be potential 'plant pests'. Until the revised regulations issued in 2020 (discussed below), this has included engineered plants produced using *Agrobacterium* in the transformation process because *Agrobacterium* is a 'plant pest'. However, USDA has exempted from oversight genetically engineered plants produced using the gene gun or other techniques that directly introduce DNA if there is no genetic material from organisms classified as plant pests. Using this same rationale, USDA has also exempted numerous genome-edited products from oversight when no organisms that are plant pests are involved. The USDA has found that the genome-editing components of CRISPR-Cas 9, TALENs, and ZFNs are not themselves plant pests (USDA, 2020a). Unlike distinctions made by regulators in Argentina, Australia, and Japan, the USDA's decisions to regulate through June 2020 have nothing to do with the presence or use of 'foreign' DNA. The USDA has exempted from regulation transgenic plants with foreign DNA, and the USDA would have exempted from oversight a plant where CRISPR was used to introduce a gene from non-plant-pest organism if such an application had been submitted to it.

In 2020, the USDA published new regulations for genetically engineered organisms with significant changes on how it regulates genetically modified organisms with specific provisions for genome-edited plants (called the Sustainable, Ecological, Consistent, Uniform, Responsible, Efficient, (SECURE) Rule) (USDA, 2020b). First, the new regulations define 'genetic engineering' to include 'techniques that use recombinant, synthesized, or amplified nucleic acids to modify or create a genome' (USDA, 2020b). This broad definition includes genetic engineering that adds one or more new genes to an organism and gene-editing techniques which edit an organism's existing genome. From the USDA's perspective, genome editing is just one type of genetic engineering.

The USDA sets forth several exemptions that eliminate any oversight for some genome-edited plants. The USDA exempts products with a single deletion of any size, a single base-pair substitution, or a single addition of a known sequence of any size (if the sequence is from the plant's gene pool). The USDA's rationale for these exemptions is that since a single deletion, substitution, or addition produces a plant that could be achieved by conventional breeding methods, and because conventionally bred plants historically have not raised plant pest risks, gene-edited plants that are the same as products that could be achieved through conventional breeding also will not pose plant pest risks (USDA, 2020b). For this reason, the USDA limits the exemption for additions to known DNA sequences that currently exist in the plant's gene pool. If an existing gene is introduced into the plant, it is exempt, but if that gene is changed in any manner (e.g. corrected or optimized), then it is not exempt. By contrast, deletions and substitutions where a random change is made are exempt, whether or not the resulting genomic sequence already exists in the plant's gene pool. The USDA rule allows developers to self-determine whether they qualify for these exemptions. It provides the developer with a voluntary option to seek confirmation of their self-determination from the USDA. There is no requirement to inform USDA when the developer self-determines their product's exempt status (USDA, 2020b).

The SECURE rule does not exempt genome-edited crops that have more than one edit to the crops' genome (USDA, 2020b). The USDA determined that, while it is possible to make more than one edit in a crop's genome through conventional breeding methods, they could not reach that conclusion for all plant species and for all types of edits, especially linked mutations or multiple mutations in a single gene. Therefore, they limited the exemption to single edits, which are 'genetic changes that could practically be achieved by conventional breeding methods in any plant' (USDA, 2020b). The USDA also clarified that for polyploid crops, a change in a gene on one chromosome is exempt but if the change is made on other chromosomes, it is considered a multi-edited crop and not exempt. This means that making a homozygously edited tomato (a deletion) would be exempt but a homozygously edited wheat with a gene deleted in all six copies would not be exempt. The USDA rule does establish a process to add exemptions for multi-edited crops, but it requires an application with scientific data that shows the multiple edits could be achieved through conventional breeding (USDA, 2020b).

For multiple edits and additions of a DNA sequence that cannot currently be found in a plant's gene pool, those genome-edited plants require some review by the USDA and possibly oversight by the USDA as a potential plant pest. The developer either voluntarily agrees to substantive oversight by seeking a permit to carry out activities with the edited plant (which includes interstate movement, importation, and outdoor activities - research and commercial plantings) or they submit an application to the USDA to conduct a Regulatory Status Review (RSR). The RSR starts with an initial 180-day process where the USDA determines if the GE plant has any plausible plant pest risks. That initial review does not require any plant-specific 'laboratory or field-test data'. If the USDA determines there are no plausible plant pest risks, the USDA sends a letter to the developer stating the plant is not regulated and publishes the letter on its website. If the USDA cannot conclude there are no plausible risks, then the USDA conducts the second part of the RSR, which is a detailed evaluation of potential plant pest risks (which can take up to 15 months). That analysis will result either in a determination that the plant is regulated and grown under permit or not regulated because it does not have any plant pest risks (USDA, 2020b). The USDA's new regulatory system set forth in the SECURE rule is summarized in Fig. 4.

The EPA regulates products that are pesticides under the Federal Fungicide, Insecticide and Rodenticide Act (FIFRA), including any pesticidal substances that have been engineered into plants. Those plant-incorporated protectants (PIPs) need to be registered, as do biological and chemical pesticides (EPA, 2020). The EPA has registered the pesticidal substance in genetically engineered Bt corn and cotton varieties as well as in crops engineered to be virus-resistant (squash and papaya). The agency has not publicly stated how it will regulate genome-edited plants. However, if a genome-edited plant has pesticidal characteristics, it is reasonable to assume that it will need to be registered as a pesticide and go through the same registration process as PIPs, especially if the pesticidal properties come from the presence of an added transgene.

The FDA regulates most food, including food from plants, under its authority in the Federal Food, Drug and Cosmetic Act (FFDCA). The FDA established a voluntary consultation process for foods derived from engineered plants in the mid-1990s (FDA, 1997). The agency has not stated that the same voluntary regulatory procedures apply to foods derived from genome-edited plants. However, the first genome-edited food crop that has been grown commercially in the United States (Calyxt's high-oleic soybeans) did submit a voluntary consultation package and the FDA reviewed the submission in the same manner as it reviews submissions for genetically engineered plants (FDA, 2019).

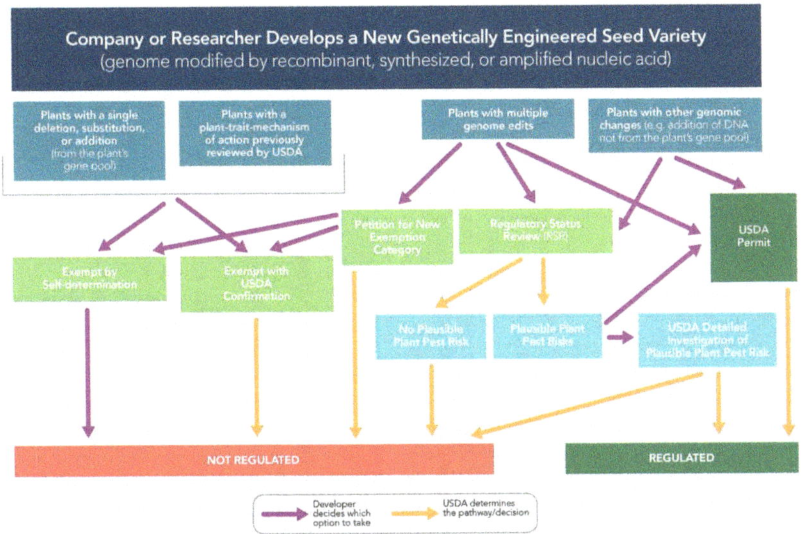

Figure 4 USDA genetically engineered plant oversight under new regulations. Source: Center for Science in the Public Interest.

The FDA also regulates genetically engineered animals and is the only federal agency with specific oversight of those animals. The FDA regulates them under the 'new animal drug' provision of FFDCA. A 'new animal drug' is anything that alters the structure or function of an animal and the FDA issued a guidance on its oversight in 2008 (Guidance #187). FDA reissued a revised version of Guidance #187 in 2017 entitled 'Regulation of Intentionally Altered Genomic DNA in Animals' (FDA, 2017). That Guidance makes clear that the same regulatory oversight that applies to genetically engineered animals applies to genome-edited animals, including single-base-pair deletions or additions (FDA, 2017).

In summary, the United States is still determining how different agencies will regulate genome-edited products. Most genome-edited products are or will be regulated under the same laws, regulations, and policies as GMOs. When those regulations are applied to genome-edited products, the result is the same oversight for animals but little or no mandatory oversight for certain genome-edited plants, depending on the number of edits and whether additions come from the plant's gene pool. For genome-edited plants, if a developer submits information to either the FDA or the USDA voluntarily for any regulatory process (such as the FDA voluntary consultation and the USDA's confirmation process), that information will be available to the public. However, if a developer chooses to market their genome-edited plant without submitting

Table 1 Summary of regulation of genome-edited crops in different countries

Country	National biosafety law and/or regulation	Definition of key terms in biosafety law/regulations	Which genome-edited crops are exempt from oversight?	Regulatory oversight procedures for genome-edited products	Which genome-edited products treated as GMOs?	Information required for regulatory process for genome-edited crop
Argentina	The regulatory process in Argentina is established by various resolutions including, Resolution 763/2011 issued by the Ministry of Agriculture	**'GMO'** is a 'novel combination of genetic material created with modern biotechnology'[a]; **'event'** is 'the insertion into the plant genome in a stable and joint form, of one or more genes or DNA sequences that are part of a defined genetic construction'	No genome edited products are exempt (every product must go through a consultative process)	**Mandatory case-by case review** of products to determine if novel DNA has been added and they are a GMO	Genome edited products with foreign DNA from a template or with genes from a sexually compatible species (unless they are a perfect allelic swap)	Information on gene editing method and changes made; confirmation of no addition of foreign DNA; phenotype and potential impact of the phenotype
Australia	Gene Technology Act of 2000; Gene Technology Regulations of 2019	**gene technology** is 'any technique for the modification of genes or other genetic material'	Genome edited products modified through a mutational event without introduction of any foreign DNA and those that did not use a template to guide homologous repair	**Regulated as GMOs** based on the genome editing process utilized	Genome edited organisms modified with oligonucleotide-directed mutagenesis, site-directed nuclease techniques involving templates to guide the repair action, or any other introduction of any new DNA (even transiently)	Same information as required for GMOs

European Union	EU Directive 2001/18 on the Deliberate Release into the Environment of Genetically Modified Organisms; EU Regulation 1829/2003 on Genetically Modified Food and Feed	**GMO** is 'an organism, . . . in which the genetic material has been altered in a way that does not occur naturally by mating and/or natural recombination'	No genome-edited products are exempt; (only crops modified through mutagenesis if it is not done using recombinant DNA and the technique has a long history of safe use)	**Regulated as GMOs** with full risk assessment and pre-market approval	All genome edited crops	Same information as required for GMOs (In 2020, EFSA is reviewing which aspects of risk assessment are needed for deletions and substitutions)
Japan	Law Concerning the Conservation and Sustainable Use of Biological Diversity through Regulations on the Use of Living Modified Organisms (Law No. 97 of 2003)	**LMO** is 'living organisms having nucleic acids obtained by utilizing a technique for processing nucleic acids outside the cell for the purpose of transferring or replicating the nucleic acids by transferring them into a cell, virus, or viroid'. **genome-editing technology** is 'technology that modifies a specific site of a base sequence on a chromosome in order to add a specific function'	No genome edited products are exempt	**Mandatory notification** for all genome edited crops. Some will be regulated as GMOs. Products not found to be GMOs are required to conduct research under GMO containment conditions and submit pre-market notifications for clearance	Genome edited crops that have added nucleic acid using a template or if they have a replicated product from any inserted nucleic acid	Products that are not GMOs are still required to submit phenotype information for added traits, information on possible influences on biological diversity, and information on food safety changes (allergens, toxicity, etc.)

(Continued)

Table 1 (*Continued*)

Country	National biosafety law and/or regulation	Definition of key terms in biosafety law/regulations	Which genome-edited crops are exempt from oversight?	Regulatory oversight procedures for genome-edited products	Which genome-edited products treated as GMOs?	Information required for regulatory process for genome-edited crop
New Zealand	Hazardous Substances and New Organism Act of 1996	**GMO** is 'any organism in which any of the genes or other genetic material – (a) have been modified by in vitro techniques; or (b) are inherited or otherwise derived, through any number of replications, from any genes or other genetic material which has been modified by in vitro techniques'	No genome edited products are exempt (exemptions in the regulations do not apply because genome editing techniques are not identified as exempt and are not sufficiently like the exempt techniques)	**Regulated as GMOs**	All genome edited crops	Same information as required for GMOs
Nigeria	The National Biosafety Management Agency Act of 2015; 2019 Amendments to the National Biosafety Management Agency Act.	**GMO** is 'any organism . . . that possesses a novel combination of genetic material obtained through the use of modern biotechnology';[b] **gene editing** is 'a type of genetic engineering in which DNA is inserted, deleted, modified or replaced in the genome of a living organism'	No genome edited products are exempt	**Regulated under the Biosafety Law** (details to be determined in forthcoming regulations/ guidance)	All genome-edited crops regulated under National Biosafety Management Agency Act	To be determined

United States Department of Agriculture[c]	The Plant Protection Act, 7 U.S.C. 7701; Sustainable, Ecological, Consistent, Uniform, Responsible, Efficient, (SECURE) Rule of 2020, 7 CFR Part 340	**genetic engineering** is 'techniques that use recombinant, synthesized, or amplified nucleic acids to modify or create a genome'	Products with a single deletion, substitution, or addition (if it is from the same plant's gene pool); multiple edits may be exempt in the future	**Regulated as GMOs** (i.e. products undergo review as potential plant pests); a streamlined regulatory status review process is used to determine if crop requires oversight	Genome edited crops with plausible plant pest risks	No laboratory or field data is needed to complete initial regulatory status review; if the more detailed 15-month review is required, then specific data needed on the crop and trait being assessed and potential to be a plant pest

[a] This definition is identical to the definition in the Cartagena Biosafety Protocol.
[b] This is the same definition as in the Cartagena Biosafety Protocol.
[c] The U.S. Department of Agriculture (USDA) is the only US agency that has established different policies and regulations for genome edited crops. The other two agencies – the U.S. Environmental Protection Agency (EPA) and the U.S. Food and Drug Administration (FDA) – currently apply their existing regulatory process for GMOs to genome-edited crops.

information to a regulatory agency, only the developer and whomever the developer provides information to will know about the product.

5 Analysis of regulations in different countries

When one reviews how different countries regulate genome-edited crops, one finds both similarities and differences among the different regulatory systems. The following analysis highlights some key issues that result from analyzing and comparing those regulatory systems.

5.1 Most genome-edited products will be subject to specific regulation in almost all countries

Many genome editing advocates argue for no special genome editing oversight for at least some products and want those products to be treated identically to conventionally bred varieties. However, virtually all the regulatory systems elaborated above, which are representative of the regulation that will likely occur throughout the world, have some level of genome-editing specific regulation. If these regulatory systems are used as models for countries to adopt, then most countries will have some regulatory oversight that uniquely applies to genome-edited products.

One can view specific regulation for genome-edited products on a continuum, with a high level of oversight and inclusivity at one end and exemptions from oversight for some or all genome-edited products at the other end. The European Union and New Zealand could be found at the end of the continuum where there is a high degree of oversight. Both countries currently treat all genome-edited products identically to GMOs, which require pre-market approval after completing a detailed risk assessment. Nigeria's recent amendment to their National Biosafety Management Agency Act also explicitly puts oversight of genome-edited products under a pre-market approval requirement. However, the details of whether the process and scientific evidence for approval of genome-edited products in Nigeria will be the same as for GMOs have yet to be decided.

Argentina and Japan - which both have 'case-by-case' oversight - are in the middle of the continuum, as they do not exempt any genome-edited products from all oversight, but products vary in the information needed by the regulators and the procedures that the regulators will follow. In those regulatory systems, the degree of oversight depends on the types of edits and the product's final composition. However, in each of those countries, there is a mandatory requirement to consult and/or submit an application for every genome-edited product, and the product cannot proceed to market unless they receive a decision by the government agency based on the application and information they submitted. As stated by an Argentinian regulator, the

Argentinian consultation process for genome-edited plants and animals results in a government-regulatory determination that the genome-edited product is or is not a GMO (Lema, 2019).

Australia's system is similar to the system in Argentina and Japan in that it separates genome-edited products into different categories based on the techniques used in the development of the products as well as its final characteristics. However, it does exempt a small subset of genome-edited products from any oversight under their Gene Technology Act (if the change is the result of random mistake caused when the cell repairs its broken DNA). Therefore, Australia and other countries (such as the United States) which exempt certain genome-edited products from any special oversight inhabit the other end of the continuum as it currently is today - where at least some genome-edited products require no specific oversight. Interestingly, the United States exemption could be perceived as narrow or large depending on whether one believes most genome-edited products will be single edits, which are exempt, or multiple edits, which require some oversight.

The conclusion from reviewing these varied oversight systems is that most countries will require some specific regulatory oversight for many genome-edited products. For many products, it will likely be less comprehensive and less burdensome than that country's oversight of GMOs. This may be adequate for some genome-edited products that have simple changes to one or a couple of genes since those products may be safer than a product with an introduced transgene. However, for some genome-edited products that most closely resemble GMOs, those products will be required to complete a detailed review and approval process with government regulators. Only rarely (e.g. for some products in Australia or the United States) will genome-edited products be treated identical to conventionally bred products and require no specific safety oversight.

5.2 The details matter - the criteria for case-by-case review vary in different countries

For countries that have decided how genome-edited crops will be regulated, the most common regulatory system involves some sort of case-by-case assessment where some products are treated identical to GMOs and others get a less thorough review. In addition to Argentina, Australia, and Japan, other countries that have adopted similar systems include Brazil, Colombia, and Chile (Schmidt et al., 2020). However, a genome-edited product could receive different degrees of oversight and require different information to be submitted in countries arguably having similar case-by-case regulatory systems. Therefore, the details around the genome-editing technique utilized, the composition of the final product, and the requirements of each country's regulatory system will determine the degree and kind of oversight in most regulatory systems. The result is that a genome-edited product's oversight could vary significantly from country to country.

Two examples show how, depending on the details of the editing process and the final product, the same product could receive different levels of oversight in different countries. The first example is a genome-edited crop where a gene from a sexually compatible species is introduced into the organism in place of the original copy. In Argentina, this product would not be considered a GMO and only needs to complete the consultation process. The developer need only show that the introduced gene is a '"perfect allelic replacement" completely indistinguishable from the outcome of a sexual crossing'. The same treatment would occur for that addition in the United States for USDA's regulatory oversight. However, in Australia, that product would be regulated as a GMO because a 'template' was used to introduce the gene - whether the gene is from a sexually compatible species is irrelevant. A similar result would occur in Japan as they consider a genome-edited product a GMO if any 'extracellular DNA is inserted, independent of whether it comes from a sexually compatible species'. In almost all country regulatory systems, if foreign DNA is introduced through genome editing and that foreign DNA remains, the resulting product is regulated as a GMO.

The second example is a genome-edited product where a gene has been knocked out through a small deletion. In Argentina, that product would only require a consultation in which the developer needs to prove that no 'foreign DNA' is present in the final organism. Whether the deletion was random or occurred from a DNA template is immaterial if the DNA template is not integrated into the final genome-edited product. In Australia, the degree of regulatory oversight depends on whether the deletion occurred randomly or was the result of a temporarily introduced DNA template. The former would be exempt from any special oversight while the latter would be treated as a GMO. That is also the case in Japan. However the regulations in Japan also distinguish for edits that are repaired randomly between edits where the DNA is cut using an artificial nuclease containing or lacking 'foreign nucleic acid'. If the artificial nuclease used to cut the DNA includes 'foreign nucleic acid', it is a GMO, and if it lacks 'foreign nucleic acid', it is not. Finally, the USDA SECURE rule would exempt that product with a single deletion because it is comparable to a product that could be developed with conventional breeding but does not exempt organisms with multiple deletions.

The amount of information in the consultation also can vary by country. For the second example above involving a simple deletion, Argentina is primarily interested in obtaining information about the techniques used and the final product to prove there is no foreign DNA in the final product. They do require an analysis of off-targets, but it is unclear from the regulations what this entails. Japan, on the other hand, requires not only information about the techniques used and the characteristics of the final product but also information on its usage, its impacts on biological diversity, as well as confirmation that it does not produce allergens or enhance toxins. Therefore, while Japan does not treat a genome-edited product with a deletion as a GMO, it still requires information similar to the information that would be required if a developer conducted a GMO risk assessment.

5.3 Many provisions of the regulatory systems are not science-based

None of the countries that have established regulatory systems for genome-edited products seem to have succeeded in making regulatory procedures and requirements that are based primarily or exclusively on scientific evidence. Instead, they integrate into their regulatory systems distinctions - sometimes arbitrary - without a scientific basis. For example, several countries, including Argentina and Japan, decide on the degree of oversight based on whether the genome-edited product has introduced foreign DNA. Anyone who has taken high school or college biology knows that DNA is the same molecule in all species and that many species have significant overlaps of common DNA (e.g. humans and other primates share over 90% of their DNA) (Cummins, 2018). There is evidence that many organisms that are considered safe contain introduced DNA from another species that occurred long ago (e.g. sweet potato with genes from *Agrobacterium*) (Kyndt et al., 2015). In addition, there is no evidence that 'foreign' DNA itself is inherently risky and needs special oversight - instead, it would depend on the introduced gene, what the gene does, the molecule it encodes, and the resulting introduced trait. Regulatory systems that establish arbitrary classifications perpetuate a view that introducing 'foreign' DNA into an organism automatically increases its potential risk, and that all genomic changes within an organism's own DNA are safe, yet neither of those two propositions is supported by scientific evidence.

Regulations in several countries also do not distinguish between different types of random deletions. In some countries (e.g. Argentina, Australia, Japan, and the United States), a single base-pair deletion that affects one gene is treated the same as a deletion of thousands of base-pairs, which might impact multiple genes and multiple cellular pathways. Most scientists would not categorize the potential phenotypic and genotypic impact of these two examples as the same, yet these regulatory systems do.

In addition, some regulatory systems assume that if a genome edit is similar to what could arise through conventional breeding, it is not risky enough to require oversight. The rationale is that conventional crop varieties have not had oversight in the past so they must not have significant risk and the same treatment should be given for genome-edited crops that could arise through conventional breeding. The USDA SECURE rule specifically exempts single-sequence deletions and substitutions because the same genotype and phenotype could arise from conventional breeding (such as chemical mutagenesis or irradiation breeding methods). Those exemptions continue 'process-based' regulation, which exempts products based on their method of development. A science-based regulatory system would not regulate based solely on how the product was made but instead determine whether a crop's introduced trait or phenotype has any potential risks based on the product's characteristics along with some consideration of how the phenotype was developed.

Could a national regulatory system for genome-edited products base its level of oversight on scientific evidence? Yes, if the oversight conducted was influenced by both the final product and the process used to achieve that product. Such a regulatory system would likely determine the degree of oversight and whether regulation is needed at all by considering criteria such as: (1) the type of edit(s) made; (2) the process for making the edit(s); (3) the resulting genotype of the organism; and (4) the organism's resulting phenotype. Some regulatory systems, such as in Argentina and Australia, consider the first three of these criteria but not the resulting phenotype, which is a key factor in determining any potential risks to humans or the environment.

5.4 Harmonization of regulatory oversight by different countries is unlikely (especially in the short run)

Many stakeholders advocate for countries to adopt the same regulatory procedures and requirements. Harmonization, if it were possible, would have numerous benefits, including cost savings and making international trade and collaborative research significantly easier. However, based on the review of the range of regulatory systems in the countries highlighted in this chapter, it is unlikely that harmonization will occur, especially in the short run. The regulatory systems in the United States and the European Union treat genome-edited products differently and harmonizing those two systems would require legislative action. Enacting and implementing legislation would take years. In addition, the regulatory systems that are considered 'case-by-case', treating some genome-edited products as GMOs and others with less oversight, may at a superficial level already seem harmonized. However, when one reviews the details, they treat identical products differently and require different levels of information, making them much less-harmonized than one might expect. Therefore, for the immediate future, global regulatory harmony for genome-edited products is a distant dream. This also has been the case for biosafety oversight of GMOs. However, it is possible that there could be regional harmonization or subregional harmonization between neighboring countries. Several Latin American countries have used the Argentinian mandatory consultation process as a model for determining how they might regulate genome-edited products (Schmidt et al., 2020; Metje-Sprink et al., 2019).

5.5 Transparency is critical but is inconsistent among regulatory systems

In today's world, information - both accurate and inaccurate - is available on the internet 24 hours a day around the world. Some individuals and stakeholders will want to know whether a product has been genome-edited. How will that farmer, grain handler, retailer, consumer, or government official find factually correct information? If that information is not available or is perceived by some people to be hidden, will that lack of information translate into concern that genome-edited products are unsafe and avoidance of them in the marketplace? Transparency about which products have been genome-edited and where they are in commerce will be important for ensuring that: (1) consumers who believe this information is important to them can find it; (2) the market can meet consumer expectations by providing non-genome-edited products for those who want them; (3) regulators know which products qualify for oversight; and (4) international commerce can occur without trade disruptions.

While the primary purpose of biosafety regulatory systems is to ensure safety, a corollary benefit is that those regulatory systems provide transparency about emerging biotech products to the public, other countries, and the marketplace. Regulatory oversight depends on transparency to identify regulated products. Information regarding regulatory decisions on GMOs has been made available throughout the world through international and national mechanisms, and this information has been critical to smooth international trade disputes and to enable stakeholders in the food chain to address the demand for information by some consumers. In contrast, the degree of transparency around regulated genome-edited products seems to vary depending on the country. Japan's regulatory system for genome-edited products is relatively transparent. The different regulatory agencies publicly provide information on genome-edited products that do not fall under the GMO regulations, including information about each organism and its modified traits. At the other extreme, Argentina conducts consultations for all genome-edited organisms, but does not release any information about those applications nor its decisions that the organism does not qualify as a GMO. In the United States, if developers self-determine that their genome-edited crops qualify for an exemption, they do not need to notify the USDA or the public, so there is no transparency. However, if a genome-edited crop developer makes a submission to the USDA to confirm its self-determination, obtains a permit, or goes through the Regulatory Status Review process (as would be required for genome-edited plants with multiple edits), the USDA will make its determinations publicly available. In other countries, such as Australia, it is unclear what information will be provided publicly on organisms that are not regulated as GMOs. For many regulators and other stakeholders, transparency regarding the regulatory status of genome-edited products is important (Eckerstorfer et al., 2019). When governments are not providing consistent and accurate information, third parties will generate their own lists of genome-edited products that vary in their accuracy. Such a result could lead to misinformation and hinder genome-edited product acceptance. Some stakeholders have called for national or international registries that identify all commercial genome-edited products in agriculture and food (Jaffe, 2019; Eckerstorfer et al., 2019).

5.6 Addressing organisms with multiple or complex genomic changes

Countries that have established regulatory policies for genome-edited products generally explain how their systems operate using simple editing examples - for example, a deletion to knock out a gene, a base-pair change to correct a gene, or the addition of a new gene. However, scientists have been successful editing multiple genes at one time in an organism. In addition, products could be developed where edited genes are stacked into a single organism to create a trait from the cumulative or synergistic effect of multiple gene changes, such as changing a metabolic pathway or causing the organism to produce a new molecule. How complex genome-edited products will be regulated in most countries is unclear.

As stated earlier, one of the arguments for exempting genome-edited products from regulation is that the same genomic allele or point mutation could be achieved from conventional breeding and should not be considered a 'novel' organism that might pose risks that need oversight. That scientific argument seldom applies to complex genome-edited products because some of them could not occur naturally or be developed with conventional breeding. It may be relatively easy to argue and prove scientifically that a single-point mutation likely could occur naturally and is not novel, but it is harder to scientifically support the same argument when multiple edits are involved, either done at the same time or stacked by crossing plants over several generations (Custers et al., 2019). The USDA SECURE rule does not exempt crops with multiple edits for just this reason - that from a practical standpoint, multiple edits are not universally achievable in all crop varieties using conventional breeding methods. Also, it is hard to argue that the potential risks of making one precise edit in one location versus multiple edits (such as making two dozen precise edits at different locations) are the same. Multiple edits increase the possibility of off-targets. The potential risk, not just the novelty, should determine the degree and type of regulatory oversight for safety.

How products with complex genome editing will be regulated in most countries will only be determined when those products move toward commercial marketing and those developers approach regulators asking what oversight is required. The exception is the USDA SECURE rule. The USDA does not approach this issue using 'novelty' as the regulatory trigger. Instead it has exempted products that could practically occur using conventional breeding and finds that it can only make that scientific conclusion for single edits within a genome.

A plain reading of the regulations in Argentina or Australia suggests that if each one of the multiple edits in a crop individually is the result of template-less repair and does not introduce new DNA, then the whole organism with multiple edits would not be considered a GMO. However, if the result of the edits is a new molecule or elevated levels of existing molecules, it is possible that the analysis of whether to regulate the organism could be different. Such a product could be more like a GMO than a conventionally bred product.

As stated earlier, many countries will look to whether these complex genome changes are GMOs under the Cartagena Protocol. While one may be able to scientifically establish that knocking out one gene or correcting a single base-pair mutation 'could occur in nature' and is not a 'novel combination of genetic material', - the trigger for inclusion under the Cartagena Protocol -- those arguments are much more difficult when many genes are altered in an organism or multiple genes are edited to change a metabolic pathway. Clearly, the more edits that are made in an organism, the harder it is to say that the organism's genomic changes are not 'novel'. Therefore, in many countries, including those with established regulatory procedures for genome-edited products, there remains uncertainty about the level of oversight for complex genome-edited products.

6 Conclusion and future trends

The science around genome editing is moving rapidly and many different crop varieties will be developed that could benefit farmers and consumers around the world. However, before those crops can be grown by farmers, they will need to comply with each country's applicable laws and regulations that ensure those products are safe. In all countries, genome-edited crops will need to comply with the same laws and regulations applicable to crop varieties developed using conventional breeding techniques. In some countries and for some genome-edited crops, compliance with those laws and regulations will be sufficient to market the genome-edited crop. For other countries, some or all genome-edited crops will need to comply with additional laws and regulations that only apply to genome-edited products (or that equally apply to GMO crops developed with genetic engineering). For most countries, whether to require specific regulations for genome-edited crops has not yet been determined.

The regulatory systems that have been established in several 'leader' countries around the world – Argentina, Australia, the European Union, Japan, New Zealand, Nigeria, and the United States – vary tremendously in their approach to regulate genome-edited crops. If one views the degree of oversight on a continuum from minimal to stringent, the EU and New Zealand would be at the stringent extreme, regulating each genome-edited product in the same manner as it regulates GMOs – requiring mandatory pre-market approval after conducting a detailed scientific risk assessment. At the other end of the continuum is the United States and Australia, which exempt certain genome-edited products (such as deletions that occur spontaneously after a DNA break) from any regulation other than what would be required for conventionally bred crops. In the middle of the continuum would be Argentina and Japan, which require mandatory government consultations for all genome-edited crops with limited scientific information and then determine if the genome-edited crop qualifies as a GMO and requires more stringent oversight.

The conclusions that can be reached from reviewing those regulatory systems are as follows:

- All countries will regulate some genome-edited products using the same regulations that are required for GMOs (primarily genome edited products that add DNA or a foreign gene).
- Most countries will require some regulation of all genome-edited products solely because they were genome edited, although that oversight could be a streamlined consultation process set up to ensure that the genome-edited crop is not a GMO.
- The regulatory systems for genome-edited products in different countries will differ in their detail, meaning that identical genome-edited crops will be treated differently in different countries (e.g. a genome-edited crop may be considered a GMO in one country and not in another).
- Many provisions in the regulatory systems are not necessarily based on science, such as treating changes made with a template different from those made randomly or using 'foreign' DNA as a trigger for oversight independent of the phenotype introduced by the added DNA.
- Global harmonization of regulations for genome-edited crops is not likely in the immediate future.
- The degree of transparency about which genome-edited products have been reviewed by regulators and which ones are commercial products will vary tremendously by country.

- The regulatory pathway for complex genome-edited products with multiple edits is much less certain than the pathway for simple edits.

A regulatory system that ensures the safety of a product should be transparent and predictable, treat products proportionally based on their potential risk, and use scientific evidence and risk assessments to ensure safety before a product is released into the market. The policy debate over how to regulate genome-edited crops has primarily focused on whether a genome-edited crop should be treated as a GMO or treated as a conventionally bred crop. Some genome-edited crops may be less risky than GMOs while other genome-edited products may be riskier than conventionally bred crops. Genome editing of crops will produce many different types of products, each of which has its own risk profile. Any regulatory system needs to consider both the genotype and phenotype of the final product as well as the process by which it was created. It should treat different categories of genome-edited products (e.g. deletions, additions, multiple edits) proportionately based on their potential risk. Instead of trying to pigeonhole new products with a range of potential risk profiles into the existing regulatory systems for GMOs or conventionally bred products, countries should establish regulatory oversight procedures that ensure safety in a proportionate, science-based and risk-based manner.

7 References

AMALF (2015). Argentina Ministry of Agriculture, Livestock and Fisheries: Resolution 173/15, Available at: http://servicios.infoleg.gob.ar/infolegInternet/anexos/245000-249999/246978/norma.htm (Accessed June 8, 2020).

AMALF (2017). Argentina Ministry of Agriculture, Livestock and Fisheries: Resolution 79-E/17/, Available at: http://servicios.infoleg.gob.ar/infolegInternet/anexos/285000-289999/287337/norma.htm (Accessed June 8, 2020).

AMALF (2019a). Argentina Ministry of Agriculture, Livestock and Fisheries: Administrative Order 1/2019. DI-2019-1-APN-DB#MPYT.

AMALF (2019b). Argentina Ministry of Agriculture, Livestock and Fisheries: Resolution 36/2019. RESOL-2019-36-APN-SAYBI#MPYT.

ASTA (2020). American Seed Trade Association Frequently Asked Questions on Plant Breeding, Available at: https://www.betterseed.org/wp-content/uploads/ASTA-Resource_NBT_Updates.pdf (Accessed July 21, 2020).

Burachik, M. (2012). Regulation of GM Crops in Argentina, *GM Crops and Food*, 3(1), 48–51.

CBD (1992). Convention on Biological Diversity (CBD), Available at: https://www.cbd.int/convention/text/ (Accessed July 20, 2020).

CBP (2000). Cartagena Protocol on Biosafety to the Convention on Biological Diversity (CBP), Available at: https://www.cbd.int/doc/legal/cartagena-protocol-en.pdf (Accessed July 10, 2020).

CBP (2020). Cartagena Protocol on Biosafety Official Website, Available at: https://bch.cbd.int/protocol/ (Accessed June 7, 2020).

CRS (2015). Congressional Research Service (CRS - USA): "Restrictions on Genetically Modified Organisms: Argentina", Available at: https://www.loc.gov/law/help/restrictions-on-gmos/argentina.php (Accessed April 23, 2020).

Cummins, E. (2018). Bananas: Your Cousin, Maybe?, *Popular Science*, August 14, 2018, Available at: https://www.popsci.com/humans-genetically-linked-to-bananas/ (Accessed June 8, 2020).

Custers, R., Casacuberta, J. M., Eriksson, D., Sági, L. and Schiemann, J. (Jan 2019). Genetic Alterations that do or do not occur naturally; Consequences for genome-edited organisms in the context of regulatory oversight, *Frontiers in Bioengineering and Biotechnology* 6, Article 213.

ECJ (2018). European Court of Justice (ECJ). In: Case C-528/16, July 25, 2018, Available at: http://curia.europa.eu/juris/document/document.jsf;jsessionid=93215E86AD2241BFA04AEEAFD64E00D7?text=&docid=204387&pageIndex=0&doclang=en&mode=req&dir=&occ=first&part=1&cid=3070137 (Accessed June 7, 2020).

Eckerstorfer, M. F., Engelhard, M., Heissenberger, A., Simon, S. and Teichmann, H. (Feb 2019). Plants Developed by New Genetic Modification Technologies - Comparison of Existing Regulatory Frameworks in the EU and non-EU Countries, *Frontiers in Bioengineering and Biotechnology* 7, Article 26.

EFSA (2012). European Food Safety Authority (EFSA): Scientific Opinion Addressing the Safety Assessment of Plants Developed with Zinc Finger Nuclease 3 and Other Site-Directed Nucleases with Similar Function, *EFSA Journal* 10(10), 2943.

EPA (2020). United States Environmental Protection Agency (EPA): 40 CFR 174. Procedures and Requirements for Plant-Incorporated Protectants, Available at: https://ecfr.io/Title-40/pt40.26.174 (Accessed June 7, 2020).

EU (2001). Directive 2001/18/EC of the European Parliament and of the Council of 12 March 2001 on the Deliberate Release into the Environment of Genetically Modified Organisms and Repealing Council Directive 90/220/EEC - Commission Declaration, Available at: https://eur-lex.europa.eu/legal-content/EN/TXT/?uri=CELEX:32001L0018 (Assessed July 31, 2020).

EU (2003). Regulation (EC) No 1829/2003 of the European Parliament and of the Council of 22 September 2003 on genetically modified food and feed, Available at: https://eur-lex.europa.eu/legal-content/EN/ALL/?uri=CELEX:32003R1829 (Accessed July 31, 2020).

EU (2020). European Commission Study on New Genomic Techniques, Available at: https://ec.europa.eu/food/plant/gmo/modern_biotech/new-genomic-techniques_en (Accessed July 20, 2020).

EUSAGE (2019). Open Statement of the European Union Sustainable Agriculture through Genome Editing, Available at: https://www.uu.nl/sites/default/files/Open%20Statement%20EU-SAGE%20July%202020_EN.pdf (Assessed July 31, 2020).

FAS (2019a). United States Foreign Agriculture Service (FAS) GAIN Report: "MAFF Guidance for the Handling of Genome Edited Organisms under the Cartagena Act," November 19, 2019.

FAS (2019b). United States Foreign Agriculture Service (FAS): GAIN Report: "Japan Invites Comments on Genome-Edited Food Handling Procedures," July 5, 2019.

FAS (2019c). United States Foreign Agriculture Service (FAS): Gain Report, "Labeling Guidance for Genome Edited Food Products," October 8, 2019.

FAS (2020). United States Foreign Agriculture Service (FAS): GAIN Report: "Japan Finalizes Handling Procedures for Genome-Edited Feed and Feed Additives," February 14, 2020.

FDA (1997). United States Food and Drug Administration (FDA): Consultation Procedures Under FDA's 1992 Statement of Policy for Food Derived from New Plant Varieties, Available at: https://www.fda.gov/food/ingredients-additives-gras-packaging-guidance-documents-regulatory-information/consultation-procedures-under-fdas-1992-statement-policy-foods-derived-new-plant-varieties (Accessed June 7 2020).

FDA (2017). United States Food and Drug Administration (FDA): Guidance #187 entitled "Regulation of Intentionally Altered Genomic DNA in Animals", https://www.fda.gov/media/74614/download (Accessed June 7, 2020).

FDA (2019). United States Food and Drug Administration (FDA): Letter to Calyxt from FDA Regarding High Oleic Soybean Dated 2/26/19, Available at: https://www.fda.gov/media/120707/download (Accessed June 7, 2020).

Friends of the Earth (2018). Gene-Edited Organisms in Agriculture: Risks and Unexpected Consequences, Available at: https://1bps6437gg8c169i0y1drtgz-wpengine.netdna-ssl.com/wp-content/uploads/2018/09/FOE_GenomeEditingAgReport_final.pdf (Accessed June 8, 2020).

FSANZ (2019). Food Standards Australia New Zealand (FSANZ) Final Report "Review of Food Derived Using New Breeding Techniques," December 2019, Available at: https://www.foodstandards.gov.au/consumer/gmfood/Pages/Review-of-new-breeding-technologies-.aspx (Accessed July 28, 2020).

GTA (2000). Gene Technology Act (Australia), Available at: https://www.legislation.gov.au/Details/C2016C00792 (Accessed July 10, 2020).

GTA (2019). Gene Technology Amendment (2019 Measures No. 1) Regulations 2019 (Australia), Available at: https://www.legislation.gov.au/Details/F2019L00573/Explanatory%20Statement/Text (Accessed July 10, 2020).

GTT (2001). Gene Technology Regulations (2001) (Australia), Available at: https://www.legislation.gov.au/Details/F2016C00615 (Accessed July 10, 2020).

Jaffe, G. (2019). Biotech Blog: A Call for a National Registry of Gene-Edited Crops April 22, 2019, Available at: https://www.pnas.org/content/112/18/5844 (Accessed June 8, 2020).

Kyndt, T., Quispe, D., Zhai, H., Jarret, R., Ghislain, M., Liu, Q., Gheysen, G. and Kreuze, J. F. (2015). The Genome of Cultivated Sweet Potato Contains Agrobacterium T-DNAs with Expressed Genes: An Example of a Naturally Transgenic Food Crop, *Proceedings of the National Academy of Sciences* 112(18), 5844–5849.

Lema, M. A. (2019). Regulatory Aspects of Gene Editing in Argentina, *Transgenic Research* 28(S2), 147–150.

Metje-Sprink, J., Menz, J., Modrzejewski, D. and Sprink, T. (Jan 2019). DNA Free Genome Editing: Past, Present and Future, *Frontiers in Plant Science*, 9, Article 1957.

MOE (2019). Ministry of Environment (MOE) (Japan): "To Genome Editing Technologies Users", Available at: https://www.env.go.jp/press/2_2_%20genome%20editing_En.pdf (Accessed August 25, 2020).

NASEM (2017). National Academies of Science, Engineering and Medicine (NASEM): Preparing for Future Products of Biotechnology, Available at: https://www.nap.edu/catalog/24605/preparing-for-future-products-of-biotechnology (Accessed June 8, 2020).

NBMA (2019). Nigeria National Biosafety Management Act Amendments, Available at: https://nbma.gov.ng/wp-content/uploads/2017/06/Amendment-of-NBMA-Act-2019.pdf (Accessed June 7, 2020).

NBMA (2015). Nigeria National Biosafety Management Act (2015), Available at: https://www.nbma.gov.ng/wp-content/uploads/2017/06/National_Biosafety_Management_Agency__Act_2015-signed.pdf (Accessed June 7, 2020).

New Zealand (1996). Hazardous Substance and New Organisms Act, Available at: http://www.legislation.govt.nz/act/public/1996/0030/latest/whole.html#DLM381222 (Accessed June 7, 2020).

NGMO (2020). Non-GMO Project Website, Available at: https://www.nongmoproject.org/blog/emerging-techniques-in-biotechnology-pose-new-risks-to-the-non-gmo-supply-chain/ (Accessed June 8, 2020).

NZ (2020). New Zealand Ministry for the Environment webpage on "about genetic modification in New Zealand,", Available at: https://www.mfe.govt.nz/more/hazards/new-organisms/genetic-modification-new-zealand/about-gm-new-zealand (Accessed July 31, 2020).

NZ EPA (2013). New Zealand Environmental Protection Agency (NZ EPA): Decision Whether or not an Organism Is a New Organism under the Hazardous Substance and New Organism Act, Available at: https://epa.govt.nz/assets/FileAPI/hsno-ar/APP201381/03e845f138/APP201381-APP201381-Decision.pdf (Accessed June 7, 2020).

NZHC (2014). New Zealand High Court, Wellington Registry (NZ Court Decision). Court Decision CIV 2013-485-877 (2014) NZHC 1067. Judgement of Mallon J.

OGTR (2019). Office of the Gene Technology Regulator (OGTR) (Australia): Overview of the Amendments to the Gene Technology Regulations 2001, September 2019, Available at: http://www.ogtr.gov.au/internet/ogtr/publishing.nsf/Content/23A002A6AC8A4908CA258478000D2031/$File/Overview%20of%20amendments%20to%20the%20Gene%20Technology%20Regulations%202001.pdf (Accessed June 7, 2020).

OSTP (1986). United States Office of Science and Technology Policy (OSTP): Coordinated Framework for the Regulation of Biotechnology, *US Federal Register*, 51(123), 23302.

Putcha, H. (2016). Using CRISPR/Cas in three dimensions: towards synthetic plant genomes, transcriptomes and epigenomes, *The Plant Journal* 87, 5–15.

Schmidt, S. M., Belisle, M. and Frommer, W. B. (2020). The evolving landscape around genome editing in agriculture, *EMBO Reports*, 21(6), e50680.

Tsuda, M., Watanabe, K. N. and Ohsawa, R. (2019). Regulatory Status of Genome-edited Organisms under the Japanese Cartagena Act, *Frontiers in Bioengineering and Biotechnology* 7, Article 387.

USDA (2010). United States Department of Agriculture (USDA): 7 CFR Part 340. Regulation Entitled: Introduction of Organisms and Products Altered or Produced through Genetic Engineering Which Are Plant Pests or Which There is Reason to Believe Are Plant Pests. https://www.aphis.usda.gov/biotechnology/downloads/7_cfr_340.pdf (Accessed June 7, 2020).

USDA (2020a). United States Department of Agriculture – Biotechnology Regulatory Services (BRS): Regulated Articles Letters of Inquiry, Available at: https://www.aphis.usda.gov/aphis/ourfocus/biotechnology/am-i-regulated/regulated_article_letters_of_inquiry/regulated_article_letters_of_inquiry (Accessed June 7, 2020).

USDA (2020b). United States Department of Agriculture (USDA): Movement of Certain Genetically Engineered Organisms Final Rule, *US Federal Register*, 85(96), 29790–29838.

Wasmer, M. (June 2019). Roads Forward for European GMO Policy – Uncertainties in the Wake of ECJ Judgment Have to be Mitigated by Regulatory Reform, *Frontiers in Bioengineering and Biotechnology*, 7, Article 132.

WEF (2018). World Economic Forum: "Bioinnovation in the Food System. Toward a New Chapter in Multistakeholder Collaborations", Available at: http://www3.weforum.org/docs/WEF_Bio-Innovation_Dialogue_WP.pdf (Accessed June 8, 2020).

Whelan, A. I. and Lema, M. A. (2015). Regulatory Framework for Gene Editing and Other New Breeding Techniques (NBTs) in Argentina, *GM Crops and Food* 6(4), 253–265.

Chapter 3

Key issues in the regulation of microbial bioprotectants in the European Union: challenges and solutions to achieve more sustainable crop protection

Rüdiger Hauschild, APIS Applied Insect Science GmbH, Germany; and Willem J. Ravensberg, Koppert Biological Systems, The Netherlands

1 Introduction

Microbial bioprotectants, also known as biopesticides, are subject to registration when used for the control of pests and diseases in crops, forestry and other situations where plants require protection. On a global scale, they are regulated by the same legislation as chemical pesticides. In this chapter, the term bioprotectant is used to describe a plant protection product (PPP) based on an active substance of natural origin, in accordance with the terminology used by the International Biocontrol Manufacturers Association (https://ibma-global.org/wp-content/uploads/2020/12/ibmadefinitionleafletweb.pdf; accessed 4 April 2021) and not the term biopesticide. Bioprotectants have many modes of action but often not a direct �-cidal� effect. In many cases, the term biopesticide is not defined in legislation, except for the USA (https://www.epa.gov/pesticides/biopesticides; accessed 4 April 2021). In European Union (EU) legislation, microorganisms are defined but the term biopesticide does not appear in pesticide regulations.

http://dx.doi.org/10.19103/AS.2021.0093.07

The legislative procedures for pesticides were designed many years ago and since the first bioprotectants were developed they have all been assessed, without exception, using these existing regulations. The resulting key issue is that data requirements for chemicals are not appropriate for living organisms that behave completely differently to chemical molecules. Moreover, the microorganisms currently used are found in the natural world and have co-evolved with all organisms. In contrast, most of the chemical molecules designed as crop protection substances are xenobiotic and unknown to nature.

The registration of pesticides involves two interconnected aspects: the administrative procedures and the data requirements that govern dossier compilation and the subsequent evaluation by means of risk assessment and risk management decisions. Both have an impact on the approval process of the microorganism as an active substance. In this chapter, we focus on data requirements as the key issue, but we will also discuss the procedures and the influence these two aspects have on each other and on microbial protectants entering the market. In the frame of the EU policy support action REBECA (EU Policy Support Action, 2006–2007), Hauschild et al. (2011) compared the data requirements and procedures between the EU and other systems. The data requirements are relatively similar due to the harmonization efforts of the OECD (Organisation for Economic Co-operation and Development), but interpretation differs between regulatory authorities. Procedures are more time-consuming in the EU due to the federal structure of the system. Guest (2018) provides an overview of regulatory procedures for biopesticides in 15 countries spread over the world and recognizes similar shortcomings and flaws in nearly all countries. Many authors have argued that the approval process for bioprotectants does not function efficiently (Villaverde et al., 2014) and that the system severely hampers placing new innovative biological products on the market (Guest, 2018; Grant and Gwynn, 2020), in contrast to the aim of our society to reduce the use of hazardous chemicals and replace them with safer alternatives.

The EU is recognized as having the most stringent approval system for active substances and the most complicated and time-consuming procedures (Balog et al., 2017; Frederiks and Wesseler, 2018; Wesseler, 2019; Guijarro et al., 2020; Sundh and Eilenberg, 2020). From the authors' personal experience, it takes approximately 5 years to obtain approval for a new active substance, and the PPPs based on that active substance to be registered, before it can be put on the market in one or more EU countries and used by farmers. One microorganism (*Metschnikowia fructicola* strain NRRL Y-27328) with almost simultaneous dossier submission in the USA and the EU was first granted product authorization after 21 months in the USA versus 60 months in the EU. Furthermore, in other parts of the world, the approval systems are also a major hurdle for (microbial) bioprotectants (Arora et al., 2016), including Australia (Hauxwell et al., 2010), India (Arjjumend and Koutouki, 2018; Keswani et al.,

2019), Asia (Singh et al., 2016) and Brazil (Bettiol, 2011). The regulatory process for biopesticides in the USA is generally recognized as the most appropriate and many authors recommend the implementation of a similar approach in other jurisdictions. Many authors worldwide argue that we need better systems that have more appropriate risks assessments and faster and cheaper procedures. In this chapter, we will outline the issues and provide suggestions on how an improved procedure can be achieved.

2 Issues of the administrative registration procedure in the European Union

2.1 Procedural and timeline issues

Administrative registration procedures vary substantially between authorities, including timelines, fees, but also dossier formats, methods of (electronic) submission, advice to applicants through pre-submission meetings, helpdesks, guidance documents, etc.; without doubt, the process is the most complicated in the EU. In most instances a national authority governs the process; however, in the EU all 27 (28 until 2020) Member States, the European Food Safety Authority (EFSA) and the European Commission are involved in the whole procedure. Moreover, it is a two-step process with approval of the active substance at EU level and product authorization at national or zonal level, for some applications (greenhouses, indoor treatments, seeds) at European level, making it a very complex, costly and lengthy process. Currently, it takes about 5 years for a product with a new active substance to get to the market; about 3–3.5 years for the approval of the active substance and 1–1.5 years for product registration. This is approximately twice as long as it takes in the USA, as concluded by Frederiks and Wesseler (2018) who compared both regulatory systems for the registration of microbial control agents. In the EU, delays are frequent due to limited resources in the relevant authorities and issues regarding dossier content. Moreover, some Member States have a dedicated authority performing dossier evaluation for the active substance as well as authorization of the PPP in-house, while in most EU countries the evaluation is performed by several institutes, specialized in, e.g. human toxicology or environmental evaluation, etc. The latter case is, as expected, frequently inefficient and contributes to delays and other issues. Another issue for applicants involves the procedural costs, which vary considerably between Member States. For instance, the fees for a Rapporteur Member State (RMS) procedure (writing the Draft Assessment Report, DAR) ranges from some tens of thousands of Euros up to a few hundred thousand Euros. The choice of a Member State as Rapporteur is up to the applicant although the choice is also guided by the availability of a time slot and the experience of an RMS. However, for a renewal procedure, the European

Commission assigns the RMS and an applicant has to accept the decision and pay the expenses of that particular RMS.

Considering the regulatory process of microorganisms in the EU, since 2015, 24 microorganisms have navigated the regulatory process in the EU, 17 as new active substances and 7 as renewals. Just 2 Member States acted as the RMS for the majority of the strains, 14 out of 24, with the other strains were spread over 6 countries. Only four countries invariably commented on the DAR during the peer-review process (Fig. 1). This shows that most EU countries do not take part in the evaluation of microorganisms, which may be due to a lack of resources, expertise or interest. Clearly, the current review process is not functioning efficiently. To improve the system, we recommend the following:

- Appoint specialist countries for groups of microorganisms (bacteria, fungi and other microorganisms, or according to ecological groups as discussed below).
- Build up expertise and, subsequently, risk assessment quality.
- Allow only a few other countries to comment, e.g. three per zone.

In this way, a more efficient review can be performed, which is faster and resource effective. All countries must then accept the evaluation of the

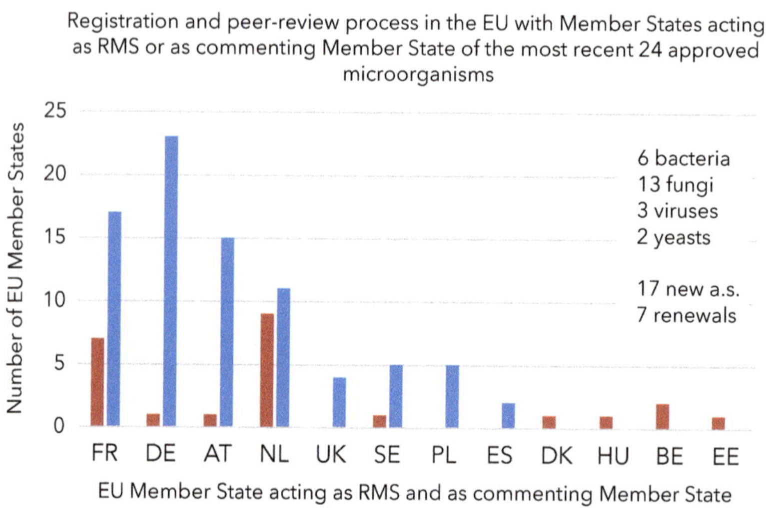

Figure 1 Registration and peer-review process of the most recent 24 approved microorganisms in the EU. Red: Country acted as RMS, blue: country provided comments. Countries: AT = Austria; BE = Belgium; DE = Germany; DK = Denmark; EE = Estonia; ES = Spain; FR = France; HU = Hungary; NL = the Netherlands; PL = Poland; SE = Sweden; UK = United Kingdom.

specialist countries; this approach should work as it is already, more or less, current practice.

The EU has evaluated its regulatory system in a so-called Refit procedure and the report (EC, 2020a) acknowledges many of these issues. The main conclusions, not only for bioprotectants, were that: ⍰Efficiency stands out as the critical area requiring attention. Due to a lack of resources and capacity in the Member States, most of the procedures set out in the regulations are suffering from severe delays, in particular, those related to the renewal of approval of active substances and the authorization of PPP containing approved substances, which in turn negatively affects their effectiveness⍰. However, it lacks recommendations for improvement, which will be developed within the Farm to Fork Strategy of the EU (EC, 2020b).

Both bioprotectants and conventional actives are processed by the same people and via the same procedures, whereas fundamental differences between these two types of products warrant a separation. In the USA Environmental Protection Agency (EPA) two separate streams were designed and have functioned well over the decades. Adopting such an approach in the EU would overcome many problems. Details and recommendations to improve the EU system for bioprotectants were detailed by Grant and Gwynn (2020), two crucial improvements include dedicated regulators at all levels with the necessary skills and harmonization and reciprocity of evaluations.

2.2 Low-risk active substances and plant protection products

EU Regulation (EC) 1107/2009 recognizes a low-risk (LR) category for active substances (a.s.) and products if they fulfill specific criteria; these were provided in an annex but designed initially only for chemicals. These criteria were later amended through Regulation (EU) 2017/1432 (EC, 2017) in order to also apply to microorganisms. Incentives to develop LR actives involve a longer initial approval period of 15 years instead of 10 years. However, approval renewal of the active substance is valid for 15 years for both chemicals and microorganisms and an LR advantage is therefore lost. Product authorization by national authorities should be within 120 days, which is ideal but in practice is almost never achieved, with the authorization process still taking approximately 1 year. Another issue is that authorities in an increasing number of cases do not automatically process a product containing a LR active substance as a LR product, which means that the reduced evaluation time of 120 days for product registration is not applied because the LR status is only assigned at the end of the evaluation. The LR status cannot be stated on the product label, which is the first information a farmer encounters before deciding to apply the product; however, it can be mentioned in any advertisement (Art. 66 of Reg. 1107/2009). Data protection for a LR product is 13 years, 3 years more than for a regular product. This means that data used for approval of an active ingredient are exclusive for the first applicant who generated the data for a longer period and cannot be used or referred to by other applicants. This period can be extended up to a maximum of 15 years when it includes extensions for minor uses. This incentive of longer data protection is seen as of little value by the biocontrol industry as much of the information on microorganisms is available in the public domain and other information is strain-specific and cannot be used by competitors anyway. LR products can benefit from reduced requirements for efficacy data: according to the European Plant Protection Organization (EPPO) Guideline PP 1/296 (1) (EPPO, 2017), less efficacy trials are needed for a LR product compared with a conventional product. However, if the LR status of the product is only assigned at the end of the evaluation, the applicant needs to provide the full set of trials anyway.

A specific problem arises from the fact that for precautionary reasons, respiratory protection is requested for products containing microorganisms due to their (theoretical) potential to induce sensitization. Implementing personal protection measures is seen as risk mitigation, which prevents registration as a ⍰low-risk⍰ PPP. Although the sensitizing potential does not explicitly preclude microorganisms from being low risk, their products are not approved as ⍰low risk⍰ because a product may not be approved as ⍰low risk⍰ if specific risk mitigation measures are applied, and although the respiratory protection derives from the precautionary phrase, it is considered a specific risk mitigation measure.

In practice, the ⍰low-risk⍰ status of the active substance has little effect on the producer or distributor of the product. The advantage of the extended approval period is relatively limited as it only reduces investment and workload at a later stage. Extended data protection is of little value for microorganisms as detailed above. Finally, the most attractive incentive, reduced evaluation times for low-risk products, is rarely achieved because either timelines are not met, or the decision if the product is ⍰low risk⍰is only taken after the evaluation for product approval.

All criteria outlined in Commission Regulation (EU) 2017/1432 refer to the hazard of a substance or a microbial strain and not to risk from their use. The term ⍰low-hazard active substance⍰ would be more appropriate than ⍰low-risk active substance⍰. Defining a ⍰low-risk active substance⍰ is impossible without determining the exposure to this active substance, and ⍰low risk⍰ cannot be a property of a substance. Defining a risk (for humans, non-target organisms or the environment) requires information on product formulation and information on the use of the product (application rates, frequency and methods, crops, etc.). For this reason, active substance evaluation is coupled to the evaluation of a representative formulation and representative use, and an active substance cannot be approved unless ⍰safe use⍰ is demonstrated. Consequently, as risk is the multiplication of hazard and the exposure to it, a product containing a ‘low-hazard active ingredient’ would in most (if not in all) cases automatically qualify as a ‘low-risk product’. Finally, the lack of implementation of the ‘low-risk’ principle results from confusion and inconsistencies in the terminology in Regulation (EC) 1107/2009.

Active substances are only approved as low risk after the full evaluation procedure has taken place. There is no faster procedure for potential low-risk active substances, only for products based on a low-risk active, but not in every case. For many years, the biocontrol industry advocated for faster approval, such as a preliminary approval based on a quick evaluation of the dossier at the point of dossier completeness check. Such an extended and adapted completeness check would also allow the provision of a temporary approval as low risk and a low-risk product (IBMA, 2018). A second option would be to grant this early approval after the DAR has been finished by the Rapporteur Member State; in this case, the low-risk status would be based on a full evaluation. If not confirmed in the peer-review process, this could be corrected after the full evaluation. The advantage is faster market access, which would give the farmer more low-risk tools as promoted in Article 1 of the Sustainable Use of Pesticides Directive 2009/128/EC (EC, 2009b) and has been in force since 2014. Currently, 10 years after the regulation came into force, only 26 active substances, of which 17 are microorganisms, have been approved as low risk according to the EU Pesticide Database (https://ec.europa.eu/food/plant/pesticides/eu-pesticides-database/public/?event=activesubstance.selection&language=EN; accessed 25 March 2020). This has led to a limited number of LR PPPs at national level and due to national differences in interpretation of mitigation measures, some products based on the LR a.s. are not authorized as LR PPP in certain countries. This illustrates the lack of harmonization within the EU Member States, which has prompted clarification. The EU also published a list of potentially LR a.s. that have been approved under former legislation EC 91/414 and are awaiting LR approval after the renewal process that is still ongoing for many of those a.s., of which 30 are microorganisms (EC, 2018).

Not all microorganisms are approved as LR. a.s. as illustrated by strains of *Bacillus*: 17 are approved as an active substance and just 2 as LR a.s. although some were evaluated before the low-risk category was established. Seven are being assessed as potential LR (EC, 2018) and await the status following the renewal decision. Another example is strains of *Beauveria bassiana*: six are approved (in October 2020), two of them under the old legislation and two in 2017 and two in 2019 under Regulation 1107/2009; however, none of them are LR. One of the reasons for not being assessed as low risk could be antibiotic resistance, but it is often data gaps and concerns leading to uncompleted risk assessments and therefore low risk is not granted. Here again, the inappropriate data requirements, often on secondary metabolites, prevent microorganisms from being assessed as low risk. However, the only low-risk criterion for microorganisms is the absence of multiple resistance to antibiotics (or antimycotics) (EC, 2020c). In cases where low risk was not granted because of data gaps for secondary metabolites, the decision was made prior to the low-risk criteria being implemented or by incorrect interpretation of the low-risk criteria. The EU is currently revising the data requirements for microorganisms, which will hopefully provide a better match between questions and answers and lead to less data gaps and concerns. As a consequence, more microorganisms are expected to be approved as low-risk a.s.

The principle of low risk in Regulation 1007/2009 is a positive initiative; however, in practice, very few LR PPPs have become available to farmers in the 10 years since this regulation came into force. The Refit report (EC, 2020a) states: “The availability of low-risk active substances, including microorganisms, has increased but is still considered by stakeholders as insufficient and procedures are considered lengthy”. The same report mentions: “The Commission acknowledged the need to find ways to accelerate placing on the market of low-risk active substances and products”. In addition, the new Farm to Fork Strategy of the Commission (EC, 2020b) only discusses a reduction of chemical pesticide use and states: “There is an urgent need to reduce dependency on pesticides...”. Furthermore, the Farm to Fork Strategy mentions: “The Commission will also facilitate the placing on the market of pesticides containing biological active substances” and “It will act to reduce the length of the pesticide authorization process by Member States”. Most likely “facilitate” refers to revision of the data requirements that will be seen as crucial; however, we doubt this will lead to a faster procedure as the Commission refers to the Member States. The latter are mainly responsible for product authorization, but the a.s. procedure takes considerably longer.

2.3 General Food Law and transparency in risk assessment

In June 2019, the “Transparency” Regulation EU 2019/1381 (EC, 2019a) on transparency and sustainability of the EU risk assessment in the food chain was published. It establishes new rules under the General Food Law (GFL, Regulation (EC) No 178/2002; EC, 2002) of the procedures for approvals of active substances and publication of information. It amended Regulation (EC) No 1107/2009, and another eight EU regulations and one directive concerning eight sectors of the food chain.

This new law details rules on transparency for products in the food chain regulated by the EFSA, which includes microorganism-based PPPs. It enables more public access to studies and limits protection of proprietary information. The most important topic in the General Food Law is the notification of studies. All studies that are initiated from 27 March 2021 onwards and intended for use in an application dossier must be notified to the EFSA by applicants and study laboratories. If studies have not been notified, or when notified studies are missing in the dossier at the time of submission, this will lead to delays or even rejection of the dossier by authorities, for both renewals as well as a first approval case. The obligation to notify all studies intended for a dossier adds further workload for applicants and requires detailed planning for several years in advance. This is particularly difficult in the case of active substances resulting from public research, where experimental data are not planned with the perspective of a regulatory submission as this may not be predictable during an early research phase.

Further to this, the EU Commission and EFSA decided, without consultation of the industry, to use the International Uniform ChemicaL Information Database (IUCLID) data format for the electronic submission of pesticide dossiers under the GFL. This is considered a step towards further harmonization between the EFSA and European Chemicals Agency (ECHA), which already uses IUCLID for Registration, Evaluation, Authorisation and Restriction of Chemicals under Regulation (EC) 1907/2006 (REACH) and Biocides. REACH is a Regulation of the EU, adopted to improve the protection of human health and the environment from the potential risks posed by chemicals (⍰except pesticides⍰) (https://echa.europa.eu/regulations/reach/understanding-reach). IUCLID is a software application to record, store, maintain and exchange data on intrinsic and hazardous properties of chemical substances. The format of dossiers is designed for REACH chemicals and biocides and not for PPPs. It needs to be adapted for chemical pesticides, and even more so for microbial active substances. The International Biocontrol Manufacturers Association (IBMA) requested a pilot project, as it was expected the implementation of this format would create difficulties for a dossier involving a microorganism. The pilot with a dossier on an entomopathogenic fungus revealed many difficulties involving how to fill in the studies and statements in the IUCLID format. Furthermore, transferring a dossier from a CADDY (⍰Computer-Aided Dossier and Data Supply⍰, a special XML format for electronic submission) or EU format takes a couple of hundred hours resulting in costs of many tens of thousands of Euros; consequently, guidance is being prepared by the EFSA. Improvements on the new format were necessary but the implementation date remained March 2021. This is another case where dossiers of microorganisms need to be squeezed into an inappropriate chemical format, accompanied by the burden of considerable extra expense. This will be extremely difficult for small biocontrol companies that develop new microbial PPPs. The new provisions in the GFL originated from issues with chemical PPPs, notably glyphosate, as the EU Parliament and Commission required more transparency in order to prevent issues on, for example, studies included or not for the risk assessments. Through the notification system, this will be prevented and transparency in decision-making improved; however, at the same time, it creates another hurdle for small niche and less harmful products that are required by our society, and will probably deter applicants from submitting their dossier in the EU.

3 Issues and solutions for the data requirements for microbial bioprotectants

3.1 The development of data requirements for microbials

The first bioprotectants were approved around 50 years ago and must have been a difficult process for both the applicant and the evaluating authority with only having data requirements for chemicals. Gradually, certain initiatives emerged that developed more appropriate requirements for microorganisms. The early development of these data requirements started in the 1970s in the USA and later in the EU. Over time, other organizations such as the International Organisation for Biological and Integrated Control (IOBC), OECD, EPPO and the Food and Agriculture Organization of the United Nations (FAO) became involved and developed guidance documents. For a historical overview on the development of data requirements in the USA and Europe see Ravensberg (2011). Despite the many improvements that have been made in this respect, many adaptations are currently still being implemented by authorities and are still requested by the biocontrol industry. Most of the adaptations are in the form of guidance documents, either updates or new documents developed by the OECD, EFSA, EU or FAO. Some of the latest examples are new guidance on secondary metabolites (EC, 2020d) and antimicrobial resistance (EC, 2020c) (published 23 October 2020) and draft guidance on whole genome sequencing and bacteriophages, a new group of biocontrol agents used in PPPs. Although the guidance was introduced, adaptations to the data requirements as laid down in the European Regulations have not been made.

We will use the EU as an example to illustrate why the registration of microbial PPPs remains a major obstacle in getting biopesticides to the market. The need for such improvements is best illustrated by the conclusions of the risk assessments of microorganisms by the EFSA, which generally contain many ‘data gaps’ and ‘areas of concern’. Although the conclusions of the EFSA on risk assessments in general contain these data gaps, the risk management decision, taken by the European Commission and Member States, leads to approval of the active substance, that is, the strain of a microorganism. In our opinion, this shows the fundamental mismatch between the legal data requirements for microorganisms and their products (laid down in Regulation 283/2013 part B and Regulation 284/2013 part B, respectively (EC, 2013a,b)), and the outcome of the registration process, resulting from different interpretations of data requirements. It would be helpful if the legal data requirements in the regulation were more appropriate to the biological characteristics of the microbial active substances, which could lead to less data gaps. However, the guidance documents would still be open to different interpretations. Historically, the questions were developed for chemicals and over time adapted for microorganisms, but that approach has not led to an appropriate and coherent set of relevant questions for a risk assessment of microbes. The European Commission has finally realized this and is currently revising the legal data requirements with the Member States and the EFSA, including the ‘Uniform Principles for evaluation and authorization of plant protection products’ (Regulation (EU) No. 546/2011 Part II) (EC, 2011). The Uniform Principles describe the harmonized way by which active substances and pesticide formulations need to be evaluated for the national authorization of pesticides. The biocontrol industry, through the IBMA, is also consulted. The IBMA is currently developing a new model based on a decision tree approach, which would basically lead to predictability of which portion of the data requirements can be addressed by public literature on the species' biology and ecology and where strain-specific experimental data are needed to answer only the relevant data requirements for a certain type of microorganism. We will describe this proposal in more detail later in this chapter (Section 3.3).

Another issue in the approval process is the knowledge and experience of the evaluators within authorities (Grant and Gwynn, 2020; Sundh and Eilenberg, 2020). They should be trained in microbial biology or ecology and have scientific experience in microbial interactions related to plants; however, many of them are trained as chemists and lack the scientific background on biology, ecology, etc. related to microorganisms. This is also the case for many evaluators within the EFSA. Living organisms behave completely differently under natural conditions than molecules and therefore this should be strongly considered during a risk assessment. The evaluation of the dossier should be performed by scientists that have a good understanding of the subject. Many chapters in this book illustrate the complex microbial interactions in the microorganism-plant system and between microorganisms. A way to improve the risk assessment is by installing a ‘Green TEAM’ as evaluators for biopesticides, as done by the Ctgb (Dutch Board for the Authorization of Plant Protection Products and Biocides) in the Netherlands. Scientists with a background in biology are trained to fulfil a risk assessment for biologically active substances. The complexity of microbial interactions is enormous and recognizing which ecological characteristics could pose a risk is a real challenge. A positive step is that the European Commission is setting up a training program for evaluators through the ‘Better Training for Safer Food’ Initiative on this topic (https://ec.europa.eu/food/safety/btsf_en, accessed on 30 March 2021). The Ctgb will be one of the trainers. Each competent authority, including the EFSA, should ideally have a ‘green team’ with specialists trained and experienced in the field of (microbial) bioprotectants. For many years, the number of new active substances applied for approval in the EU has been higher for bioprotectants than for traditional chemical active substances, and this trend will surely develop further in the direction of biological substances. In 2020, over 70% of the pending substances are of biological origin; hence, authorities must adapt to this trend. In the USA, the EPA has had a special division for the registration of biopesticides for more than 30 years, which has worked well (Leahy et al., 2014; Arthurs and Dara, 2019). This example should be followed all around the world. Brazil has recently adopted a similar approach and many biopesticides have been approved over the last 2 years, with registration roughly taking only 1 year (Togni et al., 2019; Domingues, 2019).

3.2 Data requirements: issues and solutions

The data requirements currently in place were introduced in June 2001 when Commission Directive 2001/36/EC (EC, 2001) came into force. These data requirements were transferred without change into the new Regulation (EC) No. 1107/2009 (EC, 2009a) with entry into force in April 2013 of "Commission Regulation (EU) No. 283/2013 (EC, 2013a)" and "Commission Regulation (EU) No. 284/2013 (EC, 2013b)" referring to the active substance and the PPP, respectively. This means that the experiences gained during the evaluation of microorganisms as "new" and "existing" active substances under Directive 91/414 between 2002 and 2013 were not considered for transfer to the new regulation.

Data requirements were derived from those used for chemical active ingredients and, to a certain extent, adapted to evaluate aspects that are specific for living organisms. The crucial difference between a living microorganism and every other plant protection active ingredient is the potential for survival and proliferation after product application. This is addressed with the data requirements on infectivity and pathogenicity to mammals and non-target organisms and persistence and proliferation on crops and in the environment. On the other hand, requirements for long-term toxicity, fate and behaviour in the environment and effects on non-target organisms are reduced compared with chemical active ingredients.

During the use of these data requirements over the last 19 years, interpretation has changed considerably. Data required and evaluated in, e.g. 2004 may either not be requested or would not be sufficient. This became evident during the re-approval process for more than 20 microbial active ingredients in the Renewal Programs AIR 2 to AIR4 (https://ec.europa.eu/food/plant/pesticides/approval_active_substances/approval_renewal_en (accessed February 24, 2021), when data accepted at the time of the first approval as an active ingredient were no longer sufficient for the renewal of approval, although legal data requirements were not changed. Interpretation of data requirements further varies considerably between evaluators from different EU Member States and the EFSA. This has led to a substantial loss of predicting what data would be needed, and to the costs and timelines for applicants.

One possible reason for the differences in interpretation may be due to data requirements being defined very broadly to include all microorganisms. Decisions defining which data are needed for a particular strain are taken on a case-by-case basis, meaning that some data may not be needed for the risk assessment of a particular group due to the biological properties of a microorganism. However, such data can be requested by evaluators, which leads to scientific effort and a regulatory burden for both the applicant and the regulator that is not justified.

Authorities and applicants have gained experience with the current data requirements over the last 19 years during the evaluation of more than 60 microbial strains. One key lesson learned is that new data requirements are necessary.

3.3 Biology as the basis of the risk assessment

Microorganisms should be assigned to groups, depending on their biology and ecology, to facilitate the interpretation of data requirements and the decision of what data are needed for a given microorganism. For the risk assessment, the actual biology and relationships to other organisms are more relevant than the taxonomy alone, and the risk of a biocontrol agent depends on its host range.

A decision tree is being developed by the IBMA based on ecological groups, which allows a more reliable definition of what experimental data are required and where literature information on the species is sufficient.

The decision tree (as well as the current data requirements) starts with the question of identity of the microorganism. Clear molecular identification according to up-to-date techniques and taxonomy is needed to identify the strain, to assign it to a species and to distinguish it from other related strains. Most biocontrol species (and all the species currently approved or under evaluation in the EU) are well known and consequently, their biology, ecology and host range are well described in the literature. Published information on the species and additional strain-specific information on the target organisms and mode(s) of action from the development of the biocontrol agent will allow assigning the species to an ecological group as a first approach for the risk assessment.

If the host range is well defined and limited (few related species of the same order or species from closely related families) the risk assessment is relatively easy because only host species that are exposed to the applied product need to be evaluated. In particular, obligate parasites with a limited host range have a very low risk to organisms outside their host range. Baculoviruses are a good example of this group which was acknowledged with the guidance document (EC, 2008) on the assessment of new isolates of baculovirus species already included in Annex I of Council Directive 91/414/EEC. Such organisms can only proliferate in the presence of their host, in fact, only inside the insect larval stage, which restricts their ability to multiply in the environment. For this group (also including other microorganisms with a comparably restricted host range, such as *Pasteuria* spp. and many bacteriophages), literature information should be sufficient for the risk assessment, and no studies of toxicity, infectivity or pathogenicity to mammals or other non-target organisms should be required. Such an approach is currently proposed for bacteriophages in the draft guidance based on their specificity. Formally, this group can be divided into two subgroups, the first (1.1) with a host range that only includes pest species or pathogens, and the second (1.2) with a broader host range that may also include non-pest species. In the second case, exposure of a non-target host needs to be determined. In fact, most if not all of the examples cited above would fall into the first subgroup.

If the host range is broader than just a few species of the same family, or no defined host exists, the risk assessment needs to be extended to all non-target organisms that are exposed. Such organisms may also be able to proliferate as saprophytes in the absence of a host, which might enable multiplication in the environment. This group 2 can also be subdivided into two subgroups: the first subgroup 2.1 includes specific parasites (e.g. *Ampelomyces quisqualis*, *Coniothyrium minitans*) with a restricted host range and low capacity to proliferate in the absence of their host. As they are not likely to adversely affect any organisms outside their host range, strain-specific data would be limited to survival and proliferation in the environment. The second subgroup, 2.2, includes non-specific parasites with the capacity to survive and proliferate as saprophytes. It includes most of the species commonly used in biocontrol, such as *Trichoderma* spp., *Gliocladium* spp., *Clonostachys* spp., ▯*Bacillus amyloliquefaciens* functional group▯, and *Pseudomonas* spp., which are active against plant-pathogenic fungi or bacteria, and insect or nematode pathogens such as *Beauveria* spp., *Isaria* spp., *Metarhizium* spp., *Purpureocillium* spp. or *Bacillus firmus*. For strains of this subgroup, strain-specific data are needed for the effect on non-target organisms (including mammals) and survival and proliferation in the environment.

The third group includes microorganisms that live permanently or occasionally inside a plant as an endophyte. To date, no such microorganisms have been evaluated in the EU. In addition to strain-specific data required for group 2 microorganisms, data on the potential accumulation of metabolites with adverse effects in the plant might be needed.

As described in the following chapter, metabolites produced by biocontrol microorganisms are difficult to evaluate. The decision tree proposes restricting the list of metabolites that need to be assessed to known bacterial or fungal toxins, especially mycotoxins known from plant pathogens or storage pathogens or bacterial toxins that cause problems as food or feed contaminants. These compounds are well known and a list of toxins that need to be addressed for biocontrol agents can be established based on lists for food or feed contaminants. If information detailing the biocontrol microorganism can produce such substances is available, the active ingredient or the formulated product needs to be analyzed for the presence or absence of these compounds. In case the compound is present, a risk assessment similar to the one for chemical active substances can be conducted. If there is no indication from the literature of the production of such compounds by the biocontrol species, and no adverse effects are observed in strain-specific studies, no further data on the production of secondary metabolites are justified.

3.4 Secondary metabolites as the key issue

3.4.1 Regulatory situation

One aspect of the risk assessment that probably triggers most additional requirements during the evaluation is microbial metabolites. Some microorganisms are well known for producing metabolites (primary and secondary) that may have an impact on non-target organisms including humans. Prominent cases linked to adverse effects via toxic metabolites (bacterial toxins or mycotoxins) include *Clostridium botulinum*, *Aspergillus flavus*, plant-pathogenic *Fusarium* spp. and others. The toxic metabolites produced by these microorganisms are well known and tested routinely in food or feed. None of these toxin producers has ever been considered as a biocontrol agent. If strains of these species are used in biocontrol, the absence of production of these metabolites needs to be assured, as in the atoxigenic *Aspergillus flavus* strain (see elsewhere in this book). Obviously, the data requirements for microbial metabolites are intended to prevent a toxin producer being used in plant protection. However, for microorganisms currently used in biocontrol, information is available in published literature or can be generated using genetic or biochemical methods to exclude the occurrence of toxin accumulation. Biocontrol microorganisms should not generally be regarded as toxin producers, and biocontrol companies and research institutes discard such organisms early in their screening processes due to this characteristic (Deising et al., 2017; Koch et al., 2018).

Another issue is a result of the terminology in data requirements for PPPs and their actives. Regulation (EC) 1107/2009 defines �metabolite� as follows: '"Metabolite" means any metabolite or a degradation product of an active substance, safener or synergist, formed either in organisms or in the environment�. This definition only refers to metabolites of chemical molecules and is not suitable for microbial metabolites. As described above, microbial metabolites are completely different regarding production and biology, but this is not reflected in the regulation and the data requirements for microorganisms. Moreover, EU regulations seem to overemphasize secondary metabolites compared with other countries such as Australia, the USA and Canada (Sundh and Eilenberg, 2020). We therefore suggest discussions take place regarding microbial compounds in order to emphasize their biological origin and avoid the link with chemical metabolites.

3.4.2 Microbial compounds and their biology

For species that are present in the environment such as those currently used in biocontrol (*Bacillus*, *Pseudomonas*, *Trichoderma*, *Clonostachys*, *Beauveria*, *Isaria*, *Metarhizium*, etc.), and their microbial compounds, other organisms have co-evolved and developed mechanisms to degrade or inactivate these compounds in case of adverse effects. Consequently, the effects on non-target organisms are expected to be limited and transient (Köhl et al., 2019a). However, if these compounds are contained in the PPP it must be considered that concentrations of microbial compounds of biocontrol agents can be substantially higher in the environment after product application compared with under �natural� conditions. In this case, a risk assessment is required and is possible based on the content of the compounds in the product, application rates and data on effects.

The case of *Bacillus thuringiensis* is excluded as the mode of action is at least partly based on crystal proteins (cry proteins), also called endotoxins. These proteins are pro-toxins that are only activated under the alkaline conditions of the insect midgut. The activated toxins are specific to certain insect orders (Lepidoptera, Coleoptera) and no other non-target organisms are affected; therefore, they do not present any risk to the environment or humans.

�Potentially toxic� microbial compounds are also known for some species of biocontrol agents, but to date, no adverse effects on humans or the environment linked to potentially toxic compounds from biocontrol agents have been observed or reported. Many so-called �toxins� are described in the literature for species of the genera *Bacillus*, *Trichoderma*, *Beauveria*, *Metarhizium* and others frequently used for the selection of biocontrol strains. However, the discovery and description of such toxins are in many cases driven by the search for compounds with activity for medical use, and effects in cell cultures or other *in vitro* systems are often considered sufficient to characterize a molecule as a ⍰toxin⍰. Some biocontrol agents (e.g. some *Trichoderma* strains) are able to produce compounds belonging to the same chemical group as known toxins, but the compounds from biocontrol strains are only distantly related to such known toxins and have little chemical similarity. In these cases, the compounds produced by biocontrol agents have far lower toxicity (if any at all) to mammals or other non-target organisms. Although the chemical, biological and toxicological properties of these compounds produced by biocontrol agents are different from such toxins, the fact that these compounds belong to the same chemical group causes concerns, which can trigger extensive data requirements.

In some cases, molecular phylogeny can exclude the production of a group of potentially toxic compounds if such production is limited to a phylogenetic group that does not include the biocontrol agent.

Among microbial compounds, a distinction needs to be made between those involved in the mode of action against the target organisms, and those produced independently of the target organism. For the risk assessment, compounds produced during fermentation and contained in the formulated product need to be distinguished from those produced *in situ*.

If a biocontrol strain belongs to a species known to produce ⍰potentially relevant metabolites⍰, different approaches are possible to exclude the exposure of humans and the environment to such compounds. If the genes involved in the biosynthesis are known, genome sequence information can be used to verify if these genes are present. However, the presence of such genes does not necessarily mean that these genes are functional, expressed and that they encode functional proteins. Genome data cannot reveal when and under what conditions a gene is expressed. Microbial compounds can further be identified (and possibly quantified) in the fermentation broth, the technical grade active ingredient and the formulated product.

We believe the production of microbial compounds in the environment after application of the biocontrol product (⍰*in situ* production⍰) does not represent a risk for humans or non-target organisms. Considering microbial ecology, the production and secretion of microbial compounds represent a loss of energy and nutrients for the microorganism. In environments where biocontrol organisms are applied, both are limited and many organisms compete for the same resources. Consequently, microbial compounds are produced and secreted in a highly regulated manner, depending on their function, the presence of a host and the status of the interaction with these (Köhl et al., 2019b). As many of the compounds are not stable under environmental conditions, they will not accumulate. In general, the accumulation of microbial compounds by microorganisms only occurs if substrate and energy are abundant. This is the case for plant pathogens such as *Fusarium* species in cereals that accumulate trichothecenes, *Aspergillus flavus* that accumulates aflatoxins or *Aspergillus* and *Penicillium* species that produce ochratoxins. The production and accumulation of mycotoxins can occur during plant growth or during storage if conditions are permissive for production. In these cases, the fungi have abundant substrates (energy and nutrients) and toxins are also produced to prevent other organisms from colonizing the substrate. The only situation where a biocontrol organism has (sufficient) substrate to accumulate microbial compounds is the case of entomopathogenic fungi. After infesting and killing their host, generally an insect, some strains of fungi such as *Beauveria*, *Metarhizium* or *Isaria* can produce and accumulate microbial compounds while they are growing inside and on the surface of the dead host. However, as these fungi occur naturally, other animals potentially feeding on colonized insects have co-evolved with the fungi and either avoid such feed or have mechanisms to cope with these compounds if they are toxic. Consequently, secondary poisoning in birds or mammals by such compounds produced by biocontrol agents has never been observed although it may be theoretically possible. Taken together, the *in situ* production and accumulation of potentially toxic compounds by biocontrol microorganisms is very unlikely and effects on non-target organisms, including humans, have never been observed. This was confirmed in an extensive review in Annex 1 of the OECD guidance document (OECD, 2018), which collected information on secondary metabolites from different microorganisms and demonstrated that these do not pose a risk to non-target organisms including humans.

3.4.3 Stability of microbial compounds

A technical problem of the current data requirements is the low stability of many microbial compounds. Data requirements include information on the stability and persistence of relevant compounds in the environment. As described above, most of the compounds produced by biocontrol agents are produced in small amounts at discrete sites, and accumulation does not occur (with the exception of entomopathogenic fungi, though only in hosts and thus in a limited quantity). In addition, many of the compounds have a short half-life. If data proving the absence of a certain compound from an environmental matrix (crops, soil) are required, technical difficulties make such proof impossible. Many of the compounds are not commercially available, and the development of methods is very expensive in terms of costs, time and workload. However, even if sensitive methods could be developed, proving absence is mostly impossible because many compounds will be below detection levels. Taken together, low *in situ* production rates and instability make it impossible to demonstrate the absence of a compound, and this often results in a 'data gap' that can prevent the approval of the microorganism as an active substance for plant protection. On the other hand, no reports of adverse effects from compounds produced by biocontrol organisms are known to the authors, and the risk appears to be extremely low. The probability that a biocontrol agent of a well-known species produces a yet unknown toxin can be neglected if no adverse effects are observed in studies on toxicity of the active ingredient or the product with mammals or other non-target organisms. Similarly, for a species that is not well described, studies with the active ingredient or the product should be sufficient to detect potentially toxic compounds. In addition, sufficient knowledge on the biology, ecology and mode of action needs to be provided to be able to assess the possibility of producing and accumulating potentially toxic compounds.

We believe that data requirements for microbial compounds should be limited to toxins that are well described and routinely assessed in food or feed unless data from toxicity studies on mammals or other organisms indicate that toxic compounds are present in the active ingredient or the product.

3.4.4 New EU guidance on secondary metabolites

The EU recently adopted a new guidance document on microbial metabolites (EC, 2020d), with the aim of identifying all compounds potentially produced by a strain, starting with literature information. Data on potential adverse effects and exposure need to be collected. Genome information can be used to exclude certain compounds from the assessment if it can be demonstrated that the gene(s) involved in the biosynthesis are absent. Subsequently, toxicity data are collected and compared with exposure in a risk assessment similar to those for chemical actives. Unless exposure from compounds contained in the product or produced *in situ* after application of the product cannot be excluded by experimental data (which is extremely difficult as detailed above), a risk assessment needs to be conducted. Unfortunately, this guidance starts from the assumption that all compounds are toxins, and that biocontrol microorganisms are toxin producers, similar to some fungi and bacteria cited above. Only unknown compounds produced *in situ* do not need to be considered unless adverse effects are observed in studies with non-target organisms. No practical experience with the guidance document exists to date, but it appears it would considerably increase the workload for applicants to prove the ‘innocence’ of a biocontrol organism if every compound needs to be assessed as a potential toxin.

3.5 Testing methods

Many data requirements can be addressed by the literature on the species. Others need strain- and/or product-specific experimental data and studies have to be conducted. Testing methods for studies to fulfil regulatory requirements are another aspect where improvements are needed.

3.5.1 Physical, chemical and technical properties of the formulated product

Guidelines for physical-chemical and technical properties are designed for conventional chemical products and the same guidelines are applied to microbial products. This is justified as products containing microorganisms are applied, in most cases, with the same equipment as conventional products, and compatibility and safety for the user need to be assured. However, sometimes methods are difficult to apply, especially if the content of the active substance needs to be determined in a bioassay, which involves activity testing via quantification of the amount of the biocontrol agent needed to kill the host of the biocontrol microorganism, as required for baculoviruses. For example, this is the case for testing suspensibility or spontaneity of dispersion, and many laboratories that can perform the technical test are not able to perform the bioassay, which complicates study conduct.

The determination of microbial contaminants is currently recommended to be performed using traditional plating methods, which were originally developed for food material with a relatively low microbial content. For a microbial biocontrol product consisting of a large proportion of microbial material, sensitivity of the growth media may not be sufficient and potential contaminants may still be overgrown by the biocontrol agent. The detection of contaminants according to the list provided in the Working Document on Microbial Contaminant Limits for Microbial Pest Control Products (EC, 2012) by quantitative polymerase chain reaction (qPCR) has already proven to be more sensitive and reliable in other fields but is still not widely accepted by authorities for PPPs.

Methods used in studies need to be validated, and again the relevant guidance document (EC, 2019b) was developed for chemical actives and not adapted for microorganisms. However, deficiencies in the guidance concerning microorganisms are recognized and adaptations are accepted by authorities.

3.5.2 Toxicity, infectivity and pathogenicity to mammals

Testing methods to assess toxicology, pathogenicity and infectivity to mammals are mostly acceptable. Methods developed in the USA (OCSSP 885.3050, 885.3150 and 885.3200) are widely accepted. These methods enable evaluation of the pathogenicity of the biocontrol microorganism and potential contaminants, infectivity of the biocontrol agent and contaminants, and toxicity of compounds contained in the test material that originates from the biocontrol agent, potential contaminants or impurities from growth media or co-formulants. Technically, clearance assessment for viruses or obligate parasites that do not proliferate in the absence of their host is probably the most difficult part, because their presence in the test animals can only be determined in a bioassay with the target organism. This bioassay might be inhibited by compounds from the test animal tissue matrix. For example, conducting an infectivity test with baculoviruses on their host insect with material from rat tissue may not be possible; however, the clearance assessment can be accomplished by qPCR methods.

Another point of discussion is the necessity to conduct studies for toxicology, pathogenicity and infectivity via three pathways, oral, pulmonary and injection. Pulmonary exposure is mostly achieved by intratracheal instillation instead of inhalation, which is technically difficult because of the particle nature of microbials. Injection application is preferably intravenous for bacteria and viruses and intraperitoneal for fungi and protozoa. Oral exposure is the most likely exposure scenario for humans (consumers, if the microorganism persists long enough) and wild mammals after product application, but on the other hand, both groups are continuously exposed to a plethora of microorganisms and mammalian defence systems are adapted to withstand microorganisms. Inhalation exposure is also relevant, especially for operators or bystanders, and lung infections are probably the most dangerous potential infection by biocontrol organisms, especially in immunocompromised people. Using the intratracheal application, microbial material is placed within the lungs at a site, which microbial particles would probably not access under natural conditions, because they would not stay in the airstream throughout the passage from the upper lung tracts to the trachea. Consequently, the exposure in the study represents more a worst-case scenario than a realistic scenario as the first protection of the upper respiratory tract against infection has been overcome by placing them at a site that would be unreachable under natural conditions. Intraperitoneal or intravenous injection testing represents an unrealistic exposure, and the outcome can, in most cases, be predicted from intratracheal studies. As skin infection through wounds is a more likely scenario of microbial infection than deep body infections, subcutaneous testing may be a preferred method if infectivity testing is required. We recommend a tiered approach: intratracheal testing should be performed first as the ‘realistic worst case’. If no adverse effects are observed, no further testing should be required.

In our experience, over the last 20 years, adverse effects have only been observed in one infectivity study out of more than 100 studies involving over 50 microbial strains; all the other studies revealed no adverse effects. This demonstrates that knowledge on the taxonomy and biology of the strain is crucial for the evaluation and that testing can be reduced to a single study on intratracheal toxicity, pathogenicity and infectivity per strain.

3.5.3 Exposure to consumers and the environment

Data requirements do not foresee studies for residues and environmental fate, but information is required for survival, persistence and proliferation of the microorganism on crops, food and feed, and in soil, air and water. Such information on the species can generally be found in the literature. Strain-specific data could be generated along with efficacy trials under realistic application conditions, but they would only be needed for some entomopathogens with a broad host range. However, in general, introduced microorganisms always decline due to the natural biological buffering of the environment to levels that are within common fluctuations and ranges without strongly affecting microbial communities and pose no risk to the environment (Köhl et al., 2019a).

3.5.4 Effects on non-target organisms

Data requirements on the effects on non-target organisms can also be addressed using literature about the species, and knowledge on the taxonomy and biology of the biocontrol agent can exclude a hazard for many non-target organisms. If studies are needed, the guidelines used to date do not always address the critical point appropriately. Similar to human health assessment, studies should be able to address pathogenicity and infectivity by the biocontrol microorganism and potential contaminants, and toxicity of compounds contained in the test material that originates from the biocontrol agent, potential contaminants or impurities from growth media or co-formulants.

Guidelines for birds and fish are widely acceptable, but over recent decades, adverse effects have rarely been observed. *Daphnia* tests are probably the most difficult for microbial products as *Daphnia* are very sensitive to particles in their growth media. As microbial biocontrol products always present such particles, adverse effects are frequently observed in *Daphnia* studies although the particles have more impact than microorganisms *per se*. Determination of infectivity and pathogenicity in daphnids is very difficult due to the size of the test organism and the difficulty of distinguishing between microorganisms in or on the daphnid. Such assessment requires great experience and expertise but is not always possible. The test using *Daphnia* is probably more appropriate for the detection of potentially adverse effects caused by microbial compounds or process impurities than for the assessment of infectivity or pathogenicity. On the other hand, *Daphnia* pathogens are well known. Interestingly, two biocontrol agents currently approved in the EU belong to the same genera as daphnid pathogens (*Metschnikowia* and *Pasteuria*). The biocontrol microorganisms and the *Daphnia* pathogens are only distantly related and differ in their biology, and no adverse effects were observed in the studies. This again highlights that sound information on taxonomy and biology is sufficient to exclude a risk to non-target organisms.

Testing the effects of biocontrol agents on non-target arthropods requires adaptation of methods in many cases. Tests should assess infectivity/pathogenicity rather than toxicity. Finding the right test organism is difficult because some of the established test organisms live for a shorter time than the infection period needed to show an effect. This is the case, for example, for *Aphidius*, where adults only live a few days whereas infection may need up to 2 weeks to establish. Alternative tests would be far more intensive and, in many cases, would need to be developed. On the other hand, effects are expected for entomopathogens with a broad host range and can be excluded for many biocontrol microorganisms based on literature information. Very recently, Reinbacher et al. (2021) proposed a new method to test the effects of entomopathogenic fungi on *Gaeolaelaps* (*Hypoaspis*) *aculeifer*.

No adverse effects from biocontrol microorganisms are expected for soil microorganisms and earthworms and tests are not required. Argumentation based on literature is widely accepted. In fact, these tests are obsolete and do not provide any useful information for a risk assessment; hence, this data requirement should be removed.

A further aspect is the methods determining the effects on sewage organisms or on methods for the determination of drinking water quality. Interestingly, both issues are not included in the data requirements, but in the Uniform Principles. Effects on sewage treatment organisms can be excluded for biocontrol organisms because they are not aquatic organisms. Biocontrol agents may survive in water, but seldom enter in a vegetative phase and very rarely proliferate. Furthermore, they are weak competitors compared with sewage microorganisms and only small populations reach wastewater treatment plants. A risk to sewage microorganisms can therefore be excluded. Methods for drinking water analysis are specific for potential contaminants and designed not to interfere with other organisms. Therefore, any interference is unrealistic. Both data requirements are obsolete.

4 New research, novel products and application methods

Research on the biological control of pests and diseases involves both academia and industry, and regularly delivers innovative products. A search on Google Scholar with the term ‘biocontrol agent’ presents 30 000 hits over the last 5 years, with even more hits on ‘biocontrol products’. How many of these products reach the market is impossible to say, but novel, innovative PPPs have to pass regulations and, as is well known, innovation is often hampered by regulations that were not designed for those novelties. One could even say this for biocontrol products in general that have to pass regulations designed for chemicals. New biocontrol products that we see today or in the near future include bacteriophages, weakened or mild strains of plant viruses, endophytic microorganisms, microbial consortia, amoebae (already developed as a biocide), gene-edited microorganisms, genetically modified microorganisms, etc. Research on the microbiome of plants is a new, exciting field where many microbes are being discovered as well as their functions in relation to the plant. When these relate to plant protection will they fall under the current regulations? We still need to see practical applications originating from this microbiome work, but it is bound to generate new approaches to keep plants healthy. In addition, we see new functions and applications of biocontrol agents, sometimes with multiple modes of action and multiple uses, such as biofertilizers, biostimulants, inoculants, plant defence inducers or elicitors, etc. Novel application methods are developed to deliver the microorganism to their target, for example, using honeybees and bumblebees as vectors to deliver the product at the flower, stem-injection (see elsewhere in this book), applying weak plant viruses by making small leaf surface wounds via a rub-in liquid, various types of seed treatment, and wound treatment via pruning scissors. Furthermore, formulations with two active substances can be developed, such as with two strains, or a microorganism and a chemical or a botanical. With precision agriculture, we expect a different method of controlling pests. Early applications will become more routine based on early observations of pests and diseases in distinct locations, with different application devices, such as robots and drones. What are the consequences of the risk assessment of such technologies? It can be a real challenge to get such innovative products, with their new ways of application, through the registration system. It requires a flexible approach in building a dossier, but also in the manner of risk evaluation by authorities. The product developers understand their new products well, but for evaluators it can be a challenge to reach the same level of understanding and allow a flexible approach; hence, ticking the box on all standard questions is not an appropriate approach. Knowledge of the subject should guide a risk assessment and will be key with such novel PPPs as mentioned above, and the training of evaluators will become even more relevant. The development of a decision tree will help to guide applicants and evaluators through the system and allow only relevant issues to be emphasized. Guidance is crucial to applicants, so consultation with authorities at an early phase is an absolute requirement, such as pre-submission meetings, guidance documents, workshops and training.

5 Conclusion and future trends

5.1 *The Pesticide Paradox*

Pesticides have been under societal pressure ever since the publication of '*Silent Spring*' by Rachel Carlson in the 1960s. Many organizations advocate for the decreased use of chemical pesticides because they have harmful effects on human health and the environment, even after their approval by authorities. Recently, there have been fierce ongoing debates over chemicals such as glyphosate and neonicotinoids. Switzerland will have a vote to ban synthetic pesticides in early 2021. Another example is the launch in early 2020 of an ambitious roadmap by 24 European research institutes to find alternatives to chemical pesticides (https://www.inrae.fr/sites/default/files/pdf/CP_MoU_VFinal2302_VA_0.pdf; accessed 25 October 2020). Moreover, authorities recognize this and policies are being developed to reduce the use of chemical pesticides, e.g., the Sustainable Use of Pesticides Directive (Dir. 2009/128/EC) (EC, 2009b) and, very recently, the EU Farm to Fork (F2F) Strategy (EC, 2020b) as part of the EU Green Deal. Surprisingly, little or no provision has been developed to allow alternatives to chemical pesticides to enter the market in a timely and appropriate way. Pests and diseases will always be present in agricultural crops and farmers need tools to protect their crops and maintain food security; bioprotectants can be one set of tools that can be used in a sustainable way. The F2F Strategy states: 'The Commission will also facilitate the placing on the market of pesticides containing biologicalactive substances and reinforce the environmental risk assessment of pesticides. It will act to reduce the length of the pesticide authorization process by Member States'. The word 'facilitate' most likely refers to revision of the data requirements for microorganisms. However, to what extent this will make the questions more appropriate remains to be seen. Moreover, the Commission does not appear to be planning to revise the requirements for pheromones and botanicals. Draft guidance, similar to that on secondary metabolites, does not appear to make it easier, but rather more demanding. On the other hand, we see that a risk assessment can be made more appropriate to fit the biology of a bioprotectant, the example we refer to is the draft guidance for bacteriophages. Phages and baculoviruses are the outstanding examples of the direction crop protection products should be taking, to specific solutions that target the intended pests and without harmful effects on other organisms. Many microbial bioprotectants are developed for niche markets as their biology only allows such narrow targeting. Examples are baculoviruses, but also many fungi, yeasts and bacteria that, due to their mode(s) of action, only have effects on a limited number of other organisms. The decision tree discussed above is based on biology and ecology, and a risk assessment should only address the real risks related to the biology of the microorganism. We are convinced that in this way, with specific microorganisms, safe PPPs can be developed for nearly all pests and diseases. With this approach, many small products have the potential to become available for the farmer's toolbox. However, here lies the difficulty, the regulatory system has developed in such a way that the expense and administrative burden has become so enormous that it is impossible for small-scale applicants with niche products to get through the system (Robin and Marchand, 2019); however, these are exactly the type of products that are demanded in our society. The system has become so restrictive that only broad-spectrum products with big markets and developed by multinationals can get through. The lengthy procedure of about 5 years is a compounding factor, as waiting that long for any income is difficult or even impossible to overcome for small companies; it also deters investors in this area. Product development and registration combined take about 10 years and a return on investment from the outset of a project can only be expected after 13–15 years. It is clear that getting niche products to the market is a very complex and extremely risky entrepreneurial exercise. Although our society wants safe and selective PPPs to be used by farmers, the regulatory system is a huge hurdle to achieving that goal. The GFL will only exacerbate this situation although its purpose is to have a more transparent and accepted risk assessment. This is the great paradox of pesticide registration in the EU. Therefore, we advocate for a separate legislative framework for bioprotectants, particularly for products based on microorganisms. An independent system with its own resources and expert evaluators will then be able to achieve timely market access for the innovative and sustainable tools that agriculture needs from now on. We have seen, over a number of years, that a large part or even more than half of the new active substances are based on products of biological origin, so the system should be ready imminently to deal with that. There can be no reversion to traditional chemicals and an ambitious EU agricultural sector cannot wait any longer for such a new approach.

5.2 Crop protection policy

This policy is mainly shaped via efforts to reduce the use of chemical pesticides and the dependency thereon. What we see in practice is that the Sustainable Use Directive (SUD) goal is not reached as Integrated Pest Management (IPM) is hardly used by European farmers and this reduction is not achieved, with dependency on chemicals remaining very high (ECA, 2020). A tool like the comparative assessment of chemicals for candidates for substitution has failed completely. Chemicals are withdrawn from the market when they are re-assessed and this was achieved with the implementation of Regulation 1107/2009 in the process of taking over from Directive 91/414 when approximately 400-500 actives were taken from the market. It is a process that is based solely on the risk assessment of individual chemicals. Market demand and company policies determine whether new actives are submitted for approval in Europe; however, the number of pesticides over the last 10 years has increased (Buckwell et al., 2019). Farmers have complained, and still do, about not having enough tools, but this pesticide policy does not look at whether a farmer has a sufficient set of tools for a certain crop and for an appropriate IPM programme. The European policy, in this respect, is completely insufficient, and to our knowledge, there are no programmes in individual Member States that are performing any better despite several attempts. New attempts are being developed in countries like France and the Netherlands. The Dutch plan aims for resilient agriculture, where resilient crops and minimal use of low-risk pesticides will protect the crops and have less impact on the environment. This approach demands a better and faster registration system for new, selective low-risk pesticides, mainly bioprotectants, and only a new approval regime will be able to deliver that. Farmers will need to be further trained in this field and need advice from IPM experts. The ambitious plans in the EU need to deliver sustainable agriculture within the next 10 years in order to stop the decline in biodiversity, soil fertility, pollinators and other ecosystem services, as well as the debates on the harmful effects of today's pesticides. The future of crop protection is in regenerative agriculture with bioprotectants and biostimulants as the farmers' tools to intervene in cases of abiotic and biotic stresses, in crop-specific IPM programmes. An innovative legislative framework is needed to make that possible. In addition to ambitious plans in the EU Green Deal, efforts are also required to design an appropriate regulatory system to support the transition.

5.3 Recommendations to improve the approval process and accelerate market access of microbial bioprotectants

The registration of a microbial bioprotectant is a huge challenge worldwide and assessing safety for humans and the environment is the key issue. The variety of microorganisms in nature is enormous and new ones are still being discovered, and for the majority of them, scientists do not know what their functions are. With modern technologies, the microbiome of the plant and soil are being studied and both contain a plethora of microorganisms that are crucial to the effective functioning of both plant and soil (see elsewhere in this book). On the other hand, many are known to be beneficial, while others can cause serious problems and diseases in many organisms. These features are most likely the cause of a general fear of "invisible" microorganisms as we experience today with the coronavirus pandemic.

Is the risk assessment driven by fear of the unknown? Microorganisms make all kinds of compounds to defend themselves from antagonists and predators and these substances can have beneficial as well as harmful effects. Microorganisms can adapt to new situations and even transfer genetic material;

these combined factors make them extremely complex and hard to fully understand. However, they are part of nature and have evolved in conjunction with other organisms.

In order to assess their use and safety, their biological origin must be considered in this context. This requires evaluators that understand this aspect of biological sciences and ecological principles. Living microorganisms are totally different from chemical molecules and assessing their safety is a completely different discipline. The first microorganism used as a PPP was evaluated as a chemical product, following the same regime, very similar data requirements and by the same evaluators. This was quite logical and understandable, but of course that system was not suitable for the purpose. Adaptations were made over time, over the course of decades, to improve the process. This was repeated in many countries and today that process is still ongoing. The above sections demonstrate that the current legislative system in the EU has its flaws and weaknesses, and is in fact a complete mismatch, despite all the attempts to repair it with adapted data requirements, guidance, etc. The authors' experience goes back 40 years when it was a struggle in the EU and still is today, and maybe even more so. Recommendations to improve the EU system were formulated in the EU policy support Action REBECA (Ehlers, 2011) and further detailed by Scheepmaker et al. (2019) for microbial metabolites, but very few of these recommendations have been implemented.

Based on the above arguments we conclude the following:

Recommendations for improvements in the procedure:

- Establish specialized teams with well-trained experts;
- Concentrate evaluation of microbials with a few Member States in the EU; and
- Implement conditional approval for low-risk products.

Recommendations for improvements in data requirements:

- Consider the biology and ecology of the biocontrol agent first;
- Restrict evaluation of ‘metabolites’ to known toxins;
- Conduct studies only if other information is not sufficient; and
- Adapt study protocols.

Remarkably, we are currently witnessing the same situation with biostimulants. This new category of products needs to be regulated due to concerns over their use and safety, and they have been added within legislation that was designed to cover plant nutrients and fertilizers. In particular, biostimulants based on microorganisms do not fit in and the system lacks a procedure and data requirements to assess their safety and approval. They have been placed in a legislative framework that superficially seems acceptable but in fact simply causes problems, in parallel to what happened with microbial PPPs. Sundh and Eilenberg (2020) observed that for many applications of microorganisms in food, feed, agriculture and the environment, regulations and the risk assessments vary greatly and lack proportionality. Similar issues can be expected when microbiome research leads to the use of more microbiological products in various disciplines. These issues may be justified in the early days of such microbial products but when the utilization of microorganisms increases greatly in all these different uses, a new approach is necessary. These microorganisms need to be assessed on safe use and hence, the exposure of humans and the environment to them. Their claimed functional use is hardly relevant as microorganisms are not limited to one claimed area but can have several properties, such as plant protection and biostimulation, which may be expressed depending on the environmental conditions. Their biological characteristics dictate whether or not they can be safely used, and not the administrative box they are put in. Currently, a strain of a microorganism is assessed differently on safety related to which use is claimed and under which legislation with its own data requirements it falls. This makes no sense; therefore, a specially designed legislative framework for microorganisms is required in which safety is assessed independently from claimed functions. Data requirements need to be proportionate to the actual risk, and literature information at species level should be accepted as the basis of the evaluation. Together with solid molecular identification and taxonomy, a microorganism can be assigned to an ecological group, which defines what experimental data are needed in addition to the literature on the species for the risk assessment. The new system should also have scientists educated and experienced with microorganisms, including pathology and ecology. Procedures should be set up and resources allocated specifically for this administrative process. In this way, conflicts with other products under the same regulation can be avoided as we see today with microbials in the chemical system and biostimulants in the fertilizer system. In current systems, we see that attention and resources are mainly used for the biggest groups of products with the biggest concerns. Examples are, respectively, health and environmental concerns with pesticides, and pollution concerns with fertilizers. With a separate system, this can all be avoided, the assessments will be more appropriate and market access can be achieved much faster. Products that are much more sustainable and environmentally friendly will then reach the market, and policy goals on circular economy, less pollution and increased sustainability can be achieved. Therefore, we strongly appeal for a new legislative framework for the approval of microorganisms for use as PPPs, plant biostimulants, feed additives, biocides and maybe more in the future. The EU can lead the way and be an example for other countries around the world. We advise considering the suggestion of the former Ctgb director L van Duijn (2018) who proposed using as a baseline ‘biopesticides are low risk unless…’. Many of his other suggestions are in line with many of our recommendations and solutions. His low-risk proposal was submitted to the EU via the Dutch Ministry of Agriculture as a reply to actions needed based on the Refit evaluation of Regulation (EC) 1107/2009.

The daily exposure of humans to millions of microorganisms in our food and our co-evolution can be regarded as the basis for this proposal. With every apple we consume we are exposed to 100 million (10^8) bacteria (Wassermann et al., 2019), the majority of which are unknown and have never been studied. A similar situation is found regarding fungi in strawberries and apples (Abdelfattah et al., 2016a,b). Interestingly, it is broadly recommended to eat fresh fruit every day for our health, which is illustrated by the well-known saying ‘an apple a day keeps the doctor away’. The application of a biocontrol agent would result in a population of 1.3 x 10^6 colony forming units (CFU)/apple, which is about 1% of the overall population on an apple (see Box 1 for the detailed calculation).

Box 1

To estimate the populations of a biocontrol microorganism on apples resulting from the application of a microbial biocontrol product, the following assumptions are made:

- Single application rate for the biocontrol microorganism: 3 x 10^{12} CFU/ha
- Apple yield: 28 t/ha (average for Germany, in France 38–40 t/ha)

$$3 \times 10^{12} \text{ CFU} / 28 \text{ t} = 1.1 \times 10^5 \text{ CFU/g}$$

Apple weight: 200 g: 2.2 x 10^7 CFU/apple.

- If 60% of the population end up on plants and 40% on soil, theoretical maximum population levels are 6.6 x 10^4 CFU/g fruit or 1.3 x 10^7 CFU/apple. This calculation does not take into account that the majority of cells of the biocontrol agent will end up on leaves and not on fruit because the fruit surface is relatively small when compared with leaves. If we assume that 10% of the biocontrol agent ends up on fruit, the population would be 1.3 x 10^6 CFU/apple. This is about 1% of the existing bacterial microbiome of the apple.
- In addition, decreased populations of the biocontrol agent between application of the product and harvest through UV radiation or washing off by rain is not considered, but this decrease is substantial and requires repeated applications.

How relevant is the addition of 1% of a well-characterized microbial strain to the mostly uncharacterized microbiome on a fruit and how realistic are risks associated with this additional microorganism? This example justifies, in our opinion, that the evaluation of microorganisms needs to be revised.

6 References

Abdelfattah, A., Wisniewski, M., Li Destri Nicosia, M. G., Cacciola, S. O. and Schena, L. (2016a). Metagenomic analysis of fungal diversity on strawberry plants and the effect of management practices on the fungal community structure of aerial organs. *PLoS ONE* 11(8), e0160470. https://doi.org/10.1371/journal.pone.0160470.

Abdelfattah, A., Wisniewski, M., Droby, S. and Schena, L. (2016b). Spatial and compositional variation in the fungal communities of organic and conventionally grown apple fruit at the consumer point-of-purchase. *Hortic. Res.* 3, 16047. Available at: https://doi.org/10.1038/hortres.2016.47 (accessed 5 October 2016).

Arjjumend, H. and Koutouki, K. (2018). Science of biopesticides and critical analysis of Indian legal frameworks regulating biocontrol agents. *Int. J. Agric. Environ. Biotechnol.* 11(3), 563–571.

Arora, N. K., Verma, M., Prakash, J. and Mishra, J. (2016). Regulation of biopesticides: global concerns and policies. In: *Bioformulations: For Sustainable Agriculture* Arora, N. K. and Mehnaz, S. and Raffaella Balestrini, R. (Eds), Springer, New Delhi, pp. 283–299.

Arthurs, S. and Dara, S. K. (2019). Microbial biopesticides for invertebrate pests and their markets in the United States. *J. Invertebr. Pathol.* 165, 13–21.

Balog, A., Hartel, T., Loxdale, H. D. and Wilson, K. (2017). Differences in the progress of the biopesticide revolution between the EU and other major crop-growing regions. *Pest Manag. Sci.* 73(11), 2203–2208.

Bettiol, W. (2011). Biopesticide use and research in Brazil. *Outlooks Pest Manag.* 22(6), 280–283.

Buckwell, A., de Wachter, E., Nadeu, E. and Williams, A. (2019). *Crop Protection & The EU Food System: Where Are They Going? Report*, Rise Foundation, Brussels.

Deising, H. B., Gase, I. and Kubo, Y. (2017). The unpredictable risk imposed by microbial secondary metabolites: how safe is biological control of plant diseases? *J. Plant Dis. Prot.* 124(5), 413–419. Available at: https://doi.org/10.1007/s41348-017-0109-5.

Domingues, F. (2019). Reforms to the Brazilian regulatory system, the good and the bad areas. *Presentation given at Annual Biocontrol Industry Meeting (ABIM) 2019*. Available at: https://www.abim.ch/index.php?eID=tx_securedownloads&p=124&u=0&g=0&t=1606562638&hash=749d057ae549ca14cc84b323051a91fda7b32909&file=/fileadmin/abim/documents/presentations2019/ABIM_2019_2_02_Fabio_Domingues.pdf (accessed 27 November 2020).

EPPO (2017). Efficacy evaluation of plant protection products PP 1/296 (1) Principles of efficacy evaluation for low-risk plant protection products. *Bulletin OEPP/EPPO Bulletin* 47, 297–304.

EC (2001). Commission directive 2001/36/EC of 16 May 2001 amending council directive 91/414/EEC concerning the placing of plant protection products on the market. Available at: https://eur-lex.europa.eu/legal-ontent/EN/TXT/PDF/?uri=CELEX:32001L0036&from=EN (accessed 5 December 2020).

EC (2002). Regulation (EC) No 178/2002 of the European Parliament and of the Council of 28 January2002 laying down the general principles and requirements of food law, establishing the European Food Safety Authority and laying down procedures in matters of food safety. Available at: https://eur-lex.europa.eu/legalcontent/EN/TXT/PDF/?uri=CELEX:32002R0178&from=EN (accessed 7 December 2020).

EC (2008). Guidance Document on the assessment of new isolates of baculovirus species already included in Annex I of Council Directive 91/414/EEC. SANCO/0253/2008 rev. 2; 22 January 2008. Available at: https://ec.europa.eu/food/sites/food/files/plant/docs/pesticides_aas_guidance_baculovirus.pdf (accessed 23 December 2020).

EC (2009a). Regulation (EC) No 1107/2009 of the European Parliament and of the Council of 21 October 2009 concerning the placing of plant protection products on the market and repealing Council Directives 79/117/EEC and 91/414/EEC. Available at: https://eur-lex.europa.eu/legal-content/EN/TXT/PDF/?uri=CELEX:32009R1107&from=EN (accessed 5 December 2020).

EC (2009b). Directive 2009/128/EC of the European Parliament and of the Council of 21 October 2009 establishing a framework for Community action to achieve the sustainable use of pesticides. Available at: https://eur-lex.europa.eu/legal-content/EN/TXT/PDF/?uri=CELEX:32009L0128&from=EN (accessed 2 December 2020).

EC (2011). Commission Regulation (EU) No 546/2011 of 10 June 2011 implementing Regulation (EC) No 1107/2009 of the European Parliament and of the Council as regards uniform principles for evaluation and authorisation of plant protection products. Available at: https://eur-lex.europa.eu/legalcontent/EN/TXT/PDF/?uri=CELEX:32011R0546&from=EN (accessed 2 December 2020).

EC (2012). Working document on microbial contaminant limits for microbial pest control products SANCO/12116/2012 –rev. 0 September 2012.

EC (2013a). Commission Regulation (EU) No 283/2013 of 1 March 2013 setting out the data requirements for active substances, in accordance with Regulation (EC) No 1107/2009 of the European Parliament and of the Council concerning the placing of plant protection products on the market. Available at: https://eur-lex.europa.eu/legal-content/EN/TXT/PDF/?uri=CELEX:32013R0283&from=EN (accessed 23 December 2020).

EC (2013b). Commission Regulation (EU) No 284/2013 of 1 March 2013 setting out the data requirements for plant protection products, in accordance with Regulation (EC) No 1107/2009 of the European Parliament and of the Council concerning the placing of plant protection products on the market. Available at: https://eur-lex.europa.eu/legal-content/EN/TXT/PDF/?uri=CELEX:32013R0284&from=EN (accessed 23 December 2020).

EC (2017). Commission Regulation (EU) 2017/1432 of 7 August 2017 amending Regulation (EC) No 1107/2009 of the European Parliament and of the Council concerning the placing of plant protection products on the market as regards the criteria for the approval of low-risk active substances. Available at: https://eur-lex.europa.eu/legal-content/EN/TXT/PDF/?uri=CELEX:32017R1432&from=EN (accessed 2 December 2020).

EC (2018). Commission Notice concerning a list of potentially low-risk active substances approved for use in plant protection (2018/C 265/02). Available at: https://eur-lex.europa.eu/legal-content/EN/TXT/PDF/?uri=CELEX:52018XC0727(01)&rid=7 (accessed 2 December 2020).

EC (2019a). Regulation (EU) 2019/1381 of the European Parliament and of the Council of 20 June 2019 on the transparency and sustainability of the EU risk assessment in the food chain and amending Regulations (EC) No 178/2002, (EC) No 1829/2003, (EC) No 1831/2003, (EC) No 2065/2003, (EC) No 1935/2004, (EC) No 1331/2008,

(EC) No 1107/2009, (EU) 2015/2283 and Directive 2001/18/EC. Available at: https://eur-lex.europa.eu/legal-content/EN/TXT/PDF/?uri=CELEX:32019R1381&from=EN (accessed 7 December 2020).

EC (2019b). Technical active substance and plant protection products: guidance for generating and reporting methods of analysis in support of pre- and post-registration data requirements for Annex (Section 4) of Regulation (EU) No 283/2013 and Annex (Section 5) of Regulation (EU) No 284/2013. SANCO/3030/99 rev.5 22 March 2019. Available at: https://ec.europa.eu/food/sites/food/files/plant/docs/pesticides_ppp_app-proc_guide_phys-chem-ana_3030.pdf (accessed 23 December 2020).

EC (2020a). Report from the commission to the European Parliament and the council evaluation of regulation (EC) No 1107/2009 on the placing of plant protection products on the market and of Regulation (EC) No 396/2005 on maximum residue levels of pesticides {COM(2020) 208 final}. Available at: https://eur-lex.europa.eu/legal-content/EN/TXT/PDF/?uri=CELEX:52020SC0087&from=EN (accessed 2 December 2020).

EC (2020b). Farm to fork strategy. For a fair, healthy and environmentally-friendly food system. Available at: https://ec.europa.eu/food/sites/food/files/safety/docs/f2f_action-plan_2020_strategy-info_en.pdf; accessed 2 December 2020).

EC (2020c). Guidance on the approval and low-risk criteria linked to "Antimicrobial Resistance" applicable to microorganisms used for plant protection in accordance with Regulation (EC) No 1107/2009, SANTE/2020/12260, 23 October 2020. Available at: https://ec.europa.eu/food/sites/food/files/plant/docs/pesticides_ppp_app-proc_guide_180652_microorganism-amr_202011.pdf (accessed 23 December 2020).

EC (2020d). Guidance on the risk assessment of metabolites produced by microorganisms used as plant protection active substances in accordance with Article 77 of Regulation (EC) No 1107/2009, SANCO/2020/12258 23 October 2020. Available at: https://ec.europa.eu/food/sites/food/files/plant/docs/pesticides_ppp_app-proc_guide_180653_microorganism-metabolites-concern_202011.pdf (accessed 23 December 2020).

ECA (2020). Sustainable use of plant protection products: limited progress in measuring and reducing risks. Special report 05/2020. Available at: https://www.eca.europa.eu/Lists/ECADocuments/SR20_05/SR_Pesticides_EN.pdf (accessed 7 December 2020).

Ehlers, Ralf-Udo (Ed.) (2011). *Regulation of Biological Control Agents*. Springer Science + Business Media B.V., Dordrecht, Heidelberg, London, New York.

Frederiks, C. and Wesseler, J. H. H. (2018). A comparison of the EU and US regulatory frameworks for the active substance registration of microbial biological control agents. *Pest Manag. Sci.* 75(1), 87–103.

Grant, W. and Gwynn, R. (2020). Improving regulatory approval processes for biopesticides and other new biological technologies in agriculture. In: Biopesticides for *Sustainable Agriculture* Birch, N. and Glare, T. (Eds), Burleigh Dodds Science Publishing Ltd., Cambridge, pp. 45–60.

Guest, P. (2018). Global biopesticide regulations 2018. Report, Agribusiness Intelligence Informa UK, London.

Guijarro, B., Fernández-Getino, A. P. and Alonso-Prados, J. L. (2020). Regulation and requirement challenges for the definition of a successful biocontrol strategy within the EU framework. In: *How Research Can Stimulate the Development of Commercial Biological Control Against Plant Diseases, Progress in Biological Control 21* De Cal,

A., De Cal, P., Melgarejo, P. and Magan, N. (Eds), Springer Nature, Switzerland AG, pp. 323–346. Available at: https://doi.org/10.1007/978-3-030-53238-3_19.

Hauschild, R., Speiser, B. and Tamm, L. (2011). Regulation according to EU Directive 91/414: Data requirements and procedure compared with regulation practice in other OECD countries. In: Ehlers, Ralf-Udo (Ed), *Regulation of Biological Control Agents*. Springer Science + Business Media B.V., Dordrecht, Heidelberg, London, New York, Chapter 2, S. 25-77.

Hauxwell, C., Tichon, M., Buerger, P. and Anderson, S. (2010). Registration and use of microbial biopesticide in Australia. In: *The Use and Regulation of Microbial Pesticides in Representative Jurisdictions Worldwide*. Kabaluk, J. T., Svircev, A. M., Goettel, M. S. and Woo, S. G. (Eds). IOBC Global, pp. 80–88. Available at: www.IOC-Global.org

IBMA (2018). New regulatory framework for bioprotection agents, IBMA vision on how to improve regulation in the European Union. IBMA White Paper, Brussels, IBMA.

Keswani, C., Dilnashin, H., Birla, H. and Sigh, S. P. (2019). Regulatory barriers to Agricultural Research commercialization: A case study of biopesticides in India. *Rhizosphere* 11, 10155. Available at: https://doi.org/10.1016/j.rhisph.2019.100155.

Koch, E., Becker, J. O., Berg, G., Hauschild, R., Jehle, J., Köhl, J. and Smalla, K. (2018). Biocontrol of plant diseases is not an unsafe technology! *J. Plant Dis. Prot.* 125(2), 121–125. Available at: https://doi.org/10.1007/s41348-018-0158-4.

Köhl, J., Booij, K., Kolnaar, R. and Ravensberg, W. J. (2019a). Ecological arguments to reconsider data requirements regarding the environmental fate of microbial biocontrol agents in the registration procedure in the European Union. *BioControl* 64(5), 469–487.

Köhl, J., Kolnaar, R. and Ravensberg, W. J. (2019b). Mode of action of microbial biological control agents against plant diseases: relevance beyond efficacy. *Front. Plant Sci.* 10(845), 845. Available at: https://doi.org/10.3389/fpls.2019.00845.

Leahy, J., Mendelsohn, M., Kough, J., Jones, R. and Berckes, N. (2014). Biopesticide oversight and registration at the U.S. Environmental Protection Agency. In: *Biopesticides: State of the Art and Future Opportunities* Gross, A. D., Coats, J. R., Duke, S. O. and Seiber, J. N. (Eds), American Chemical Society, 1172(16), 3–18.

OECD (2018). Working document on the risk assessment of secondary metabolites of microbial biocontrol agents, series on pesticides no. 98, ENV/JM/MONO(2018)33; Annex 1, ENV/JM/MONO(2018)33/ANN1.

Ravensberg W. J. (2011). Critical factors in the successful commercialization of microbial pest control products. In: Ravensberg W. J. (ed) A roadmap to the successful development and commercialization of microbial pest control products for control of arthropods, Vol 10, Progress in biological control. Springer Science Business Media, Dordrecht, pp 295–356

Reinbacher, L., Bacher, S., Praprotnik, E. and Grabenweger, G. (2021). Standard non-target tests for risk assessment of plant protection products are unsuitable for entomopathogenic fungi—a proposal for a new protocol. *J. Soils Sediments* 21(6), 2357–2368. Available at: https://doi.org/10.1007/s11368-021-02919-w.

Robin, D. C. and Marchand, P. A. (2019). Evolution of the biocontrol active substances in the framework of the European pesticide Regulation (EC) No. 1107/2009. *Pest Manag. Sci.* 75(4), 950–958.

Scheepmaker, J. W. A., Busschers, M., Sundh, I., Eilenberg, J. and Butt, T. M. (2019). Sense and nonsense of the secondary metabolites data requirements in the EU for beneficial microbial control agents. *Biol. Control* 136, 104005.

Singh, H. B., Sarma, B. K. and Keswani, C. (2016). *Agriculturally Important Microorganisms. Commercialization and Regulatory Requirements in Asia*, Springer, New Delhi.

Sundh, I. and Eilenberg, J. (2020). Why has the authorization of microbial biological control agents been slower in the EU than in comparable jurisdictions? *Pest Manag. Sci.* 77(5), 2170–2178. Available at: https://onlinelibrary.wiley.com/doi/pdf/10.1002/ps.6177 (accessed 15 December 2020).

Togni, P. H. B., Venzon, M., Lagoa, A. C. G. and Sujii, E. R. (2019). Brazilian legislation leaning Towards fast registration of biological control agents to benefit organic agriculture. *Neotrop. Entomol.* 48(2), 175–185.

van Duijn, L. P. (2018). A member state proposal to improve the regulatory approach for biological PPPs. *Presentation given at Annual Biocontrol Industry Meeting (ABIM) 2018*. Available at: https://www.abim.ch/fileadmin/abim/documents/presentations2018/1_03_Luuk_van_Duijn.pdf (accessed 23 December 2020).

Villaverde, J. J., Sevilla-Morán, B., Sandín-España, P., López-Goti, C. and Alonso-Prados, J. L. (2014). Biopesticides in the framework of the European pesticide Regulation (EC) No. 1107/2009. *Pest Manag. Sci.* 70(1), 2–5.

Wassermann, B., Müller, H. and Berg, G. (2019). An apple a day: which bacteria do we eat With organic and conventional apples? *Front. Microbiol.* 10, 1629. Available at: https://doi.org/10.3389/fmicb.2019.01629 (accessed 24 July 2019).

Wesseler, J. (2019). Perspective: regulation of pest and disease control strategies and why (many) economists are concerned. *Pest Manag. Sci.* 75(3), 578–582.

Chapter 4

Improving regulatory approval processes for biopesticides and other new biological technologies in agriculture

Wyn Grant, University of Warwick, UK; and Roma Gwynn, Biorationale, UK

1 Introduction

The emergence of a regulatory state is a phenomenon that has attracted extensive commentary by analysts. It has been driven by demands for more enhanced forms of social and environmental protection, supported by campaigns by a range of non-governmental organisations. The limited fiscal powers of the European Union (EU) have encouraged it to rely on regulation to expand and deepen its sphere of competence. The innovative nature of biological technologies has encouraged regulatory interventions by governments across the world and these will be reviewed here. These regulatory interventions take the form of schemes of regulation specifically designed to cater to the particular requirements of biopesticides or reduced-fee schedules such as the Biopesticides Scheme launched in the United Kingdom in 2006.

Pesticides have been subjected to a particularly rigorous regulation. Conventional chemical plant protection products are synthesised: they are

http://dx.doi.org/10.19103/AS.2020.0073.04

not generally found in nature (though some synthetic pesticides are derived from molecules occurring in nature). Regulation therefore has to be strict to anticipate and prevent potential harm to human health or the environment. All players in the system have been trained and embedded into this precautionary approach. There is an unavoidable tension between regulation and the agrochemical industry yet the linkages between them are strong in terms of exchanges of expertise and personnel so they are co-reliant. What is changing is that some regulators are choosing to break these linkages, giving less priority to the views of the agrochemical industry, and moving to support substances that intrinsically cause less harm than some long-established pesticides.

For a product and its producers, regulation is a double-edged sword. It provides a means of protection from producers who make exaggerated or scientifically unsound claims for their products, evident in relation to 'grey area' products like plant enhancers or strengtheners. In that sense it provides reputational protection. It also ensures that there are no harmful side effects from new products, for humans (operators, consumers, bystanders and residents), the environment (air, soil and water) and non-target organisms (aquatic organisms, birds, soil organisms, invertebrates including bees). In addition, compatibility of new products with existing methods and product that may be used in crop protection is evaluated. However, it can also provide a barrier to new entrants and in that sense can serve as a dampener on valuable innovation.

2 Establishing a regulatory framework for biopesticides

The challenge is to design a regulatory system for biological technologies that is robust and scientifically rigorous, but not totally inflexible as biological technologies are a diverse group of substances that have their origins in nature and consequently can often be accommodated in the agroecosystem with no or limited perturbations or unintended consequences. The unintended consequences of some conventional pesticides such as the well-documented detrimental human and environmental effects resulting from (over)use of organophosphates have rightly created a cautious approach to their regulation (Hertz-Picciotto et al., 2018).

This overall precautionary approach is carried over to biological technologies mainly because of the regulator's unfamiliarity with a new field of research, for example, with microbial ecology rather than synthetic chemistry. It means less-experienced regulators of biological technologies revert to a very precautionary approach and demand more data and effectively raise standards for a set of substances for which this is usually not needed. However, it should be noted that some biopesticides can be toxic to non-target organisms, for example, rotenone to fish, amphibians and macro-invertebrates. Thus there are specific cases where precaution is warranted for natural product biopesticides. Moreover, biopesticides need to be used with appropriate concentrations, timings and frequencies of application to avoid adverse effects on non-target organisms and consumers, for example, residue issues relating to human toxicity or flavour taint.

It is not a question of non-regulation, but ensuring that the regulations are well-designed and appropriate. The regulatory system must not impose undue transaction costs on the regulator and the regulated, while ensuring that all appropriate checks relating to toxicity and other issues are made. It must not degenerate into a bureaucratic box-ticking exercise that inhibits scientific progress or unduly delays the availability of new products that may be of real benefit. It must be open to dialogue with those it seeks to regulate. Responding effectively to these challenges is not easy as bureaucrats are typically risk averse and successful regulatory innovation requires a fortuitous combination of circumstances including both exogenous and endogenous pressures (Greaves, 2009). This is in contrast to developers of biological technologies who tend to be innovative and less risk averse. This leads to a tension for regulatory development, not least in how the different parties communicate.

For a new regulatory approach, data requirements themselves are not necessarily the primary issue, but deciding which data are relevant, how they are interpreted and the experience and expertise of the person asking the questions. For example, the European Food Safety Agency (EFSA) is meant to perform this role, but it still does not employ or retain the specialist services of microbial ecologists. It still believes that general skills in food microbiology are sufficient even, for example, in assessing environmental effects. Examination of the peer review reports and dialogue between applicants, regulatory agencies, the Commission and EFSA demonstrates a frequent mismatch between the parties in understanding where the relevant risks might lie. For example, regulators often state that the dossiers submitted by applicants are poor and propose to resolve this by offering training to applicants. However, in some cases the dossiers may be sufficient but, to correctly understand them, the evaluators also need training in the new technologies. To be effective, all parties are best served if there is trust and a good partnership. There can also be poor levels of trust between national authorities.

3 Pesticide regulation in the European Union (EU)

As noted in the previous section, in addition to the need for mutual understanding between regulators and product manufacturers, effective regional and international regulation relies on trust and cooperation between national regulatory agencies. After the implementation of directive 91/414/EEC it quickly became evident that mutual recognition between different EU Member States was not working (Chandler et al., 2011). Implementation of EC/1007/2009 had a specific aim to improve mutual recognition but, in the experience of the authors, the lack of trust between regulatory authorities and bodies continues eight years after its coming into force, inhibiting the system from working well. Trust is eroded because some regulatory bodies consider that they apply better standards than others. For example, when becoming a member of the European Economic Area Agreement in 1994 Norway decided not to be a part of the EU pesticide regulations, because the Norwegian regulations were assumed to provide higher protection to human health and the environment. In June 2015 Norway finally implemented Regulation EC 1107/2009 and a new Norwegian regulation came into force. In the Norwegian Parliament it was argued that it would be wise to implement the EU regulation, because the criteria for pesticide assessment would be very similar to current practice in Norway (Kvakkestad et al., 2020). The consequence of this precautionary approach and supposed raising of standards is that Norway has four biological technologies approved. This is in contrast with neighbouring Sweden which has engaged more in building capacity in biological technologies, and which has 20 biological products approved.

Trust is eroded between applicants and regulatory bodies when applicants are communicating to evaluators who are clearly not sufficiently skilled in the art. This probably explains why applicants tend to use the same evaluators for microbials, based on countries which are more supportive of biopesticide development (Fig. 1), while some EU Member States are never or rarely used (e.g. Bulgaria, Czech Republic, Greece, Finland, Hungary, Ireland, Lithuania, Latvia, Romania and Slovakia).

The EU approval system for plant protection products was originally designed to cater for synthetic or conventional chemical pesticides. The effect

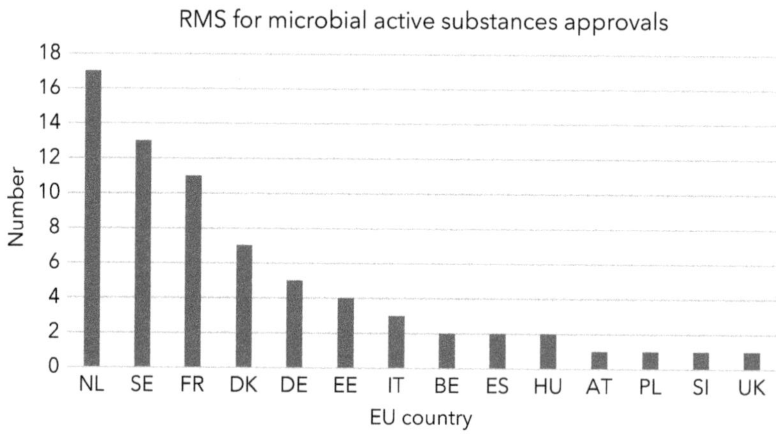

Figure 1 EU Rapporteur Member States responsible for approving microbial active substances (new, renewals and pending).

of the influence of conventional pesticides is found at all layers of the system because it was developed by and for them. Many of its procedures, or the questions that could be asked of applicants, were irrelevant for biocontrol products, for example, asking questions about the physical-chemical properties such as the boiling point of a microorganism but not asking about their biological properties. The majority of the companies researching and developing these products are micro companies, typically a spin-off from a university on a science park or innovation start-up. Staffed by scientists, they have limited understanding of the political process and few spare resources to devote for lobbying. They often find the checks and balances of regulation baffling and restrictive. They have lacked the political influence of the big agrochemical companies. Developing an effective representational structure for the emerging biopesticides industry took some time and continues to do so. Hence, a regulatory barrier was created to the entry of more of these sustainable products on to the market.

4 The development of the biopesticide sector and new regulatory requirements

The market share for biological technologies has grown from virtually nothing in 1993 to 4 billion US$ in 2018, in a span of 15 years, and is growing exponentially; forecast reveals the share to double and reach 8 billion US$ within the next 5 years. In terms of market share in 2018, Europe and the United States each hold 32%, Latin America 18% and Asia-Pacific 16% (Dunham, 2018). Of the different technologies globally available in 2018, microorganisms continue to dominate with 53% of the market, biochemicals 39% and macrobials 7.4% (Dunham, 2018). There has been some consolidation in the industry with multinational agrochemical companies such as Bayer, BASF and Syngenta and smaller multinational companies such as Certis having been brought into the sector by acquiring smaller companies. Individual spin-outs and start-ups have also been acquired by larger companies or by venture capital funders. The involvement of the larger pesticide companies in industry representation has strengthened the International Biocontrol Manufacturers Association (IBMA).

However, it can also be argued that the involvement of larger agrochemical companies is shifting the agenda, potentially increasing regulatory hurdles by making guidance documents more precautionary. This approach is informed by its existing experience from the conventional chemical regulatory framework. Well-resourced companies, embedded in using familiar conventional pesticide regulations, are used to a 'tick-box' approach to regulation where they, more or less, address every data requirement with a study, following established protocols. There is little consideration in presenting a reasoned-case approach for non-provision of study data. This is even when the nature of the substance being registered may make this applicable. This often ignores the explicit information in relevant guidance documents that a reasoned-case approach is accepted. This increases the information provided in a dossier of sometimes extraneous data and thereby increases the workload of assessors who are obliged to evaluate all information presented, relevant or not. The cost of providing this extra information is not a burden to a large company so there are fewer consequences to them from taking this approach.

The traditional approach of established pesticide manufacturers can be contrasted with the newer generation of biopesticide companies. They could develop products that are toxic or pathogenic (because these products work best against target pests) but they actively choose not to; they focus on substances which cause minimal harm. Those regulators who have developed expertise in relation to biological technologies have chosen a similar path: they are motivated to approve products that cause less harm. This approach is reinforced where there is a strong government policy to facilitate the availability of biological technologies such as the Netherlands and French governments' 'Green' agendas. One should add a note of caution. At present biological technologies are, by definition, seen to pose minimal harm to people or the environment. However, this may change if there are unforeseen adverse environmental impacts or harm to insects, animals or humans.

Nevertheless, biopesticide manufacturers have to overcome a framing of the crop protection dialogue around an approach that a pest controlled (> 80% efficacy) by a conventional chemical, leads to a good crop. This way of thinking has been embedded into the mental maps of farmers as well as support industries and regulators. Advocates for biopesticides are asked why products do not provide 80% efficacy, even though they do not need to. A 50% pest reduction twice equals 100%, so two applications are able to achieve an acceptable reduction in the target pest population. The aim is to reduce the pest population below a damage threshold, rather than eliminate the entire population. Expensive plant breeding programmes focus on overcoming resistance to conventional chemicals when they should be asking a different and more appropriate question about alternatives to conventional chemicals. Breeding programmes could focus more usefully on crop plants that have better natural defences and retain their natural ecology and relationship with microorganisms and so develop plants that have more resilience that many present varieties currently have.

5 Challenges in improving the regulatory framework for biopesticides

The biggest challenge that biopesticide manufacturers have faced in moving the debate forward has been the lack of a clear political constituency of support and hence of cohesive advocacy for biological technologies. Relatively few broadacre farmers have used them so there has been relatively little pressure from farming organisations for their wider availability (although the importance of the greenhouse sector creates a somewhat different political context in countries such as the Netherlands). In the United Kingdom, for example, the National Farmers Union (NFU) policy for integrated pest management (IPM) appears to be strongly influenced by and embedded in the use of agrochemicals. Farmers are subtly persuaded that they will fail if they do not have conventional chemistry and this is used to influence the conversation about what crop protection solutions will look like. The potentially influential organic sector is not really a source of support as they are suspicious of biological technologies or 'biopesticides' on the grounds that they are 'still pesticides'.

Retailers are a potential source of support because of their interest in food safety and quality issues. There has been a trend towards the privatisation of the regulation of such issues with big retail chains having their own supplementary systems operated by approved growing protocols which may prohibit their suppliers from using plant protection products approved by the state system. Many retailers certainly present themselves as 'greener' than their rivals as a useful means of market differentiation. However, they find it easier to give a message by promoting organics than by saying that their suppliers are users of IPM and biological technologies which does not convey a clear meaning to consumers.

In the United Kingdom at least, the biological technologies policy network has been weakly developed and relatively immature, particularly when compared with the agrochemicals network. An incomplete and loosely integrated policy network 'results in an underperforming network ... As one may expect, a very loose and incomplete network has little impact on observable policy outcomes' (Greaves and Grant, 2010: p. 35).

Nevertheless, some progress has been made at both the Member State level and the EU level. How has this been achieved? Analysis suggests that five factors have been important:

1 Interventions by global organisations such as OECD, IOBC, FAO and WHO which encourage international cooperation to develop harmonised, pragmatic and proportional regulatory requirements that are adapted to biological technologies. Certain EU agencies (EPPO[1] and MUCF[2]) and Member States (the Netherlands, France, Sweden, Germany and formerly the United Kingdom) are regular and active participants. Such activities support communication and shared experiences and

1 European Plant Protection Organisation: www.eppo.int.
2 Minor Use Coordination Facility: www.minoruses.eu.

engender trust between countries and develop guidelines that can be used by applicants and evaluators in the EU providing support where currently none exists.

2 Interventions by national governments who wish to enhance the contribution made by biological technologies to sustainability goals. The Biopesticides Scheme in the United Kingdom made a positive contribution in facilitating the successful completion of biopesticides registrations.
3 A greater commitment by EU regulatory agencies to biological technologies, in part stimulated by national governments policies to reduce chemical pesticide use and increase biological technology alternatives, but also in pursuit of 'first mover' advantage, although this commitment has its fragilities. Unfortunately, such commitments are often reliant on particular champions within agencies who may leave and they may also be affected by cutbacks in resources. However, this can work both ways. In some EU organisations there have been significant changes of personnel which bring in a new approach. There are early signs of a willingness to consider new ways of regulating biological technologies because of a desire to support food production that has less environmental impact.
4 The role of regulatory consultants both as a resource for the industry and as intermediaries between the industry and regulatory agencies. Their particular advantage is that they can look across the whole industry and have insight into its diversity whereas individual companies look inwards to seek solutions to show that their idea or technology can be proved to work. Consultants have also argued for greater political sophistication and the mobilisation of external expertise to assist this process.
5 Improvements in the capacities and sophistication of the European trade association, the IBMA which has had some impact on the policy-making process at the EU level. It has influenced the debate about regulatory changes and the guidance documents which inform their implementation. It has also raised the profile of biopesticides among decision-makers although this is an area in which more work needs to be undertaken.

The extent of this progress can be seen, for example, in current EU regulation in this sector.

6 Current EU regulation of biopesticides

In the EU, microbial, biochemical and semiochemical biological technologies are registered under the same regulatory framework as chemical pesticides although some special provisions have been developed. The legislation provides for measures that favour the registration of products defined as 'low risk' (most of the biological technologies) but the process has not worked well in practice. Although 'biological technologies or biopesticides' do not exist as a regulatory category, the pesticide categories 'basic substances' and 'low risk substances' were introduced in August 2017, as defined in Regulation 2017/1432, amending Regulation 1107/2009.

Before plant protection products (PPP) could be introduced to the European market, the active substance must be approved for the EU, after which authorisation of the product can be sought in individual Member States. For the approval of an active substance, the applicant submits a dossier to an individual Member State, the Rapporteur Member State. The Rapporteur Member State produces a Draft Assessment Report, which is submitted to the European Commission and the EFSA which is responsible for peer reviewing the risk assessments of active substances. Finally, the European Commission presents a review report to the Standing Committee for Food Chain and Animal Health (SCFCAH), who votes on approval or non-approval of the active substance.

It should be noted that the resources available to regulatory bodies in the Member States vary considerably and this in turn affects their competence and reputation. As a broad generalisation, those in Northern Europe and one or two Member States in Central Europe are the most innovative, while those in the rest of Central and Southern Europe have more limited capacity. This is reflected in the Member States that have invested capacity for evaluating microbial active substances (Fig. 1).

Under the current regulatory system (in Reg. (EC) 1107/2009), and depending on the complexity and completeness of the dossier, it takes 2.5-3.5 years to approve a new active substance. The subsequent authorisation of a product is carried out by the Member States, with the applicant specifying the Member State carrying out the evaluation which takes a further 1.5-2 years. Following authorisation, a company may apply for mutual recognition within one of the three geographic zones (North, Central and South) for products with the same use(s) under similar agricultural conditions. The approval process takes the same time regardless of the type of active substance or the dossier size (e.g. the majority of biopesticide dossiers contain about 10% of the paperwork of a conventional pesticide). Biopesticide dossiers are smaller because they usually do not trigger the need for higher level studies, for example, for many biopesticides a reasoned case can be provided to justify the lack of relevant residues whereas extensive studies are needed for most conventional pesticides.

The IBMA White Paper comments, 'At the latest count, 60% of new submissions to the EU are bioprotection technologies but evidence indicates that, the existing approach is not working well in Europe, despite guidance produced for use in Reg. (EC) 1107/2009 for these types of technologies' (IBMA, 2018). It suggests a single EU body for regulating bioprotection technologies. This would coordinate registration of active substance and products integrally (together) and would ensure evaluations are provided by experts with the proper field of expertise and relevant scientific knowhow. However, the chances of securing such a body are not good at this time with the EU distracted by so many issues.

7 Global national initiatives in biopesticide regulation

Within the developed world, the United States has attracted praise for its proactive approach to the encouragement of biological technologies or biopesticides. In the United States, in 1994, the Environmental Protection Agency (EPA) established the Biopesticides and Pollution Prevention Division (BPPD) in the Office of Pesticide Programs to facilitate the registration of biopesticides. The BPPD aims were to promote the use of safer pesticides, including biopesticides, as components of IPM programmes. The success of a biopesticide-specific registration programme is demonstrated by the United States having the highest number of active substances available worldwide. In early 2019 there were 366 registered biopesticide active ingredients compared to 135 approved and 16 pending in the EU. A 'white paper' from the Europe-based IBMA states:

> The USA EPA process for regulating biopesticides is the best example of a system which has been set up solely to regulate bioprotection products. It has a track record of delivering new product approvals through robust rigorous review within a year of submission since it was established in 1994. Importantly it follows a tiered approach and reviews are conducted by dedicated evaluators with relevant expertise (IBMA, 2018).

More recently, national initiatives on biological technologies registration have been assisted by the production of a joint Food and Agriculture Organization (FAO) and World Health Organisation (WHO) document 'Guidelines for the registration of microbial, botanical and semiochemical pest control agents for plant protection and public health uses' (FAO and WHO, 2017) which provides a guidance framework for national policy and replaces an earlier 1988 document. The FAO and WHO documents emphasise that, in relation to the registration of biological pest control agents, 'Harmonization of the different approaches is important for streamlining and facilitating the research, development, commercialization and use of biopesticides for plant protection and public health. Using similar data requirements and evaluations should make it easier for applicants to submit applications to different countries and for regulatory agencies to benefit from each other's reviews' (FAO and WHO, 2017).

A number of countries outside Europe have been taking initiatives to boost the use of biological technologies, foremost among them China, Kenya and Cuba. Over the past five years, the number of biological technologies registered in China has enjoyed a 16% increase while growth of conventional chemical products grew just 3%. They have adopted the FAO and WHO guidelines. The Chinese government is 'eager to find safer alternatives to the conventional chemicals currently used domestically. Food safety is a massive concern in China and the government has tried to drive substitution of hazardous conventional chemicals domestically by subsidizing biopesticide users and offering favorable registration policies'. The authorities started to draw up a supplementary document for data requirement on biological technology registration in 2010 and a revised draft of the Data Requirement on Pesticide Registration was released for public consultation the following year. However, the consultation never progressed any further, partly because of delays in approving the overarching regulation being developed for the sector, the 'Regulation on Pesticide Administration'. The biggest obstacle facing China's biopesticide registration policy remains the delay in the legislative process for approving this overarching regulation.

Horticulture is a vital industry for Kenya with its products accounting for 92% of exports. The adoption of maximum residue levels (MRLs) by the EU in 1994 provided a stimulus for the extensive use of biological control agents. About 30 products derived from plants or microbes are registered in Kenya. 'In 2003, in response to horticulture industry needs, Kenya's pesticide regulatory authority, the Pest Control Products Board (PCPB), supported by a range of stakeholders such as the UK Department for International Development (DFID) Crop Protection Programme, developed biological technology-specific registration pathways, including one for microbial pest control products' (Gwynn and Maniania, 2010). The total number of biological technology products available in Kenya in 2008 was 25. By 2018 this had doubled to 51. Thirty-four products were active against insects: 12 microorganisms based on *Bacillus thuringiensis* (Bt), 12 other microorganism non-Bt and 10 botanical. There were 13 products active against diseases and 4 against nematodes and 3 with other activity.

Cuba is characterised by intensively cropped shareholdings, often located in residential areas, which use biomanagement strategies for crop protection, although there is a shortage of supply of the required biological technologies. However, it has achieved a substantial reduction in imports and made suitable products available for small-scale farmers. 'The regulation system has to be flexible and not too costly or too lengthy but it must have the full and justified trust of the public. As current legislation stands, there are certain categories of microbial pesticides that have an easier and quicker passage to registration than others' (Diaz, 2004).

8 Developing good regulatory practice

In the EU over 30% of approved active substances are now biological technologies. Yet, with one or two Member States that are good exceptions, the problems within the EU regulatory systems for 'alternative' active substances and products are similar to those in many other regions or countries:

- Lack of specific data requirements for this new class of products;
- Limited knowledge and experience within government regulatory agencies;
- Lack of government financial support to regulatory agencies to build their capacity; and
- No mechanisms for a fast track system for products that are, in the main, inherently less harmful to humans and the environment.

Organisations such as the IBMA and its sister organisations globally have, for at least the last 15 years, publicised these shortcomings and barriers and proposed well thought-out solutions. The continuing inertia, to reduce the regulatory barriers to allow better crop protection solutions to get onto the market, is a disservice in achieving a more sustainable future.

However, the activities of international bodies such as OECD, FAO and WHO, and good national programmes, show that there are good examples of initiatives pushing for good regulatory practices that support disciplined and robust regulation but that remain proportional to the risks that biological technologies may present to consumers and the environment. As outlined in this chapter, there are some countries which have been proactive to reduce the barriers and consequently, they have more biological technologies available for farmers.

It is all too easy for any activity on biological technologies regulation to remain embedded in the detail of how to prepare guidance. This creates the illusory sense of progress but, without a framework for implementing this guidance, it does not substantially change timelines for regulatory approvals. What would good regulatory practice be? Certainly, it would be one that balances properly the need to be safe with the need for resilient crop protection tools that fit good quality IPM and supports farmers to grow a good quality crop with good yields at economically viable prices.

To better regulate biological technologies and emerging technologies is to ensure all participants in the process are well skilled in the relevant technologies. This needs investment because of the diversity of technology types. These skills are needed by the applicants, the contract research organisations running the testing (with appropriate testing methods), the evaluators, the peer review organisations and finally the decision-makers (risk assessment and risk management). Any break in this chain can lead to difficulties and delays, for example, if the lack of expertise acts as a barrier and leads to delays and usually additional costs through requests for extra studies and delays in getting products onto the market.

Technology transfer and knowledge exchange for biological technologies need to be actively pursued by governments to resource their representatives adequately so they can acquire the skills to fulfil their regulatory roles. Governments and public policies are requiring that food is grown following IPM practices and that alternative technologies are the mainstay of this more environmentally friendly approach. Yet they are failing to resource their regulatory agencies sufficiently to allow them to fulfil this obligation so combined with the inherent inertia progress to adopt new technologies progress is slow. There need to be additional efforts and resources for regulators to build their capacity in new technologies. Many regulatory agencies work on a full-cost recovery basis within which there is no mechanism to fill the funding gap for capacity building in new technologies.

More precisely 'Good Regulatory Practice Guide for biological technologies (alternatives)' is one that upholds the intent of government policy for IPM which promotes alternatives and indicates that good practice involves:

1 Dedicated regulators (at all levels) that have the necessary skills for biological solutions (microbial, natural substances-botanical, semiochemical).
2 Having a strong partnership using a clear, structured pre-submission process that engages equally both evaluators and applicants for example:
    a initial enquiry with evaluators, ideally evaluators visiting the applicants research facility, to discuss broad principles while the applicant is still developing new active substances
    b early-stage meeting once the active substance specification is established
    c mid-stage meeting during initial dossier development
    d late-stage meeting once studies have been done and the dossier has been prepared
    e completeness check

   During this process there should be interim exchanges of emails to confirm studies proposed and so on.
3 Providing a high-level framework for the principle that data are excluded 'except when ...' and the exceptions are explained: for example, are

ecotoxicology studies for microbials needed when it is known from their ecology they are specifically plant pathogen hyperparasites? In principle it could be stated that microbials are of concern only for arthropod non-targets when they are insect pathogenic.

4 Trusted and respectful partnership between evaluators for quick resolution of any issues and for promoting harmonised outcomes.
5 Harmonisation of evaluations that the same standards are applied across evaluations.
6 Reciprocity of evaluations, if a dossier is accepted in one country at a suitable standard then it is automatically accepted by another, without re-evaluation. There are already agreements on acceptable standards so this can be implemented quickly if there is a political will.
7 Reciprocity for efficacy: data from one region should be accepted in another area, provided there are no major differences in uses and agronomic practices.

The unrealised advantage of regulating biological technologies better is that the focus can be on where the real risks are and it frees up resources to pay correct attention to these. This in turn should speed up the evaluation process supporting innovative biological technology companies and provide farmers with more much-needed IPM tools. The consequent benefits also accrue to society.

9 Conclusions

Biological technologies have a key but specific role to play in crop protection. Synthetic chemical pesticides are a precious resource for crop protection and should be treated as such. However, the occurrence of human and environmental harm because of misuse or overuse of pesticides, problems of resistance to pesticides and withdrawal of some products means that fewer chemical pesticides are available in the market. Alternative crop protection tools are needed and biological technologies have an increasing and important role to play in crop protection, especially as part of IPM programmes.

Biological technologies should be regulated. Just because something is natural does not mean it is safe. However, the regulatory system has tended to follow a chemical pesticide model that does not facilitate the efficient registration of biological technologies. An international 'epistemic community' has been formed among experts, facilitated by organisations such as FAO, WHO and OECD, which has been able to identify the key unresolved issues, recommend solutions to technical challenges, promote harmonisation of regulation and suggest ways forward.

However, this agenda for more effective forms of regulation will only be fulfilled with sufficient political support. There is greater interest in biological technologies than in the past and an increased recognition of the contribution they can make to environmentally sustainable solutions to plant protection. This reflects a number of factors, including greater effectiveness on the part of the industry representative body, the IBMA, including a willingness to build coalitions with environmental groups. It also reflects the greater involvement of major agrochemical companies in the sector as market opportunities improve. However, public knowledge and understanding of biological technologies remains limited and this means that there is no substantial constituency of public support for their greater use. This in turn influences the priority that political decision-makers give towards improving the responsiveness and speed of regulation that allows new products to enter the market.

10 Where to look for further information

Each country maintains information and resources (usually internet-based) for the countries' regulations for plant protection produces, depending on the countries expertise in biopesticides, these resources can provide good guidance, for example, the Netherlands regulatory agency has guidance on how to prepare a dossier for biopesticides. Details of any guidance or other provisions for biopesticide registration are provided by organisations such as the EU department for the environment (DG SANTE), by the OECD Pesticides facility and by the UN agency, FAO on their IPM webpages.

Organisations such as IBMA and its sister organisations worldwide, actively participate to get provisions made for biopesticide regulation and for regulations to be adapted for them.

Research that relates to the regulatory risk areas for biopesticides supports the efforts of industry and provides good-quality scientific evidence to underpin the development of appropriate and proportional regulation. There is a need to consider the impact of Brexit on regulation in the UK and how this interacts with EU regulation. More generally, the responsiveness of regulation to a variety of stakeholders requires analysis and drawing an appropriate balance between risk avoidance and facilitating innovation.

11 References

Chandler, D., Bailey, A. S., Tatchell, G. M., Davidson, G., Greaves, J. and Grant, W. P. 2011. The development, regulation and use of biopesticides for integrated pest management. *Philos. Trans. R. Soc. Lond., B, Biol. Sci.* 366(1573), 1987–98. doi:10.1098/rstb.2010.0390.

Diaz, L. H. 2004. Registration of biological pesticides in Cuba. In: Wabule, M. N., Ngaruiya, P. N., Kimmins, F. K. and Silverside, P. J. (eds), Registration for Biocontrol Agents in Kenya. Kenya Agricultural Research Institute, Nairobi, pp. 117–34.

Dunham, W. 2018. Biocontrol's position in the global marketplace. Presentation to *ABIM Conference*, Basel.

FAO and WHO. 2017. International Code of Conduct on Pesticide Management: guidelines for the registration of microbial, botanical and semiochemical pest control agents for plant protection and public health uses.

Greaves, J. 2009. Biopesticides, regulatory innovation and the regulatory state. *Public Policy Admin.* 24(3), 245–64. doi:10.1177/0952076709103810.

Greaves, J. and Grant, W. 2010. Underperforming policy networks: the biopesticides policy network in the United Kingdom. *Br. Pol.* 5(1), 14–40. doi:10.1057/bp.2009.15.

Gwynn, R. L. and Maniania, J. N. K. 2010. Africa with special reference to Kenya. In: Kabaluk, J. T., Svircev, A. M., Goettel, M. S. and Woo, S. G. (Eds), *The Use and Regulation of Microbial Pesticides in Representative Jurisdictions Worldwide*. IOBC Global, 99p.

Hertz-Picciotto, I., Sass, J. B., Engel, S., Bennett, D. H., Bradman, A., Eskenazi, B., Lanphear, B. and Whyatt, R. 2018. Organophosphate exposures during pregnancy and child neurodevelopment: recommendations for essential policy reforms. *PLoS Med.* 15(10), e1002671. doi:10.1371/journal.pmed.1002671.

IBMA. 2018. A new EU regulatory framework for bioprotection agents: IBMA vision on how to improve regulation in the European Union. IBMA White Paper.

Kvakkestad, V., Gwynn, R., Sundbye, A. and Klingen, I. 2020. Authorization of microbial plant protection products in the Scandinavian countries: a comparative analysis. *Environmental Science and Policy* 106, 115–24. doi:10.1016/j.envsci.2020.01.017.

Chapter 5

Developing effective product dossiers for regulatory approval of new animal feed products

Manfred Lützow, saqual GmbH, Switzerland

1 Introduction: what are feed products?

Feed is part of the food supply chain on which society relies for production not just of food for humans but also for pets and non-food animal products such as wool from sheep. The scale of feed production (as part of the overall food supply chain) can be seen in the 164.811 million tons of compound feed used in the 28 member states of the European Union (EU) in 2018 (FEFAC, 2019). In 2018, 321.3 kg of compound feed was produced for each citizen of the EU.

Countries differ in their food supply systems and therefore have different regulatory frameworks to ensure that sufficient safe food of the required quality and quantity is available. Regulatory differences potentially hinder trade and, as a result, there have been significant efforts to harmonise food legislation. The Codex Alimentarius Commission, an international non-governmental

http://dx.doi.org/10.19103/AS.2020.0083.15

organisation that is managed jointly by the Food and Agriculture Organisation of the United Nations (FAO) and the World Health Organisation (WHO), has been developing food standards for almost 60 years to address key aspects of food quality and safety. Many countries have implemented Codex standards, their food legislation and regulatory frameworks are similar, they use the same definitions and legal terms and have comparable approaches. We have for example internationally accepted definitions for food additives or the requirements for the composition and labelling of dairy fat spreads.

Such international regulatory harmonisation is less advanced in the area of feed. The Codex Alimentarius Commission started the discussion of animal feed much later than for food and, so far, has not developed any binding overall rules governing the international trade in feed. A Codex task force has developed general principles for the risk assessment of animal feed (Codex, 2013). However, no international feed product-specific standards have been developed (apart from the adoption of common standards for certain contaminants in feed e.g. for some mycotoxins). In the absence of internationally accepted definitions, the feed product categories and terms applied by the relevant jurisdiction need to be used. Those are, unfortunately, quite heterogeneous (Smedley, 2013). In part they reflect differences in responsibility for feed safety. In the United States for example the Food and Drug Administration (FDA) regulates food additives. This means there are no feed additives as such but, rather food additives intended for animal feed. There are also differences in whether a product constitutes feed. Animals do not only receive feed but are also treated with various other products such as veterinary products and biocides which may or may not be defined as feed. The main categories defined by Codex Alimentarius, the United States and the European Union are compared in Table 1.

Prior to the development of a dossier for a feed product, the regulatory framework of the target market needs to be understood, as requirements differ significantly between jurisdictions. The following outline is based on the regulatory framework of the European Union that has been developed over the past 20 years and revised recently from 2017 to 2019. No other jurisdiction has gone so far in incorporating current risk assessment principles and

Table 1 Comparison of categories and terms for feed products (Codex, 2013; Smedley, 2013)

Codex Alimentarius	United States	European Union
Feed (Feeding stuff)	Animal Feed	Feed
Feed additive	Food Additive	Feed additive
Feed ingredient	Feed Ingredients	Feed material
Veterinary drug	Animal Drugs (Medicated feeds)	Veterinary medicinal product
-	Utility, claim	Effect, function, claim

state-of-the-art scientific methodology in their official guidance documents. This does not mean that other agencies involved in the evaluation of feed ingredient dossiers are not using state-of-the-art methodology, but their published guidance is often less up-to-date. It should be noted that the life-cycle approach to the evaluation of feed ingredients discussed in the following section is used by most jurisdictions.

2 The food chain: the product life cycle for feed ingredients

Animal feed is composed of agricultural raw materials such as wheat middlings, rice bran, soybean meal, minerals such as calcium phosphate, trace minerals like manganese salts, vitamins such as retinyl acetate (vitamin A), sources of proteins and amino acids, carbohydrates and lipids (oils and fats). Feed also includes ingredients that serve a specific purpose such as preservatives (benzoic acid), emulsifiers and anticaking agents. For each of these ingredients, safety and efficacy along the food chain needs to be assured (Fig. 1).

As an example, aflatoxin B1 present in grains fed to dairy cows may end up as aflatoxin M1 in milk products consumed by humans. Likewise, *Salmonella* present in chicken feed could mean that eggs are unsafe for consumption if contaminated with this bacterium. As well as issues of safety and quality, a life-cycle approach to feed production must also include other aspects such as workers' safety during feed processing and potential environmental pollution. A holistic feed-safety approach starts at the very beginning of the production process of a feed ingredient; continues with the investigation of the ingredient's properties and its behaviour when mixed into feed; and assesses safety for the worker exposed to the ingredient, the safety of the target animals when ingesting it, environmental safety when digested feed materials are excreted by the animal, and eventually the safety of the consumer who is possibly

Primary production, extraction, synthesis

↓

Feed mill, processing, storage, transport

↓

Animal, farm, stable, pasture, cage, home

↓

Meat, milk, eggs, fat, wool, ...

↓

Consumer

Figure 1 Fate of feed ingredients

exposed to feed constituents, their metabolites, or feed contaminants when consuming animal products.

The extent to which these aspects are covered by specific regulatory requirements differs. However, any feed producer must take legal responsibility and is liable for the wholesomeness and safety of the feed marketed. Hence the development of a new feed product, whether a single ingredient or a ready-to-consume feed, should take into consideration all aspects of quality, safety and efficacy across the life-cycle of the product.

3 Safe feed and safe food

Animal feed is an important element of the human food chain. In one way or another most feed ends indirectly on the plate of the consumer. Several food hazards originate from animal feed. Some of these resulted in a number of prominent food scares in the European Union between 1995 and 2001. Among these were incidents involving BSE, mycotoxins, dioxins, residues of banned veterinary drugs, and fears concerning the development of antimicrobial resistance. These food scares were a major driver for new food legislation in the European Union which integrated fully all aspects of feed production and animal feeding into one horizontal food safety framework (EU, 2002). This framework applies not only to feed for food-producing animals, also to feed for companion animals or for non-food-producing animals (e.g. in zoos). Until that time the safety of animal feed was seen as requiring less rigorous safety regulation compared to requirements for food for direct human consumption.

The new EU food legislation meant that the safety of feed for animals now attracted the same level of scrutiny as the safety of food for consumers in many respects, resulting in a significant increase in the data needed for feed products requiring authorisation. This increased scrutiny of safety issues relating to animal feed was accompanied by the development of guidance documents and new scientific evaluation procedures by the FEEDAP panel, the expert group at EFSA responsible for the evaluation of feed additives. The revised food law of 2002 triggered a re-evaluation of all feed additives that had been authorised by the European Union after 1970. This included a large group of flavourings with several thousand substances listed. This re-evaluation started in 2010 and created a considerable number of case studies of specific problems which allowed the experts (and the applicants) to collect additional information on the range of issues relevant to the life-cycle of a feed ingredient. Based on this information, the EFSA revised the relevant guidance documents between 2017 and 2019 and developed a set of up-to-date guidelines for the evaluation of feed ingredients (EFSA 2017a,b,c, 2018a,b, 2019a).

4 Where do you start?

As discussed, we will, for the purpose of this chapter, describe the data needed for a feed additive dossier in the European Union. When developing a dossier for a feed product, the first step is to identify the legal category it will fall into, and to understand the requirements for that legal category in the jurisdiction of choice. In the European Union feed ingredients may be either feed additives which require premarket authorisation, or feed materials which may be put into circulation without such authorisation. A health claim may be developed for a feed material or a compound feed. No authorisation is required but a dossier documenting the claim is needed which must be shared with the authorities if they request it. In addition such data may be necessary for an effective marketing strategy. In principle, similar datasets are needed in other jurisdictions such as Australia, Canada, Japan and the United States, though the emphasis of the dossier may differ.

Once you have identified the legal category and you have decided your product is a feed additive, you must draft the proposed conditions of use which will be part of the authorisation process. In that sense you start at the very end of the project (i.e. the intended use). The nature of the eventual authorisation is what drives the need for data. The authorisation of a feed additive will comprise characterisation of the additive, its function in the feed or in the animal, the categories of animals that shall receive feed containing the additive, the duration of administration, levels of the additive to be incorporated into feed, analytical methods for the additive present in the feed, and other relevant aspects, for example interaction of the additive with other feed ingredients and feed components.

All key data that support the feed product to be authorised need to be generated with the final product in mind. If studies and data were originally generated for a different product (different degrees of purity or active principle), specific arguments need to be presented that justify their use for the product that is subject of the application. These aspects of product development are familiar to those who develop pharmaceutical products, where standard project schemes are applied for single phases of product development. However, projects for new feed ingredients are often managed ad hoc without taking full account of the eventual regulatory framework and its requirements.

Another important aspect of developing a new feed product or a new use for an existing feed product is understanding the baseline of current knowledge. For this purpose, a systematic literature review is recommended that makes it possible to identify and collect all data that would be potentially relevant when discussing the application with experts at the relevant authority. For literature-based dossiers, for example for health claims, the literature review may even be the main part of the dossier, and corresponding guidance for undertaking such work is available (Glanville et al., 2014). In the context of food safety and of the regulatory framework for the evaluation and authorisation of a feed product, a literature review must be free from any bias and provide an objective summary of the already available data. Such literature reviews may be valuable steps in the generation of data on the characterisation, the safety, and the efficacy of a feed additive, and the claims for a feed product.

5 Building the dossier

Regulatory authorities provide templates for the submission of data, now usually in electronic form. These dossiers support the application of a feed ingredient for market authorisation. In addition to the template, guidelines are issued for the main chapters of a dossier. Both the template structure and guidelines allow feed producers to develop a project plan that comprises all data that need to be generated, the experiments that need to be run, and further additional work required. The data needed can be put into a logical sequence of experiments and studies. As an example, for any studies with animals at least one analytical method is needed to determine the quantity of the additive in the feed. The development of a suitable analytical method therefore needs to precede testing the additive with animals. Likewise, the safety for target animals depends on the intended use level. You first need to identify (using a dose range finder) the efficacious level of the feed product before you can test multiples of this level of feed for the safety of the target animal. Another example is consumer safety which depends on potential residues in animal products. Residues could result from the ingredient itself or after metabolization by the animal. As a result, investigations into the metabolism of the feed additive by target animals must precede the discussion of consumer safety. In reality, the picture is more complex, as data are often generated in order to justify the development of a product, which means that frequently efficacy data are produced at an early stage. However, one may then need to differentiate clearly between data generated to prove the concept of a product idea and data generated for a regulatory purpose.

For efficient project management, a right-first-time approach is advisable, as applied for human pharmaceutical products. All steps of product development are completed to a standard that can be included directly in the dossier. This minimises repeating experiments because they have not met regulatory standards and accelerates the assembly and completion of the final dossier. It is important to note that regulatory authorities discourage submission of incomplete datasets, a practice which may be characterised by the attitude of 'wait and see whether the experts say this data is are really needed'. The EFSA, for example has a unit dedicated to checking the completeness of a dossier as described in the relevant guidance documents. Incomplete dossiers are considered invalid and are therefore rejected (EFSA, 2007).

6 Characterisation and properties

The chemical composition of a feed product informs us directly about the potential hazards that may occur during the product's life cycle. Toxic impurities could endanger the health of the animal, proteins present could trigger allergenicity in workers, active substances may be sensitive to oxygen and transform into products that are less active or toxic, solubility in water may be relevant to whether the excreted additive could end up in groundwater.

First, the feed additive's composition needs to be fully characterised (EFSA, 2017a). This may be easy for substances of high purity such as crystalline vitamin C but may be a challenge if a more complex extract from a plant is the material of choice. In such cases a systematic literature review of possible constituents makes it possible to narrow down the substances that need to be characterized. In addition to the active substance that is responsible for the effects of the feed product, all other major and minor constituents typically or potentially present need to be analysed. This obviously depends on the raw material, the manufacturing process and eventually the purification steps. Certain impurities, in the European Union called 'undesired substances', need to be tested by default. Examples include dioxins in products of mineral or industrial origin, mycotoxins in products derived from plant materials, organic solvents in extracts and heavy metals in all feed products.

Besides composition other properties need to be investigated. Degradation products need to be part of stability studies since they may constitute impurities to be accounted for as part of the specification of the product. Other physicochemical properties that need to be studied are particle size, dusting potential, solubility in water and organic solvents, stability under normal and exceptional storage conditions and reactivity with other components. For specific product categories such as microorganisms, enzymes and plant extracts, separate additional datasets are needed.

For microorganisms, whether used as feed ingredients or in the production of feed ingredients (e.g. in fermentation processes), there is a need for data that address issues such as taxonomy (i.e. unambiguous identification of the species), the risk of antimicrobial resistance or the production of toxins. This is now often most easily achieved by investigating the whole genome sequence (WGS) of the organism which makes it possible to allocate the organism to its proper taxonomic species and to search for genes that indicate the presence of antimicrobial resistance or metabolic pathways leading to the formation of toxic compounds.

Secondly, data are needed to understand the properties of the product when added to feed. The most obvious properties are mixing behaviour and stability. Experimental evidence is needed that the product can be mixed homogeneously into feed of different types for example mashed or pelleted feed. The stability of the ingredient needs to be investigated during all processing stages including mixing, pelleting (at elevated temperatures) and storage of the feed. Should an additive be used in an intermediate product, for example as a premixture (which is a commercial product in its own right), data may also be needed for the premixture. All such data should reflect commercial practice. As an example, stability in premixtures should be studied for a period of 6 months, whereas stability in feed for food-producing animals can be studied for 3 months only. The required period of study depends on the expected storage recommendation for the final feed product. In the case of pet food, guaranteed stability is frequently for 1 or 2 years, and any ingredient that serves a nutritional need or is designed to have specific health effects needs to provide those benefits until the end of the 'best before' period (which is authorised for the inclusion on labelling).

Product characterisation requires analytical methods to determine each ingredient and confirm that it is present in animal feed. Analytical methods may also be needed to determine residues in animal tissues (e.g. muscle, fat or liver) and animal products (e.g. eggs or milk). In case an analytical method mentioned in the authorisation is not an official method, it needs to meet criteria that indicate that it has been sufficiently validated (EU, 2008a: Section 2.6. in Annex II).

7 Safety data: animals, consumers, workers and the environment

The first step in any safety assessment for a feed product is an inventory of available data to allow the development of a preliminary toxicological profile. This will help identify the need for new data and the sequence of experiments and studies to be undertaken to provide that data. As feed products for food-producing animals enter the food chain, some basic questions need to be addressed and understood before undertaking studies that involve major costs and resources. Substances added intentionally to the food chain must not be genotoxic, must not carry antimicrobial resistance genes or enable their occurrence and should not have hormone-like activities which might classify them as endocrine disruptors. When developing new feed products these risks should be investigated at the beginning of the process. For many processed feed ingredients these questions may be easy to answer, but natural extracts for example are complex and may contain potentially genotoxic compounds, microorganisms which may carry and spread antimicrobial resistance genes, and molecules able to bind to hormone receptors.

Once these basic questions have been addressed, the further need for classical toxicological studies depends on the composition of the compound and the chemistry of its constituents. For substances to which consumers may be exposed, the EFSA usually requests a minimum 90-day oral feeding study in laboratory rats (EFSA, 2017b). Over the period of 90 days the test substance is added to the diet of animals, usually being allocated to a control group and three groups with increasing doses (e.g. 0, 100, 500, 1000 mg/kg bodyweight per day). This study serves two purposes: to identify organs, tissues and physiological parameters that may be affected or damaged by the substance, and to assess the dietary levels at which such effects are observed or not yet observed using a dose-response approach. The level at which no adverse effects are observed is used as the point of departure to establish safe intake levels for humans and, where appropriate, animals. As an example, at the doses quoted above we might see increased liver weight in rats at 1000 mg/kg bodyweight per day, but we do not make this observation at the medium dose level of 500 mg/kg bodyweight per day. The value of 500 mg/bw*d is determined to be the 'no observed adverse effect level' (NOAEL); it is divided by a default factor of 100 to establish the intake by humans considered to be without appreciable risk. The acceptable daily intake (ADI) would in this case be 5 mg/bw, which translates for an adult of 60 kg bodyweight to a daily intake of 300 mg.

This approach of using toxicological studies to establish an acceptable daily intake for consumers was developed more than 60 years ago. The European Food Safety Authority (EFSA) has recently extended this concept to animals, whether food-producing or pet animals, applying the same default factor of 100 when deriving acceptable intakes for a broiler or dog. It is possible to establish acceptable feed levels for all animals in a species using one study in rats. However, this approach requires rather high no observed adverse effect levels in a laboratory animal study. Table 2 compares the corresponding maximum safe feed levels for two feed additives, a natural and a synthetic flavouring, as calculated by the EFSA.

It is clear from both examples in this table that substances with a low safe level in rats will lead to low save feed levels for the target animals in question. Should the intended and technologically justified level in feed be higher, the safety of the feed product needs to be studied in the target animals themselves. The strategy for safety testing in target animals will depend on the range of animal species and categories (e.g. age range) for which the substance shall be added to feed. If just one animal category is involved, for example piglets, safety is studied in this category.

Table 2 Maximum safe feed concentrations for selected target animals

Feed additive	Essential oil from Origanum vulgare ssp. hirtum (Link) letsw.	1-Octen-3-ol
Reference	EFSA, 2019b	EFSA, 2020
NOAEL (mg/kg bw*d)	200	6.7
Maximum safe concentrations (mg/kg feed)		
Chickens for fattening	22	1.1
Piglets	40	1.3
Cattle for fattening	88	2.9
Dogs	106	3.5

If the additive is to be used in a broader range of animal species a specific strategy needs to be developed. The EFSA recently systematised safety assessments by taking into account the life stage (e.g. young/growing, adult/reproductive) and the broad classifications of animals as mammals (ruminants, monogastric mammals), birds (poultry) and fish (EFSA, 2017c). A safety study in piglets is applicable to young pigs in the fattening stage, but not to adult sows used for piglet production. If an ingredient is to be used in all pig categories, separate safety studies in piglets and sows are needed. Similarly, in the case of ruminants, two separate studies in calves and in dairy cows are required. In the case of poultry, separate studies in broilers (for meat) and laying hens (for eggs) are needed. If the intention is to develop a feed additive for all animal species and categories (including pets), a minimum of four studies in piglets, broilers, dairy cows and one fish species is needed. These general principles of extrapolation between growing and reproductive animals are applied on a case-by-case basis for specific products in particular situations and should not be taken as a default as sufficient in all cases.

Safety studies in target animals are called tolerance studies. They are not to be compared with toxicological studies in rats or other laboratory animals for which detailed and extensive guidelines by the OECD are available. The EFSA has developed a set of principles to be considered for tolerance studies (EFSA, 2017c). The design of the study is driven mainly by the margin of safety that is envisaged: beside a negative control group, the ingredient is given at the highest proposed dose to the treatment group, and multiples of this dose are administered to one or more tolerance groups. Should the tolerance be hundredfold higher, a simple study observing animal growth and performance plus health monitoring is sufficient. Where the tolerance group needs to be administered a dose at a factor of 10 or more than the highest proposed dose, detailed data including monitoring of blood chemistry and haematology are needed. At tolerance levels below 10 it is expected that selected representative animals from all dose groups are subjected to histopathological examination.

Specific attention needs to be paid to ensure a sufficiently robust statistical methodology for the study which must include statistical evaluation and comparison of the parameters observed in the various groups of animals. The EFSA formally requests that the protocol for a tolerance study is a signed document that is added as an annex to the final study report. It is a good idea, as is practiced in GLP studies, to document any change with respect to the protocol and to formally sign off those changes. It is important to note that following Good Laboratory Practice (GLP) is essential if the study is intended for a product dossier for the FDA. GLP is recommended by the EFSA but not mandatory.

In assessing the safety of the feed additive for consumers, the first question to be answered is whether the feed ingredient or its metabolites will be carried over into food products. In case of colourants this carryover may be intentional (e.g. to give egg yolk or salmon a desirable appearance). For the assessment of these residues basic toxicological data packages are necessary, with a minimum requirement for mutagenicity studies and a 90-day rat study (see this section, second paragraph). The acceptable daily intake (ADI) for residues must ensure their safety for the consumer.

To understand if metabolites are deposited in animal organs or tissues, specific metabolism studies may be needed. No specific designs and protocols for such studies for feed ingredients are available since different types of materials may need to be studied. The guidance provided by the VICH is an appropriate starting point, suitably adapted to the properties of the substance being studied (EMA, 2011). Clearly plant ingredients consisting of typical nutrients or fibres eaten by animals will not raise concerns. There may also be data from laboratory animals or humans that can be used. However, since no general approach is available, investigations need to be developed on a case-by-case basis. They will depend on the data available for similar substances in predicting likely metabolism. One obstacle is that metabolism of chemicals is often only studied in laboratory animals (rats, mice or dogs), raising the question of whether the same metabolic pathways are present in other animal categories such as birds or fish.

Data on worker safety are needed in the evaluation of feed additives. Workers at feed ingredient production sites, feed mill workers and farmers may be exposed to the feed product, on the skin, in the eyes, the oral cavity and the gastrointestinal (GI) tract or by inhalation into the lungs. Corresponding guidance is available from the EFSA (2012). Of specific interest is accidental exposure at higher levels and a set of OECD study protocols are available to investigate this. Important considerations include whether the substance is solid or liquid, whether it generates dust that may be inhaled and whether it contains specific substances of concern such as endotoxins. Within the European Union any ingredient that is marketed between businesses needs to be assessed and classified for its safety with respect to workers and the environment, and needs to be labelled accordingly using the appropriate warning statements and pictograms (EU, 2008b).

Feed ingredients pass through the GI tract of animals and are excreted as such or in a modified form. They may also be spread as manure on farmland. In aquaculture feed ingredients may leak out of feed pellets or be excreted by fish or shrimps into water and end up in the sediment of a lake or below a sea cage used to contain the fish being farmed. Environmental safety evaluations of feed additives follow the approach for pesticides and veterinary drugs. The requirement may be waived if the substance occurs naturally in the environment (this excludes ingredients derived from exotic species) or is fully metabolised and not excreted at all. Hazards to be investigated are toxic effects on organisms in environmental compartments where the feed additive could be present: soil, groundwater, surface water (lakes, rivers), freshwater sediments, seawater or sea water sediments. A full evaluation requires a considerable amount of data. However, it is also possible to model the behaviour of substances in different conditions. Extensive guidance is available from the EFSA to help develop a product-specific strategy (EFSA, 2019a).

8 Effects and efficacy

Feed ingredients are developed for a wide variety of reasons. They may affect the properties of feed (e.g. a preservative will allow storage for a given period), change the behaviour of animals (e.g. an appetiser may increase feed consumption), improve gut health (e.g. pre- and probiotics), support digestion of feed constituents (e.g. enzymes), improve productivity parameters such as feed conversion ratio (e.g. increase of milk output per dry matter input), reduce environmentally hazardous materials (e.g. phosphate in faeces or exhaled methane), change the appearance of animal tissues and products (e.g. colouring egg yolks) and provide macro and micro nutrients (e.g. carbohydrates in corn flour or selenium-enriched yeast). There is no single way of studying and proving the efficacy of such a broad range of functions and effects of feed ingredients.

The number of studies required, their design and the endpoints to be investigated all depend on the specific target effect of the specific substance (EFSA, 2018a). The common denominator is that evidence of efficacy needs to be generated using scientific protocols that reflect state-of-the-art knowledge and methodology. The same general considerations apply as for animal safety. Efficacy may differ for each target animal category i.e. ruminating and monogastric mammals, poultry, fish and non-vertebrates. It may also differ for growing and reproductive animals. An ingredient that can be used for different species will require separate studies for each species. Where an ingredient is targeting the feed rather than the animal (e.g. a preservative), it must be tested on the different types of feed to which it will be added.

In the European Union, a specific group of feed additives are classified as so-called zootechnical additives, that is additives designed to optimise the performance of healthy animals. Their authorisation resembles that for veterinary products. Authorisation is holder specific and depends on "guaranteed" effects on performance, growth and health of the animal. Three independent efficacy studies are needed for each target animal category in order to demonstrate that beneficial effects are likely to be achieved regularly in the marketplace. It is important to justify the intended level of use, and an appropriate dose range finding study should be included in the study programme. Such a study may serve as one of the three required efficacy studies. Study planning also needs to take into account the number of animals required to provide statistically reliable results for both tolerance and efficacy studies.

As European agricultural practices are diverse, studies should be carried out in at least two different locations. Studies from other countries (e.g. the United States) can be accepted, but only if feed compositions and animal husbandry practices are sufficiently similar to those in the European Union. Diets offered to broilers or pigs may differ significantly between the US and Europe with respect to major ingredients used as sources for carbohydrates and proteins. Efficacy data generated in the United States may not be applicable to the European Union if such differences occur (and vice versa).

There is a considerable difference between the European Union and United States with respect to the planning of efficacy studies. The EFSA in the European Union provides detailed written guidelines and expects the applicant to read them and then prepare a final report for assessment by the EFSA (with limited opportunity to prior consultation). In contrast the FDA in the United States encourages companies to approach FDA experts and to discuss with them all relevant aspects of study protocols up front. There is a formal review process for draft study protocols which results in detailed written comments supported by further clarifications in meetings. This applies specifically to the application procedure for food additives (for animal feed) but can also be used for the generally recognised as safe (GRAS) routes.

Although it is difficult to provide extensive guidance on efficacy studies, as the endpoints of efficacy differ significantly, some elements are common to all studies. Animal studies must record all details with respect to origin of animals (health status), feed composition, analysis of ingredients in feed, dosing of the feed ingredient, feeding regime, husbandry conditions in general, specific husbandry practices (e.g. layout of stables and pens), water supply any diseases and medication used to treat them. Animals should be in good health throughout the study and receive an appropriate diet in accordance with recommendations by breeders.

It is widely accepted that it is difficult to establish efficacy under optimal conditions which may not reflect real commercial conditions. However, since findings may not be reproducible, field studies under practical conditions are not accepted by EFSA. Results from such studies may be used to corroborate the dossier, but are not considered to provide the pivotal data that allow a decision to be made about the effects and the efficacy of a feed additive.

9 Substantiating claims

In the previous section we discussed those feed ingredient effects that are assessed by an authority as part of pre-market authorisation. Some claims about the effects of a feed ingredient or feed product (including a complete feed/complementary feed) can be made outside the regulatory procedure required for authorisation of a feed additive. Such claims are accepted in the European Union for feed materials (which are not feed additives) and complete feed, and they are possible in many other jurisdictions as long as they do not mislead the buyer/user of the feed. Being not misled about a claimed effect means that the statement is true as far as common understanding among scientists applies. Such claims are often acceptable under freedom of speech rules and can be made on the product label itself as well as on leaflets, webpages, at exhibitions and in direct discussions with customers.

In the European Union the following conditions need to be met when making a claim. It should be objective, verifiable by the competent authorities and understandable by the user of the feed. In order to meet the second condition, the feed producer should be able to produce scientific substantiation of the claim that can be shared on request with the relevant authorities (EU, 2009). The regulation suggests either publicly-available scientific evidence or documented company research as scientific substantiation for a claim. No further specific requirements are given. In that sense anybody making claims about feeds is free to document them as they think best.

One important limitation is claims for veterinary medicinal products, products treating or preventing disease in animals, or products restoring, correcting or modifying physiological functions by pharmacological, immunological or metabolic means (EU, 2019). However, there is an overlap in the area of diseases caused by improper nutrition; there are diseases or disease-like conditions such as milk fever that may be addressed by dietary intervention. In the European Union, such dietary claims related to specific diseases are regulated under the feed law for specific ingredients and target animals. The recently updated list of such claims for particular nutritional uses should be taken into account when developing new claims (EU, 2020).

Two federations representing animal feed manufacturers have recently developed more detailed guidance on the evidence files that can be used to document and justify claims for animal feeds (COPA-COGECA and FEFAC, 2018; FEDIAF, 2018). In both documents the guidance for making claims is part of more comprehensive good practice covering labelling guidelines developed by the feed sectors for food producing animals and for pet animals in close consultation with government authorities. They are recognised by these authorities and are therefore a valuable and recommended point of reference.

For a specific claim to be made for a complementary feed it will be important to consider the available evidence and to phrase carefully the actual statement. There are differences between statements such as 'plant extract XYZ improves feed conversion by 15%' and 'extracts from plants used for XYZ have been studied for their possible contribution to improved feed conversion'. To substantiate the first statement requires at least one study that demonstrates significant quantifiable effects on feed conversion. The second statement refers to studies from the published literature that may be relevant for a product but does not make any claim about a guaranteed effect.

As discussed elsewhere, a comprehensive scientific literature review is always a preferred starting point. Studies relating to the effects of plant extracts in humans (e.g. for use as traditional herbal medicines) may help to support a product, providing the feed company avoids any specific claims reserved for veterinary medicinal products which require a stronger evidence base. It is possible to use results from field studies undertaken with customers or research institutions without implementing all the requirements needed for studies which need to be evaluated by authorities like the EFSA or the FDA. Field studies are typically available in larger numbers. A meta-analysis based on a number of field studies can result in a more robust data set to support a claim. It is also possible to investigate the mode of action *in vitro* and to predict effects *in vivo* without having data from studies in target animals. However, robust claims require developing study designs and protocols similar to those for feed additives described in Section 8.

10 Dossier submission and follow-up

A successful dossier needs to comply with all legislative requirements and guidelines from the relevant authorities. In addition it should be written in a concise language that is easy to understand and which is free of company and lab jargons. The narrative parts of the submission should be clear and to the point. This requirement reflects the fact that national expert bodies are composed of individuals with different technical, scientific and linguistic backgrounds. It is important that you use the terminology and definitions laid down in the legal text and guidelines of the procedure under which you will submit your application. It is always a good idea to study reports for similar products so that you can phrase your documents in a similar way to experts.

Dossiers for EFSA are now submitted in electronic form, with the current submission by CD or USB stick, to be replaced by online submission. It is important that the dossier, its main parts and the annexes and references are submitted in an 'easy-to-use' format, for example with a clearly labelled structure and the ability to move between sections and documents via hyperlinks (EFSA, 2007). Incomplete dossiers are prevented from progressing until the missing parts have been submitted. Additional questions may be raised by experts, in which case a deadline to answer them is set by the authority in question. Follow-up of such additional requests is critical and should be prioritised within the company.

The European Union adopted recently specific rules for the notification of studies commissioned with a view to support an application for authorisation of a feed additive. Such notifications will be mandatory for industry and for contract laboratories, whether they are located in the European Union or in a third country. Failure to notify will result in a default six-month delay of the evaluation of the dossier by EFSA. These new rules will come into force on 27 March 2021. Before then the EFSA will publish more detailed guidance on the notification procedure and the publicly-accessible database where notified studies need to be filed.

11 Continuously updating the file

The processes of generating evidence for the efficacy of a feed product and assembling an evidence file are not activities that should stop once authorisation is granted and the product is launched. All feed businesses have the obligation to monitor any unexpected or unwanted effects resulting from the use of their products. As science progresses, relevant research relating to efficacy should be monitored and the evidence file updated as appropriate.

In the European Union there is currently an obligation to renew authorisation of feed additives after 10 years. The dossier for a renewal application needs to be submitted one year before the end of authorisation. In practice preparation for such a dossier starts 6-8 years after the first authorisation. The focus of the renewal dossier depends on the substance category. In all cases possible changes in the manufacturing and quality of the additive need to be discussed, and the impact of any change on safety and efficacy of the additive needs to be assessed. Relevant progress in science related to the additive needs to be documented by a systematic literature search.

12 Where to look for further information

European Union	European Food Safety Authority: www.efsa.europa.eu.
	European Union Reference Laboratory for additives for use in animal nutrition - EURL for Feed Additives: ec.europa.eu/jrc/en/eurl/feed-additives.
	European Commission - DG SANTÉ: ec.europa.eu/food/safety/animal-feed_en.
United States	US Food and Drug Administration: www.fda.gov/animal-veterinary/products/animal-food-feeds.
	Association of American Feed Control Officials: www.aafco.org/.
Sector guidance	FEFAC - European Feed Manufacturers' Association: www.fefac.eu. FEFANA - EU Association of Specialty Feed Ingredients and their Mixtures: www.fefana.org.
	Fediaf-the European petfood industry: www.fediaf.org.

13 References

Codex Alimentarius Commission (2013). *Guidelines on the Application of Risk Assessment for Feed*. CAC/GL 80-2013.

COPA-COGECA and FEFAC (2018). *EU Code of Good Labelling Practices for Compound Feed for Food Producing Animals*. European Feed Manufacturers Federation.

EFSA (2007). Administrative guidance to applicants on the preparation and presentation of applications for authorisation of additives for use in animal nutrition. *EFSA Journal* 5(5): 1447. [63 pp.] https://doi.org/10.2903/j.efsa.2007.1447, Updated on 22 March 2019.

EFSA (2012). Guidance on studies concerning the safety of use of the additive for users/workers. *EFSA Journal* 10(1): 2539. [5 pp.] https://doi.org/10.2903/j.efsa.2012.2539.

EFSA (2017a). Guidance on the identity, characterisation and conditions of use of feed additives. *EFSA Journal* 15(10): 5023. [K been the 12 pp.] https://doi.org/10.2903/j.efsa.2017.5023.

EFSA (2017b). Guidance on the assessment of the safety of feed additives for the consumer. *EFSA Journal* 15(10): 5022. [17 pp.] https://doi.org/10.2903/j.efsa.2017.5022.

EFSA (2017c). Guidance on the assessment of the safety of feed additives for the target species. *EFSA Journal* 15(10): 5021. [19 pp.] https://doi.org/10.2903/j.efsa.2017.5021.

EFSA (2018a). Scientific opinion on guidance on the assessment of the efficacy of feed additives. *EFSA Journal* 16(5): 5274. [25 pp.] https://doi.org/10.2903/j.efsa.2018.5274.

EFSA (2018b). Guidance on the characterisation of microorganisms used as feed additives or as production organisms. *EFSA Journal* 16(3): 5206. [24 pp.] https://doi.org/10.2903/j.efsa.2018.5206.

EFSA (2019a). Guidance on the assessment of the safety of feed additives for the environment. *EFSA Journal* 17(4): 5648. [78 pp.] https://doi.org/10.2903/j.efsa.2019.5648.

EFSA (2019b). Safety and efficacy of an essential oil from Origanum vulgare ssp. *hirtum* (Link) Ietsw. for all animal species. *EFSA Journal* 17(12): 5909. [14 pp.] https://doi.org/10.2903/j.efsa.2019.5909.

EFSA (2020). Scientific opinion on the safety and efficacy of oct-1-en-3-ol, pent-1-en-3-ol, oct-1-en-3-one, oct-1-en-3-yl acetate, isopulegol and 5-methylhept-2-en-4-one, belonging to chemical group 5 and of isopulegone and a-damascone belonging to chemical group 8 when used as flavourings for all animal species. *EFSA Journal* 18(2): 6002. [16 pp.] https://doi.org/10.2903/j.efsa.2020.6002.

European Medicines Agency (2011). VICH GL46: Studies to Evaluate the Metabolism and Residue Kinetics of Veterinary Drugs in Food-Producing Animals: Metabolism Study to Determine the Quantity and Identify the Nature of Residues. 14 March 2011. EMA/CVMP/VICH/463072/2009. Committee for Medicinal Products for Veterinary Use (CVMP).

European Union (2002). Regulation (EC) No 178/2002 of the European Parliament and of the Council of 28 January2002 laying down the general principles and requirements of food law, establishing the European Food Safety Authority and laying down procedures in matters of food safety. Official Journal of the European Communities, 1.2.2002, L31/p. 1.

European Union (2008a). Commission Regulation (EC) No 429/2008 of 25 April 2008 on detailed rules for the implementation of Regulation (EC) No 1831/2003 of the European Parliament and of the Council as regards the preparation and the presentation of applications and the assessment and the authorisation of feed additives. Official Journal of the European Union, 22.5.2008, L133/p. 1.

European Union (2008b). Regulation (EC) No 1272/2008 of the European Parliament and of the Council of 16 December 2008 on classification, labelling and packaging of substances and mixtures, amending and repealing Directives 67/548/EEC and 1999/45/EC, and amending Regulation (EC) No 1907/2006. Official Journal of the European Union, 31.12.2008, L353/p. 1.

European Union (2009). Regulation (EC) No 767/2009 of the European Parliament and of the Council on the placing on the market and use of feed, amending European Parliament and Council Regulation (EC) No 1831/2003 and repealing Council Directive 79/373/EEC, Commission Directive 80/511/EEC, Council Directives 82/471/EEC, 83/228/EEC, 93/74/EEC, 93/113/EC and 96/25/EC and Commission Decision 2004/217/EC. Official Journal of the European Union, 1.9.2009, L229/p. 1.

European Union (2019). Regulation (EU) 2019/6 of the European Parliament and of the Council of 11 December 2018 on veterinary medicinal products and repealing Directive 2001/82/EC. Official Journal of the European Union, 7.1.2019, L4/p. 43.

European Union (2020). Commission Regulation (EU) 2020/354 of 4 March 2020 establishing a list of intended uses of feed intended for particular nutritional purposes and repealing Directive 2008/38/EC. Official Journal of the European Union, 5.3.2020, L67/p. 1.

FEDIAF (2018). *Code of Good Labelling Practice for Pet Food*. European Pet food Industry.

FEFAC (2019). *Feed & Food Statistical Yearbook* (vol. 2018). European Feed Manufacturers Federation.

Glanville, J., Wood, H., Arber, M., Varley, D., Frampton, G. and Brazier, H. (2014). Technical Manual for Performing Electronic Literature Searches in Food and Feed Safety. Annex to Inventory of Sources of Scientific Evidence Relevant to EFSA's Risk Assessments and Information Sessions on Literature Searching Techniques (CFT/EFSA/SAS/2011/03). EFSA Supporting Publication.

Smedley, K. O. (2013). Comparison of Regulatory Management of Authorized Ingredients, Approval Processes, and Risk-Assessment Procedures for Feed Ingredients Jurisdictions Covered: Brazil, Canada, China, European Union, Japan, South Africa, and United States. July, 2013. Published by the International Feed Industry Federation.

www.ingramcontent.com/pod-product-compliance
Ingram Content Group UK Ltd.
Pitfield, Milton Keynes, MK11 3LW, UK
UKHW021843100926
8556IPUK00019B/1026

9 781835 450062